AF615407

OXFORD
POLYTECHNIC
LIBRARY

0031212600

TELEPEN

THE EVOLUTION OF SEDIMENTARY BASINS

THE EVOLUTION OF SEDIMENTARY BASINS

PROCEEDINGS OF
A ROYAL SOCIETY DISCUSSION MEETING
HELD ON 3 AND 4 JUNE 1981

ORGANIZED AND EDITED BY
SIR PETER KENT, F.R.S., M. H. P. BOTT, F.R.S.,
D. P. McKENZIE, F.R.S., AND C. A. WILLIAMS

LONDON
THE ROYAL SOCIETY
1982

Printed in Great Britain for the Royal Society
at the
University Press, Cambridge

ISBN 0 85403 184 7

First published in *Philosophical Transactions of the Royal Society of London*,
series A, volume 305 (no. 1489), pages 1–338

Published by the Royal Society
6 Carlton House Terrace, London SW1Y 5AG

CONTENTS

PAGE

CONTINENTAL MARGIN BASINS

THE NORTH SEA

INTRACONTINENTAL BASINS

Phil. Trans. R. Soc. Lond. A **305**, 3 (1982)
Printed in Great Britain

Introduction

BY SIR PETER KENT, F.R.S.

The proposal for a Discussion Meeting on sedimentary basins arose from the realization that although there is a wealth of new concepts on horizontal movement of the Earth's crust, vertical movements had been relatively neglected, despite their great importance as a source of fluids – water, oil, and hydrocarbon gases – which are of such high importance to mankind. Basin development as a function of sedimentary loading has long been discussed; other basins are related to major tension faults; others again are suspected of being due to deep flow compensating for orogenic crustal uplift.

During the last two decades deep drilling by industry, particularly offshore, has provided a wealth of accurate statistical information on sedimentary basins, and has provided also reference sections to which seismic data can be related. A review of the new information was felt to be overdue.

The selection of authors and subjects was designed to provide statistics and a history of a range of sedimentary basins that were known to considerable depths, the aim being to illustrate the history of subsidence, depositional rates, sedimentary facies, thermal régime and also their relation to crustal structures. A considerable geographical range was attempted at the same time, recognizing, however, that at a two day meeting only a minimal sampling was possible.

The first two papers describe the accumulation of the sedimentary prism on the Atlantic margin west of the French coast, the relations of oceanic and continental crust, and the effects of oceanic opening. The third and fourth deal with the Mesozoic and Tertiary sedimentational history in Aquitaine and North-west Australia, followed by two papers on the North Sea, an internal sedimentary basin developed from the Upper Palaeozoic to the present time, and a general review of the fault control of sedimentary basins in northwest Europe.

The Upper Palaeozoic basins of the United States, the Permian–Miocene Zagros basin of Iran, the Palaeozoic–Mesozoic Alberta basin and the Arctic basin of northern Canada provide a range of cases of specific intracontinental sedimentary basins, all of them of major economic importance. More general papers cover Carboniferous basins of northern Europe and the less well documented Precambrian basins.

The final group of papers review the problem of the mechanism of subsidence, particularly in relation to modern concepts of global structure.

The organizers, on behalf of the Royal Society, are grateful to the authors for this account of a very varied collection of sedimentary basins, and of the problem of their development in time and space.

Phil. Trans. R. Soc. Lond. A **305**, 5–25 (1982) [5]
Printed in Great Britain

A seismic refraction and reflexion study of the continent–ocean transition beneath the north Biscay margin

By F. Avedik†, A. L. Camus‡, A. Ginsburg§,[1] L. Montadert‡, D. G. Roberts§[2] and R. B. Whitmarsh§

† *Centre Océanologique de Bretagne, B.P.* 337, 29273 *Brest cédex, France*
‡ *Institut Français du Pétrole, Rueil-Malmaison, France*
§ *Institute of Oceanographic Sciences, Brook Road, Wormley, Godalming, Surrey GU*8 5*UB, U.K.*

The structure of the northern margin of the Bay of Biscay consists of a series of tilted and rotated blocks bounded by prominent listric faults whose polarity is consistently down toward the continent–ocean boundary. These blocks formed by rifting in late Jurassic – early Cretaceous time and are now thinly covered by post-rift sediments of Aptian to Recent age.

Seismic refraction profiles were occupied on the shelf, on either side of and across the continent–ocean transition to the shelf, using Pubs and Obs with explosives and a $4 \times 1000\ in^3$ ($4 \times 16400\ cm^3$) airgun array. Two-ship expanding spread multichannel (48-trace) seismic reflexion profiles and 30 km fixed offset reflexion profiles were located along the seismic refraction profiles on either side of the transition. A two-ship 30 km fixed offset multichannel profile was located across the transition as well as a 5 km fixed offset multichannel profile extending from the ocean crust to the shelf. Conventional 48-trace single ship multichannel profiles were located along all the refraction and two-ship reflexion lines.

Interpretation of the refraction profiles has been made by using ray tracing as well as synthetic seismograms. Conventional seismic processing techniques have been used to prepare the two-ship multichannel seismic data for interpretation. The survey is believed to be the first attempt to apply two-ship multichannel seismic data to the study of the change in crustal structure of a rifted passive margin from the shelf to the ocean crust.

The results from the experiment led to the identification of a zone of transition between continental and oceanic crust about 8 km wide. The seismic refraction data show progressive thinning of the continental crust from 33 km to about 5 km close to the transition zone. However, extension values calculated in the upper crust from the rotation of fault blocks are much less (1.1–1.4) and suggest that the majority of the thinning is achieved by extensive attenuation of the lower crust.

Introduction

Theoretical models of lithospheric extension, as well as the geological consequences of such extension in the form of rifted cratonic basins and of passive continental margins, have attracted considerable recent interest. In the European area, examples of rifted cratonic basins include the North Sea (Ziegler 1978, and this symposium) and the adjacent margins of the North Atlantic Ocean. It has become clear that extension may also occur in strike-slip and convergent settings (Bally & Snelson 1980), and examples include the Western Mediterranean (Biju-Duval

[1] On leave from Tel-Aviv University, Tel-Aviv, Israel.
[2] Present address: BP, Britannic House, Moot Lane, London EC2Y 9BU, U.K.

et al. 1977), the Pannonian Basin (Royden & Sclater 1981) and the Great Basin (Profett 1977). The occurrence of crustal extension, expressed by rifting, in these different geological settings emphasizes that very little is known of the mechanisms producing the rifting process, including its initiation, that cause the profound changes in crustal structure revealed by attenuation and ultimately by the formation of a spreading ocean basin.

Several innovative theoretical models (McKenzie 1978; Royden *et al.* 1980; Royden & Keen 1980; Sclater & Christie 1980; Turcotte, this symposium) have proposed mechanisms for lithospheric extension and have attempted to evaluate quantitatively the mechanical and thermal consequences of each model in terms of histories of margin and basin subsidence. These models may have a potential practical value in predicting the thermal history and thus the degree of hydrocarbon maturity in sedimentary basins formed in response to a lithospheric extension (Royden & Keen 1980; Royden *et al.* 1980; Sclater & Christie 1980). However, a rigorous description of the deep crustal structure beneath a cratonic sedimentary basin or a passive continental margin that would provide sufficient data to quantify the extension in order to constrain these models has been lacking, as has a comprehensive study of the geological and thermal history of basins that have evolved in response to lithospheric extension.

In this paper, we present some of the results of a detailed seismic refraction and reflexion study of the northern or Armorican margin of the Bay of Biscay. The study examined the deep crustal structure of the rifted and attenuated continental crust and of the transition to the adjoining oceanic crust. It is believed to be the first time that such a detailed seismic description of a rifted passive margin has been obtained. A comprehensive discussion of the seismic data and their interpretation will be published elsewhere (Ginzburg *et al.*, in preparation).

Regional structure of northern Biscay

The continental margin in the Southwestern Approaches to the English Channel (figure 1) consists of a broad shelf, a relatively steep slope and a broad rise that merges imperceptibly with the Biscay Abyssal Plain. The shelf is underlain by the NE–SW trending Western Approaches Basin, which seems to intersect the margin. Further north, the Cornubian High separates the Western Approaches Basin from the Celtic Sea Basin (Montadert *et al.* 1979; Roberts *et al.* 1981).

Beneath the slope and rise, multichannel seismic reflexion profiles show that the margin consists of thin Tertiary and Cretaceous sediments that overlie a series of large tilted and rotated fault blocks. Rifting apparently took place in late Jurassic – early Cretaceous time in a pre-existing marine basin, and was completed by late Aptian time (Montadert *et al.* 1977, 1979). The overall tectonic style consists of tilted and rotated fault blocks whose downthrow is consistently down toward the ocean. The pattern of faults is probably *en-echelon* or anastomosing. Many of the fault blocks are bounded by listric normal faults (figure 2*a*) with throws of up to 3000 m. The listric faults flatten out with depth close to the top of a subhorizontal group of reflectors called 'S' (figure 2*b*). Comparison with early refraction data suggested that this reflector corresponded to the boundary between a 4.9 and 6.3 km s^{-1} layer (Avedik & Howard 1979). De Charpal *et al.* (1978) and Montadert *et al.* (1979) interpreted the 6.3 km s^{-1} refractor as the boundary between the upper brittle and the lower ductile continental crust. Refraction data on the shelf (Holder & Bott 1971) suggested that the Moho shoaled from 25 km beneath the shelf break to 12 km near the continent–ocean transition, where, from the available refraction data, the observed ductile crust was estimated to be only 3 km in thickness. These

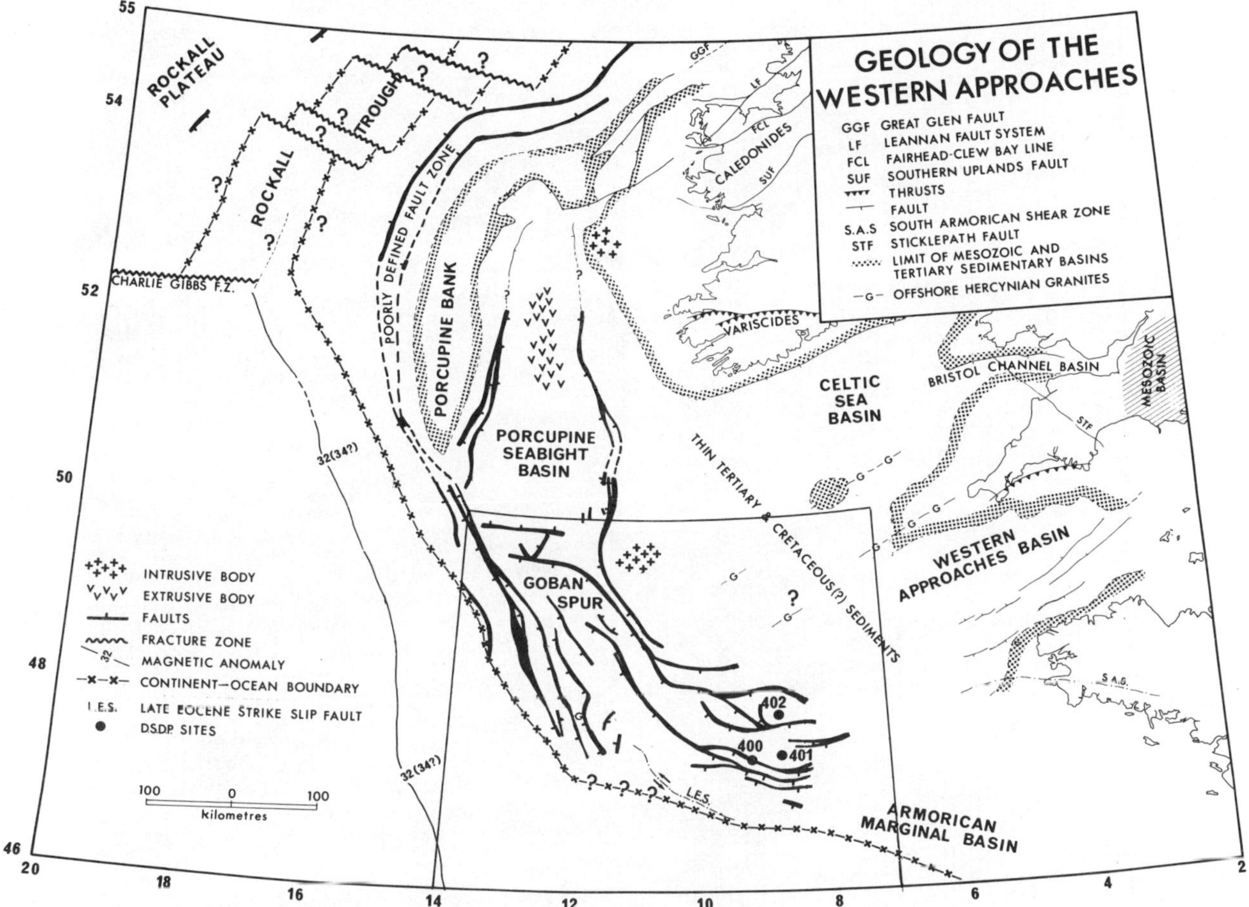

FIGURE 1. Regional structure of the continental margin of north Biscay (after Roberts *et al.* 1981).

early results showed that the observed thinning of the crust could not be fully explained by the extension of 10–15 % in the upper crust and resulted in the suggestion that the lower ductile had thinned by a greater amount (de Charpal *et al.* 1978; Montadert *et al.* 1979). The drill, dredge and reflexion data suggested that at the end of rifting a system of half-graben about 2.5 km below sea level existed adjacent to the continent–ocean transition (Montadert *et al.* 1977, 1979). After rifting, regional subsidence took place in response to cooling of the lithosphere of which the continental crustal part had been thinned during the rifting process. This interpretation of attenuation, although based on good-quality multichannel seismic data, was dependent on earlier refraction profiles whose directions are now known to cross underlying fault blocks.

EXPERIMENTAL DESIGN

In June 1979, a number of single-ship refraction profiles and two-ship reflexion and refraction profiles were located on the margin. The purpose of the experiment was to define the variation in the deep structure of the continental crust from beneath the shelf, to define the continent–ocean transition and to examine the relative amounts of thinning in the upper and

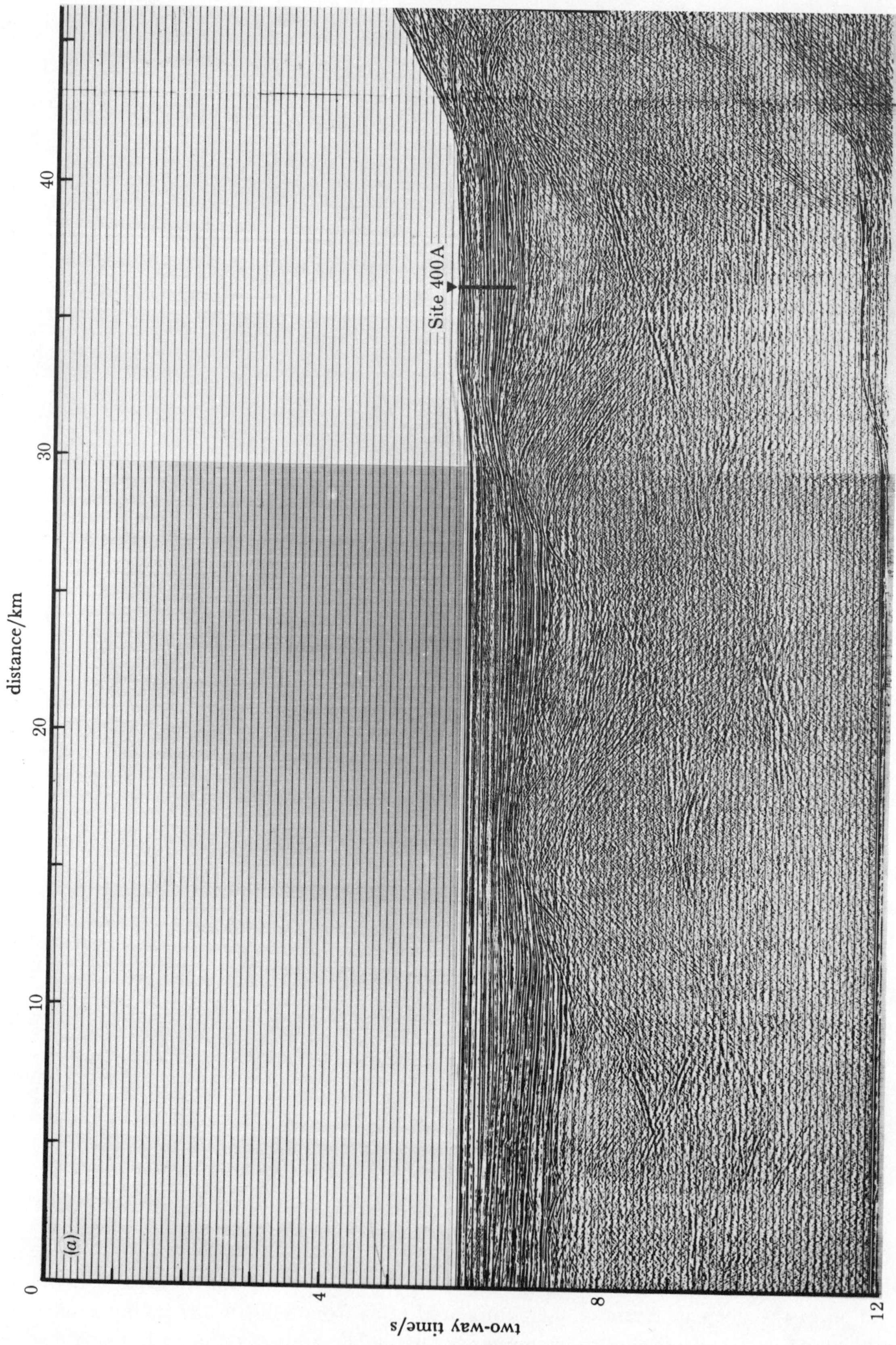
distance/km
0
10
20
30
40
Site 400A
(a)
two-way time/s
4
8
12

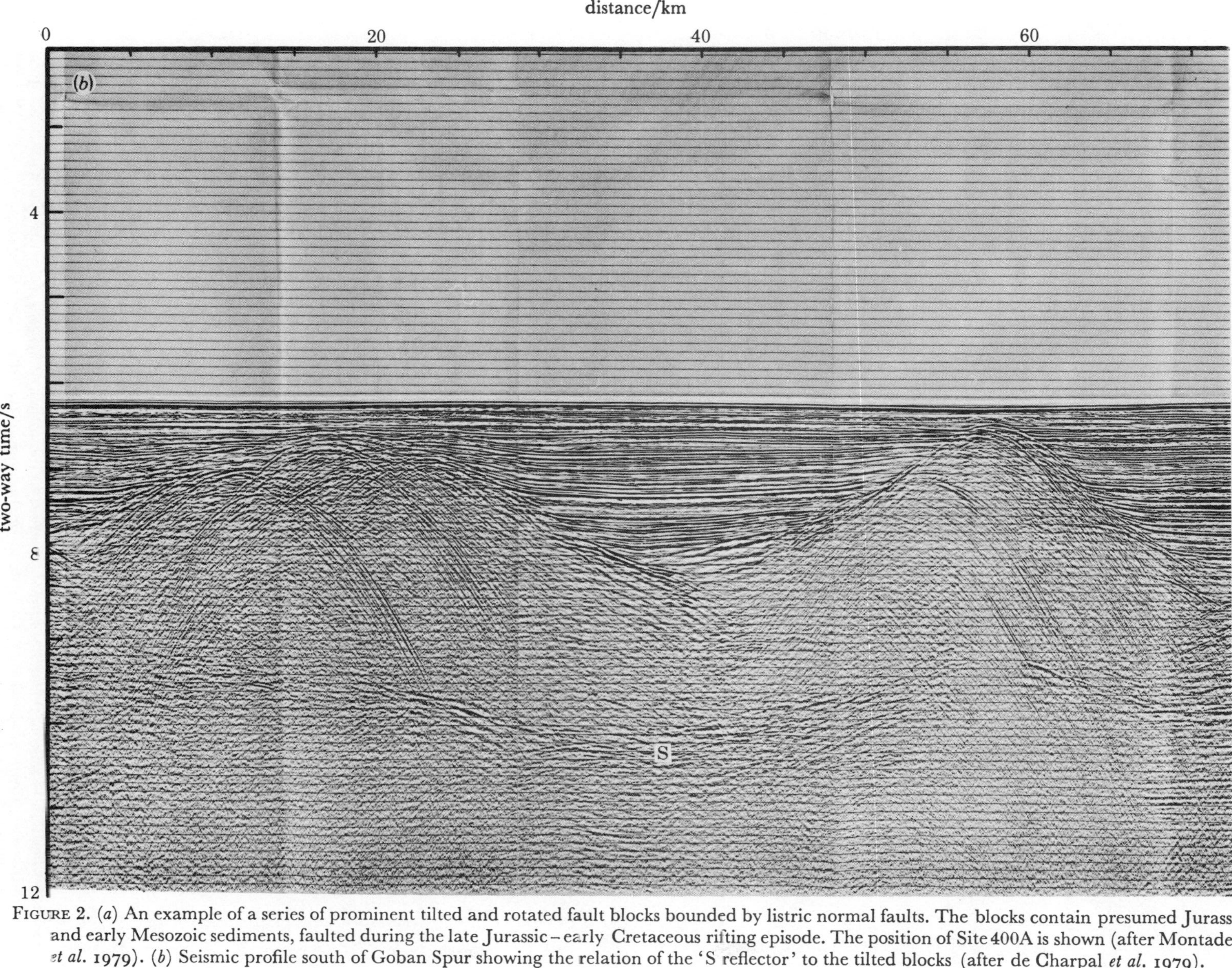

FIGURE 2. (*a*) An example of a series of prominent tilted and rotated fault blocks bounded by listric normal faults. The blocks contain presumed Jurassic and early Mesozoic sediments, faulted during the late Jurassic – early Cretaceous rifting episode. The position of Site 400A is shown (after Montadert *et al.* 1979). (*b*) Seismic profile south of Goban Spur showing the relation of the 'S reflector' to the tilted blocks (after de Charpal *et al.* 1979).

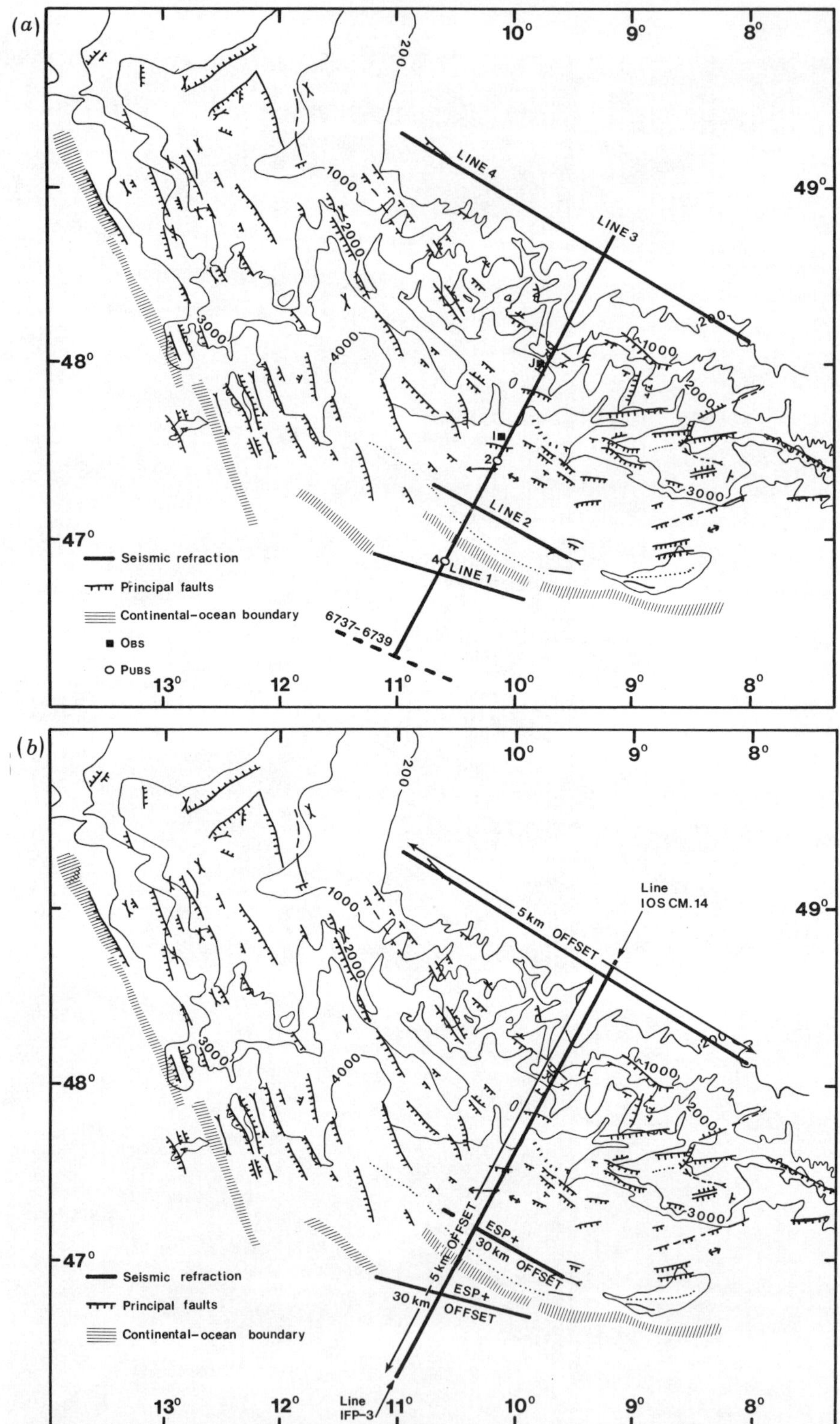

FIGURE 3. (*a*) Position of seismic refraction lines in relation to the major structural elements of the margin (after Montadert *et al.* 1979). I, OBS ISIS, J, OBS JULIE. (*b*) Position of two-ship expanding spread (e.s.p.), 30 and 5 km fixed offset profiles in relation to the major structural elements of the margin. Conventional 24-fold 48-trace multichannel single-ship profiles were occupied along all offset profiles with the exception of line 4. The position of line IOS CM-14 is also shown.

lower parts of the crust. The experiment was also designed to examine the width and nature of the transition from continental to oceanic crust.

The study was carried out in two parts. During the first phase refraction profiles were shot by R.R.S. *Shackleton* using the Pop-Up Bottom Seismographs (PUBS) of the Institute of Oceanographic Sciences (Kirk *et al.* 1982) and the Ocean Bottom Seismographs (OBS) of the Centre Océanologique de Bretagne (Avedik *et al.* 1978) along the lines shown in figure 3*a*. The study area, some 30 km west of the region studied in detail around Site 400A (Montadert *et al.* 1979), was selected to avoid complications that might be posed by the prominent uplift associated with late Eocene–Oligocene shearing in the vicinity of the Trevelyan escarpment. Line 1 was situated on the oceanic crust and line 2 on the presumed thinned continental crust. As far as was possible these lines were oriented parallel to the continent–ocean transition and, for line 2, parallel to the strike of the tilted blocks. An additional unpublished reversed refraction line (figure 3*a*) situated on the ocean crust south of line 1 and made with PUBS (Discovery Station 6737–6739) was reinterpreted and incorporated in the study. Line 3 consisted of a long refraction profile from the ocean crust to the shelf encompassing the continent–ocean transition and the progressive change from attenuated crust to thick crust beneath the shelf. PUBS and OBS were deployed at various points along the line to give a series of overlapping reversed profiles. The energy source used along lines 1, 2 and 3 was a 4×1000 in^3 (4×16400 cm^3) airgun array. Profile 4 along the shelf was unreversed and was shot by using explosive charges of up to 270 kg. It was intended to use explosives along line 3 but this was not possible for reasons beyond our control.

During the second phase of the experiment (figure 3*b*), R.R.S. *Shackleton* and the R.V. *Résolution* of the Institut Français du Pétrole occupied a series of two-ship multichannel seismic reflexion profiles. R.R.S. *Shackleton* acted as the shooting ship with a 4×1000 in^3 airgun array, and R.V. *Résolution* acted as receiving ship with a 2.4 km 48-trace streamer. Relative positioning was maintained by using a PULSE-8 shore-based radio navigation system (accuracy ± 25 m) and radar. Shot breaks were transmitted from R.R.S. *Shackleton* to R.V. *Résolution* by using a time-break transmitter. The following types of two-ship profile were made. Along refraction lines 1 and 2 on either side of the continent–ocean transition, two-ship expanding spread profiles were made to ranges of 35 km. In addition, along both these lines and part of line 3, 30 km fixed offset profiles were made; 30 km was chosen as the offset from a preliminary Moho depth estimate to ensure that only Pn arrivals would be obtained as first arrivals. Reflexion profiles at 5 km fixed offset were also made along the whole of line 3 and line 4. Conventional 24-fold 48-trace reflexion profiles were also obtained along all refraction profiles by R.V. *Résolution*. Bathymetric and gravity data were acquired on all profiles by R.R.S. *Shackleton*.

To aid the seismic and gravity modelling, a multichannel seismic reflexion profile extending from the continental shelf to the ocean crust with a recording length of 12 s (I.O.S. line CM-14) was migrated by the Institut Français du Pétrole. We discuss below interpretation of the expanding spread (Camus 1980) and refraction data. The fixed offset data have been discussed in part by Camus (1981) and will be presented in Ginzburg *et al.* (in preparation).

Expanding spread seismic profiles

The principles of two-ship expanding spread seismic reflexion profiling have been discussed by Stoffa & Buhl (1979), and a detailed discussion of the way in which our data was processed can be found in Camus (1980).

In this experiment, R.R.S. *Shackleton* acted as shooting ship with a 4×1000 in^3 airgun and R.V. *Résolution* as receiving ship with a 48-trace (2400 m) streamer. The time interval between shots was 80 s, resulting in a shot–receiver spacing increment of 400 m at a speed relative to the sea bed of 9.3 km h^{-1} maintained by each ship. The position of the expanding spread profiles (e.s.p.) on either side of the continent–ocean transition is shown in figure 3*b*. The recorded data were subjected to wave number and frequency filtering before sixfold stacking to ensure maximum enhancement of signal.

Interpretation of the expanding spread data has been made with reference to individual shots (48 traces per shot), as well as to the stacked section.

Individual shots in expanding spread seismic profiles

The slope(s) and intercept(s) of refracted arrivals were measured for each individual shot. This method was used to take advantage of the precision with which slopes could be measured. The multiple measurements of velocities along the line were used to assess variations in apparent velocity within a given layer. Examples of some traces displayed by using different reduction velocities are shown in figure 4.

The range of apparent velocities and intercept times are given in table 1.

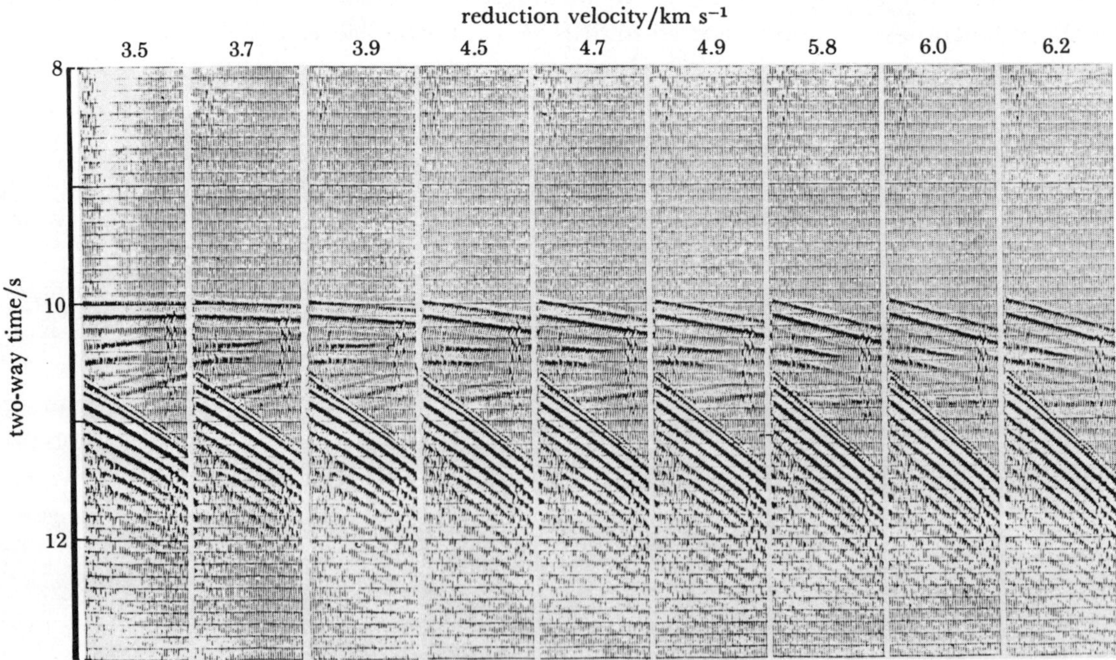

Figure 4. Expanding spread profile 1: examples of 48 traces per shot data displayed at different reduction velocities (from Camus 1980).

TABLE 1. APPARENT VELOCITY AND INTERCEPT DATA (INDIVIDUAL SHOTS), EXPANDING SPREAD LINES 1 AND 2

line 1 (oceanic crust)		line 2 (thinned continental crust)	
velocity/(km s^{-1})	intercept/s	velocity/(km s^{-1})	intercept/s
2.0	—	2.0	—
3.65–3.8	6.4–6.75	3.5–3.7	6.3–6.4
4.4–5.3	7.6–8.4	4.5–4.7	7.2–7.3
6.5–6.8	8.5–8.8	6.1–6.7	8.1–8.4
7.3–7.8	9–9.3	8.1–8.7	8.7–8.7

In the line 1 (ocean crust) data, the 6.5–6.8 km s^{-1} arrivals have come from the top of the oceanic layer 3 and the 7.3–7.8 km s^{-1} arrivals are from the Moho.

In the line 2 (thinned continental crust) data, the 4.5–4.7 km s^{-1} velocity is a refraction from the top of the tilted blocks. The 6.1–6.7 km s^{-1} refractor is the top of the crystalline basement within the tilted blocks, and the 8.1–8.7 km s^{-1} refractor is the Moho.

In comparing the two lines, the scatter of the intercepts is greater on line 1 situated on the oceanic crust. Although the scatter may be due in part to the existence of strong velocity gradients within the oceanic basement, some of the scatter of velocities and intercepts on both lines probably results from the oceanic basement topography on line 1 or the tilted blocks on line 2. For example, a dip of 10° on a 8.2 km s^{-1} refractor will reduce the apparent velocity to 7.55 km s^{-1}.

Stacked expanding spread data

Results from the expanding spread profile along line 2 are shown in figure 5*a*. Each e.s.p. was shot twice, with the two ships first approaching and then going away from each other. Intercept times were measured on both sides of the stacked sections to give averaged apparent refraction velocities and intercept times (Camus 1980).

The results calculated assuming horizontal layering are given in table 2.

TABLE 2

velocity/(km s^{-1})	velocity accuracy (%)	intercept/s	depth/km
line 1 (oceanic crust) from expanding spread data			
1.51	—	—	—
2.35	—	4.95	4.6
3.5	±5	6.65	6.1
4.8	±8	7.8	7.95
6.7	±8	8.85	10.2
7.8	—	9.4	12.5
line 2 (thinned continental crust) from expanding spread data			
1.51	—	—	—
2.4	—	4.7	4.52
3.5	±5	6.4	6.1
4.6	±8	7.3	7.4
6.4	±8	8.25	9.2
8.2	—	8.9	11.2

From these results, travel-time curves for lines 1 and 2 have been calculated and both the refracted and reflected branches are in good agreement with the observed travel-time data (figure 5*b*) (Camus 1980).

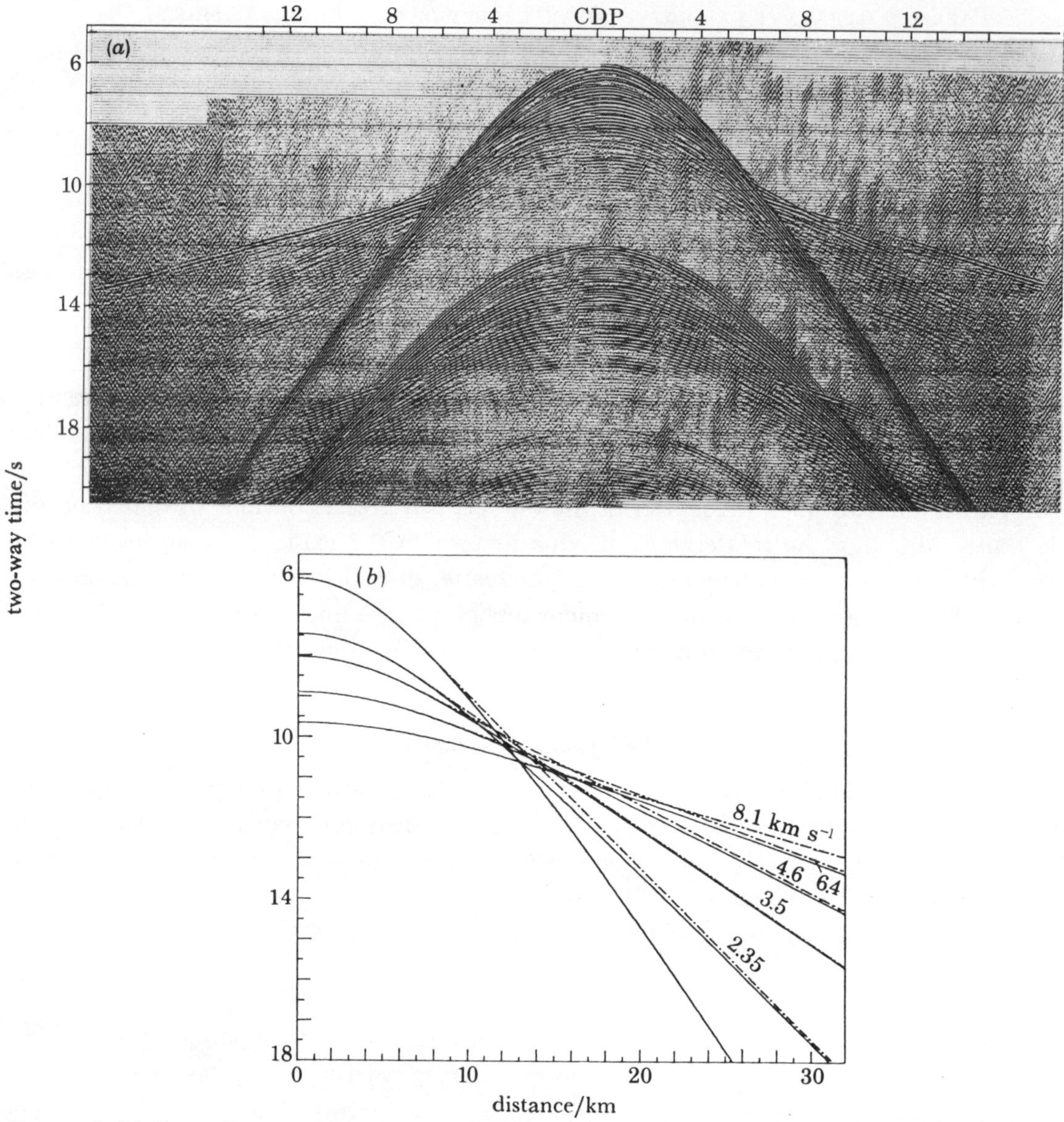

FIGURE 5. (*a*) Expanding spread profile 2. (*b*) Hodochron of expanding spread profile 2. ——, Primary P-type reflexions; —·—, P-type refractions.

SEISMIC REFRACTION PROFILES

Seismic refraction profiles were occupied along lines 1, 2, 3 and 4 by using PUBS and OBS. Data were recorded in analogue form in the PUBS and by pulse-coded modulation on the OBS. A 4×1000 in^3 airgun array firing every 2 min at 2000 lbf in^{-2} (*ca.* 13.8 MPa) was used as the sound source on lines 1, 2 and 3. Line 4 was shot by using explosive charges of 11.3 kg for the near shots and 135 or 270 kg for the far shots. Ranges were computed for each trace for the PUBS data and every 15 shots for the OBS by computer modelling of water-wave travel-times. Sound velocities in the sea down to 2000 m were obtained from velocimeter data (I.O.S., unpublished data) and to greater depths from Fenner & Bucca (1971).

The PUBS data were digitized at 100 samples per second and the digitized data were plotted

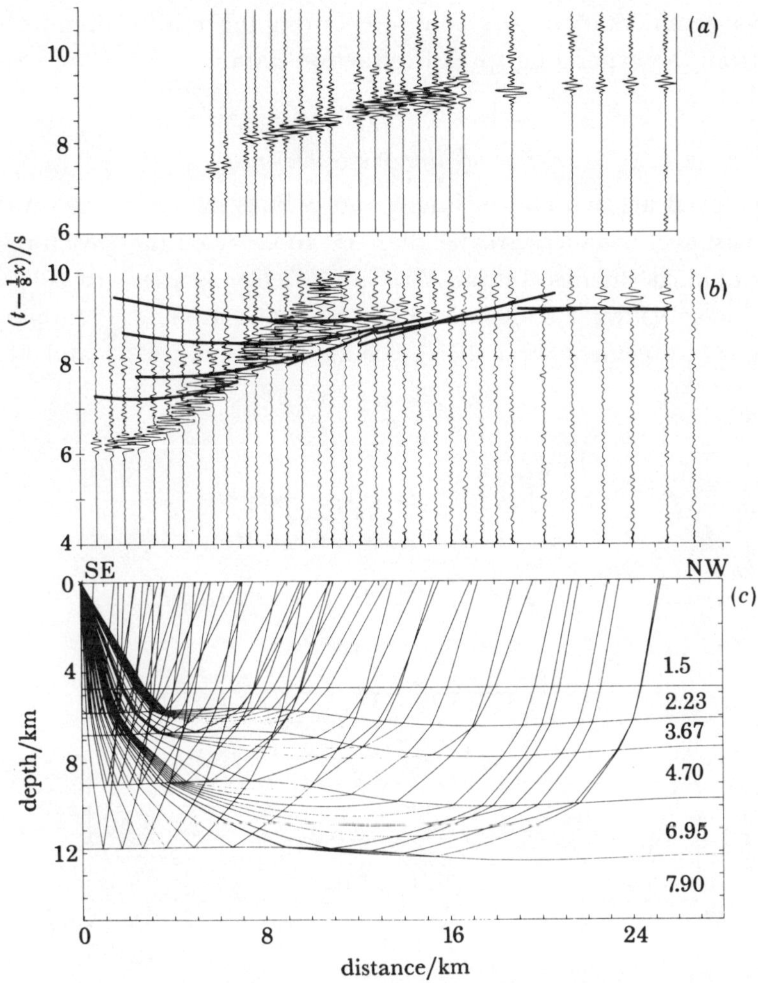

FIGURE 6. (*a*) Synthetic seismogram of Station 6739 (reduction velocity 8 km s^{-1}). (*b*) True amplitude record section (reduction velocity 8 km s^{-1}) of Station 6739 (for location see figure 3*a*). Hodochrons derived from the model in (*c*). (*c*) Ray trace model of the structure along reversed refraction line 6737–6739.

as filtered reduced travel-time record sections. Because of the low signal: noise ratio, particularly at ranges greater than 15 km from the PUBS, three-trace running averages were applied to the PUBS traces and the averaged sections were used for picking arrival times. The OBS data were displayed as unreduced, unaveraged variable-area record sections on which apparent velocities, but not arrival onsets, could be read with confidence.

In interpreting the data, a preliminary model was constructed based on estimates of true velocities from updip and downdip apparent velocities and from depths calculated by standard velocity intercept methods. This model was then fed into a two-dimensional ray tracing programme (Makris 1977), which was used to develop the final models. A timing correction was applied to all the arrival times to extend the ray paths to the sea surface vertically above the PUBS and OBS. The correction was based on a refraction velocity of 8 km s^{-1}. Errors in this assumption will only be significant for arrivals from the uppermost crust not studied in detail here. For Station 6739, amplitudes were modelled by using synthetic seismograms computed by the Fuchs & Muller (1971) reflectivity method to estimate gradients within the oceanic crust (figure 6). In addition, the models were constrained by thickness and interval velocity data for the upper sedimentary layers computed from the multichannel seismic lines and, where available, from the expanding spread refraction data.

Refraction line 1 (*oceanic crust*)

An OBS was deployed at each end of line 1 and a PUBS in the centre. Although refracted arrivals do not persist over distances greater than 35–40 km from the OBS, a reversed refraction profile was thus obtained (figures 3*a* and 7). Calculated velocities are 2.35 and 3.32 km s^{-1} (sediments), 4.44 km s^{-1} (layer 2), 6.6 km s^{-1} (layer 3) and 7.9 km s^{-1} (upper mantle). These velocities are in good agreement with those calculated for lines 6737 and 6739, some 30 km

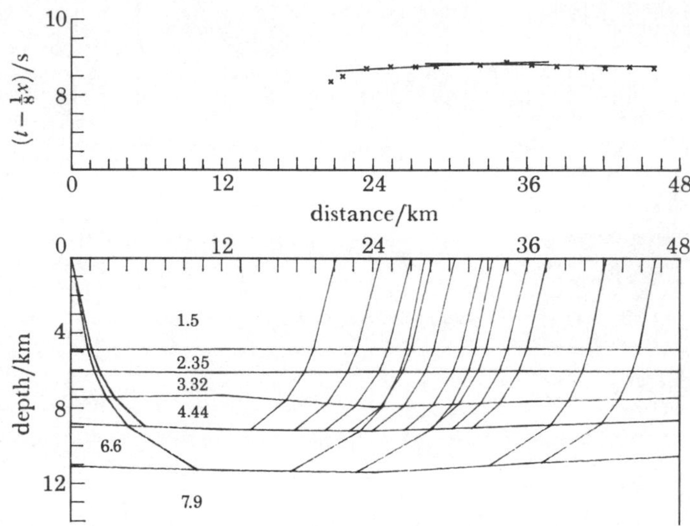

FIGURE 7. Ray trace model of refraction profile 1 (oceanic crust) obtained by using OBS and PUBS. Velocities are in kilometres per second. For location see figure 3*a*.

south of line 1. The 4.44 km s^{-1} refractor corresponds to the 3.5 and 4.8 km s^{-1} refractors observed on the unreversed e.s.p. line 1 (table 2). The existence of velocity gradients was deduced from the 6737 and 6739 data and is confirmed on line 1 by the concentration of energy near the 6.7–6.9 km s^{-1} crossover distance in the PUBS data. The thicknesses of the various layers are: sediments 2.6–3.0 km, layer 2 1.6 km, layer 3 varied from 2.4 km in the west to 2 km in the east (figure 7). The model calculations were based on the arrival times and apparent velocities obtained for the PUBS data because of the poor quality of the OBS data on this line.

Refraction line 2 (*thinned continental crust*)

Two PUBS deployed at each end of the line (figure 3*a*) yielded record sections that have been used to calculate the model in figure 8. The velocities in this model are 3.43 km s^{-1} for the second sediment layer, 4.52 km s^{-1} (Palaeozoic or Mesozoic? sediments), 6.2 km s^{-1} (crystalline basement) and 8.05 km s^{-1} (upper mantle). The 6.2 km s^{-1} layer is 2.5–3.0 km thick

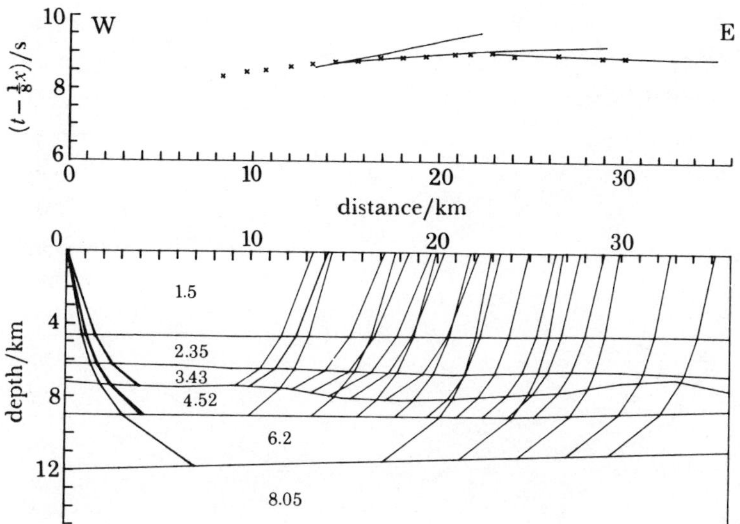

FIGURE 8. Ray trace model of refraction profile 2 (thinned continental crust) obtained by using PUBS and OBS. Velocities are in kilometres per second. For location see figure 3*a*.

and dips gently westward, as does the Moho. The configuration of the interfaces above the crystalline basement was obtained from the multichannel reflexion profiles.

Refraction profile 3

Refraction profile 3 (figure 3*a*) extends across the entire margin from the shelf to the Biscay Abyssal Plain. The profile was designed to examine the thinning of the continental crust as well as the transition from thinned continental crust to oceanic crust. The profile connects the two transverse refraction profiles discussed above, and profile 4. Three PUBS and three OBS were deployed along the line (figure 3*a*). Although it was originally intended to use explosive charges to obtain good arrivals at long ranges, this proved impossible and the airgun array was used perforce, thereby severely limiting the quality of the data.

The velocity structure used to constrain the refraction models at the south end of line 3 is derived from the reversed refraction profiles 1 and 2. After allowing for variations in sediment thickness obtained from multichannel seismic profiles along line 3 (lines IFP-3 and IOS CM-14, figure 3*b*), all variations in apparent velocity have been attributed to dip variations and not to lateral velocity variations within layers. Each reversed refraction profile between adjacent PUBS or OBS along line 3 was interpreted independently to obtain a section-by-section model for the whole line.

PUBS 4 was located on oceanic crust at the southern end of line 3 (figure 3*a*). The split refraction profile recorded by this PUBS covers the segment of line 3 from just south of line 2 to the Southern end of the line. The velocity distribution used in computing the model was that of line 1. The segment of the split profile south of PUBS 4 is shown in figure 9. Layer 3 apparently thins southward from 3.0 to 2.5 km.

The next segment of line 3 was recorded from PUBS 2 situated on thinned continental crust 15 km north of refraction profile 2 (figure 3*a*). Readable arrivals were observed to a range of 30 km to the south of PUBS 2 (figure 10). Four distinct arrivals are present and correspond to

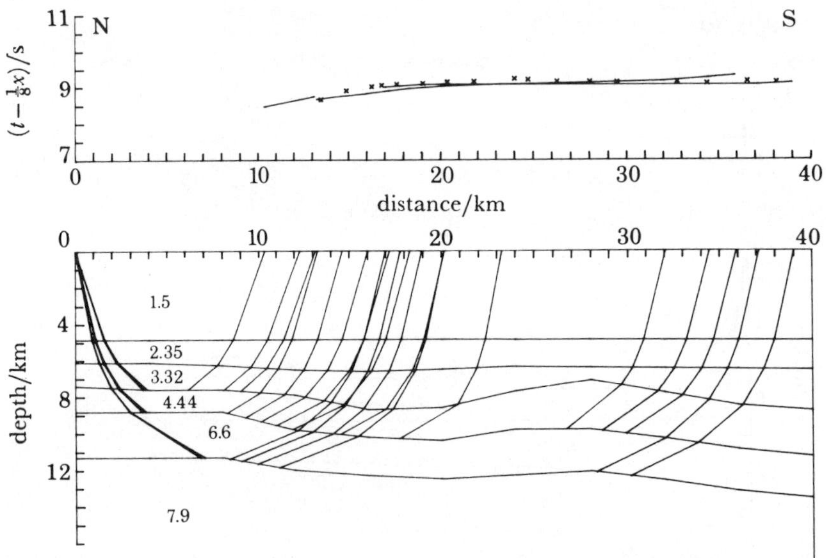

FIGURE 9. Ray trace model of reversed segment of line 3 from PUBS 4 to the south situated on oceanic crust. Velocities are in kilometres per second. For location see figure 3*a*.

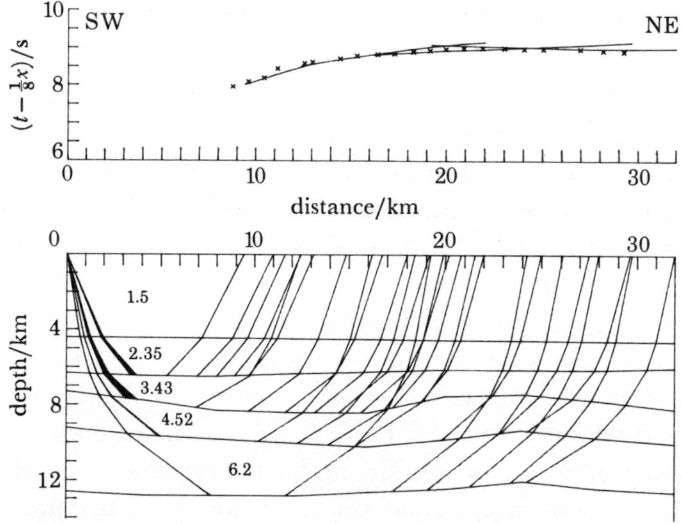

FIGURE 10. Ray trace model of reversed segment of line 3 from PUBS 2 to the south on thinned continental crust immediately north of the continent–ocean transition (cf. figure 9). Velocities are in kilometres per second. For location see figure 3*a*.

the 3.43 and 4.52 km s^{-1} sedimentary layers, the 6.2 km s^{-1} crystalline basement and the Moho. North of PUBS 2, however, arrivals were recorded only out to a range of 25 km and the Pn arrivals were thus not observed. However, the Pg arrivals define the configuration of the crystalline basement. The velocity distribution used in the models is derived from refraction line 2.

The results of modelling the refraction profiles from PUBS 2 and 4 provide some constraint on the position of the continent–ocean boundary. Modelling of the observations made at

PUBS 2, with the use of the velocity distribution calculated for refraction profile 2, gave good agreement between calculation and observation. However, use of the velocity distribution obtained for line 1 situated on ocean crust did not give good agreement. The converse was found in attempting to apply the results of refraction line 1 to the data obtained at PUBS 2. The discrepancy indicates that a major lateral change in crustal structure, here interpreted as the continent–ocean transition, is present within the region between PUBS 2 and PUBS 4 and probably is located between PUBS 4 and refraction profile 2 (figure 3*a*).

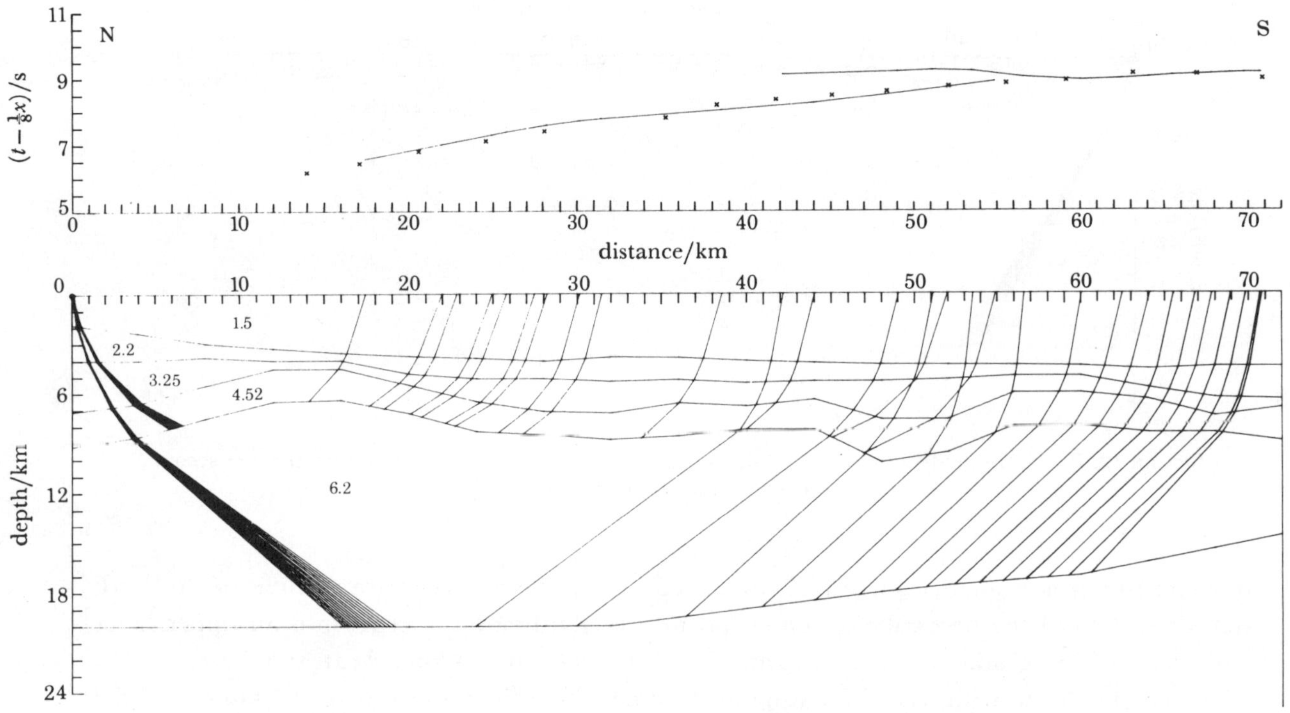

FIGURE 11. Ray trace model of reversed segment of line 3 from OBS JULIE to the southward. Note the progressive deepening of the Moho toward the continent and thickening of the crystalline crust (V_P = 6.2 km s^{-1}). Velocities in kilometres per second. For location see figure 3*a*.

The record section south of OBS JULIE (figure 3*a*) has been modelled with the use of the results from PUBS 2, refraction line 2, and the multichannel reflexion profiles IFP-3 and IOS CM-14. The signal:noise ratio of the record section is low and it is fairly certain that picks at ranges over 50 km do not represent the onset of the event. None the less, the model, which is fairly well constrained by the above data, shows the progressive thickening of the crust, and particularly the lower crust, towards the shelf (figure 11). Unfortunately, because of the poor quality of the data, P* and PIP arrivals could not be identified in the later arrivals so that the position of the discontinuity between the upper and lower crust and its behaviour in relation to the crustal thickening could not be determined.

Refraction profile 4

The refraction profile 4 (figures 3*a* and 12) is unreversed and was shot on the shelf across line 3 by using explosives. No reflexion data were available along this line and there is thus very little control on the depth and attitude of the sedimentary layers. However, a prominent

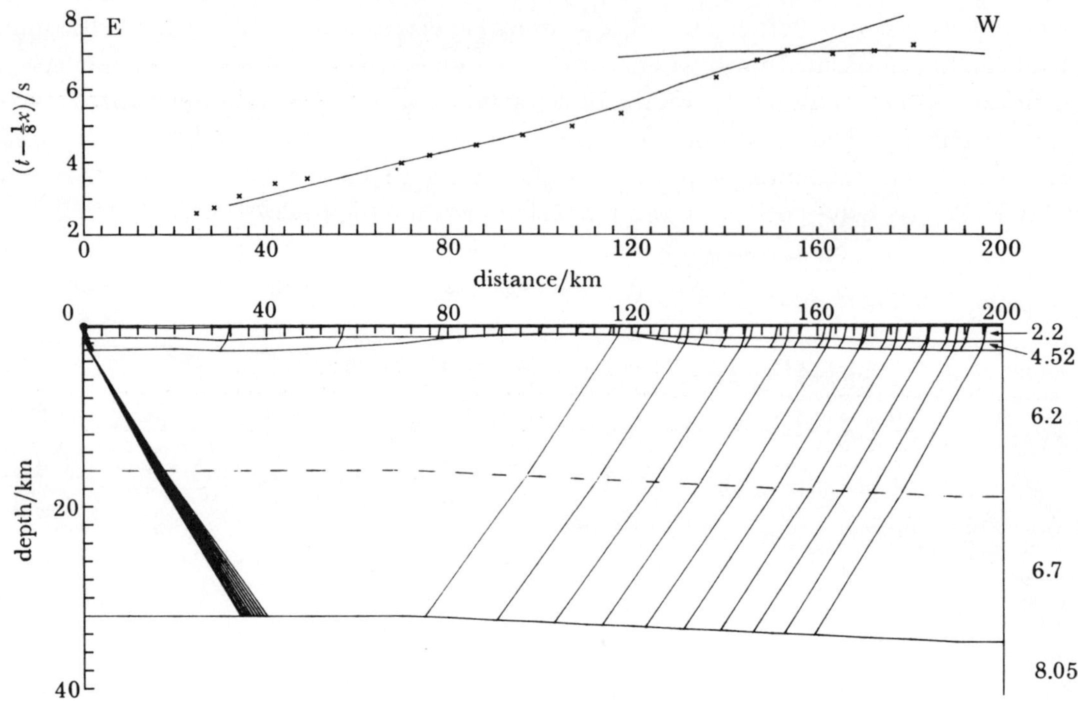

FIGURE 12. Ray trace model of unreversed line 4 shot on the shelf. Velocities in kilometres per second. For location see figure 3*a*.

basement high is evident from travel times. To satisfy the observed travel-time of the PmP reflexion, it has been necessary to somewhat arbitrarily divide the crust into an upper layer with a velocity gradient from 6.2 km s^{-1} at the top to 6.4 km s^{-1} at the base, and a lower crustal layer with velocities ranging from 6.4 km s^{-1} at the top to 6.7 km s^{-1} at the base.

Survey of the refraction interpretation

A summary true-scale section of the distribution of the principal refracting interfaces along line 3 is shown in figure 13, together with the positions of the OBS and PUBS. Depths of the interfaces have been interpolated between OBS JULIE on line 3 and line 4 on the shelf. Uncertainties in the data are shown by dashed lines. The continent–ocean transition is inferred to lie within a narrow zone 8–10 km wide, immediately south of refraction profile 2, from the following three lines of evidence. Firstly, ray trace modelling of crustal structure south of PUBS 4 made with the use of the velocity distribution determined for the thinned continental crust along line 2 does not provide satisfactory agreement with the observed refracted arrivals. This indicates that the transition must occur north of PUBS 4 and south of refraction line 2 and PUBS 2. Secondly, the position of the transition in this region is independently suggested by the marked difference in the velocity gradients observed in the basement along lines 1 and 2. Thirdly, the refracted arrivals observed on the 30 km fixed offset profile recorded along the southern part of line 3 also show a significant change in this region. South of the inferred transition, only mantle arrivals were observed, but to the north arrivals from both basement

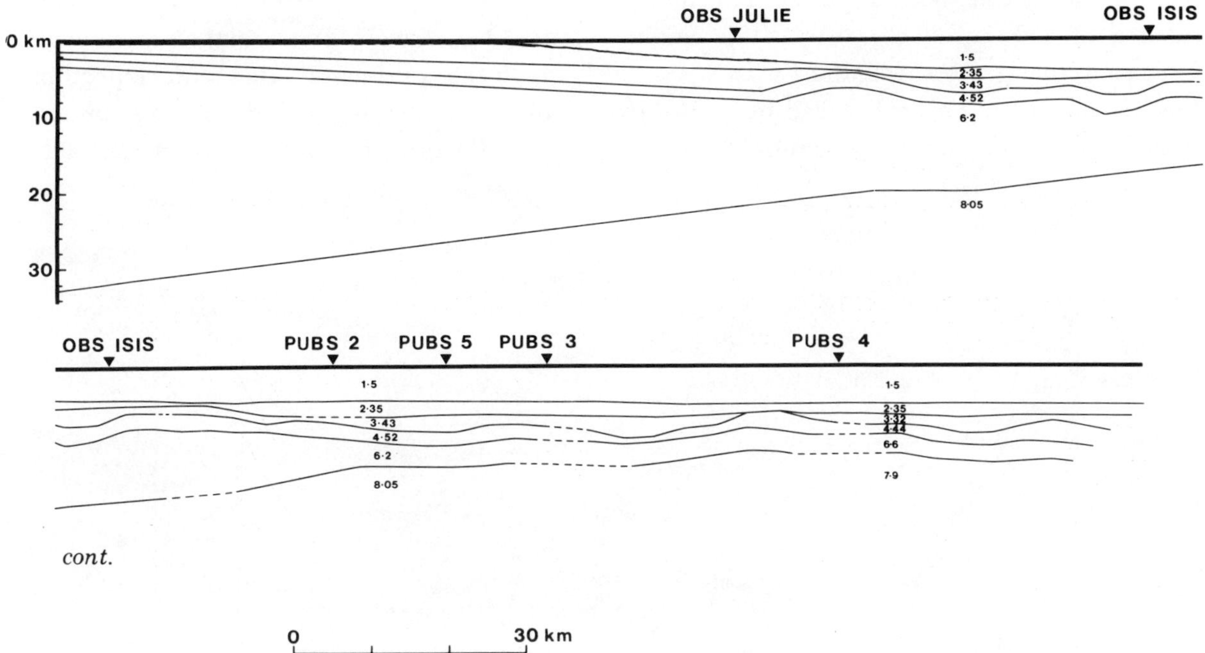

FIGURE 13. Summary true-scale section of refraction results along line 3. No vertical exaggeration. Velocities in kilometres per second. Interfaces have been interpolated between OBS JULIE and the shelf. Uncertainties are shown by dashed lines. The continent–ocean transition lies between PUBS 4 and PUBS 2. For location of section see figure 3*a*.

and mantle are present. A distinct zone of about 10 km width from which no arrivals were recorded coincides with the transition zone identified from the above data (Camus 1981).

The composite profile has a number of interesting features that will be noted here and discussed in more detail in the conclusions. Moho depths are very similar on either side of the transition, but mantle velocities are different (cf. 8.05 and 7.9 km s^{-1}). Within the continental crust, the 4.52 km s^{-1} layer remains relatively constant in thickness in contrast to the underlying crystalline basement, which thickens from 3 to 12 km progressively toward the continent.

GRAVITY MODELS

Although the refraction data provided a very good description of the variation in crustal structure across the entire margin, the absence of information on the behaviour of the upper crust – lower crust boundary, as revealed near the foot of the slope by the 'S reflector' on multichannel seismic profiles, was particularly disappointing. As an alternative approach, an attempt was made to model the observed gravity profile by using the results from the refraction experiment.

Initially, density boundaries were not introduced within the continental crust in order to test the agreement between calculated and observed free-air gravity anomalies by using the refraction data alone. Interfaces were taken directly from the seismic models, and the continent–ocean boundary was positioned from the criteria discussed above. Sediment densities were picked from figures in Hamilton (1978), continental crustal densities from the Nafe–

Drake (1963) curve, oceanic crustal densities from Christensen & Salisbury (1975), and mantle densities from Christensen (1966). The model was computed by using the two-dimensional method of Talwani *et al.* (1959). The results (figure 14*a*) show a substantial discrepancy, indicating that greater mass must exist within or beneath the continental crust with respect to the oceanic crust. This excess can be accounted for by lateral density variations in the upper mantle or within the lower continental crust, or both. Further modelling showed that the

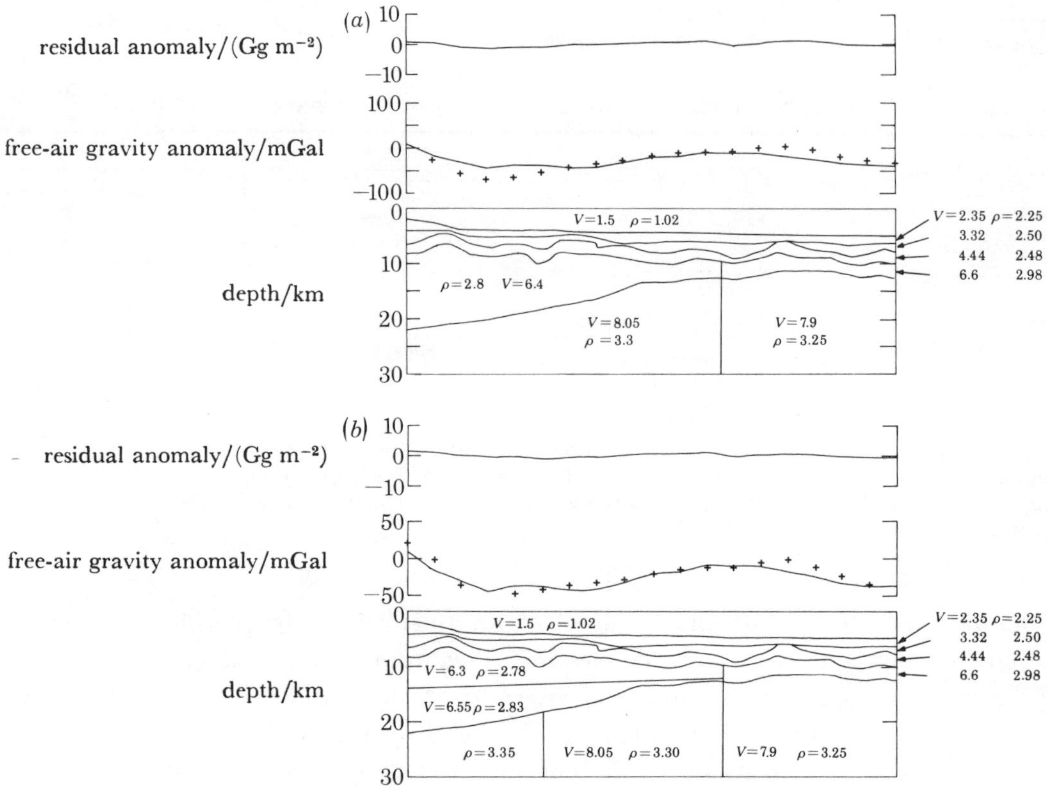

FIGURE 14. (*a*) Calculated (+) and observed (continuous line) free-air gravity profile along line 3. Interfaces have been taken from the summary section in figure 13, and densities from various publications (see text). The density model extended far beyond the limits of the figure and was constrained by the whole section in figure 13. The density contrasts in the upper mantle persisted to 40 km depth. (*b*) Calculated and observed free-air gravity profile along line 3. A density contrast in the crystalline crust at about the level of the 'S reflector' and a lateral variation in mantle density have been used to produce the agreement between calculated and observed anomaly. Otherwise the model is as in figure 13*a*. Note the thinning of the lower crust towards the continent–ocean transition. The gravity residuals can probably be explained by a lack of two-dimensionality in the real world. Note the different gravity scale with respect to (*a*).

required density variations in the upper mantle violate the refraction data and that more reasonable values do not fully account for the observed mass excess. A third model was computed by using a density interface within the crystalline part of the continental crust. The depth of this interface was arbitrarily chosen to correspond to the 'S reflector'. A small lateral variation in mantle density also needed to be included in the model (figure 14*b*). Agreement between calculation and observation is good. In particular, the model supports substantial oceanward thinning of the lower continental crust from 7 km to almost zero at about 100 km from the foot of the slope. Indeed, the possibility that the Moho and the base of the upper crust (S reflector) are coincident cannot be excluded from the gravity interpretation. In marked contrast, the upper part of the crystalline continental crust remains relatively uniform in thickness between the foot of the slope and the continent–ocean transition.

Discussion and conclusions

The results from the seismic refraction study, multichannel seismic reflexion profiles and gravity modelling have been combined in the true scale section across the entire north margin of Biscay displayed in figure 15. The geometry of the fault blocks has been reconstructed from migrated seismic sections (lines IFP-13 and IOS CM-14), subsequently converted to depth by using both interval and refraction velocity data. Depths of the deeper interfaces have been taken from the refraction models. The position of the 'S reflector' is approximate and has been computed from interval velocity data in the sedimentary layers and the 6.2–6.4 km s^{-1} refraction velocity in the crystalline basement.

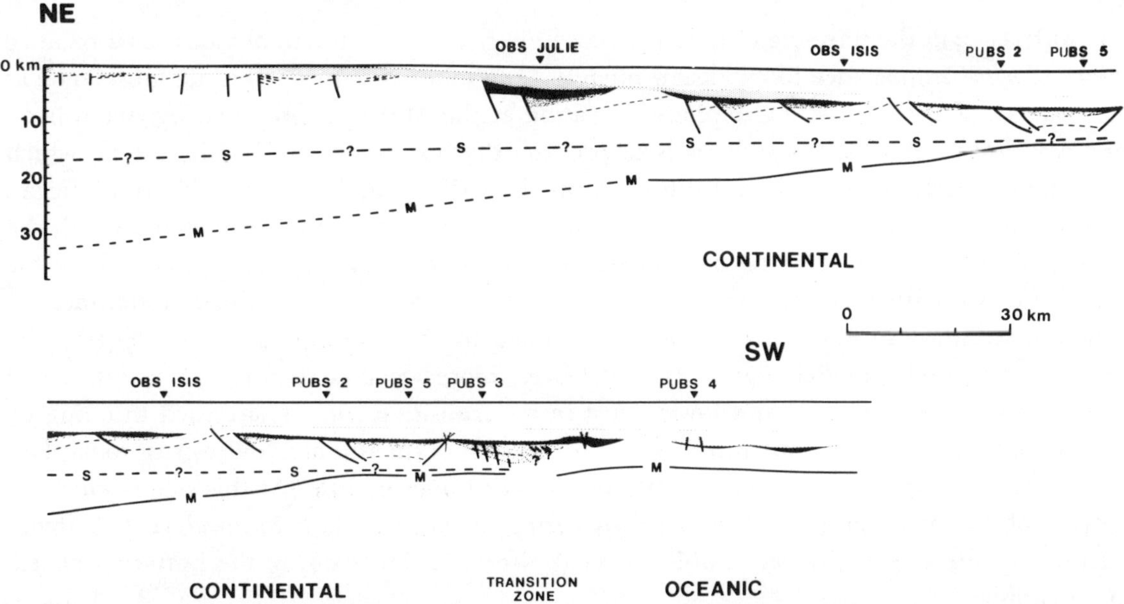

Figure 15. Summary section across the continental margin of north Biscay incorporating the results of the seismic refraction profile along line 3, migrated 24-fold 48-trace multichannel seismic profiles IPF-3 and IOS CM-14, and the gravity interpretation. For discussion see text.

The section has several features that have an important bearing on the problem of the measurement, as well as the mechanism, of crustal attenuation along rifted margins.

The observed position of the 'S reflector' indicates differential thinning within the crystalline continental crust toward the continent–ocean boundary, and the gravity model does not exclude this possibility. The lower crystalline crust thins oceanward near Obs Julie and to a value near zero close to the continent–ocean transition. In this region, the 'S reflector' observed on seismic profiles may be a complex event arising from the intracrustal discontinuity and the immediately subjacent Moho. The section indicates that the larger part of the lower crust has been lost. Thinning is also evident in the upper crystalline part of the crust, but in contrast it has a global β value of between 2.5 and 3 from the shelf to the continent–ocean transition. Over the region of extreme thinning in the lower crust, the upper crust remains relatively uniform in thickness. A third estimate of extension can be derived from the rotation of the tilted blocks. By using a true-scale section computed from the migrated multichannel seismic profile, the quantity of extension β has been found for the blocks between OBS JULIE and the continent–ocean transition. Values range between 1.1 and 1.45 and are similar to the range of values computed by Chenet *et al.* (1981) but are significantly lower than values of about 3 found for the tilted blocks by Le Pichon & Sibuet (1981; see also Foucher *et al.*, this symposium).

Estimates of the extension parameter β found by Le Pichon & Sibuet (1981) for the upper crust on the north margin of Biscay have been interpreted by them in terms of the model of uniform extension of the whole lithosphere developed by McKenzie (1978). However, the results of the refraction survey and these calculations indicate that there has been substantially greater thinning of the lower crust than of the upper. The extension parameter derived for the tilted blocks is not representative of the overall extension of the crust as was earlier noted by de Charpal *et al.* (1979) and Montadert *et al.* (1979). Indeed, according to our results, the global thinning of the crust (4.8) is much greater than that estimated by Le Pichon & Sibuet (1981).

The difference in the thinning of the upper and lower crust creates an obvious mass balance problem of some importance in extension models. One solution is to invoke rift subsidence as a consequence of a gabbro–eclogite phase transition at the Moho, as proposed by Artyushkov & Sobolev (1981). Another hypothesis is to propose that extension of the lithosphere, which perhaps had already been weakened during a previous rifting in Permo-Triassic time, began by brittle failure above and ductile failure below the 'S reflector'. As extension continued, the lower ductile part of the crust became progressively heated and gradually modified by the addition of new mantle that effectively converted the lower crust into a modified upper mantle. The lateral variation in upper mantle density implied by this hypothesis is also suggested by the best-fitting gravity model (figure 14*b*). Clearly, more work is required to assess lateral variations in the upper mantle and lower crust in the transition zone. Continued thinning by brittle and ductile flow with contemporaneous conversion of the thinned lower crust took place until the thinned upper crust rested directly on the hot upper mantle. At this stage, when the outer part of the rifted margin had subsided by rifting to about 2.0 km (Montadert *et al.* 1979), the asthensophere might have been able to break through, thus causing the initiation of sea-floor spreading.

We acknowledge support for this study from the U.K. Department of Energy, the Natural Environment Research Council, the Centre Océanologique de Bretagne, the Institut Français du Pétrole, the Centre Nationale pour l'Exploitation des Océans and the Comité des Études Petrolières Marins. The Masters and ships' crews of R.V. *Résolution* and R.R.S. *Shackleton* are thanked for their cooperation. The work could not have been done without the help of R. E. Kirk, J. J. Langford, P. R. Miles, S. Smith, G. Aubert, R. Conogan and R. Paron on R.R.S. *Shackleton* and M. Cassand on R.V. *Résolution*. The time-break transmitter was kindly lent by Société Nationale des Pétroles Aquitaine.

References (Avedik *et al.*)

Artyushkov, E. V. & Sobolev, S. F. 1981 *Mem. Am. Ass. Petrol. Geol.* (In the press.)

Avedik, F. & Howard, D. 1979 In *Initial reports of the deep sea drilling project*, vol. 48, pp. 1015–1023. Washington, D.C.: U.S. Government Printing Office.

Avedik, F., Renard, V., Buisine, D. & Cornic, J.-Y. 1978 *Mar. geophys. Res.* **3**, 357–379.

Bally, A. W. & Snelson, S. 1980 In *Facts and principles of world petroleum occurrence* (ed. A. D. Miall), pp. 9–94. Canadian Society of Petroleum Geologists.

Biju-Duval, B., Dercourt, J. & Le Pichon, X. 1977 In *Structural history of the Mediterranean basins* (ed. B. Biju-Duval & L. Montadert), pp. 143–164. Paris: Éditions Technip.

Camus, A.-L. 1980 *Rep. Inst. fr. Pétrole*, vol. 28, no. 310. (100 pages.)

Camus, A.-L. 1981 *Rep. Inst. fr. Pétrole*, vol. 28, no. 989. (63 pages.)

de Charpal, O., Guennoc, P., Montadert, L. & Roberts, D. G. 1978 *Nature, Lond.* **275**, 706–711.

Chenet, P. Y., Montadert, L., Gairaud, H. & Roberts, D. G. 1981 *Mem. Am. Ass. Petrol. Geol.* (In the press.)

Christensen, N. I. 1966 *J. geophys. Res.* **71**, 5921–5931.

Christensen, N. I. & Salisbury, M. H. 1975 *Rev. Geophys. Space Phys.* **13**, 57–86.

Fenner, D. F. & Bucca, P. J. 1971 *U.S. Naval Oceanographic Office Informal Report* no. 71–13. (86 pages.)

Fuchs, K. & Muller, G. 1971 *Geophys. Jl R. astr. Soc.* **23**, 417–433.

Hamilton, E. L. 1978 *J. acoust. Soc. Am.* **63**, 366–377.

Holder, A. P. & Bott, M. H. P. 1971 *Geophys. Jl R. astr. Soc.* **23**, 465–489.

Kirk, R. E., Langford, J. J. & Whitmarsh, R. B. 1982 (In preparation.)

Le Pichon, X. & Sibuet, J. C. 1981 *J. geophys. Res.* **86**, 3708–3720.

Makris, J. 1977 *Geophys. Einzelschr.* **34**, 1–124.

McKenzie, D. P. 1978 *Earth planet. Sci. Lett.* **40**, 25–32.

Montadert, L., Roberts, D. G. *et al.* 1977 *Nature, Lond.* **268**, 305–309.

Montadert, L., Roberts, D. G., de Charpal, O. & Guennoc, P. 1979 In *Initial reports of the Deep Sea Drilling Project*, vol. 48, pp. 1025–1060. Washington, D.C.: U.S. Government Printing Office.

Nafe, J. & Drake, C. L. 1963 In *Physical properties of marine sediments of the sea*, vol. 3 (ed. M. N. Hill), pp. 794–815. London: Interscience.

Profett, J. M. 1977 *Bull. geol. Soc. Am.* **88**, 247–266.

Roberts, D. G., Masson, D. G., Montadert, L. & de Charpal, O. 1981 In *Petroleum geology of the continental shelf of northwest Europe*, pp. 455–473. London: Institute of Petroleum.

Royden, R. & Keen, C. E. 1980 *Earth planet. Sci. Lett.* **51**, 343–361.

Royden, L. & Sclater, J. G. 1981 *Phil. Trans. R. Soc. Lond.* A **300**, 219–222.

Royden, L., Sclater, J. G. & Von Herzen, R. P. 1980 *Bull. Am. Ass. Petrol. Geol.* **64**, 173–187.

Sclater, J. G. & Christie, P. A. F. 1980 *J. geophys. Res.* **85**, 3711–3739.

Stoffa, P. L. & Buhl, P. 1979 *J. geophys. Res.* B **84**, 7645–7660.

Talwani, M., Worzel, J. L. & Landisman, M. 1959 *J. geophys. Res.* **64**, 49–59.

Ziegler, P. A. 1978 *Geologie Mijnb.* **57**, 589–626.

Phil. Trans. R. Soc. Lond. A **305**, 27–43 (1982) [27]
Printed in Great Britain

The ocean–continent transition in the uniform lithospheric stretching model: role of partial melting in the mantle*

By J.-P. Foucher†, X. Le Pichon‡ and J.-C. Sibuet†
† *Centre Océanologique de Bretagne, B.P.* 337, 29273 *Brest cédex, France*
‡ *Université Pierre et Marie Curie, Laboratoire de Géologie dynamique,*
4 *place Jussieu,* 75230 *Paris cédex* 5, *France*

The role of partial melting in the uniform lithospheric stretching model of continental margin formation is explored. It is shown that the transition from continental lithosphere stretching to oceanic accretion is most probably controlled by the production of a significant amount of partial melting in the asthenosphere immediately below the lithosphere, which requires stretching factors larger than 3. It is also shown that, at stretching factors exceeding 2, the law of subsidence is significantly changed by the presence of partial melt in the underlying asthenosphere. The implications for the existence of deep continental margin basins on thinned continental crusts are examined. The Armorican deep continental margin basin is taken as an example.

Introduction

A simple uniform stretching model of subsidence (McKenzie 1978*a*, *b*) has been applied at various continental margins (see, for example, Royden & Keen 1980; Royden *et al.* 1980; Cochran 1981; Le Pichon & Sibuet 1981). We use this model to discuss more specifically the nature and mode of transition from the stretched continental lithosphere to the accreted oceanic lithosphere. In particular, we examine whether the stretching model can account for the existence of deep continental margin basins. The Armorican deep continental margin basin is taken as an example.

The method proposed by Le Pichon *et al.* (1982) for the simple stretching model is adopted, which allows us to ignore the density stratification of the lithosphere. This is because, as a first approximation, the lithosphere is floating on top of the asthenosphere, and hence subsidence is controlled by the existence of two reference levels, one near 3.6 km and the other near 7.8 km water depth. These are the levels that would be reached by the asthenosphere in the absence of lithosphere and of formation of oceanic crust. The first one is for hot asthenosphere; the other one for asthenosphere cooled to thermal equilibrium. The instantaneous (Z_i) as well as the total subsidence after an infinite time (Z_t) can then be expressed simply as a function of the difference of elevation between the starting level and the 3.6 and 7.8 km reference levels, respectively. Thus confining ourselves for simplification to basins below water, we have

$$Z_i = \gamma(3.6-E)$$

and

$$Z_t = \gamma(7.8-E)$$

where E is the starting water depth, $\gamma = 1-1/\beta$ and β is the stretching factor (see figures 1 and 2).

In the continental margin model discussed by Le Pichon & Sibuet (1981) and Le Pichon *et al.* (1982), the amount of stretching increases from no stretching ($\beta = 1$, $\gamma = 0$) on the continental shelf to a maximum value, β_{max}, beyond which oceanic lithosphere accretion starts. It was argued by Le Pichon & Sibuet (1981) that, although instantaneous subsidence can theoretically

* Contribution no. 754, Centre Océanologique de Bretagne.

reach a limit value of 3.6 km for infinite stretching, oceanic accretion will probably start much earlier and will be increasingly likely to occur once the water depth exceeds 2.5 km, which is the level reached by new oceanic crust at mid-ocean ridge crests. However, it was proposed that stretching may occasionally exceed the value corresponding to 2.5 km subsidence, thus producing stretched continental crust that is deeper than the adjacent oceanic crust. As a result, a deep continental margin basin will be created.

The amount of uniform stretching necessary to bring the surface of the lithosphere in isostatic equilibrium from sea level to 2.5 km water depth is 3.2 ($\gamma = 0.69$; see figure 2). It has been shown by Le Pichon & Sibuet (1981) and Le Pichon *et al.* (1982) that such large values of stretching are

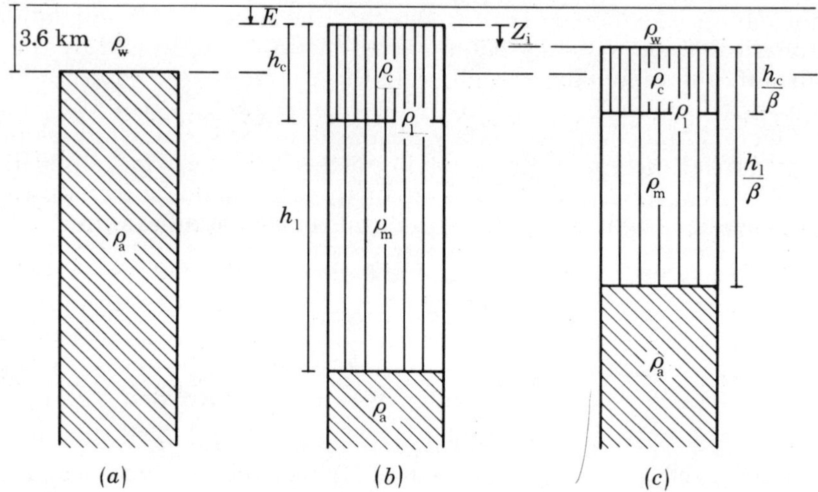

FIGURE 1. Initial stretching phase and isostatic equilibrium. Oblique hatched pattern, asthenosphere; vertical hatched pattern, lithosphere. Average densities are ρ_a (asthenosphere), ρ_m (mantle portion of lithosphere), ρ_c (crust), ρ_l (whole lithosphere), ρ_w (water). (*a*) Hypothetical column with lithosphere entirely replaced by asthenosphere (which is isostatically equivalent with $\rho_a = \rho_l$). This enables definition of the reference levels (mantle geoid and asthenosphere geoid) at 3.6 and 7.8 km respectively. (*b*) Lithosphere before stretching: h_c and h_l are thicknesses of crust and lithosphere respectively. E is the starting water depth. (*c*) Lithosphere just after instantaneous stretching by a factor β. The subsidence relative to (*b*) is Z_i.

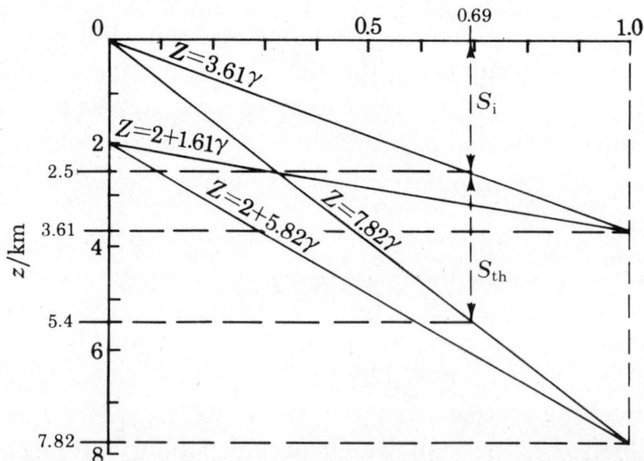

FIGURE 2. Initial instantaneous subsidence and total subsidence as a function of relative thinning of the lithosphere for two different starting elevations: 2 km above sea level and sea-level. S_i and S_{th} are the initial and thermal subsidences. The 2.5, 3.61 and 7.82 km levels are identified.

indeed measured on the deeper portion of the Armorican continental margin and that consequently the model could be accepted at least as a first approximation.

The reason why the transition from continental lithosphere stretching to oceanic accretion should occur for a given stretching factor β was, however, not discussed in detail by Le Pichon & Sibuet (1981). They did mention the probable role played by the increase in the amount of partial melting as the base of the lithosphere is raised by stretching but did not try to discuss it quantitatively. That this transition does not occur randomly at widely different stretching factor values is proved by the fact that the transition from stretched continental crust to oceanic crust, where this is documented, does not seem to be marked by a significant topographic step (see, for example, Montadert *et al.* 1971*a, b*). Thus, it can be concluded that in general the stretched continental crust had reached a level within 500 m or less of the level of emplacement of early oceanic crust when this transition occurred. Such a coincidence is unlikely to be fortuitous and should be controlled by a physical mechanism. Small differences of 500 to 1000 m, when they occur, however, are geologically very significant since they control the existence of early deep continental margin basins. It is consequently necessary to examine the mechanisms that control the transition from the stretching mode to the accreting mode to be able to discuss possible variations at the origin of the basins. In the following, we briefly present the geology and geophysics of the Armorican deep continental margin basin, then we discuss how partial melting may be the mechanism playing the dominant role in provoking the transition from stretching to accretion. We show that taking into account this partial melting introduces significant modifications in the curves of figure 2. We then discuss the possible origin of deep continental margin basins based on this mechanism.

The northern Bay of Biscay continental margin

Over the past 10 years, a considerable number of data have been acquired including those of *Glomar Challenger* Legs 48 and 80. The structure of the northern Bay of Biscay margin is featured by a series of horsts, graben and tilted fault blocks buried beneath a thin sedimentary cover that has been only slightly affected by post-rifting tectonics (Roberts & Montadert 1980). The fault blocks parallel to the margin occupy the area between the shelf edge and the ocean–continent boundary. They affect either the Hercynian basement in the western part of the Celtic margin (Pautot *et al.* 1976) or a pre-existing continental basin as shown by the presence of reflectors within the blocks (Montadert *et al.* 1979*a, b*) and the results of Hole 401 drilled in a tilted fault block (Montadert *et al.* 1979*a*). The present structure of the margin is mainly the result of a tensional phase 20–40 Ma long occurring in a submarine environment during Lower Cretaceous time (Montadert *et al.* 1979*b*; Sibuet & Ryan 1979). Le Pichon & Sibuet (1981) and Le Pichon *et al.* (1982) have shown that extensional values as large as 3 are calculated from the geometry of tilted fault blocks. The calculated amount of thinning for the brittle portion of the crust is comparable with the tinning of the whole continental crust deduced from seismic refraction measurements and required by the uniform stretching model for the whole lithosphere.

A deep margin basin, located on the thinned continental crust, is observed along the northern Bay of Biscay (figure 3) except in the western part of the Celtic margin, where the oceanic crust is directly in contact with the continental slope, a 0.5–1.0 km vertical offset being often observed. From refraction data (Avedik & Howard 1979) and density data from Leg 48, the interpreted profile of figure 3 is shown in figure 4. If the whole sedimentary cover, taken to be in local isostatic

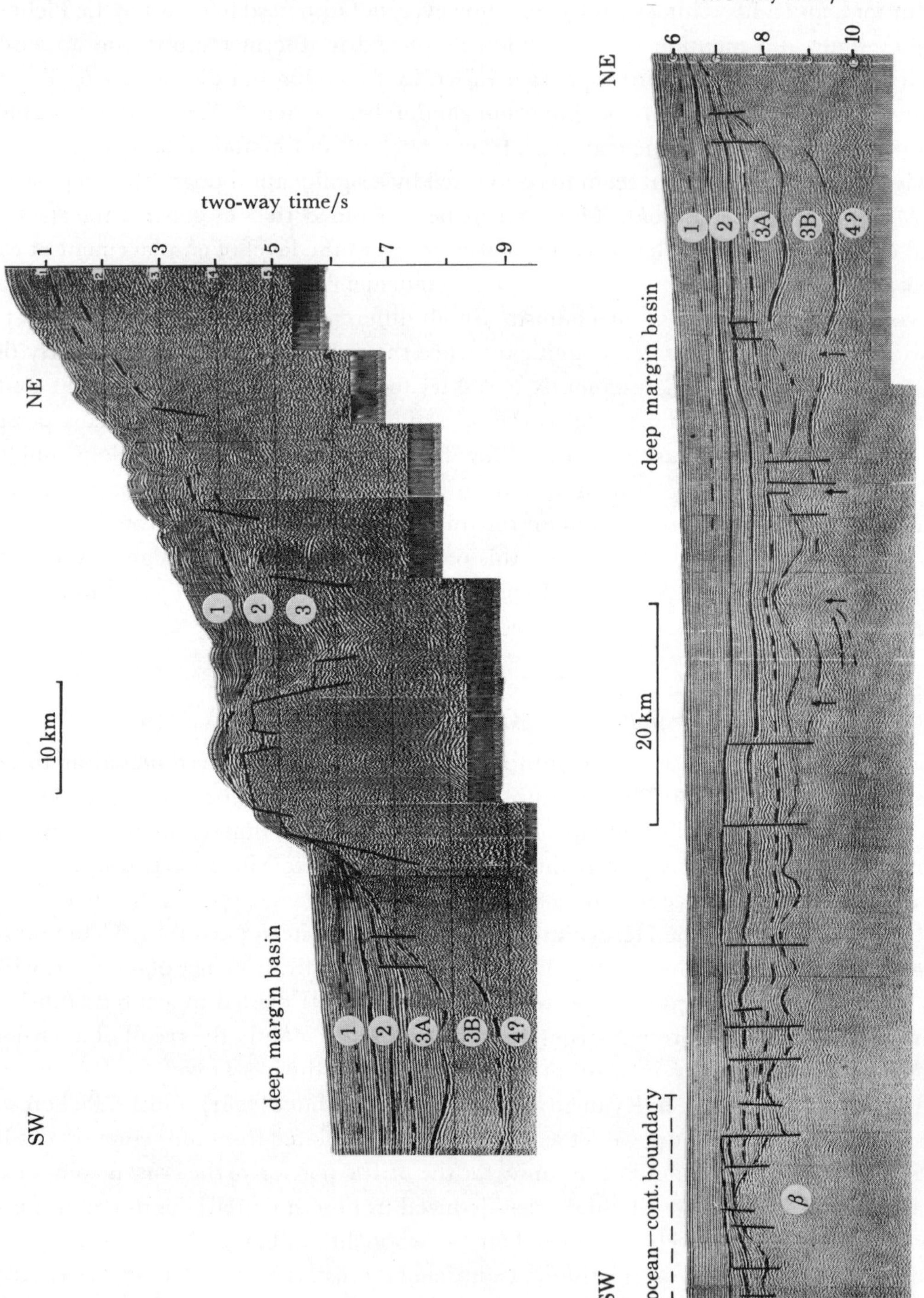

FIGURE 3. C.E.M. 017 seismic reflexion section across the Armorican margin (from Montadert *et al.* 1971 *a, b*). The location of the seismic section is shown in the inset to figure 4. Note the rough topography of the reflector at the top of the lower sedimentary unit 3 B.

equilibrium, is removed, a slight deep-margin basin, about 0.5 km deep, remains on the thinned continental crust and is fringed by the oceanic crust or by a small dam generally less than 0.5 km high. Below the deep margin basin, the inferred thickness of the thinned continental crust, based on the interpretation of free-air anomalies in terms of local as well as regional isostatic equilibrium (Lalaut 1980), is about 3 km. This value is in agreement with the interpretation of a refraction line shot in the deep margin basin about 50 km southweastwards of the seismic section of figure 3 (Limond *et al.* 1974). In summary, the deep margin basin, located on a greatly thinned continental crust about 3 km thick, is, in the absence of sediments, a slight geological feature 0.5 km or less deeper than the adjacent oceanic crust.

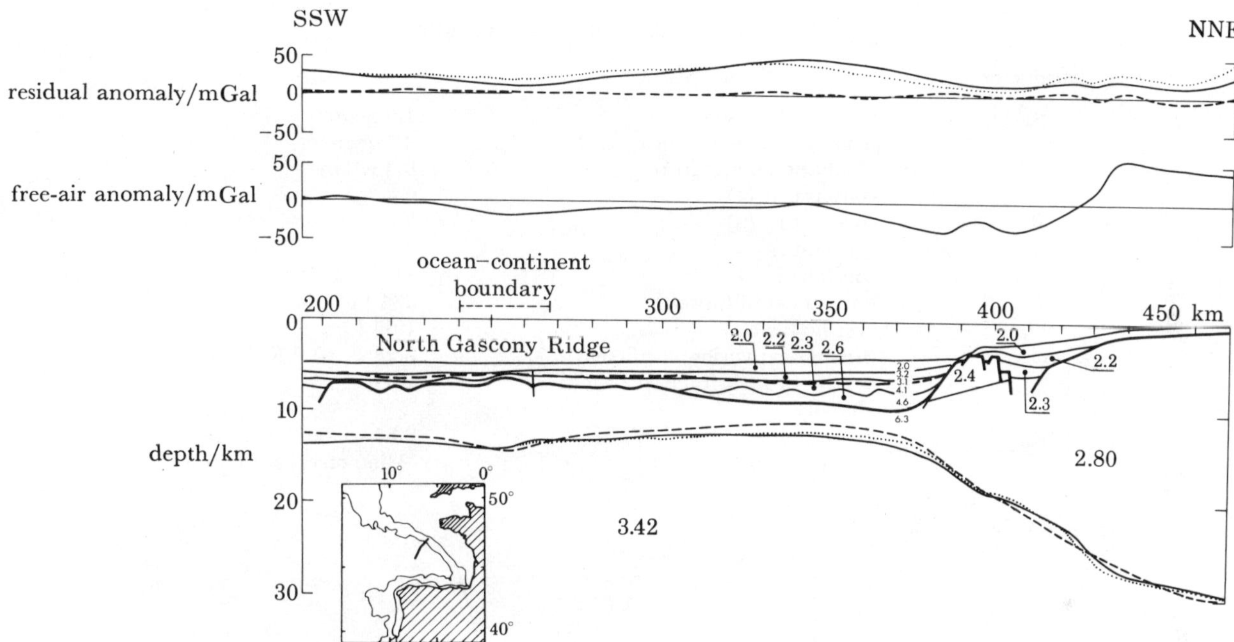

Figure 4. Free-air anomaly profile corresponding to the seismic reflexion profile of figure 3. Densities in grams per cubic centimetre. Oblique numbers are seismic refraction velocities (Avedik & Howard 1979). The base of the crust is obtained assuming local (solid line) and regional (dotted line, with flexural parameter $\alpha = 50$ km) isostatic equalibrium. The broken line corresponds to the best-fitting model. The position of the basement after correction for the effect of the sedimentary load is shown by a broken line.

Partial melting in the upper mantle at large stretching factors

The base of the lithosphere rises from h_l to h_l/β during extension by a factor β, which results in the ascent between h_l and h_l/β of hot upper mantle material at the temperature of the asthenosphere, T_a, which we take as the temperature at depth h_l in the mantle (figure 1). At small stretching factors, the pressure drop in the ascending hot upper mantle remains insufficient to produce partial melting, which means that the ascending upper mantle does not cross its solidus. This is not so at large stretching factors when the ascending upper mantle crosses the solidus, then producing partial melting. The critical extension value β_c, beyond which the stretching process produces partial melting as well as the amount of partial melting in the upper mantle at a given stretching factor β, $\beta > \beta_c$, is primarily controlled by the law assumed to describe the dependence of partial melting on temperature and pressure in the upper mantle. Although this law is poorly known, for discussion purposes we adopt that proposed by Ahern & Turcotte (1979) as it has a simple analytical expression:

$$f = A\{\exp[B(T - Cz - D)] - 1\}, \tag{1}$$

where f is the degree of partial melting, T is temperature, z is depth and A, B, C, D are constants with values given in table 1. The empirical equation (1) accounts to a fairly good approximation for the experimental data with 0.01 % water reported by Ringwood (1975) (see also fig. 2 in Ahern & Turcotte 1979). Thus, in the following section, we propose to apply the equation of Ahern & Turcotte (1979) to illustrate quantitatively how partial melting in the upper mantle varies as a function of the stretching factor β.

Table 1. Values of parameters

parameter		value
ρ_w	density of water	1.0 g cm^{-3}
ρ_a	density of upper mantle at T_a	3.3 g cm^{-3}
G	adiabatic temperature gradient	0.3 K km^{-1}
A	constant in (1)	0.4
B	constant in (1)	3.65×10^{-3} K^{-1}
C	constant in (1)	3.0×10^{-5} K cm^{-1}
D	constant in (1)	1100 °C
L	latent heat of fusion	334 J g^{-1}
C_p	specific heat	1.05 J g^{-1} K^{-1}
α	thermal expansion coefficient	3.28×10^{-5} K^{-1}
ρ_{al}	density of melt fraction (liquid basalt)	2.6 g cm^{-3}
ρ_{as}	density of melt fraction after solidification	$2.85 + 0.00833z$ (z in kilometres; see figure 6)

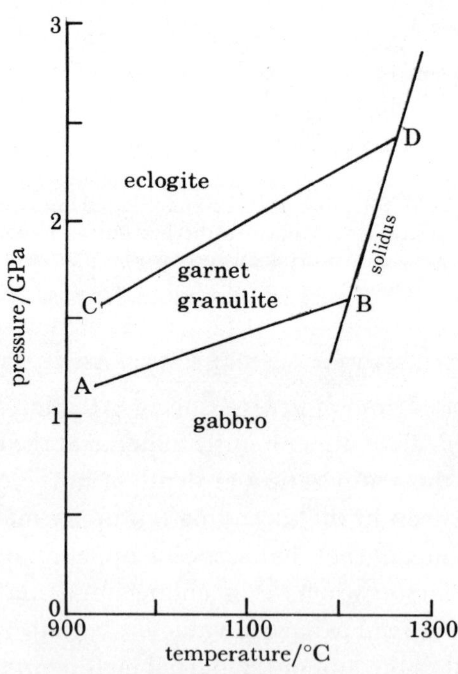

Figure 5. Experimental P–T diagram showing stability fields of eclogite, garnet granulite and gabbro (from Green & Ringwood 1967). The transition from gabbro to eclogite is not sharp but occurs through a 400–800 MPa garnet granulite interval.

Given a stretching factor β and assuming no partial melting ($\beta < \beta_c$), the initial temperature distribution is given by

$$\left.\begin{aligned} T &= (T'_a\beta/h_1)\,z \quad \text{for} \quad 0 \leqslant z \leqslant h_1/\beta \\ \text{and} \qquad T &= T'_a + G(z - h_1/\beta) \quad \text{for} \quad h_1/\beta \leqslant z \leqslant h_1, \end{aligned}\right\} \tag{2}$$

with $T'_a = T_a - G(h_1 - h_1/\beta)$.

In (2) we have introduced the adiabatic temperature gradient G, generally ignored in previous developments of the stretching model, but which should be considered in an approach including a description of partial melting effects because of the high sensitivity of the degree of melting to small changes in temperature. This high sensitivity is illustrated by the steep slope of the solidus (figure 5).

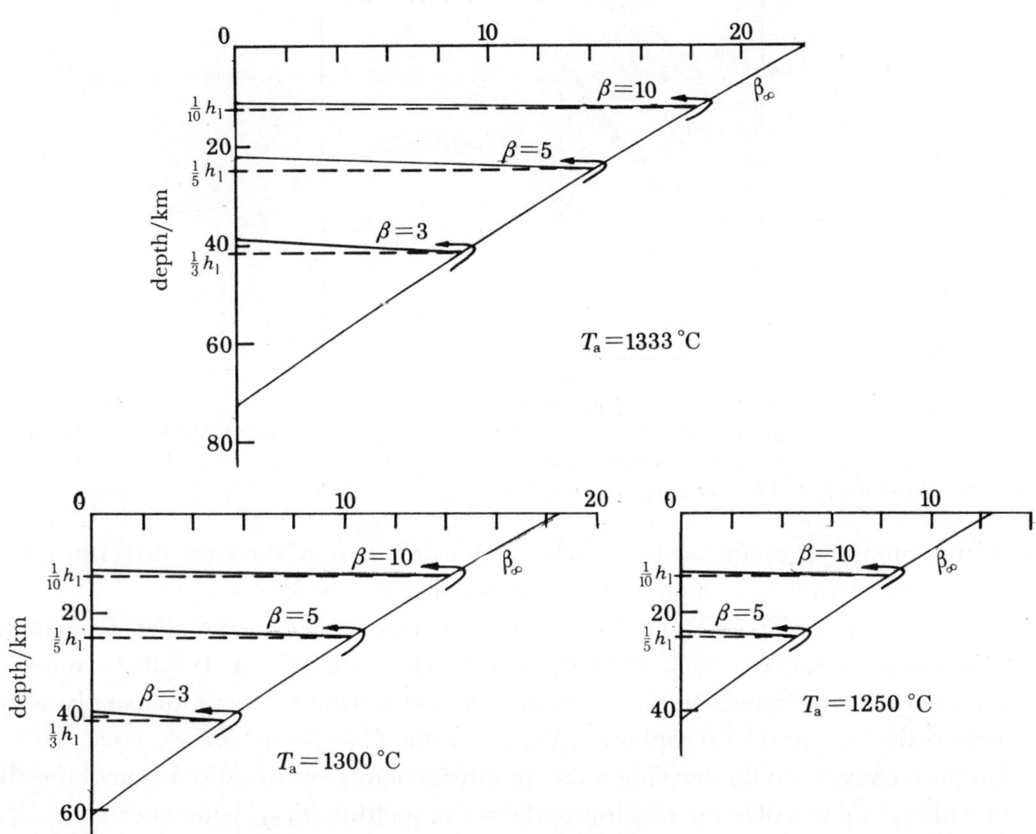

FIGURE 6. Partial melting as a function of depth at various stretching factors (3, 5 and 10) and for three different values of T_a, 1250, 1300 and 1333 °C.

Introducing now partial melting ($\beta > \beta_c$), equations (2) become

$$\left.\begin{aligned} T &= (T'_a\beta/h_1)\,z - fL/C_p \quad \text{for} \quad 0 \leqslant z \leqslant h_1/\beta \\ \text{and} \qquad T &= T'_a + G(z - h_1/\beta) - fl/C_p \quad \text{for} \quad h_1/\beta \leqslant z \leqslant h_1, \end{aligned}\right\} \tag{3}$$

where the term fL/C_p is the temperature decrease due to the extraction of latent heat of fusion, L is the latent heat of fusion and C_p is the specific heat capacity at constant pressure (table 1). Thus, the degree of melting, f, can be calculated from (1) with T in (1) given by (3). The solution can be obtained from simple numerical calculations. Figure 6 displays the results obtained. Partial melting is nearly restricted to the upper mantle underlying the stretched lithosphere. The top of the zone of melting is near h_l/β, its base is at depth z_f obtained from (1), making f equal to zero.

$$z_f = \frac{T_a - Gh_l - D}{C - G} \quad \text{for} \quad \beta > \frac{h_l}{z_f}. \tag{4}$$

Through the zone of melting, the degree of melting f decreases nearly linearly with depth so that to a good approximation the degree of melting f is given by

$$f = f_0(1 - z/z_f), \tag{5}$$

where f_0 is the degree of melting in upper mantle material ascending adiabatically from the base of the lithosphere to the surface.

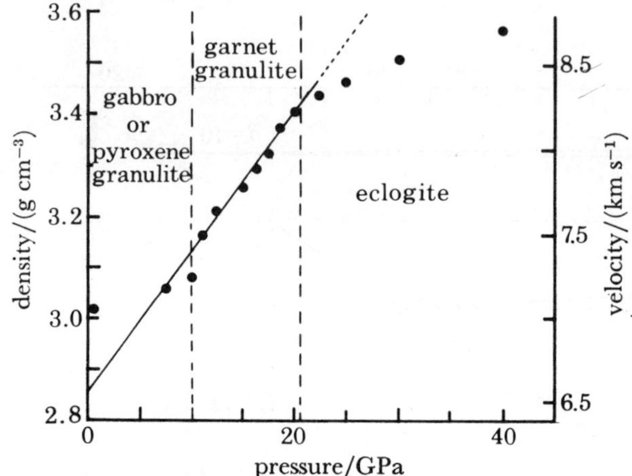

FIGURE 7. Experimental data points reported by Green & Ringwood (1967), showing the increase in density with pressure of alkali basalts. We approximate the density variation with $\rho = 2.85 + 0.00833\ z$ (solid line; z in kilometres, ρ in grams per cubic centimetre).

Thus, partial melting is predicted to occur as soon as the base of the stretched lithosphere is raised to z_f, i.e. corresponding to a critical extension value $\beta_c = h_l/z_f$.

Taking $h_l = 125$ km and $T_a = 1300$ °C, then $z_f = 60.2$ km and $\beta_c = 2.1$. For $T_a = 1333$ °C, $z_f = 72.4$ km and $\beta_c = 1.7$. For $T_a = 1250$ °C, $z_f = 41.7$ km and $\beta_c = 3$. It will be noted that partial melting begins at relatively moderate extension values, from 1.7 to 2.1 for usually assumed temperatures at the base of the lithosphere, 1333 and 1300 °C respectively.

Equation (5) is convenient for deriving some quantities of interest to a discussion of the effects of partial melting, in particular the maximum degree of melting, f_{max}, which occurs at h_l/β, is given by

$$f_{max} = f_0(1 - \beta_c/\beta), \tag{6}$$

and the total equivalent thickness, h_b, of liquid basalt produced is given by

$$h_b = \frac{\rho_a}{\rho_{al}} \int_{h_l/\beta}^{z_f} f_0(1 - z/z_f)\,dz;$$

$$h_b = \frac{\rho_a}{\rho_{al}} f_0 z_f/2(1 - \beta_c/\beta)^2. \tag{7}$$

From (6), taking $T_a = 1333$ °C, f_{max} reaches 5 % at $\beta = 2.9$, and 10 % at $\beta = 3.1$. For $T_a = 1300$ °C, the 5 and 10 % levels are reached at $\beta = 2.9$ and 4.6 respectively. For $T_a = 1250$ °C, $\beta = 5.0$ and 16.0 respectively. Again, the high sensitivity of the results to the assumed temperature T_a at the base of the lithosphere is illustrated. However, it must be emphasized that for reasons of continuity in the application of (1) at the transitions from the stretched continental crust to the oceanic crust, the value of T_a is constrained by the requirement that at β tending to infinity the amount of liquid basalt produced by partial melting is capable of accounting for the creation of a normal 5.5 km thick oceanic crust. From (7), taking $T_a = 1333\ °C$, $h_b = 10.4$ km when β tends to infinity, while for $T_a = 1300\ °C$, $h_b = 7.0$ km, and for $T_a = 1250\ °C$, $h_b = 3.2$ km. Thus, we suggest that realistic descriptions of the partial melting in the upper mantle in the present approach are for values of T_a between 1333 and 1300 °C, the exact value depending on whether it is considered that nearly all of the liquid basalt produced migrates to the surface to form the oceanic crust, or whether only part of it is involved (Ahern & Turcotte 1979; Sleep 1974).

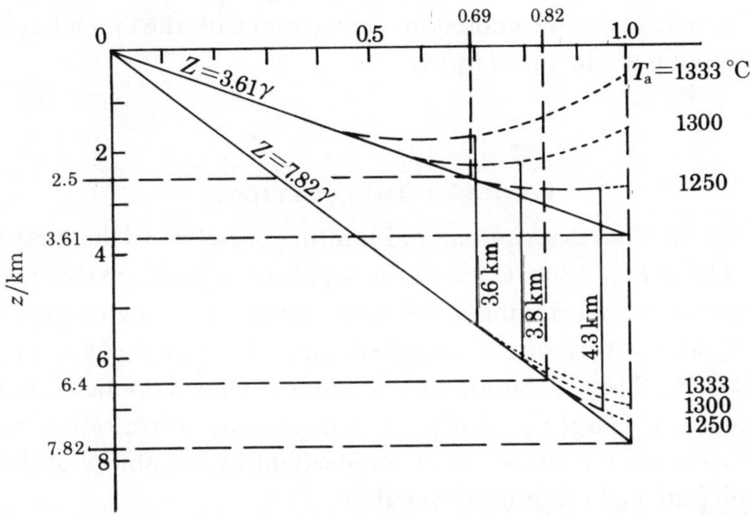

FIGURE 8. Subsidences of figure 2 (solid lines) corrected for the effects of partial melting (broken lines, then dotted lines). Broken lines are for degrees of melting less than 10 %. Dotted lines are for degrees of melting exceeding 10 %. Vertical bars show the amount of thermal subsidence at β corresponding to a degree of melting of 10 %.

Further partial melting introduces significant changes in the mean density of the upper mantle, which result in considerable modifications of the subsidence curves of figure 2. If ρ_{al} is the density of the melt fraction (liquid basalt), the mean density change $\Delta\rho_i$ at $t = 0$, introduced over the whole thickness of the melting zone, i.e. from h_l/β to z_f, is given by

$$\Delta\rho_i = \frac{\displaystyle\int_{h_l/\beta}^{z_f} \rho_{al} - \rho_a/\rho_{al} f\,dz}{z_f - h_l/\beta}\,\rho_a, \tag{8}$$

which implies a correction ΔS_i ($\Delta S_i < 0$ since $\rho_{al} < \rho_a$) to be applied to the initial subsidence S_i:

$$\Delta S_i = \frac{\rho_a}{\rho_a - \rho_w}\frac{\Delta\rho_i}{\rho_a}(z_f - h_l/\beta),$$

or

$$\Delta S_i = \frac{\rho_a}{\rho_a - \rho_w}\int_{h_l/\beta}^{z_f} \frac{\rho_{al} - \rho_a}{\rho_{al}} f\,dz. \tag{9}$$

In (9) we have assumed that local isostatic conditions prevail at depth h_l. Similarly, if ρ_{as} is the density of the melt fraction after cooling and solidification, a correction ΔS_∞ is predicted at thermal equilibrium, which is given by

$$\Delta S_{\infty} = \frac{\rho_a}{\rho_a - \rho_w} \int_{hl/\beta}^{z_f} \frac{\rho_{as} - \rho_a}{\rho_{as}} f \, dz, \tag{10}$$

where $\rho_{as} = 2.85 + 0.00833\, z$, taking into account the change in density due to pressure (see figure 7 and table 1).

Figure 8 shows the resulting subsidence curves after corrections have been applied by adopting numerical values in table 1. We have neglected, in the calculation of the corrections above at $t = 0$, the effects of the temperature drop ΔT due to the extraction of the latent heat of melting,

$$\Delta T = Lf/C_p,$$

which introduces a density change, $\rho_a \alpha \Delta T$, that is $\rho_a \alpha Lf/C_p$, about $0.03\, f$, adopting values in table 1. This density change can be ignored in comparison with the much larger density change $(\rho_{al} - \rho_a)\, \rho_a/\rho_{al} f$, about $0.9\, f$, involved in (8).

General implications

As is seen from the previous description, and limiting ourselves to the most realistic range of cases of $T_a = 1333\ °C$ and $T_a = 1300\ °C$ within our set of assumptions, partial melting is predicted to occur in the upper mantle beginning at relatively small stretching factors, of the order of 2, then reaching 10 % at the base of the stretched lithosphere at $\beta = 3.1$ for $T_a = 1333\ °C$ or $\beta = 4.6$ for $T_a = 1300\ °C$. Thus, a considerable amount of melt is predicted to be present in the upper mantle at moderately large values of β. We suggest that the transition from the stretching process to oceanic accretion is controlled to a large extent by the ability of the large volume of melt produced to migrate and erupt to the sea floor.

Melt migration models have been proposed to account for the segregation of an oceanic crust from a partly molten mantle at mid-oceanic ridges (see, for example, Sleep 1974; Ahern & Turcotte 1979). The first melt produced forms on grain boundary intersections (Waff & Bullau 1979), then with the amount of melt increasing, intersections become interconnected, which makes the matrix more permeable and favours migration due to the buoyancy of the melt fraction. It has been inferred from geochemical studies that migration begins above about 10 % melting at mid-oceanic ridges (Kay *et al.* 1970). A 10 % melt fraction is produced at extensions from 3.1 ($T_a = 1330\ °C$) to 4.6 ($T_a = 1300\ °C$), as discussed above. Such extension values may then be minimum estimates required to initiate oceanic accretion. It should be noted from (7) that, with β increasing beyond the critical value β_c at which partial melting begins, the volume of melt produced will increase rapidly, as illustrated by the fact that at $\beta = 2\beta_c$, one-quarter of the total volume of melt predicted at t_{∞} would be produced. As a result, the probability for the transition from stretching to oceanic accretion process will increase rapidly with β, which suggests that the transition will probably not occur at β greatly in excess of the minimum estimates given above. However, it will be noted that unlike at mid-oceanic ridges the zone of melting in the stretching model does not extend up to the surface but remains confined to the upper mantle underneath the stretched lithosphere. Presumably then, migration will bring the melt material to the base of the lithosphere, where it may accumulate. Large accumulations, however, would be unstable because of the high buoyancy of the melt material. Possible mechanisms for the ascent of the melt material through the thinned lithosphere are propagation along fractures or elastic cracks, and diapiric intrusion. The melt material could ascend and erupt rapidly to the sea floor once it has reached the base of the lithosphere. It follows that the transition to oceanic accretion would then be primarily controlled by the volume of partial melt produced in the upper mantle and its ability to migrate to the base of the lithosphere. If so, previous estimates of minimum extension values required to initiate oceanic accretion remain valid.

Important implications arise from the modified subsidence curves of figure 8, which permit a discussion of the relative subsidence of the stretched continental crust and the adjacent oceanic crust at large stretching factors, when accounting for the effects of partial melting. As is seen from figure 8, for the values of T_a between 1300 and 1333 °C that we considered to be realistic, the stretched continental crust is not expected to subside below 2.5 km during extension, which means, if 2.5 km is taken as the depth of the zero age oceanic crust (Parsons & Sclater 1977), that the stretched continental crust remains shallower than the level of emplacement of the oceanic crust. This is because it is assumed that the melted portion does not solidify rapidly as long as it does not erupt to the surface to form the oceanic crust. Its cooling is controlled by the overall cooling of the lithosphere. This seems to remove the possibility that deep continental margin basins may develop on a stretched continental lithosphere deeper than the adjacent oceanic lithosphere in the early stages of oceanic accretion. Assuming that the maximum depth reached by the oceanic crust is 6.4 km (Parsons & Sclater 1977), it is also seen that the stretched continental crust remains shallower than the oceanic crust for $\beta \lesssim 5$. Further, the total amount of thermal subsidence can be inferred directly from the curves in figure 8. At the 10 % melt fraction produced at the base of the lithosphere, which we considered as a minimum degree of melting required to initiate oceanic accretion, the total thermal subsidence predicted on the stretched continental crust is 3.6 km assuming that $T_a = 1330$ °C, and 3.8 km assuming that $T_a = 1300$ °C. These values would be slighly increased for slightly higher β, but in all cases remain close to the expected thermal subsidence of the oceanic lithosphere, which is about 3.9 km. An important consequence is that no major differential vertical movements at the transition from the stretched continental crust to the oceanic crust are expected to occur during the return to thermal equilibrium after the phase of active extension.

Extension of finite versus instantaneous duration

In the preceding discussion, we have assumed that stretching is instantaneous, i.e. no significant cooling of the stretched lithosphere occurs during the tectonic phase. Jarvis & McKenzie (1980) have shown that this is a reasonable approximation provided the duration of this tectonic phase does not exceed $60\gamma^2$ Ma for $\beta \geqslant 2$. With large stretching factors, as involved in the deeper parts of the stretched continental crust, and for a duration of extension of the order of megayears as may be realistically assumed, the cooling occurring during the tectonic phase cannot be ignored if topographic differences as small as 500 m are significant to the creation of deep basins.

Let us consider, for example, a phase of active extension Δt Ma long. If the extension were instantaneous, the subsidence after a time t would be S_i plus a small portion of the thermal subsidence $S_{th} (S_{th} = S_t - S_i)$, which can be approximated by

$$S_t = S_i + S_{th}(1 - e^{-t/62.8}),$$

assuming an exponential decay and a thermal time constant of 62.8 Ma for the lithosphere (Parsons & Sclater 1977). As the extension is not instantaneous, cooling starts before the end of the tensional phase. At constant extensional rate, one might approximate the effect by taking a new time $t' = t+\frac{1}{2}\Delta t$ for computing S_{th}. As the stretching factor actually changes continuously from 1 to a maximum value, the effect will be about one-half and an order of magnitude estimate can be obtained by choosing $t' = t+\frac{1}{4}\Delta t$. This was checked by numerical computations by Angelier *et al.* (1982) for the Aegean sea, where $\Delta t = 13$ Ma and $\beta = 1.5$ to 2.

As a consequence, the actual curves computed in the preceding section for instantaneous stretching should be corrected at $t = 0$ by adding a term ΔS of the order of

$$S_{th}(1-e^{-\Delta t/4\times 62.8}).$$

With $\Delta t = 20$ Ma, $S_{th} = 3900$ m and $\Delta S = 300$ m; with $\Delta t = 40$ Ma, $\Delta S = 600$ m.

Thus the subsidence curves as a function of γ now reach a level deeper than the level of implacement of new oceanic crust because the stretched lithosphere has already began to cool when the oceanic lithosphere is emplaced. Such an effect is thus necessarily implied if the stretching is not instantaneous. Simultaneously, another mechanism, the transfer of heat by lateral conduction, also generates a relative cooling at the base of the continental slope with respect to the seaward stretched continental crust and results in the creation of a depression (Watremez 1980).

Note, on the other hand, that the effect on the amount of melting produced will be much smaller and can probably be neglected. This is because cooling by conduction from the surface is unable to penetrate deep to the lower boundary of the lithosphere, below which most of the melting occurs. Consquently, if the transition to accretion is controlled by the amount of partial melting, it will occur at the same value of β but for a larger subsidence; thus a basin can be created. This difference of level will disppear progressively with time as cooling proceeds, but the basin may be preserved if it has been loaded with sediments.

Application: the origin of the deep Armorican basin

As seen previously, if our density estimates are correct, a slight unloaded basin 0.5 km deep exists on the greatly thinned continental crust of the deep Armorican margin. The boundary of the basin is marked by a little basement offset 0.5 km high in the vicinity of the ocean–continent transition. The uniform stretching model, including the effects of partial melting, accounts for the formation of an early basin if we include the effect of an extensional phase about 40 Ma long. However, this model does not explain that, at infinite time and after isostatic correction, the basement depth appears to be larger than that of the oceanic crust.

At this point, it should be noted that the law of partial melting in the mantle is poorly known and that, in any case, it is extremely sensitive to variations in temperature. Thus, one could obtain rather different results with relatively slight changes in the physical parameters. Further, we have not taken into account the compressibility of the melted portion. Stolper *et al.* (1981) argue that because the compressibility of basaltic melt is much higher than that of mantle minerals, the density contrast between melt and the solid phase should decrease with increasing source region depth. This effect would tend to increase the likeliness of formation of deep early basins on thinned continental crust.

On the other hand, if the igneous crust is indeed only 3 km thick, this model cannot account for such a large extension factor ($\beta = 10$). A possible explanation, compatible with the stretching model, could then be that the Neocomian phase of extension affected a pre-existing basin (the double rifting phase model).

Montadert *et al.* (1979*a*, *b*) have clearly shown the existence of such a pre-existing basin on the Celtic margin northwest of the Meriadzek Terrace, for example, at Site 401 (Montadert *et al.* 1979*a*), which was drilled in 2500 m water depth through a tilted fault block, where the existence of a shallow water Jurassic carbonate platform is suggested. On the other hand, several authors (e.g. Winnock 1971; Dardell & Rosset 1971; Mattauer & Séguret 1971; Olivet 1978) have proposed that the Permian to Lias tensional phase led to the formation of a depression partly filled with Jurassic sediments along the present day Northern Bay of Biscay margin. Without entering into a debate about the nature and origin of the lowest sedimentary layer identified as 3 B on figure 3 and in which velocities of 4.4 km s^{-1} (Bacon *et al.* 1969) and 4.6 km s^{-1} (Avedik & Howard 1979) have been found, we suggest that, at the level of the Armorican margin, this layer 3 B could correspond to the infilling of a pre-existing depression that has been stretched during the Neocomian phase. If the thickness of the continental crust is reduced to 3 km beneath the deep Armorican Basin, an explanation could be that this portion of the crust has been stretched twice by a global stretching factor reaching 10. In that case, if the interval between the two phases of extension is long enough to cool the lithosphere sufficiently, the base of layer 3 B after the second tensional episode should be at a depth greater than that of the new oceanic crust emplaced at 2.5 km deep, more or less at the level of the top of the stretched sediments. The resulting basin would consequently be created at the end of the second tensional phase, the newly emplaced oceanic crust acting as a dam for already emplaced sediments.

Conclusion

Adopting the simple analytical expression proposed by Ahern & Turcotte (1979) for the degree of partial melting in the mantle as a function of temperature and depth, we have determined the amount of partial fusion produced in the formation of a continental margin by using the uniform lithospheric model. Although the law is poorly known and is highly sensitive to slight changes in the physical parameters, our study demonstrates the importance of this phenomenon and indicates that the transition from stretching to accretion most probably occurs once the amount of melt produced in the asthenosphere below the stretched lithosphere becomes large enough.

Using an adiabatic temperature of 1262.5 °C (actual temperature 1300 °C) and 1295.5 °C (actual temperature 1333 °C) at the base of the 125 km thick lithosphere before thinning, we have derived the corrected curves of subsidence. As long as the melt does not migrate to the surface to form the oceanic crust, it will cool very slowly within the lithosphere and consequently will take several tens of megayears to solidify completely. Thus, the difference of density between liquid and solid phases leads to a smaller subsidence, the difference in the initial subsidence being several hundred metres, depending on the actual amount of stretching reached. Melting begins at a stretching factor of about 2 and becomes significant (maximum of 10 %) at a stretching factor of 3–4. Thus it is at this large stretching factor that the transition to oceanic crust becomes possible.

A significant feature of the new curves of subsidence is that the thinned continental lithosphere, in the initial subsidence stage, is always shallower than new oceanic lithosphere. However, as the process of extension is not instantaneous but takes several tens of megayears, cooling has already affected the thinned continental lithosphere when accretion starts. As a result, the depth reached may be larger than the depth of the mid-ocean ridge and a deep continental margin basin may thus exist in the initial stage of sea-floor accretion. However, the difference of elevation should progressively disappear with age, although sedimentary loading will maintain the depressed basement.

Although this study was initiated to explain the presence of the deep Armorican continental margin basin, the process just described does not seem to be able to explain it. This is because the crust of the basin appears to be extremely thin according to gravity estimates (3 km, requiring $\beta = 10$). Such a large thinning, if it is confirmed, may perhaps be explained within the framework of this model by a double-rifting stage.

N. Guillo-Uchard and A. Grotte helped with the preparation of the paper and illustrations.

References (Foucher *et al.*)

Ahern, J. L. & Turcotte, D. L. 1979 *Earth planet. Sci. Lett.* **45**, 115–122.
Angelier, J., Lyberis, N., Le Pichon, X., Barrier, E. & Huchon, P. 1982 *Tectonophysics* (In the press.)
Avedik, F. & Howard, D. 1979 *Init. Rep. D.S.D.P.* **48**, 1015–1024.
Bacon, M., Gray, F. & Matthews, D. H. 1969 *Earth planet. Sci. Lett.* **6**, 377–385.
Cochran, J. R. 1981 *J. geophys. Res.* **86**, 263–287.
Dardell, R. A. & Rosset, R. 1971 In *Histoire structurale du Golfe de Gascogne* (ed. J. Debyser, X. Le Pichon & L. Montadert), vol. 4, pt 2, pp. 1–28. Paris: Éditions Technip.
Green, D. H. & Ringwood, A. E. 1967 *Contr. Miner. Petr.* **15**, 103–190.
Jarvis, J. G. & McKenzie, D. P. 1980 *Earth planet. Sci. Lett.* **48**, 42–52.
Kay, R., Hubbard, N. & Gast, P. 1970 *J. geophys. Res.* **75**, 1585–1613.
Lalaut, P. 1980 Thèse de 3ème cycle, Université Pierre et Marie Curie, Paris (131 pages.)
Le Pichon, X. & Sibuet, J.-C. 1981 *J. geophys. Res.* **86**, 3708–3720.
Le Pichon, X., Sibuet, J.-C. & Angelier, J. 1982 In *A.A.P.G. Proceedings of the Hedberg Conference, Galveston*, January 1981. (In the press.)
Limond, W. Q., Gray, F., Grau, G., Fail, J. P., Montadert, L. & Patriat, P. 1974 *Earth planet. Sci. Lett.* **23**, 357–368.
McKenzie, D. 1978*a* *Earth planet. Sci. Lett.* **40**, 25–32.
McKenzie, D. 1978*b* *Geophys. Jl R. astr. Soc.* **55**, 217–254.
Mattauer, M. & Séguret, M. 1971 In *Histoire structurale du Golfe de Gascogne* (ed. J. Debyser, X. Le Pichon & L. Montadert), vol. 4, pt 4, pp. 1–24. Paris: Éditions Technip.
Montadert, L., Damotte, B., Delteil, J. R., Valéry, P. & Winnock, E. 1971*a* In *Histoire structurale du Golfe de Gascogne* (ed. J. Debyser, X. Le Pichon & L. Montadert), vol. 3, pt 2, pp. 1–22. Paris: Éditions Technip.
Montadert, L., Damotte, B., Fail, J. P., Delteil, J. R. & Valéry, P. 1971*b* In *Histoire structural du Golfe de Gascogne* (ed. J. Debyser, X. Le Pichon & L. Montadert), vol. 6, pt 14, pp. 1–42. Paris: Éditions Technip.
Montadert, L., Roberts, D. G., de Charpal, O. & Guennoc, P. 1979*a* *Init. Rep. D.S.D.P.* **48**, 1025–1060.
Montadert, L., de Charpal, O., Roberts, D. G., Guennoc, P. & Sibuet, J.-C. 1979*b* In *Deep-drilling results in the Atlantic Ocean: continental margins and paleoenvironment* (ed. M. Talwani, W. W. Hay & W. B. F. Ryan) (Maurice Ewing series, vol. 3), pp. 164–186. Washington, D. C.: American Geophysical Union.
Olivet, J. L. 1978 Thèse de Doctorat d'État, Université de Paris 7. (200 pages.)
Parsons, B. G. & Sclater, J. G. 1977 *J. geophys. Res.* **82**, 803–827.
Pautot, G., Renard, V., de Charpal, O., Auffret, G. A. & Pastouret, L. 1976 *Nature, Lond.* **263**, 669–672.
Ringwood, A. E. 1975 *Composition and petrology of the Earth's mantle*. McGraw-Hill. (618 pages.)
Roberts, D. G. & Montadert, L. 1980 *Phil. Trans. R. Soc. Lond.* A **294**, 97–103.
Royden, L. & Keen, C. E. 1980 *Earth planet. Sci. Lett.* **51**, 343–361.
Royden, L., Sclater, J. G. & von Herzen, R. P. 1980 *Bull. Am. Ass. Petrol. Geol.* **64**, 173–187.
Sibuet, J.-C. & Ryan, W. B. F. 1979 *Init. Rep. D.S.D.P.* **47** B, 761–775.
Sleep, N. H. 1974 *Bull. geol. Soc. Am.* **85**, 1225–1232.
Stolper, E., Walker, D., Hager, B. H. & Hays, J. F. 1981 *J. geophys. Res.* **86**, 6261–6271.
Waff, H. S. & Bulau, J. R. 1979 *J. geophys. Res.* **84**, 6109–6114.
Watremez, P. 1980 Thèse de 3ème cycle, Université de Bretagne Occidentale. (108 pages.)
Winnock, E. 1971 In *Histoire structurale du Golfe de Gascogne* (ed. J. Debyser, X. Le Pichon & L. Montadert), vol. 4, pt 1, pp. 1–30. Paris: Éditions Technip.

Discussion

D. G. Roberts. It is difficult to estimate the amount of extension in an area. To carry out this type of investigation you need migrated seismic sections with very good control on the interval velocities, as well as refraction control. To estimate the extension we have concentrated on those blocks where we can see the basement reflexion. The values of the extension (β) that we obtain vary from about 1.1 to 1.45. The block to the west of the one that Foucher showed has extended by 1.45, and the basement reflexion is clear. But it is by no means straightforward to interpret the section, and the value obtained for the extension depends on your assumptions about the geometry of the faulting.

J.-P. Foucher. Why I explained in some detail how we obtained our estimates of the extension was to answer some of the points that Roberts has raised. The estimates in question were all obtained from the interpreted seismic section of Montadert *et al.* They are the measured ratios between the present total length of the profile and the lengths of different layers. Where the geometry is ambiguous we have used the minimum amount of extension.

P.-Y. Chénet. Dr Foucher has remarked that the estimate of the amount of extension obtained from the shallow brittle deformation differs from that at depth. In the Meriadzek–Trevelyan region it is about 1.5 near the surface and increases to 2 at a depth of 6–8 km. I have obtained the same values for the same blocks, which is encouraging. But Dr Foucher said that this difference should be produced by greater brittle deformation at the shallow levels. This would mean a thinning of the upper layers by about 30 % by brittle failure, and none at depth. I have never seen an outcrop that shows 30 % thinning produced by internal deformation. I think a more reasonable model is one in which the extension throughout is about 1.5, which is taken up by brittle failure at shallow depths and by creep at deeper levels. The creep may cause lateral offsets in deep layers and therefore may be responsible for the difference between the extension determined from the geometry of shallow and deep layers respectively.

A. W. Bally. I should like to comment on the cross section showing layers marked A to D. Judging from the slide, it looks as if the cross section is not balanced, since the length of layer A appears to be much longer than that of layer D. Balanced cross sections have served as useful approximations in the reconstruction of folded belts. The same method can be used on sections displaying normal faulting. Thus you can only speculate that layer A has been thinned. Such thinning would prevent you from making an accurate estimate of the extension, if you do not know what the original thickness of the layer was.

J.-P. Foucher. As I said, if you wish to balance the cross section you have to allow layer A to have been thinned by internal deformation. I agree that this introduces some difficulty in measuring the extension from the geometry of the uppermost strata, which we assume in our interpretation to have been disturbed.

A. W. Bally. But then the reasoning may be circular. First you draw a cross section that is not balanced, then you conclude that because the section is not balanced, stretching must occur. However, you cannot show which layers have been thinned during the extension and which other layers have not. Also, there is no way that such thinning can be detected on the reflexion section only.

J.-P. Foucher. It seems to us reasonable to assume that the uppermost layers could deform during the gliding of the blocks to their present position, especially if these layers were poorly consolidated, as seems likely. The available seismic data do not permit the detection of such small-scale deformation, though field observations support this interpretation. Clearly, when measuring extension, care should be taken to measure it over the whole thickness of the brittle layer, and not just over the uppermost layer, which is often disturbed.

E. R. Oxburgh, F.R.S. It is only possible to carry out the type of reconstruction shown in both of the previous papers if the sections are obtained normal to the strike of the faults. There has recently been considerable debate on how the Bay of Biscay opened, and it is not yet generally agreed in what direction the relative motion occurred. I would therefore like to ask the authors of both papers how well they can constrain the strikes and the dips of the faults in the region? If these constraints are not good, then the types of calculations that they have described may only put rather broad limits on the amount of extension.

D. G. Roberts. There is a very great deal of seismic data available in the area, and as far as possible the lines have been run normal to the faults separating the blocks. We believe that a more important source of error is the absence of accurate seismic velocities. It is not possible to estimate the extension until the diffractions have been removed and the time section converted into a depth section by using the interval velocities. Since the dip of both the faults and the blocks is controlled by the velocities used in this conversion, the extension estimates are dependent on accurate velocities.

Sir Peter Kent, F.R.S. One of the maps shows a considerable amount of dip faulting, as well as strike faulting, which presumably makes extension estimates even harder to obtain.

D. G. Roberts. On the floor of the Bay of Biscay and close to the ocean junction between the ocean and the continent there are a number of inversion structures related to the late stages of the Pyrenean orogeny in the Eocene and Oligocene. One of the sections clearly shows such a inversion on the ocean side of the continental ocean boundary. In making estimates of the extension we have tried to avoid such places and have concentrated on those areas dominated by listric normal faults. For this reason we chose to carry out the experiment west of the Trevelyan Escarpment because of the clear evidence of inversion there in the Eocene and Oligocene.

P.-Y. Chénet. I believe that it is important to test whether a simple stretching model, with the same extension at all depths, can produce the main features of this continental margin. I do not believe that it can account for the observed geometry of the listric normal faults, and think that we need a model with at least two layers, each of which undergoes different amounts of extension.

J. F. Dewey. I do not believe that it is possible to use the geometry of the listric faults alone to make accurate estimates of the amount of extension. Where such features have been studied in the field, there is clear evidence for a considerable amount of internal deformation within the blocks on either side of the faults, by movement on joint planes, faults, fissures and other processes, none of which would be visible on a seismic record, and all of which would contribute to the strain.

D. G. Roberts. I agree that such processes would not be visible on a seismic section, which can only resolve the large structures. But none the less the estimates of extension that we have obtained are considerably smaller than those obtained by Le Pichon and Sibuet using the same seismic lines. In any case our values are maximum ones.

J. F. Dewey. But I do not believe that it is possible to use the geometry of the brittle faulting of the upper part of the crust to make accurate estimates of the extension. All that such studies can provide is a lower limit on the amount of extension. A better way to estimate the extension is to use the change in crustal thickness, which avoids the problems caused by internal deformation of the sediments.

M. H. P. Bott, F.R.S. It is only possible to stretch the upper crust by jointing by about 5 % before the density becomes too low to be compatible with the gravity observations. But this is far from the 100 % required to make the geometric deformation compatible with the crustal thinning.

J. F. Dewey. But this argument does not apply if the holes are filled with carbonate or some other cement!

J. A. Jackson. Another type of observation that is relevant to Dewey's suggestion is the observed strain release following large normal faulting earthquakes. Though many small aftershocks occur within the blocks on either side of the main fault plane, the displacements involved in these shocks are small compared with that on the main fault plane. These observations suggest that the internal deformation occurs because of the geometry of the main fault.

If the dip of the main fault changes with depth, motion can only occur if the blocks on one or other side deform internally. But there is no seismic evidence that this internal deformation makes an important contribution to the total strain.

M. F. Osmaston. The lithosphere-stretching hypothesis, discussed in both the preceding papers to account for the observed faulting, has the essential property that the duration of lithosphere stretching at depth must exactly match the period or periods of apparently extensional surface faulting. If the latter were to occur without the former, even for a short time, décollement of fault slices would be implied and this would put in question whether *any* of the faulting is a measure of lithosphere stretching. A duration of stretching from the Triassic until some time in the Cretaceous seems plate-tectonically unlikely; so too would be a continuance of stretching after the locus of ocean floor genesis had moved away from the margin. Therefore, how precisely at present can one define the period or periods of apparent extensional faulting and how does the date of its final cessation compare with independent evidence of the age of the ocean floor at this margin?

D. L. Turcotte. There are two other mechanisms of thinning crust that have received rather little attention today. The first is uplift followed by erosion. There is no doubt that this process has recently occurred along the East African rift system, where there is also evidence of crustal thinning. Presumably this region will later subside to form a basin. The other possibility is that phase changes occur in the lower crust. It is important that these alternatives to the crustal stretching model should continue to be discussed.

Phil. Trans. R. Soc. Lond. A **305**, 45–62 (1982) [45]
Printed in Great Britain

The northwest Australian continental margin

By D. E. Powell
Hudson's Bay Oil and Gas Company Limited, 700 *Second Street, S.W., Calgary, Alberta, Canada T2P 0X5*

The Northwest Shelf of Australia offers a typical example of a 'passive' continental margin. Major intra-cratonic basins of Permian to Middle Jurassic age developed along the present coastline, superimposed to either orthogonally trending Palaeozoic basins or Precambrian basement rocks.

In each of these depocentres distinct lithotectonic units can be recognized that are related to phases of rifting and subsequent continental break-up. The pre-break-up rift valley and intra-cratonic basin stages are represented by a very thick Permian to Middle Jurassic series of mainly fluvio-deltaic sediments but with occasional marine incursions. Break-up took place near the end of the Middle Jurassic and was accompanied by large-scale block faultings with associated uplift and sub-areal erosion. Gradually late Jurassic to early Cretaceous marine sediments transgressed over the eroded surface: within the general transgressive episode, late Callovian, late Oxfordian to Kimmeridgian, late Tithonian to early Cretaceous marine incursions may be singled out. Open marine conditions, related to the breakup of Gondwanaland and opening of the Indian Ocean, became widespread during the Albian in the southern part of the Australian Northwest Shelf and during the Cenomanian in the northern part. The deposition of a thick prograding wedge of mainly carbonate sedimentation since the mid-Eocene resulted in a northwesterly regional tilt of the Shelf.

Hydrocarbon occurrences are related to the tectonic evolution. Early Triassic, early Middle Jurassic, late Oxfordian–Kimmeridgian and early Cretaceous marine incursions are directly related to the deposition of potential source rocks in restricted basins. A regressive phase led to the deposition of Triassic fluviatile sediments with excellent reservoir potential. Break-up tectonism and subsequent marine transgression provided the relevant trapping mechanism and probably the migration paths for the major gas–condensate discoveries of the Rankin Platform. The prolonged high rate of subsidence and accompanying thick sedimentation have ensured that hydrocarbon generation occurred, despite the low geothermal gradient.

Introduction

The Northwest Shelf of Australia is regarded as a good example of a passive 'Atlantic-type' margin that evolved as a result of major rifting on a continental scale, starting in the late Palaeozoic and continuing into the early Cretaceous. These pull-apart movements resulted in the complete disassociation of eastern Gondwanaland with the separation of a western landmass from the Australian Plate.

Geologically the area as a whole comprises a number of Permian and Mesozoic epicontinental basins lying offshore from the Australian craton and developed parallel to the principal rift system. These extend westwards across a series of marginal plateaux to the base of the present continental slope. In the north the margin has been deformed within the Cainozoic orogenic belt of the Indonesian island arc system. The portion of the margin considered in this paper extends over 2000 km north from Northwest Cape towards Darwin and the adjacent portions of the Indonesian Archipelago. Seawards the width of the continental margin varies considerably but averages about 600 km (figure 1).

Geological and geophysical information on this vast area has been obtained from a number of sources. On the inner part of the shelf it has resulted from extensive offshore oil exploration carried out over the past 18 years by several large groups who have acquired over 300 000 lines of the seismic data and drilled close to 100 exploratory wells.

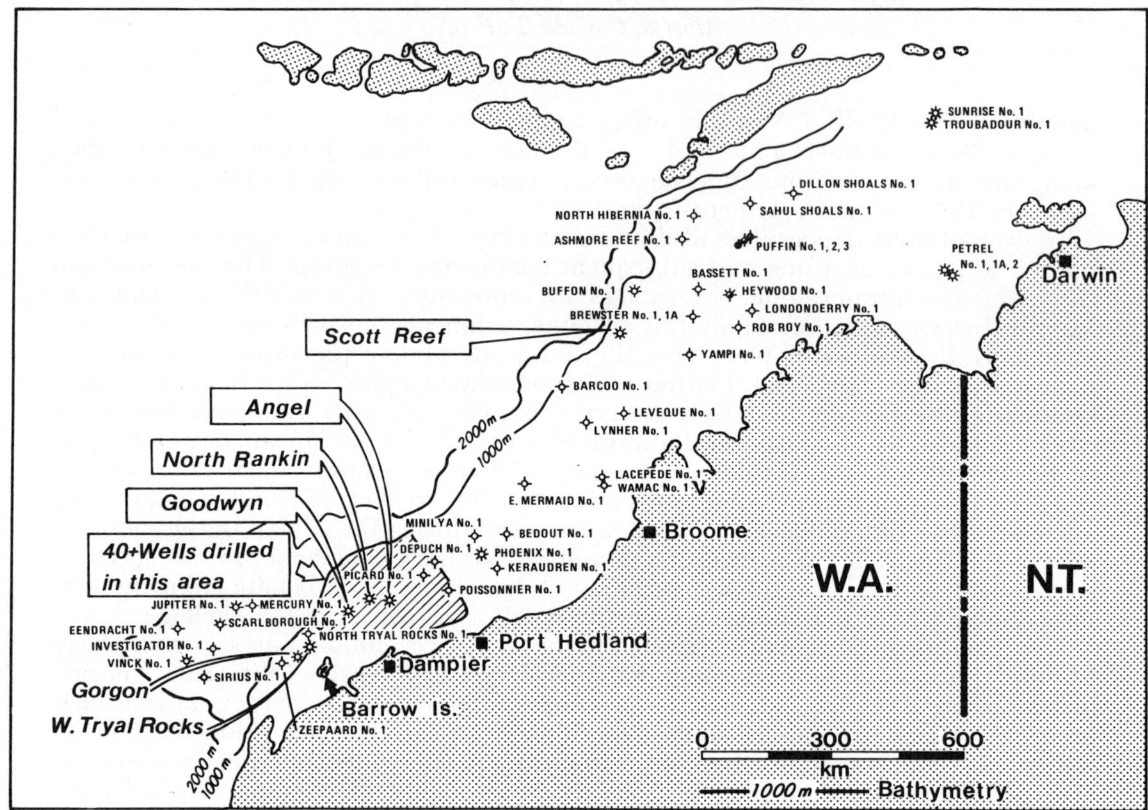

FIGURE 1. Location map with key exploration wells.

The outer part of the continental margin, extending to water depths of over 4000 m, has been investigated by the Australian Bureau of Mineral Resources who have carried out extensive gravity, magnetic and sparker surveys. In addition, several holes drilled by the *Glomar Challenger* in 1972 as part of the Deep Sea Drilling Project provided valuable information on the age both of the sea floor in the adjacent ocean basins and of the sediments in the Timor Trough. During 1977–9 the B.G.R. of West Germany conducted geoscientific investigations on the outer continental shelf, and recently, oil exploratory drilling has commenced in deep water on the Exmouth Plateau towards the outer continental margin where five wells have now been completed (von Stackelberg *et al.* 1980).

This paper outlines the geological evolution of the continental margin off Northwest Australia from the late Palaeozoic into the Upper Cretaceous and relates this to phases of crustal accretion.

Geological setting

Figure 2 shows the major geological elements, and figures 3 and 4 represent structural cross sections extending across the Browse Basin and Dampier Sub-Basin respectively, and out into the adjacent deep ocean. Gross regional isopachs of the Jurassic, Cretaceous and Tertiary within the main basinal areas are included as figures 5–7.

The onshore geology of northwest Australia is dominated by three large Archaean–

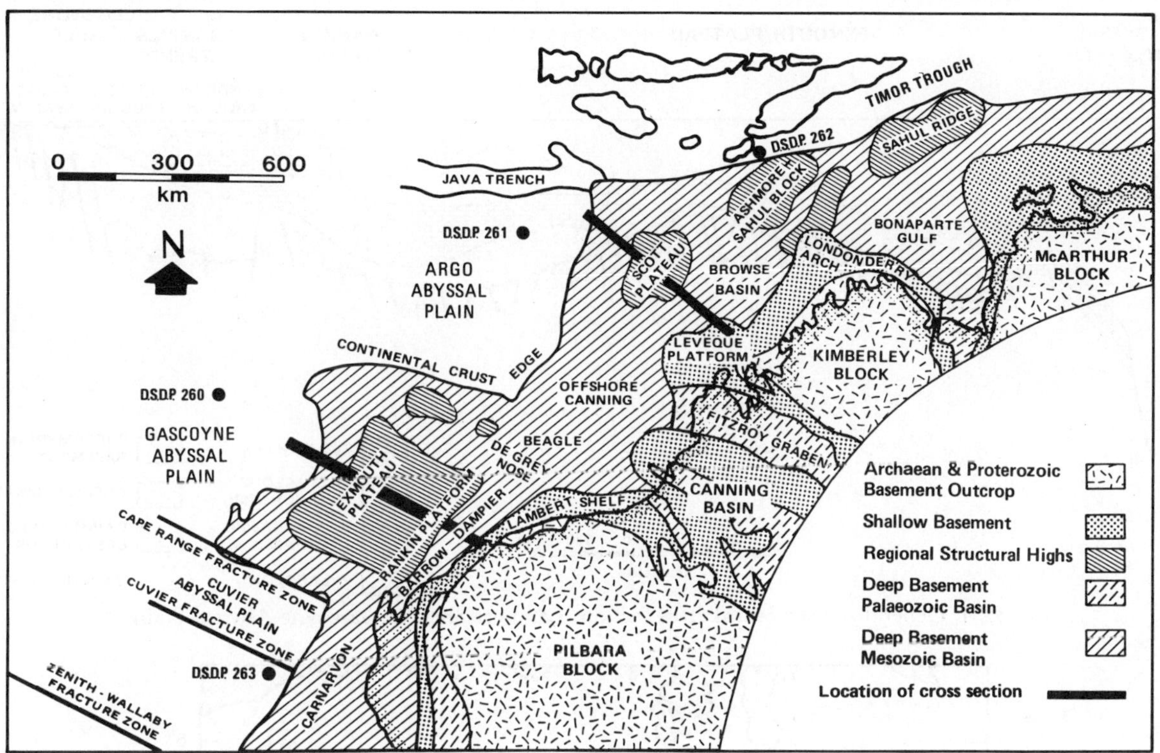

Figure 2. Tectonic elements.

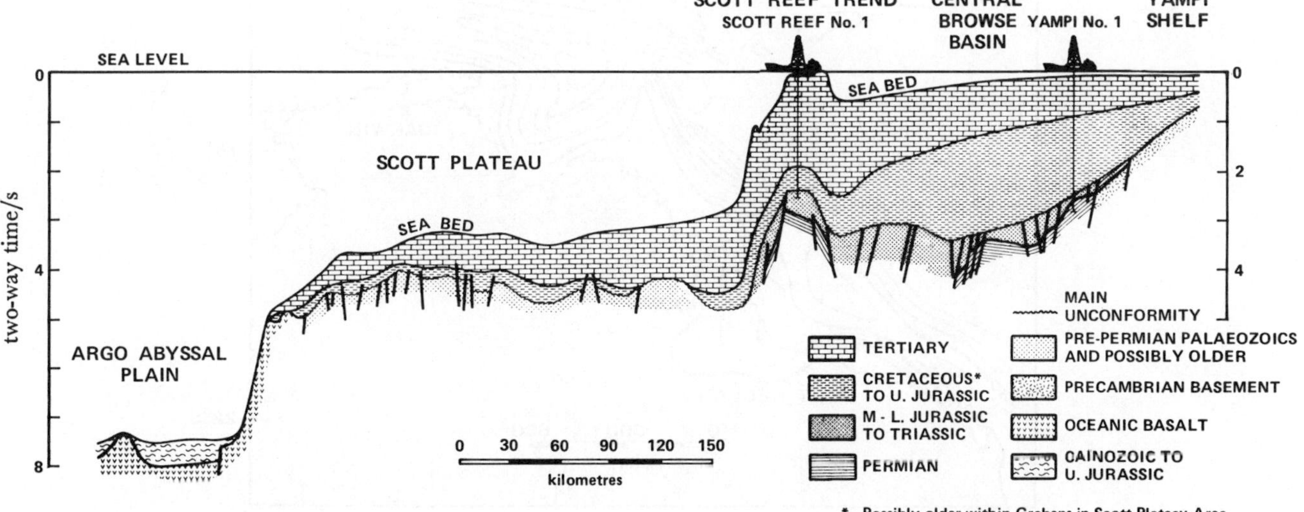

Figure 3. Generalized structural cross section, Yampi Shelf to Argo Abyssal Plain.

Proterozoic shield areas: the Pilbara, Kimberley and McArthur Blocks. These are separated by two northwest-trending intra-cratonic basins of mainly Palaeozoic age, namely the Canning Basin and the Bonaparte Gulf Basin, which are believed to have been initiated in the late Proterozoic as a result of crustal fragmentation.

Offshore a number of northeast-trending Permian to Tertiary epicontinental basins and sub-basins are located on the inner continental margin. The epicontinental basins developed off

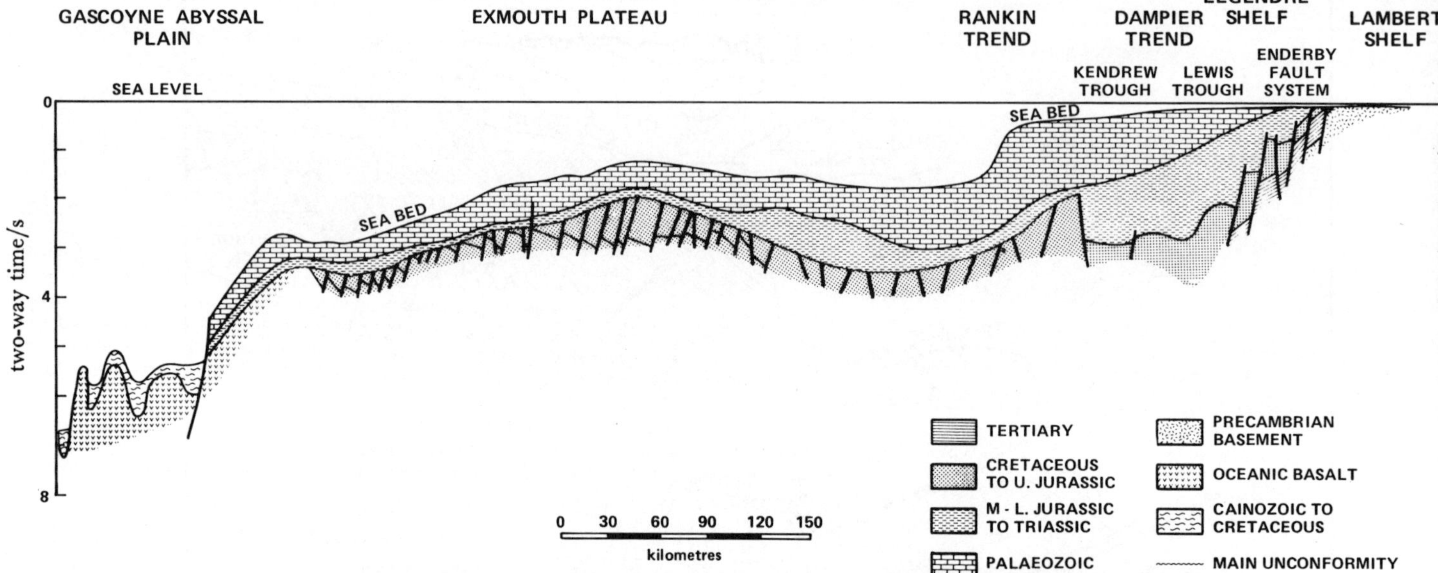

FIGURE 4. Generalized structural cross section, Lambert Shelf to Gascoyne Abyssal Plain.

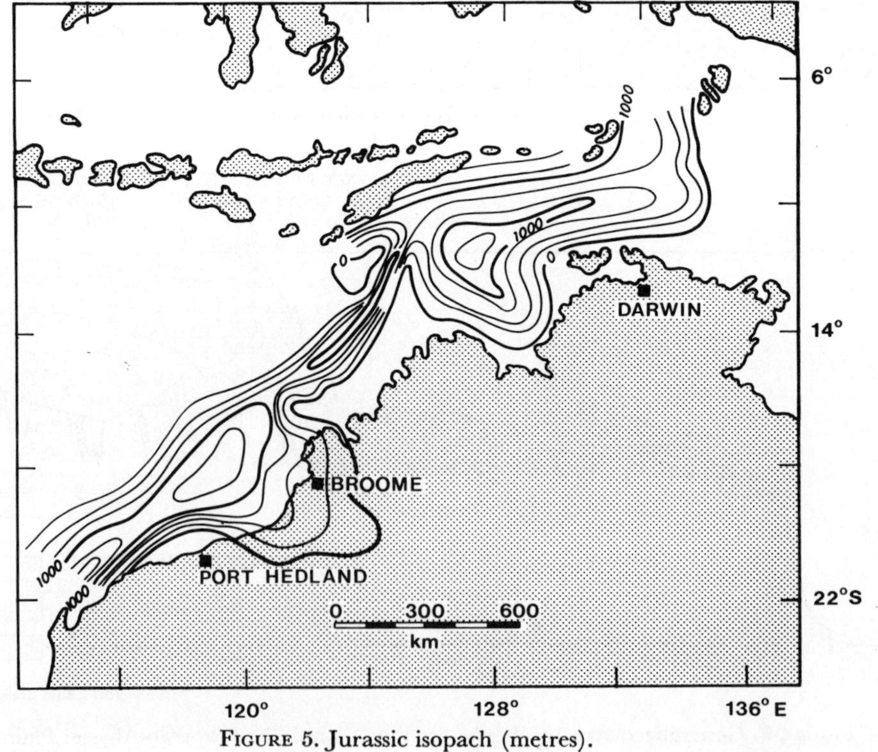

FIGURE 5. Jurassic isopach (metres).

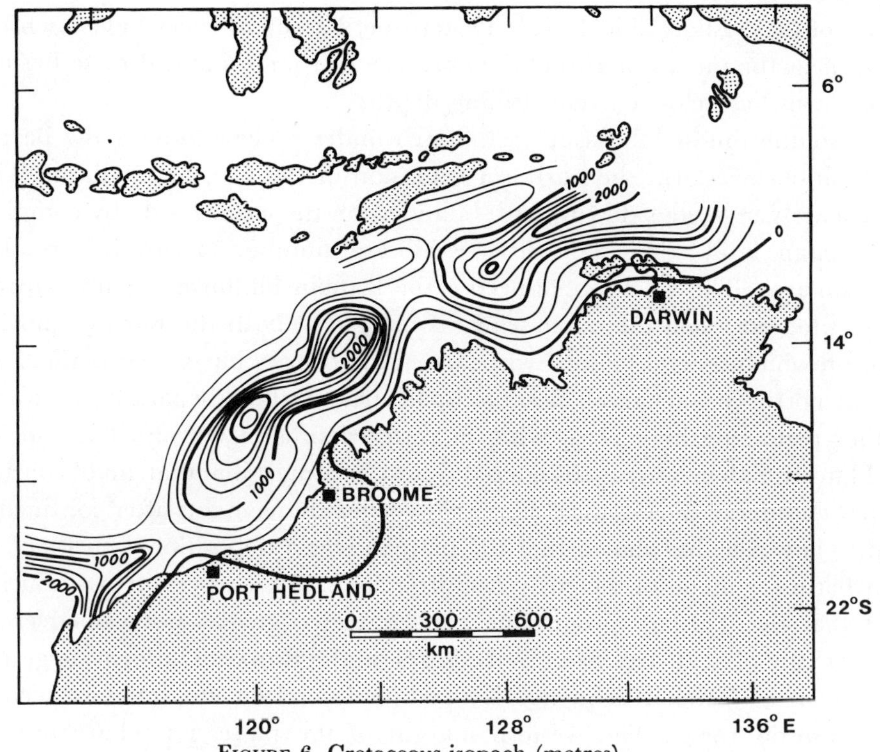

FIGURE 6. Cretaceous isopach (metres).

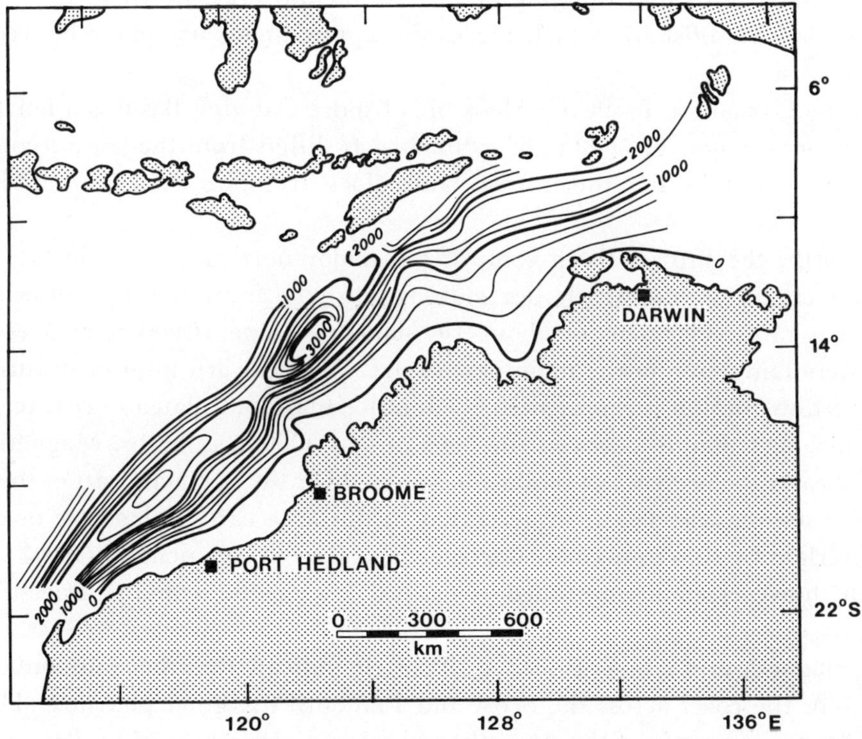

FIGURE 7. Tertiary isopach (metres).

northwest Australia were formed entirely over continental crust, with basement representing an extension of the onshore shield area. Proterozoic basement rocks have been penetrated by several wells near the inner margins of the northern Carnarvon and Browse Basins, but farther offshore basement lies below current drilling depths.

The three southernmost sub-basins in the area under review, namely the Barrow, Dampier and Beagle sub-basins, form the northern extension of the Carnarvon Basin. They represent major Triassic to Cretaceous downwarps, bounded on the eastern side by complex, essentially down-to-the-basin fault zones. All three exhibit a number of positive trends, paralleling a central basin area. The most pronounced is the Rankin Platform, a fault-controlled regional high feature situated along the outer seaward margin of both the Barrow and Dampier sub-basins, and on which several major hydrocarbon discoveries have been made.

West of the northern Carnarvon Basin a broad structural low separates the Rankin Platform from another regionally high area, lying in water depths of 900–2000 m and known as the Exmouth Plateau. Falvey (1972) considered this to be a marginal plateau of continental crustal material that foundered during the differential vertical movements after continental break-up. Recent work, however, has shown that the plateau was a depositional entity of the Northern Carnarvon Basin for much of the early Mesozoic and owes its present position in deep water to the fact that the Tertiary prograding carbonate wedge has not yet covered it. Seismic correlation into the plateau area from well control on the Rankin Platform suggests that a thick, faulted, Permian to Triassic sequence is overlain by relatively thin Cretaceous to Recent sediments (Branson 1974). The western margin of the plateau is characterized by strong northeast-trending fault blocks downthrowing westwards to the Gascoyne Abyssal Plain. This fault zone represents a rapid thinning of the continental crust down to the 4000 m isobath, located some 600 km offshore, which marks the approximate boundary between continental and oceanic crust.

North of the Carnarvon Basin the Mesozoic offshore Canning Basin is a northeast-trending linear basin containing over 5000 m of sediments. It differs from the Carnarvon Basin in that it generally does not exhibit the intense fault block tectonics that characterize the former feature.

Farther north, the Browse Basin comprises a major northeast-trending late Palaeozoic to Recent sedimentary basin, which lies entirely offshore from the Precambrian Kimberley Block and has an areal extent of about 100000 km^2. Some 10000 m of Mesozoic–Tertiary sediments were laid down in the main depocentre. The seaward limit of substantial Jurassic–Cretaceous sedimentation is represented by another marginal plateau area, termed the Scott Plateau, which lies mainly between the 2000 m and 3000 m isobaths. Magnetic and gravity evidence indicates the presence of shallow basement over the northern part of the Scott Plateau and tentative seismic correlation suggests that in this area early Palaeozoic or older rocks, or both, are overlain by thin uppermost Cretaceous to Recent sediments (figure 3). Elsewhere on the plateau, however, fault-controlled structural lows could contain Triassic and possible Jurassic rocks.

Gravity, magnetic and seismic evidence all clearly indicate that the continental crust extends westward from the coast across the Scott and Exmouth marginal plateaux. The geophysical data off the western margin of the Australian continent was reviewed by Branson (1974), who calculated, from the Bouguer gravity values, a westward thinning of the crust from a standard 35 km under the continent itself to about 22 km at the outer edge of both the Exmouth and

Scott plateaux. West of the marginal plateaux, geophysical evidence indicates a relatively rapid thinning of continental crust, generally through a block-faulted transitional zone, plunging to the abyssal plains, which are floored by oceanic crust. The outer limit of the continental margin off Western Australia is approximately defined by the 4000 m isobath. Both the Scott and Exmouth plateaux are excellent example of starved passive margins with low sedimentation rates after break-up.

Some 6000 m of Palaeozoic sediments comprising Cambrian–Ordovician clastics, Silurian–Devonian evaporites, Devonian reefal and basinal facies and a Carboniferous–Permian deltaic complex filled the northwest-trending Bonaparte Gulf Basin. The outer portion of the basin, which underlies the Timor Sea, also exhibits the same northeast structural grain and has a substantially similar Mesozoic evolution to that of the other marginal basins.

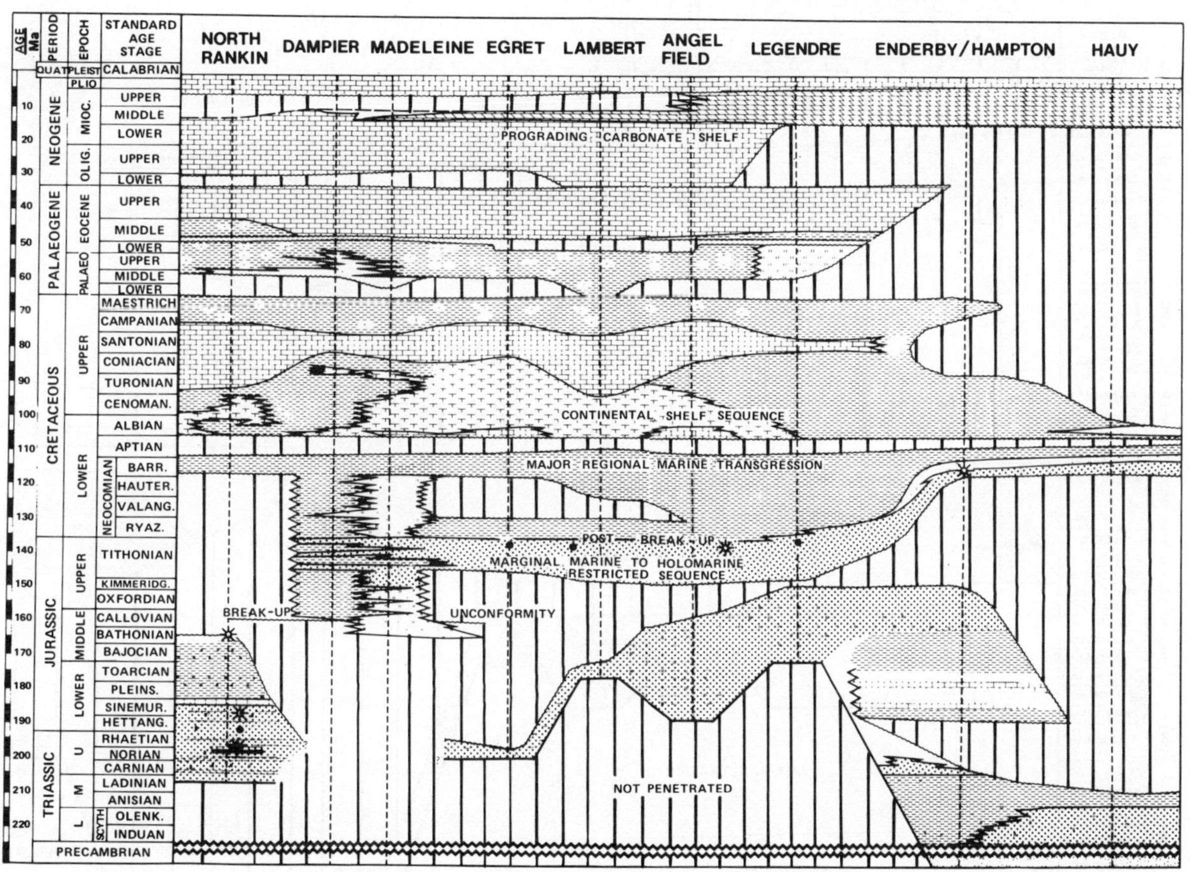

FIGURE 8. Dampier Sub-Basin lithostratigraphy.

Major tectonic features of the Bonaparte Gulf Basin include an uplifted area of thick Triassic sediments referred to as the Ashmore–Sahul Block and the Sahul Ridge, which is a basement high feature forming the northern limit of the Palaeozoic Bonaparte Gulf Basin.

STRATIGRAPHY

Figures 8, 9 and 10 summarize the lithostratigraphy of the Dampier Sub-Basin, the offshore Canning and the Browse basins. Figure 11 extends westwards from the Dampier Sub-Basin over the Exmouth Plateau. Sedimentation in all cases has been controlled by phases of rifting movements of which the most important occurred at the end of the Permian, during the late Triassic, towards the end of the Middle Jurassic and during the Lower Cretaceous. The repeated tensional stresses were responsible for producing the regional structural trends which exercised a dominant control on subsequent sedimentation. As a result four major lithotectonic units are recognized and can be correlated with sequential stages of the continental rifting process.

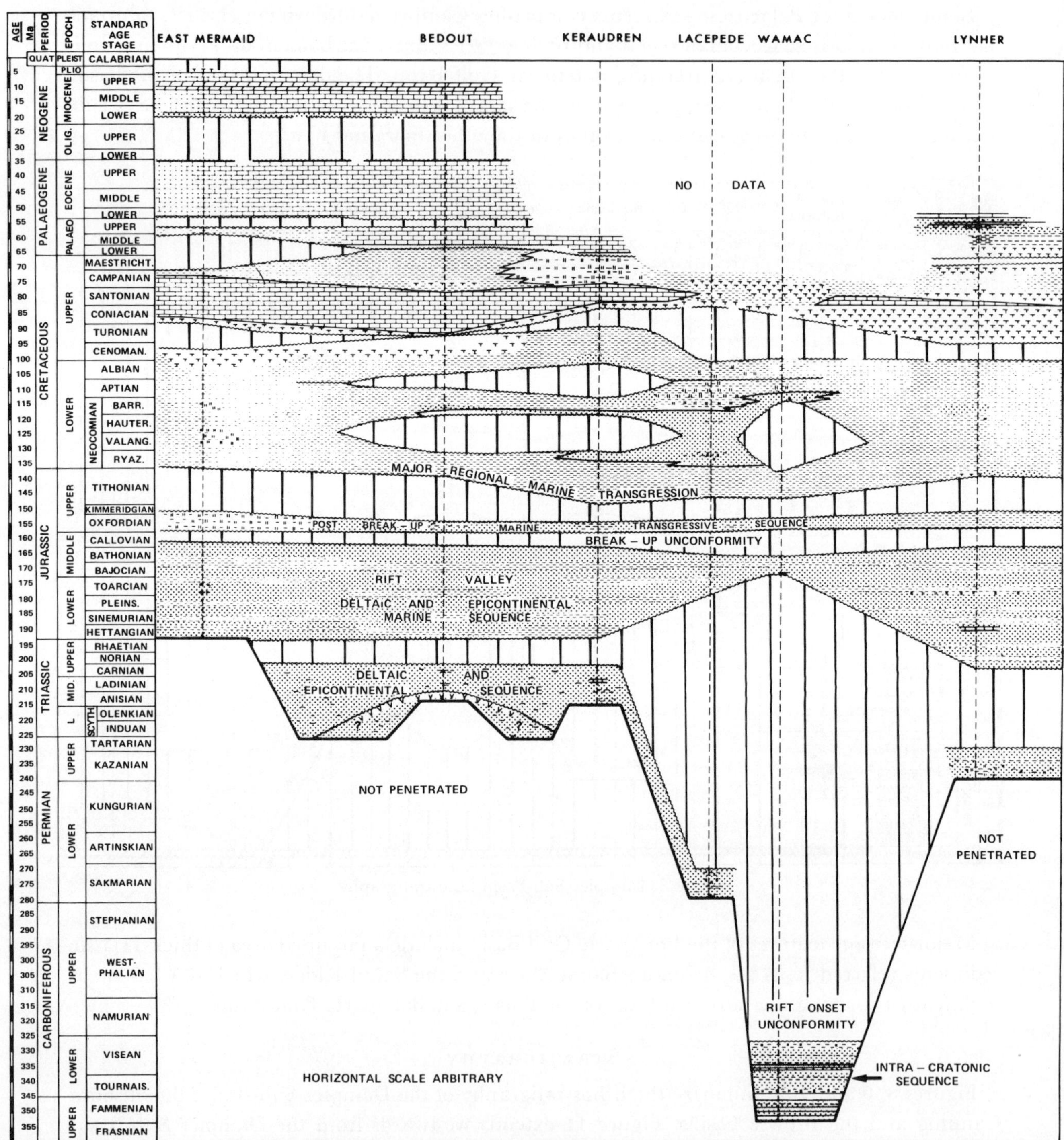

FIGURE 9. Offshore Canning Basin lithostratigraphy.

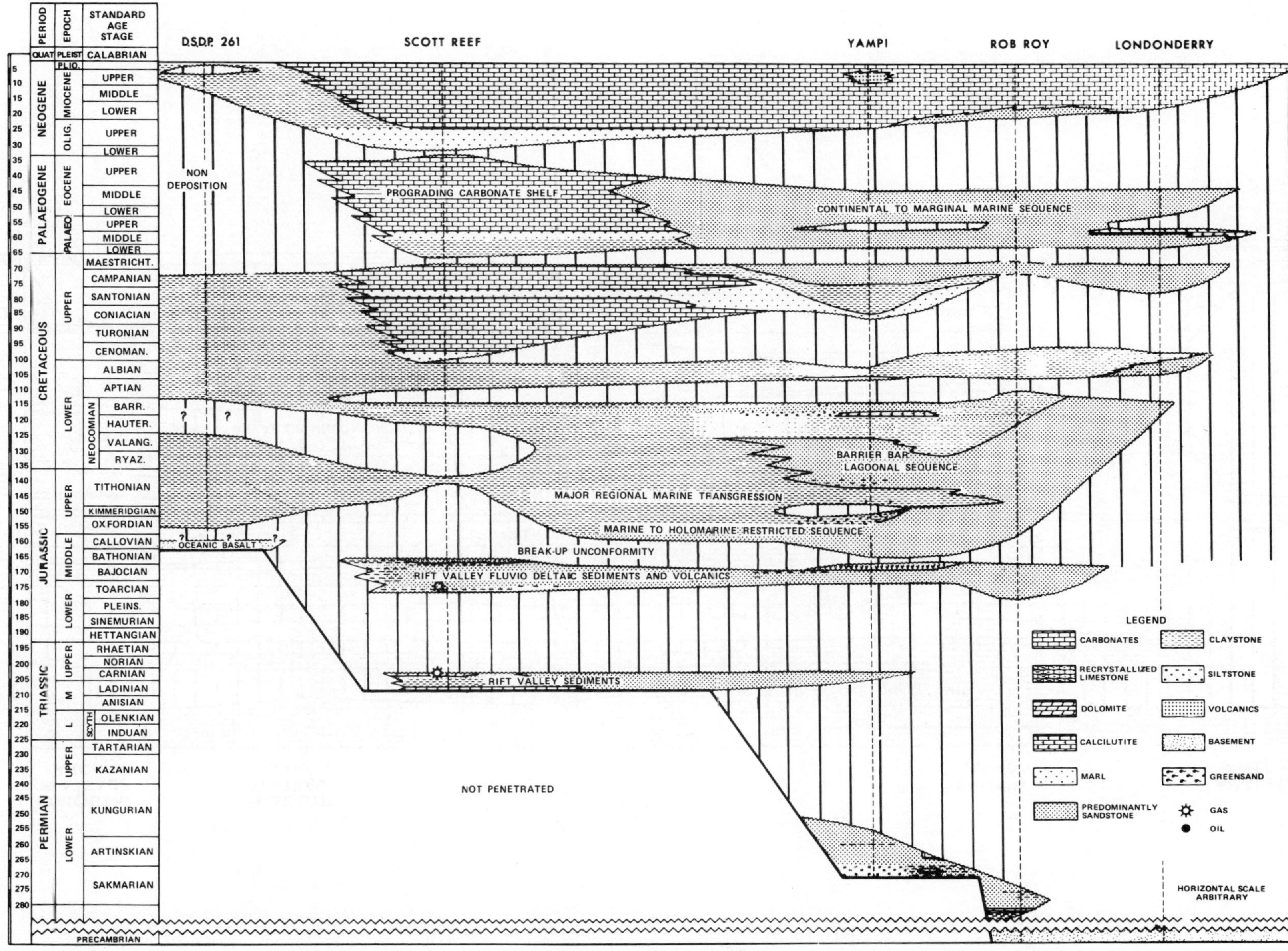

FIGUR 10. Browse Basin lithostratigraphy.

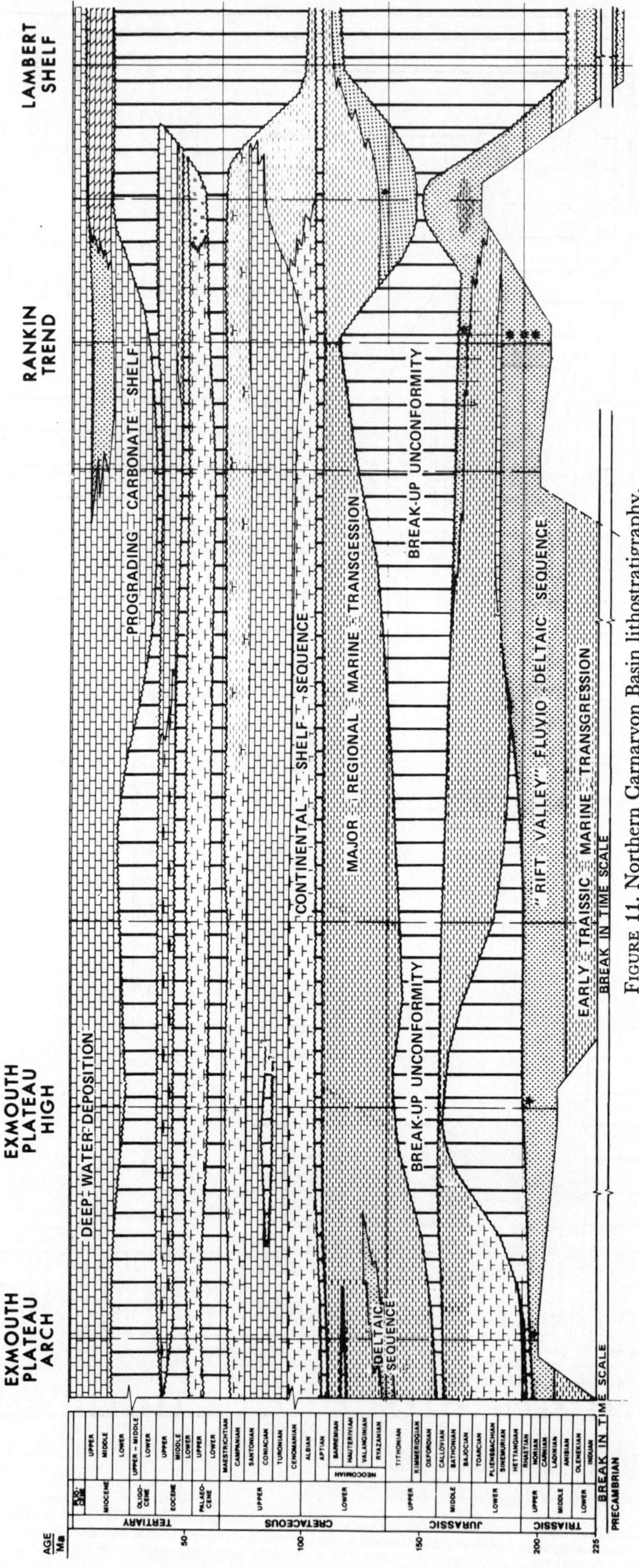

FIGURE 11. Northern Carnarvon Basin lithostratigraphy.

Permo-Carboniferous to Upper Triassic

Regional uplift of the Archaean–Proterozoic shield area accompanied by rifting was responsible for the opening of an elongate trough near the coastline of present-day Western Australia. In the offshore Northern Carnarvon Basin, little is known of this earlier depositional cycle as Permian and older rocks have been penetrated only in a few wells located near the Pilbara Shelf. From the limited subsurface data, together with outcrop evidence to the south, it would appear that the Permian is composed of over 4000 m of marine and marginal marine siltstones, sandstones and shales forming an overall regressive cycle.

Farther north, late Carboniferous to Permian rocks rest with angular unconformity on older sediments in both the offshore Canning and Browse Basins. Sparse well control indicates that sedimentation throughout was predominantly clastic in a non-marine fluviatile environment, with occasional marine incursions. The basin configuration is largely unknown, but the rapid thickening away from the basin margins observed on seismic lines is suggestive of deposition in a continually subsiding downwarp. On both well and seismic evidence a maximum of about 4000 m of Upper Carboniferous to Permian sediments appears to be present in this area.

In the Bonaparte Gulf Basin some 4000 m of Upper Carboniferous to Upper Permian fluvio-deltaic sediments are again present. Upper Permian marine shales and biomicrites in Sahul Shoals no. 1 on the Ashmore–Sahul Block reflects an increasing marine influence to the north-west. This, as well as strong faunal affinities, is consistent with the depositional trough's having been in open communication with the Tethyan Sea.

Tensional movements at the end of the Permian are evident in some areas, such as in the inner Browse Basin where they were sufficiently pronounced to cause block faulting. Elsewhere the movements are less evident, and locally in the northern Carnarvon, offshore Canning and Bonaparte Gulf basins the boundary between the Permian and the overlying Triassic seems to be merely disconformable.

The late Permian movements were accompanied by general subsidence, which led to a wide-spread transgression. Thus early Triassic deposits in the south are composed of marine shales, namely the Locker Shale of the Carnarvon Basin and the Blina Shale of the Canning Basin. Seismic evidence suggests that the Triassic and Permian sequences of the Carnarvon Basin extend westwards without break into the Exmouth Plateau area where a very thick equivalent section is also believed to be present.

Apart from the basal unit, the Triassic sequence is one of gradual regression so that by the late Middle Triassic a fluvio-deltaic environment was established throughout the Northwest Shelf area. In the Northern Carnarvon Basin, the Exmouth Plateau and the Canning Basin, the Middle and Upper Triassic sequence consists of 3000 m of fluvio-deltaic sands and muds. Sedimentation rates were about 50 m/Ma.

Although not encountered by the drill to date, the Lower to early Middle Triassic section in in the Browse Basin is predicted to be similar to that of the Bonaparte Gulf Basin, where Lower and early Middle Triassic marine shales and silts are overlain by a late Middle and early Upper Triassic fluvio-deltaic to shallow marine sequence. More than 2000 m of Upper Triassic marginal marine sediments, together with volcanics and evaporites, accumulated on the Ashmore–Sahul Block.

Lower to Middle Jurassic

A major stage in the fragmentation of the stable crustal unit of Gondwanaland occurred in the late Triassic, when tensional stresses resulted in extensive block faulting. These movements were particularly severe in the northern area (offshore Canning, Browse and Bonaparte Gulf basins), where they were accompanied by widespread erosion, so that subsequent sediments ranging in age from Lower Jurassic to Cretaceous are markedly unconformable on the older sequence. The post-erosion sedimentary cycle commenced in the early Lower Jurassic with red beds, non-marine and deltaic sediments that are widespread in the offshore Canning and Bonaparte Gulf basins. On seismic lines the lowermost beds can be seen to occur within depressions in the irregular palaeotopographic surface, but by the end of the Lower Jurassic most of this palaeotopography had been infilled. The remainder of the Lower to Middle Jurassic sequence in the three northern basins is predominantly non-marine and consists of thick bedded fluvio-deltaic sandstones with interbedded claystones, siltstones and occasional coals. Lower to Middle Jurassic fluvio-deltaic sands belonging to this sequence form the main reservoirs in the Scott Reef Troubadour and Sunrise gas–condensate discoveries.

Late Triassic to early Jurassic movements have also been observed in the Northern Carnarvon Basin, accompanied by uplift along the basin margins and over parts of the Exmouth Plateau. Locally, however, as on parts of the Rankin Platform, fluvio-deltaic sedimentation was continuous from the Triassic through into the Lower to Middle Jurassic. In this area an uninterrupted uppermost Triassic to late Middle Jurassic sequence of some 3000 m in thickness is present, with extensive sandstone development in the Upper Triassic passing upwards into predominantly claystones in the Lower to Middle Jurassic. Seismic and well evidence shows this thick Triassic to Jurassic section extends over the Exmouth Plateau, except for those parts of the Plateau that were uplifted during the Late Triassic movements. Early and Middle Jurassic sedimentation rates were 50–100 m/Ma.

Upper Jurassic to early Cretaceous

Pull-apart movements at the end of the Middle Jurassic initiated the principal period of tectonism affecting the Northern Carnarvon Basin. During this period the major positive trends were enhanced and strong uplift occurred on the Rankin Platform and the Exmouth Plateau. The succeeding marine transgression took place over a highly irregular palaeotopographic surface. Upper Jurassic to Lower Cretaceous marine sedimentation commenced in the deeper parts of the basin and, with continued subsidence, gradually submerged the upstanding scarps of the Rankin Platform, which were all inundated by the Upper Neocomian. Sedimentation rates reached a maximum of 70 m/Ma in the central part of the Barrow and Dampier sub-basins where the sequence is over 5000 m thick, consisting predominantly of claystones. These were deposited in a restricted marine environment, as evidenced by lack of planktonic fauna and the common occurrence of pyrite. In the northeast Dampier Sub-Basin, however, a thick Tithonian sand section was deposited in a nearshore, shallow marine environment. By the late Neocomian, claystone deposition was widespread, and this continued into the Aptian. During the late Jurassic to Neocomian, a delta 2000 m thick built northward onto the Exmouth Plateau and into the Barrow Sub-Basin. In the Canning Basin, Upper Jurassic sediments are generally absent, but a thick Lower Cretaceous claystone sequence of up to 1000 m was laid down on the continental shelf.

Major tensional movements during the Callovian–Oxfordian period are also evident in the Browse and Bonaparte Gulf Basins, being preceded in the former area by outpourings of volcanic basalt. These are recorded from the Scott Reef no. 1, Yampi no. 1, and Lombardina no. 1 wells. In the transgressive series that followed, early paralic sandstone sedimentation was superseded by thick marine claystone deposition, as marine conditions became widespread. A similar depositional environment to that of the Dampier Sub-Basin is generally envisaged for the Upper Jurassic to lowermost Cretaceous sequence of the Browse and Bonaparte Gulf Basins, where thick restricted marine claystones were initially deposited in the troughs, passing laterally into marginal marine to deltaic facies in the more topographically elevated areas.

Late Lower Cretaceous to Tertiary

Throughout the Northwest Shelf a marked change, from a restricted to an open marine facies, is evident during the early Albian. The later Cretaceous consists of deep water marls and calcilutites, passing upwards into a mainly claystone sequence.

An unconformity at base Tertiary level is observed regionally, and is indicative of initial subsidence and general deepening of the depositional environment followed by a gradual regressive cycle. Palaeocene to Lower Eocene sediments are predominantly deep water claystones and calcilutites, but by the Middle Eocene thick calcarenite deposition had been established in an inner-shelf to nearshore environment. Continued regression has resulted in a thick predominantly carbonate wedge prograding over the entire inner shelf. Tertiary sediments at the shelf edge attain a thickness of over 3000 m.

Geological evolution

The geological evolution of the Northwest Shelf can be related closely to a series of continental rifting movements which began in the late Palaeozoic, and which ultimately led to complete break-up of eastern Gondwanaland in Upper Jurassic – Lower Cretaceous times. The lengthy rifting history, associated with the fact that the most intensive tensional phases are not time-equivalent over the entire area, has caused a relatively complex continental margin development, in part characterized by the development of parallel rift valleys that are considered to be aborted rifts. This appears to contrast with the simpler development of the southern Australian rifted margin during the Tertiary (Boeuf & Doust 1975).

In the late Palaeozoic an elongate trough was initiated near to and paralleling the edge of the present-day Western Australian landmass. The presence of such a trough was first recognized by Teichert (1939). Its extent and orientation suggests that its inception was caused by major tensional movements normal to the approximate northwest trend of the onshore Palaeozoic intra-cratonic basins. The rift thus formed probably constituted a structurally simple trough extending westwards to at least the limit of the present continental crust, and northwards along the length of the Australian coast. It appears to have been in open communication with the Tethyan Sea, the post-Sakmarian deposits of Western Australia having close faunal similarities with Timor and the Salt Range of northern India (Dickins 1963).

It is considered that many of the major structural trends, which continued to be active throughout the Mesozoic, were initiated during a block faulting phase at the end of the Permian which caused fragmentation of the original trough. It is possible that uplift of the northwestern edge of the Scott Plateau occurred at this time. This is in contrast to the Ashmore–

Sahul Block and Scott Reef areas where a thick Permo-Triassic section, accompanied locally by volcanism, suggests graben fill. It is significant that volcanics are found only in the area that lay closest to the location of the evolving rift axis. Data from wells located east and northeast of the Scott Plateau indicate a west to northwest sediment source during the Triassic. It is considered that an uplifted area adjacent to the rift axis on the western edge of the Scott Plateau could have provided this source.

At the end of the Triassic a further major phase of the rifting process occurred, and was responsible for the main period of structural activity in the northern basins. Uplifted areas at the outer edge of the present-day margins may have included the Ashmore–Sahul Block and parts of the Exmouth Plateau. Erosion of high areas and the deposition of red beds followed by a thick non-marine sequence marks the continuation of a widespread typical rift valley stage.

The effects of the late Triassic movements were less pronounced in the Northern Carnarvon Basin, where they were probably restricted to epeirogenic movements along previously established major northeast trends accompanied by uplift along the basin margins. The observed southward migration of tectonism with time is also observed in the timing of actual break-up of the western margin of the Australian continent, which again began earliest in the northern areas (indicated by results of D.S.D.P. drillholes 261, 260 and 259). Contrary to earlier interpretation, evidence from drilling to date currently suggests that the entire Triassic sequence in the southern basins, which is largely of a fluvio-deltaic nature, was derived from the ancestral Pilbara Block to the south and east. The thick Triassic section present in the Exmouth Plateau area, however, may well have been derived in part from an unidentified landmass that lay farther to the west.

The major period of tectonism in the Northern Carnarvon Basin resulted from the rifting phase that took place at the end of the Middle Jurassic, and which represented the onset of complete continental break-up. Northeast-trending large-scale block faulting established a complex horst and graben infrastructure. General subsidence followed the tensional movements, although the positive areas such as the Rankin Platform remained topographically high. The ensuing Upper Jurassic transgression consequently took place over a palaeotopographic surface of varied relief, with sedimentation that was both local in distribution and environmentally restricted. Upper Jurassic faunas and floras still, however, have a Tethyan aspect. As subsequent rupture of the continental margin took place later and farther out to sea, the Northern Carnarvon Basin can be regarded as an aborted rift that in itself never evolved beyond this stage.

Creation of new oceanic crust occurred to the northwest of the Exmouth Plateau and to the west of the Browse Basin during the late Middle to early Upper Jurassic. Magnetic lineations on the Argo Abyssal Plain, which trend almost east–west north of the Exmouth Plateau (Heirtzler *et al.* 1978) range in age from 148 to 153 Ma, and suggest that that margin formed by a combination of rifting and transform faulting in the Callovian. This dating is also supported by the existence of Oxfordian sediments immediately above oceanic basement at D.S.D.P. Site 261 (Veevers *et al.* 1974). The Middle Jurassic basaltic lavas recorded from several wells in the Browse Basin also probably correlate with this initial formation of new oceanic crust.

Sea floor spreading began later in the southern area. Magnetic anomalies in the Cuvier and Gascoyne abyssal plains range in age from 108 to 123 Ma and suggest that the southwestern and northwestern margins of the Exmouth Plateau formed at the same time in the late Neocomian, i.e. 120–125 Ma. This is confirmed by D.S.D.P. Site 263. The southwestern margin formed as a transform fault (Cape Range Fracture Zone), whereas the northwestern margin is a rifted margin. The separation of Greater India from these two margins drastically reduced the clastic supply to the plateau, and thereafter deltaic sedimentation ceased and subsidence rates exceeded sedimentation. A period of strong north–south orientated faulting is recorded in the late Upper Jurassic or early Cretaceous of the Dampier Sub-Basin (Powell 1973), and is probably also related to this late spreading phase. Regional subsidence followed by a rapid transgression in the Lower Cretaceous marks the dispersal phase of eastern Gondwanaland.

A post-dispersal mature ocean stage, characterised by the establishment of a free circulating system, evolved in the late Lower Cretaceous. The Lower Cretaceous generally exhibits a transitional sequence from a marginal marine-deltaic environment with a virtual absence of Foraminifera to that of an open shelf facies with a diverse planktonic–benthonic assemblage. The change to an unrestricted marine environment ranges from Cenomanian in the Browse Basin to Middle Albian in the Dampier and Beagle sub-basins.

By Upper Cretaceous time an open marine environment was widespread over the entire area and, with minor variations, has persisted to the present day. Continued subsidence of the continental shelf has occurred from the late Cretaceous. A prograding carbonate wedge 3000 m thick has been deposited over the entire inner shelf area since mid-Eocene, as a result of which the shelf now has a general northwesterly regional tilt. This prograding stage is of course characteristic of a fully evolved 'Atlantic-type' margin. By contrast, during the late Cretaceous and Cainozoic the Exmouth Plateau sank 2000–3000 m, and sedimentation did not keep pace; the amount of sediment laid down averages 800 m, representing a sedimentation rate of less than 10 m/Ma.

Hydrocarbon occurrences

All of the prerequisites for major hydrocarbon accumulations can be recognized within the various development stages of the northwest Australian continental margin. The pre-break-up rift valley stage is characterized by rapid deposition of thick, mostly fluvio-deltaic clastics constituting excellent potential reservoirs. The tectonic phase accompanying rifting produced a block faulted substructure over which the initial post-break-up transgression occurred. The resulting fine-grained clastic sequence is an excellent hydrocarbon source rock and in addition is draped over the upstanding horsts to complete the trapping mechanism. Finally, the thick prograding sedimentary wedge deposited during the mature ocean stage depressed the older sediments to levels at which thermal generation of hydrocarbons could take place.

The validity of this model for the mechanism of hydrocarbon generation and entrapment is reflected in the overall results so far obtained from exploration drilling along the continental margin. To the south, in the Barrow Sub-Basin, the Barrow Island Field is a major oil producer with the main production being obtained from Cretaceous sands contained within the post-break-up sequence.

The greatest concentration of discoveries occurs in the adjoining Dampier Sub-Basin, where some large gas–condensate fields and also several smaller accumulations of both oil and gas–condensate have been found to date. These are shown in figure 12, and the stratigraphic positions of some of the hydrocarbon reservoirs are indicated in figure 8. In three of the main fields (North Rankin, Goodwyn and West Tryal Rocks) the reservoir is represented by thick fluvio-deltaic Upper Triassic to Lower Jurassic sands contained within a tilted horst block trap and sealed unconformably by Lower Cretaceous claystones. North Rankin is the largest discovery with estimated recoverable reserves of about 230 km^3 gas together with condensate in the ratio of 0.16 $m^3/10^3\,m^3$ gas. The sandstones have excellent reservoir characteristics with porosities averaging over 20% and generally very high permeabilities.

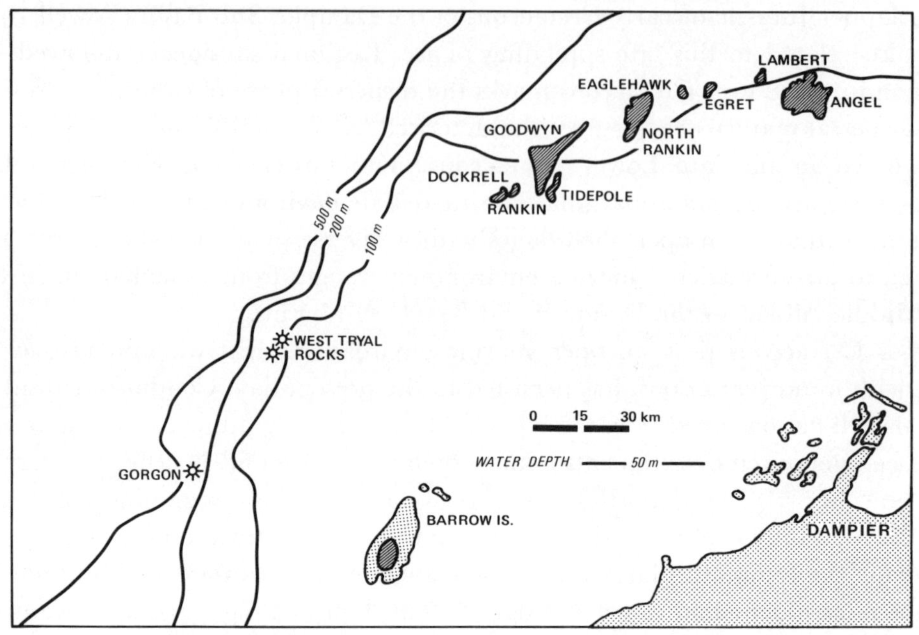

FIGURE 12. Dampier and Barrow sub-basins discoveries.

Fluvio-deltaic sands of the Upper Triassic 'rift valley stage' also constitute the main reservoirs for the gas–condensate and oil flows from the Rankin, Dockrell and Eaglehawk structures. In the Angel field, the gas–condensate accumulation is contained in sandstone reservoirs within the Upper Jurassic transgressive sequence. Beach, offshore bar and paralic sand facies within this stratigraphic interval also provide the reservoir for the small oil accumulations at Egret and Lambert while in the Legendre area, southeast of Angel, a small oil pool is present in Lower Cretaceous sands.

Despite the variations in reservoir age, the oils and condensates of many of the various Dampier Sub-Basin discoveries show a remarkable similarity in composition and a common source has been invoked for these (Powell & McKirdy 1973). The most likely sources are marine claystones deposited during the Upper Jurassic transgression, which followed the late Middle Jurassic tectonic phase. These claystones, which are rich in both cuticle and sapropel, were deposited in a restricted environment, conducive to organic preservation, and constitute excellent potential source rocks. The transgressive series filled the palaeotopographical lows so that the hydrocarbons generated later were able to migrate into structurally higher traps, such as the Triassic horsts on the Rankin Platform. The Upper Cretaceous marine claystones, which finally inundated the upstanding horsts, have provided a reservoir seal on a regional scale.

The Browse and Bonaparte Gulf basins have been less extensively explored than the Dampier Sub-Basin due mainly to excessive water depths. However, the Scott Reef no. 1 well drilled near the Browse Basin outer margin was another gas–condensate discovery and in the Bonaparte Gulf Basin two gas–condensate discovery wells, Troubadour no. 1 and Sunrise no. 1, were drilled on the Sahul Ridge.

BIBLIOGRAPHY (Powell)

Beck, R. H. & Lehner, P. 1974 *Bull. Am. Ass. Petrol. Geol.* **58**, 376–395.
Boeuf, M. G. & Doust, H. 1975 *APEA Jl* **15**, 33–43.
Branson, J. C. 1974 Structures of the Western Margin of the Australian Continent. Bur. Miner. Resour. Aust. Rec. 1974/64. (Unpublished.)
Dickins, J. M. 1963 *Bur. Miner. Resour. Aust. Bull.* no. 63.
Falvey, D. A. 1972 Ph.D. thesis, University of New South Wales.
Falvey, D A 1974 *APEA Jl* **14**, 96–106.
Heirtzler, J. R., Cameron, P. J., Cook, P. J., Powell, T., Roeser, H. A., Sukardi, S. & Veevers, J. J. 1978 *Earth planet. Sci. Lett.* **41**, 21–31.
Laws, R. A. & Kraus, G. P. 1974 *APEA Jl* **14**, 77–84.
Lofting, M. J. W., Crostella, A. & Halse, J. W. 1975 In *Proc. 9th World Petrol. Congr., Tokyo*, 1975.
Powell, D. E. 1973 Paper presented at Australasian Inst. Min. Metallurgy, Western Australia Conference, 1973, pp. 193–204.
Powell, T. G. & McKirdy, D. M. 1973 *APEA Jl* **13**, 81–85.
Schneider, E. D. 1972 *Mem. geol. Soc. Am.* **132**, 109–118.
Teichert, C. 1939 *Aust. J. Sci.* **2**, 84–86.
Veevers, J. J. & Heirtzler, J. R. 1974 *Woods Hole Oceanographic Contribution* no. 3217.
Veevers, J. J., Heirtzler, J. R. *et al.* 1974 *Init. Rep. Deep Sea Drilling Project* **27**, Washington, D.C.: U.S. Government Printing Office.
von Stackelberg, U., Exon, N. F., von Rad, U., Quilty, P., Shafik, S., Beiersdorf, H., Seibertz, E. & Veevers, J. J. 1980 *B.M.R. Jl Aust. Geol. Geophys.* **5**, 113–140.

Discussion

SIR PETER KENT, F.R.S. Though there appears to be a general similarity between the style of faulting off the Australian margin and in the North Sea, the dip of the faults shown on Australian profiles appears to be greater than those in the North Sea. Is this a consequence of the vertical exaggeration in the sections or is there a real difference between the two regions?

D. E. POWELL. The steep dips on the profiles are largely a result of the vertical exaggeration. The dips of the faults in the inner part of the basin are probably between 50° and 60°. However, the larger faults on the southwestern side of the Exmouth Plateau have dips as great as 70°, where the margin probably had a large transform component.

SIR KINGSLEY DUNHAM, F.R.S. The profiles that Mr Powell showed seem to differ from those on the northern margin of the Bay of Biscay in showing considerably less crustal attenuation. Is this difference supported by more detailed studies?

D. E. POWELL. In the Exmouth plateau area in the south, the thick middle to lower Jurassic and also the thick Triassic beds continue westwards, probably as a continuation of the North Carnarvon basin. Further north in the Scott Plateau area the situation is different. The Browse and Bonaparte Gulf basins appear to be limited by an uplifted region of Proterozoic rocks on which no Mesozoic sediments were deposited. The thickness of the continental crust obtained from gravity and magnetic surveys in this area varies from about 30 km near the present Australian coast line to about 22 km beneath the Exmouth Plateau, and 17 km beneath the Scott Plateau. Since this thinning takes place over 600 km the geometry is very different from the Biscay margin.

G. KARNER. The rifting style around the Australian margin is characterized by the creation of offshore plateaux, for example the Exmouth and Scott Plateau, by movement on widely separated normal faults. Realizing that the data presented do not directly address this problem, could Mr Powell please comment on this peculiar rifting style perhaps in light of the previous Bay of Biscay papers, where the faults are more closely spaced?

Secondly, I noticed in Mr Powell's cross sections across the Exmouth and Scott Plateaux that the ocean–continent boundary was on the plateau proper. The oldest oceanic crust is therefore elevated at a height of 1–1.5 km relative to the younger oceanic crust. What is the character of the ocean continent boundary? i.e., what is the basalt geochemistry, are they stratified, what was the depositional environment, etc.?

D. E. Powell. There is an abrupt change from oceanic to continental crust. The strata overlying the continental crust can be followed out almost to the continental edge, both by using the drill sections and seismic reflexion, and the D.S.D.P. holes are close to the boundary on the ocean side. The southern and southwestern margins of the Exmouth Plateau must have been produced by transform faulting. The northern margin was probably produced by a combination of rifting and transform faulting. The two boundaries were also produced at different times; that in the north was formed during the Callovian whereas the southern boundary did not form until the early Cretaceous. I do not understand why the lithosphere should deform in different ways around the North Atlantic and off Northwestern Australia.

M. F. Osmaston. What are the ages and character of the oldest drill-sampled rocks on the various offshore structural highs mentioned, notably the Exmouth and Scott Plateau? From reflexion seismic data at these sampled sites how much sedimentary section, if any, lies beyond the depth reached by drilling and can any marked unconformities be discerned therein?

D. E. Powell. The only areas where pre-Permian rocks have been reached by drilling are on the inner parts of the continental margin. On the Sahul ridge the Troubador well reached a granite which produced a pre-Permian date. On the Exmouth plateau I doubt if any of the wells have reached Permian rocks, but I have not seen all the available information. (Dr Murphy of Esso agreed with Mr Powell.)

A. S. MacKenzie. Have the proposed Jurassic origin of the oils and gases, and the Tertiary timing of oil and gas generation, been tested by some organic geochemical or isotopic measurements?

D. E. Powell. There is extensive geochemical evidence for a Jurassic origin of the oils. I originally believed that the Upper Jurassic claystones, laid down during a transgression when the circulation was restricted, formed the principal source rocks in the area. However, it has now been demonstrated that a number of other Lower Jurassic marine incursions also produced deposits that have acted as source rocks. It is more difficult to determine where the gas originates.

Phil. Trans. R. Soc. Lond. A **305**, 63–84 (1982) [63]
Printed in Great Britain

The Mesozoic–Tertiary evolution of the Aquitaine Basin

By R. Curnelle, P. Dubois and J. C. Seguin

Société Nationale Elf Aquitaine (Production), Boussens, 31 360 *St Martory, France*

The Aquitaine Basin, situated in southwest France, with an area of about 60000 km^2, has the form of a triangle which opens towards the Atlantic (Bay of Biscay) and is limited to the north by the Hercynian basement of Brittany and the Massif Central, and to the south by the Pyrenean Tertiary orogenic belt. Beneath the Tertiary sequence (2 km thick, and which outcrops over much of the basin) a Mesozoic series, up to 10 km thick, rests generally on a tectonized Hercynian basement but locally it covers narrow (NW–SE-trending) post-orogenic trenches of Stephano-Permian age.

The Mesozoic history can be subdivided into four major structural–sedimentary episodes:

(1) during a Triassic taphrogenic phase a continental–evaporitic complex developed with associated basic magmatism;

(2) throughout the Jurassic, a vast lagoonal platform developed, initially (Lower Lias) as a thick evaporitic sequence followed by a uniform shale–carbonate unit, indicating a relative structural stability;

(3) the end of the Jurassic and the Lower Cretaceous saw a fragmentation of this platform, due to an interplay between the Iberian and European tectonic plates, resulting in an ensemble of strongly subsident sub-basins;

(4) during the Upper Cretaceous and until the end of the Neogene, the evolution of the Aquitaine Basin was influenced by the Pyrenean orogenic phase, with the development, towards the south, of a trench infilled by flysch which, from the Upper Eocene, is succeeded by a thick post-orogenic molasse complex.

The main hydrocarbon objectives in the basin are situated in the Jurassic platform (e.g. the Lacq giant gas field) and the Cretaceous sub-basins (e.g. the Cazaux and Parentis oil fields).

To date, production has been about 4×10^7 m^3 of oil, and about 15×10^{10} m^3 of gas since the first gas discovery (St Marcet) in 1939.

1. Introduction

The Aquitaine Mesozoic–Tertiary basin is situated in southwest France and has the form of a triangle of 60000 km^2, which opens towards the Atlantic (Bay of Biscay) and is limited to the north by the Hercynian basement of Britanny and the Massif Central and to the south by the Pyrenean Tertiary orogenic belt (figure 1).

Beneath the Tertiary cover (on average about 2 km thick) exists a Mesozoic sedimentary sequence which can reach a thickness of up to 10 km, and which rests generally on a metamorphic and tectonized Hercynian basement. Locally the Mesozoic section covers several NW–SE-trending post-orogenic grabens of Stephano-Permian age.

The pre-Tertiary substratum only outcrops on the flanks of the basin. All the present knowledge of the deeper parts of the basin has resulted from the search for hydrocarbons which began after the commercial discovery of gas at Saint-Marcet in 1939. Since then over 10000 km of reflexion seismic profile have been shot and over 1000 wells drilled, the deepest of which is Lannemezan 1 with a final depth of 6900 m. All this work, realized mainly by Elf Aquitaine and by Essorep, has resulted in the discovery of about 20 economic hydrocarbon accumulations,

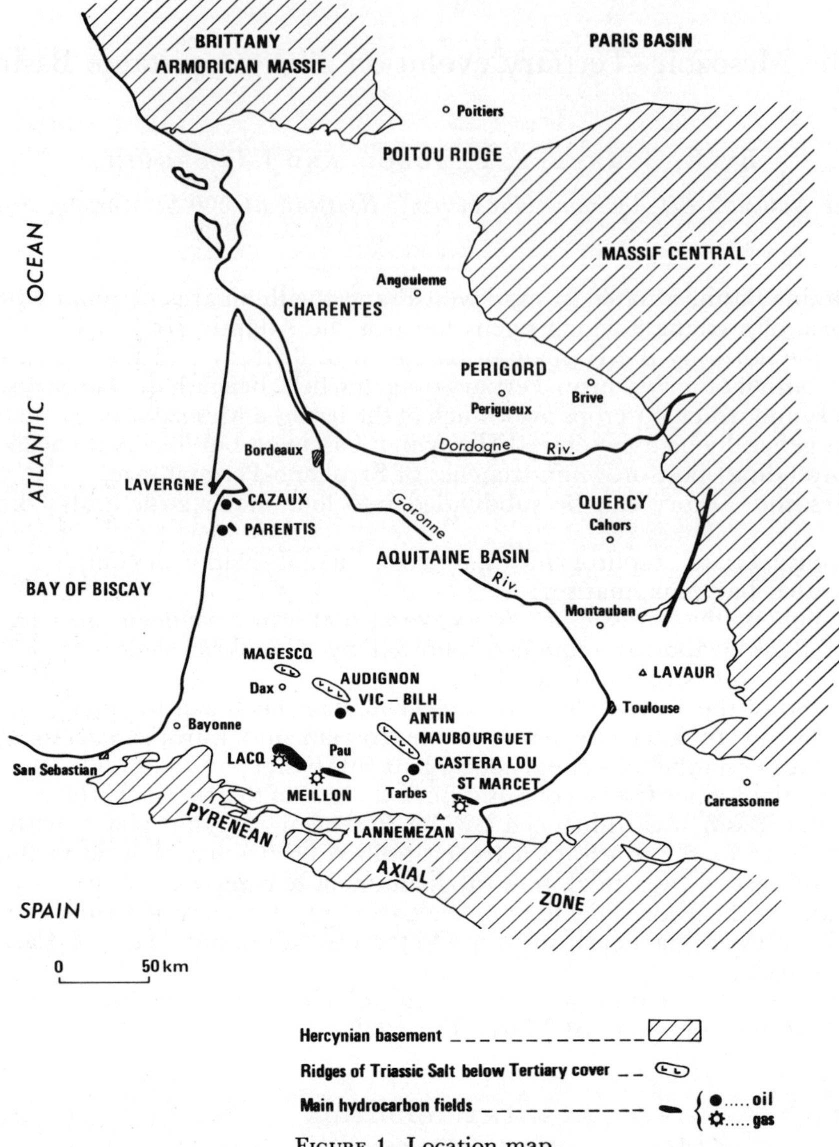

FIGURE 1. Location map.

the most important of which are the giant gas fields of Lacq and Meillon, and the oil fields of Parentis and Cazaux.

2. THE BASEMENT

The make-up of the pre-Mesozoic basement is only partly known, having been reached by only 70 wells; thus the sub-crop map (figure 2) remains very hypothetical. However, the geometry of the base-Mesozoic can be followed, albeit intermittently, from seismic reflexion data. It reaches its maximum burial of 8–10 km along the axis of the Parentis Basin and along the frontal edge of the Pyrenean orogenic belt.

The basement can be divided into two units:

(*a*) *Hercynian basement*, which seems to be of regional extent; it is composed of crystalline Precambrian, with sedimentary series of Cambrian to Carboniferous age with plutonic and

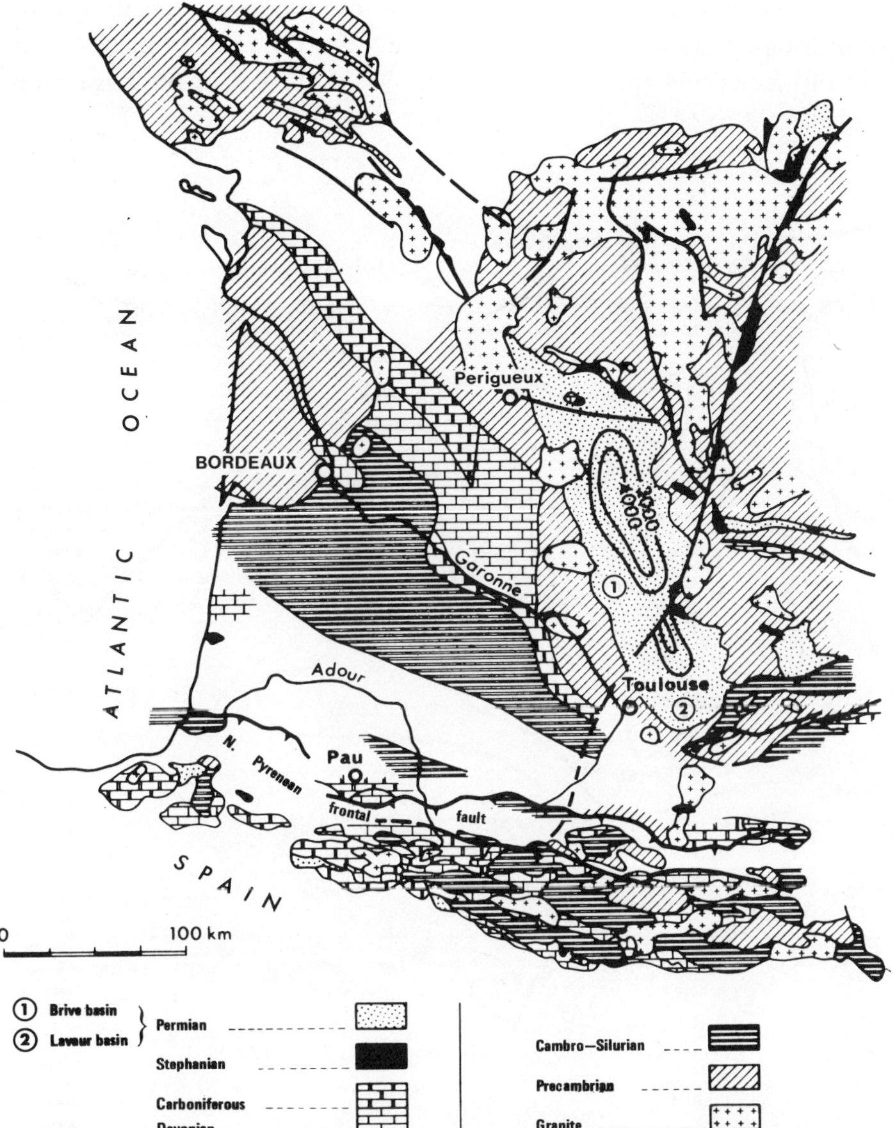

FIGURE 2. Subcrop at the base of the Mesozoic (older than 230 Ma).

hyperbyssal intrusives and also volcanic units; all have been deformed and metamorphosed during the various phases of the Hercynian orogeny.

(*b*) *Stephano-Permian* deposits of sandstone and shales that sometimes reach a considerable thickness in narrow graben, which developed after the main Hercynian orogenic phase. The best developed of these graben is that of Lavaur-Brive, situated to the north of Toulouse, which contains more than 3 km of sediment; it seems that other graben occur beneath the central part of the Aquitaine basin but their geometry is unknown.

3. THE MAJOR EPISODES OF MESOZOIC SEDIMENTATION

After the beginning of the Mesozoic, the Aquitaine Basin developed by subsidence as a vast depressed area, the present limits of which are between the Armorican (Brittany) and Central

Massifs and the Pyrenean axis. It seems probable that the southern limit during the Mesozoic extended beyond the present Pyrenees and somewhat onto the Iberian block.

The variations in facies and thickness during successive phases of the basin's evolution are summarized in figures 2 to 11.

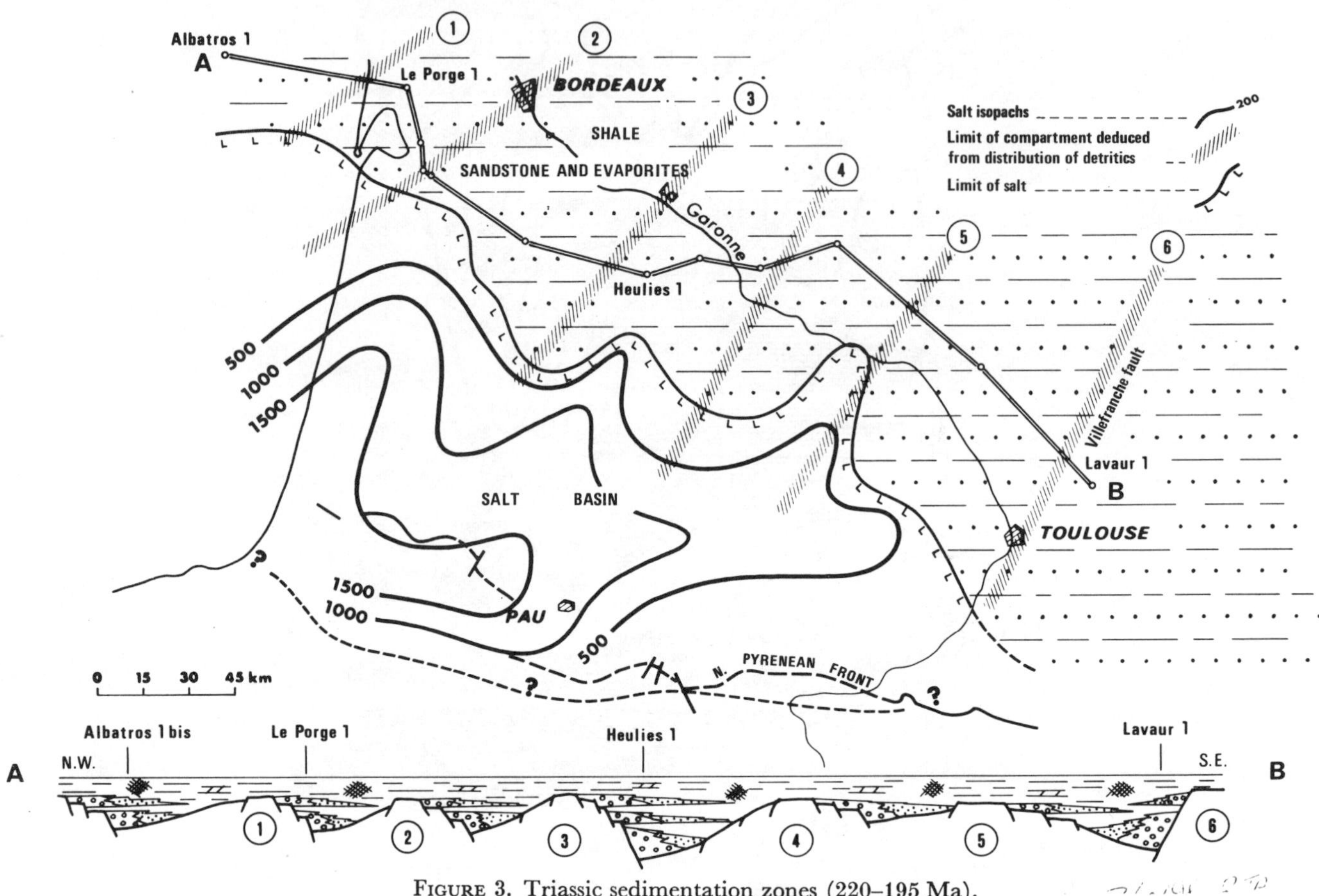

FIGURE 3. Triassic sedimentation zones (220–195 Ma).

(*a*) *The Trias* (*figure* 3)

Knowledge of the Trias is rather imprecise owing to three factors: (1) the small number of wells that have traversed it, (2) the almost total lack of good palaeo-dating (rendering difficult any correlation of the reduced series at the edge of the basin with the thick basinal series), and (3) difficulties of identifying the Trias on seismic lines.

The following three palaeogeographic units can be distinguished.

(i) A central basin, defined by the extension of halite deposits. Two main lithological units occur. The first mainly detritic unit comprises, after local development of sands at the base, a silt and evaporitic shale sequence, which can vary in thickness from 0 to 500 m. The environment is interpreted as alluvial cones developing close to active relief, passing laterally into an alluvial plain with fluviatile channels (braided stream) and finally a sabkha–playa environment. The geometry of these deposits, probably a function of tilted fault blocks, is poorly defined. The age is imprecise, the greater part is Lower Triassic, though the lower member could be of Permian age.

The upper unit comprises a number of evaporitic sequences – dolomite, anhydritic shales, anhydrite and halite – in a total thickness of more than 1500 m; the unit may have an accumulated salt thickness of up to 600 m. Within this unit are numerous intercalations of 'ophite' (basaltic lava flows or intrusions, or both). The lower sequences, which are richer in carbonate, are assigned to the Muschelkalk, whereas the top of the series is of Keuper age.

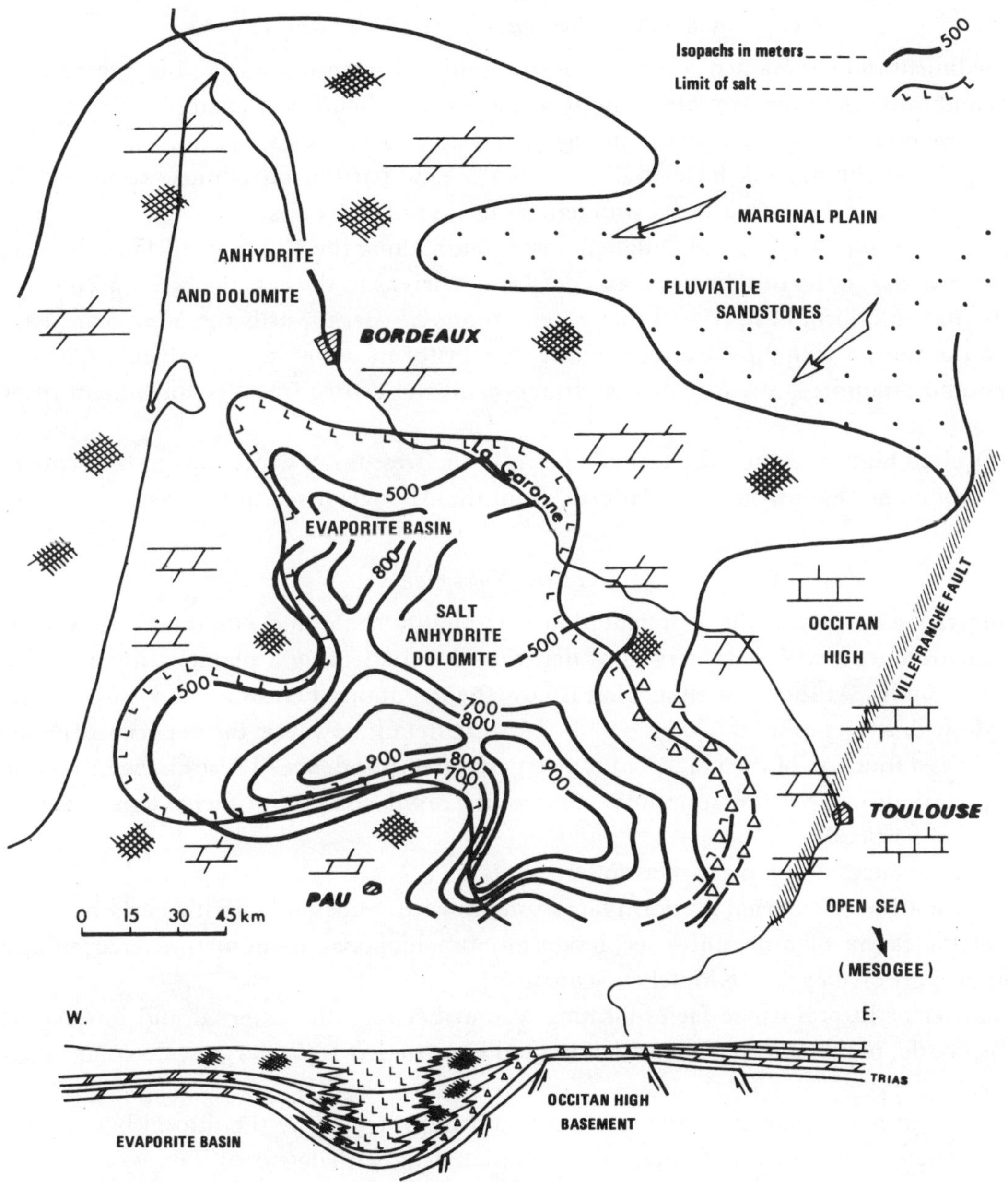

FIGURE 4. Lower Liassic sedimentation zones (195–190 Ma).

(ii) A wide northern border on which are found the same types of deposits as above – sandstones, silts, carbonates and evaporites – but, with considerably reduced thickness (100–300 m) and salt being absent. The sedimentation corresponds to similar alluvial fan, flood plain and sabkha environments, but developed in a series of small graben as indicated in the section on figure 3. The age is mainly Keuper, at least for the upper parts of the sequence.

(iii) The sedimentation of the Pyrenean zone remains poorly studied. In general the series, which consists of basal detritics with thin evaporites above (perhaps a result of tectonic thinning), leaves the supposition that the Pyrenean axis showed a tendency to be a high, thus limiting the basin within Aquitaine during the Trias.

(*b*) '*Basal Lias*' – *Rhaetian–Hettangian* (*figure* 4)

The sedimentation remained argilo-evaporitic with carbonates, resembling the end-Trias deposits and thus rendering any simple limit to the Trias difficult. For example the basal unit, the Dolomite of Carcans, is dated Middle Rhaetian in the south and Hettangian in the north. However, the distribution of facies indicates that a new pattern of sedimentation had been established with a polarity markedly different from that of the Trias:

(i) to the east, the Occitan high, aligned north–south along the Villefranche fault, is covered on its western margin by thin breccias and detritics, whereas to the east the facies becomes that of a carbonate platform of restricted marine environment (i.e. towards the Mesogean sea);

(ii) to the west, a tectonic basin containing evaporites in dolomite–anhydrite sequences at the edge and dolomite–anhydrite–halite sequences in the centre, reaches thicknesses of 800–900 m.

The Occitan high separated the tectonic basin to the west upon which was to be located the Jurassic Aquitaine shallow-marine platform, from the Mesogean sea to the east.

(*c*) *Jurassic–Neocomian*

During this long period the sedimentation was mainly shale and carbonates in a shallow marine environment, with only very local detritics, thus indicating a planification and retreat of the basin limits. At the same time what is now the Aquitaine Platform gently subsided over the 75 Ma with a maximum thickness of only 2 km. In detail, however, the depocentre changed frequently as a function of the palaeobathymetry and the subsidence. These changes took place on a topography of very low amplitude thus quite rapid changes of facies took place both laterally and vertically.

The various facies developed are as follows:

(i) open marine or external platform facies: shales with mudstone micritic carbonates (with characteristic fauna of ammonites, echinoderms, brachiopods) in numerous coarsening-up, infilling or regressive cycles (Klupfelian sequences);

(ii) barrier or littoral-fringe facies forming a limit between the external and internal platform: bioclastic, oolitic and reefal carbonates and constructed reefs (large-scale coarsening-up sequences);

(iii) confined marine facies on the internal platform protected by the littoral barrier: mudstone carbonates with benthic fauna, often laminated or with lagoonal varves, cut by tidal channels infilled with oolites or bioclasts. These carbonates pass laterally into evaporitic deposits with evaporitic breccias and continental deposits of shales with lignites often showing palaeosols and indications of emersion. The distribution of these various facies in time and space gives the following palaeogeographic evolution.

Lower Lias: progressive submersion of the evaporitic basal Lias by bioclastic and oolitic limestones.

Middle and upper Lias: the Aquitaine Platform was completely submerged and covered by ammonitic marls of external platform facies. The Atlantic and Mesogean oceans are linked over the Occitan High.

Middle Jurassic – Oxfordian (figure 5): re-establishment of the Occitan High with reappearance of internal platform deposits and even indications of emersion (lignites). On the western fringe of the High, an oolitic offshore bar with occasional constructed reefs developed (the formation of Meillon and Saint-Marcet). This barrier was continuous in a N–S sense from Pau to Perigord. Further west open marine conditions persisted in a Marl–limestone facies with ammonites.

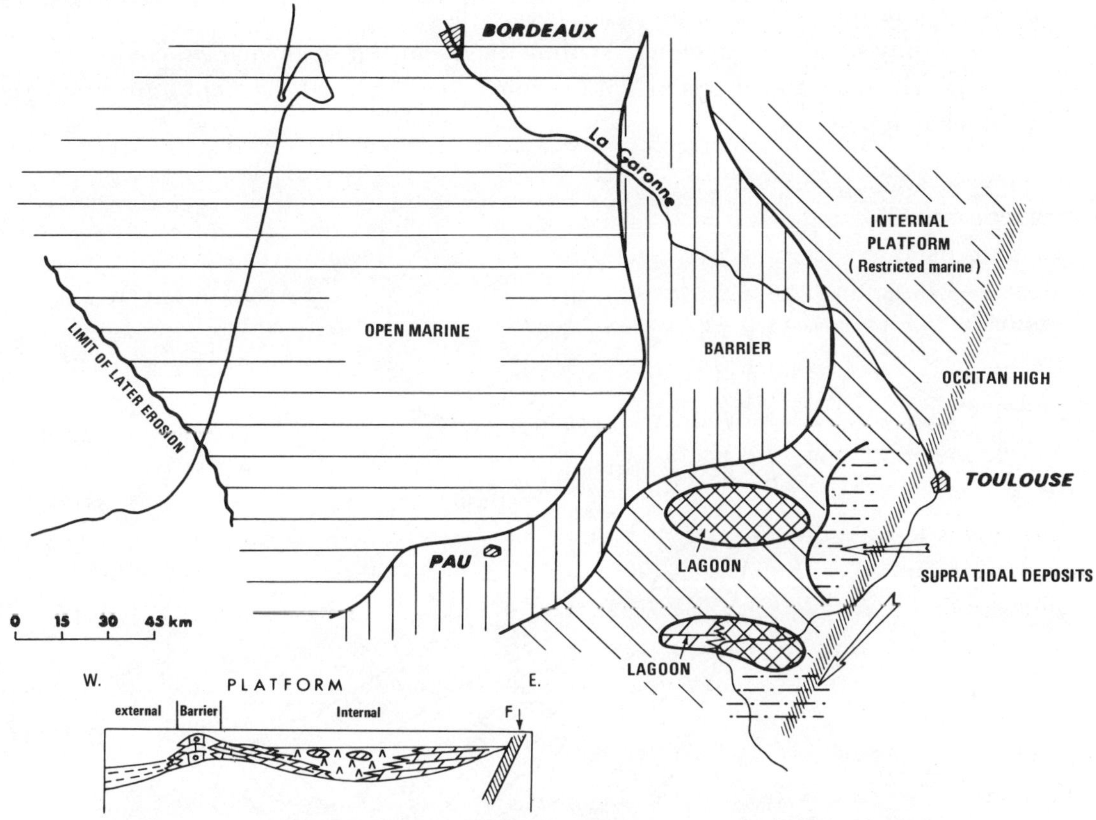

FIGURE 5. Middle Jurassic – Oxfordian sedimentation zones (170–140 Ma).

Kimmeridgian: the Occitan High was again submerged, with deposition of open marine facies composed of shaly limestone with ammonites. This was the last time that there was a connection between the Atlantic and the Mesogean oceans through the Aquitaine basin. Within the basin three depocentres developed with WNW–ESE trends: the Parentis sub-basin and the Mirande and Arzacq troughs.

Portlandian–Purbeckian (figure 6): the Occitan High reappeared and a calcareous offshore bar, which was locally reefal (in the Charentes region), migrated westwards as far as the present Bay of Biscay. The open marine environment persisted only to the northwest in the Parentis sub-basin. At the end of the Jurassic almost the entire Aquitaine platform was covered by the 'Dolomite de Mano' Formation deposited in a lagoonal platform environment. However, in the depocentres of Mirande and Arzacq thick calcareo-evaporites with anhydrites developed passing into sedimentary breccia ('Brèche de Garlin') at the edges of the basins.

Neocomian–Barremian (figure 7): this period saw the first uplift and erosion of the margins of the basin since Liassic times, with development of fluviatile sands both on the southern flank and deltaic sands to the north. Over the rest of the area shale–carbonate internal-platform facies, similar to that of the end of the Jurassic, persisted. During the Barremian, Aquitaine was a vast and very shallow lagoon. Beneath this landscape the Mirande and Arzacq sub-basins collected considerable thicknesses (up to 800 m) of fine algal limestone with characean algae ('calcaire à Annelides') and some beds with an evaporitic or sabkha tendency.

On the edge of the Parentis sub-basin one sees the change to a littoral environment with deposition of limestone and detritics. Within the basin, an open marine environment developed with siltstone and limestones into which, on the northern flank, are intercalated several thick deltaic units.

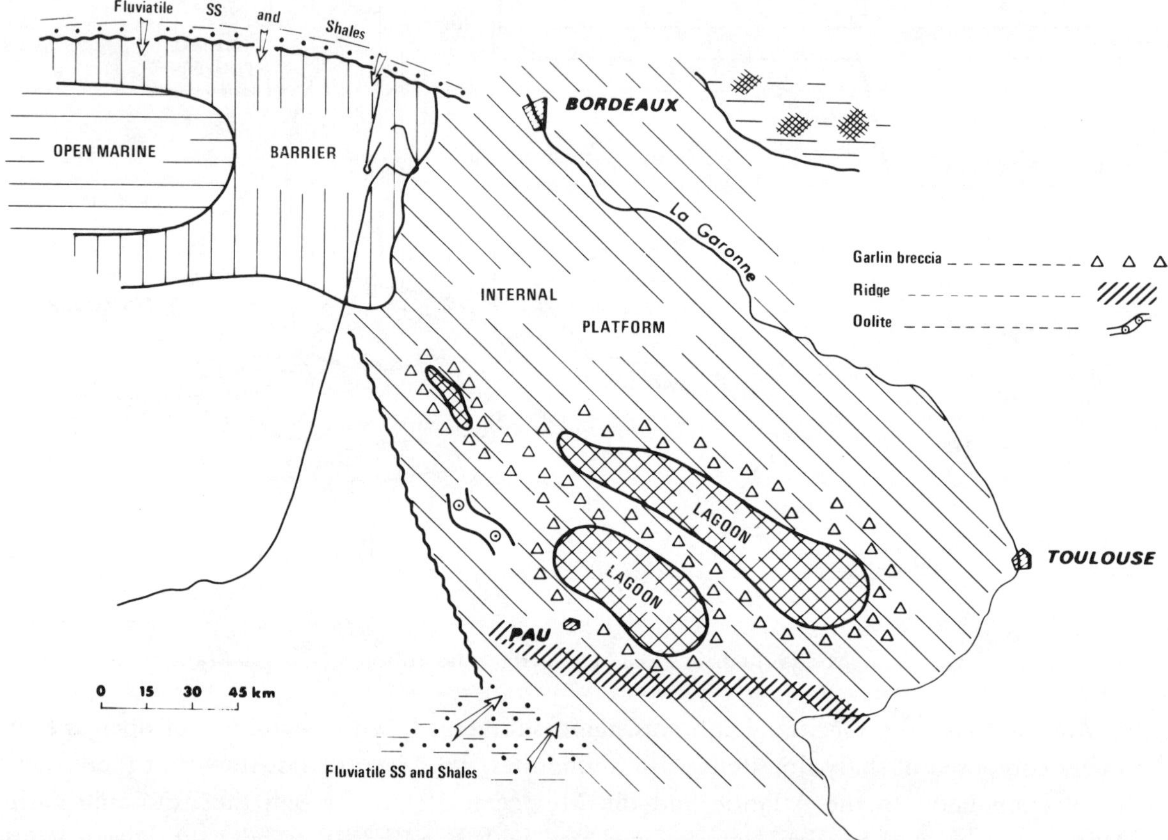

FIGURE 6. Sedimentation zones of the Portlandian–Purbeckian (140–135 Ma).

(*d*) *The Aptian–Albian and Upper Cretaceous*

From the Aptian onwards, the Aquitaine Platform, which had existed since the Lias, was completely modified by the development of two strongly subsiding zones, the sub-basin of Parentis to the north, and the North Pyrenean Basin complex to the south (itself divided into two sub-basins). The main basins were separated by an east–west-trending high, the Central Aquitaine platform. The basins and high all trend E–W (figures 8 and 9), and were to influence the sedimentation until the Senonian.

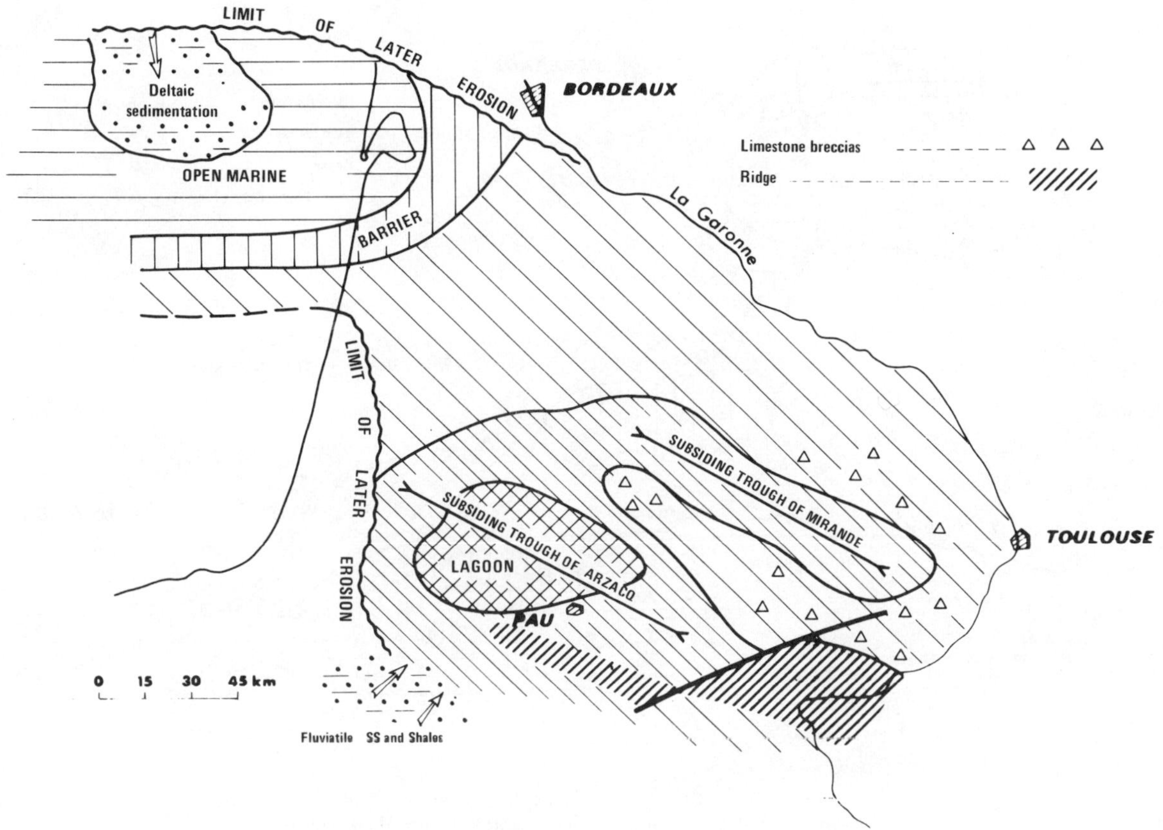

FIGURE 7. Neocomian–Barremian sedimentation zones (135–115 Ma).

(i) *The Parentis sub-basin*

This is a narrow trough that began to subside slightly in the Upper Jurassic. Subsidence continued slowly through the Neocomian–Barremian but during the Upper Aptian and Albian it accelerated. Despite a very high sedimentation rate (more than 2.5 km in 10 Ma), the subsidence outstripped the sedimentation, and the water depth continued to increase to a maximum of about 2 km at the end of the Albian. Along the margins of the basin, submarine canyons, which cut down to the Upper Jurassic, acted as conduits for detritic material that resedimented at the break of slope in lobes with grain or debris flow channel deposits that are separated by interlobe deposits of hemipelagic shale with thinly bedded turbiditic silty 'Bouma' units. The shelf edge, which was cut by the canyons, was a bioclastic offshore bar with more or less continuous bioconstructed reefs.

The Upper Cretaceous had much the same palaeogeography but sedimentation conditions were modified:

the differential subsidence was much reduced;

the whole of the Aquitaine margin was transgressed, causing a retreat of the zone of detritic deposition, and thus very little carbonate sediment reached the trough until the middle of the Senonian;

after the Middle Senonian the borders of the basin started to uplift, initiated a regressive and infilling cycle with a sedimentary (carbonate and siliceous detritic) prism prograding towards the centre of the basin. This cycle was not completed until during the Lower Tertiary.

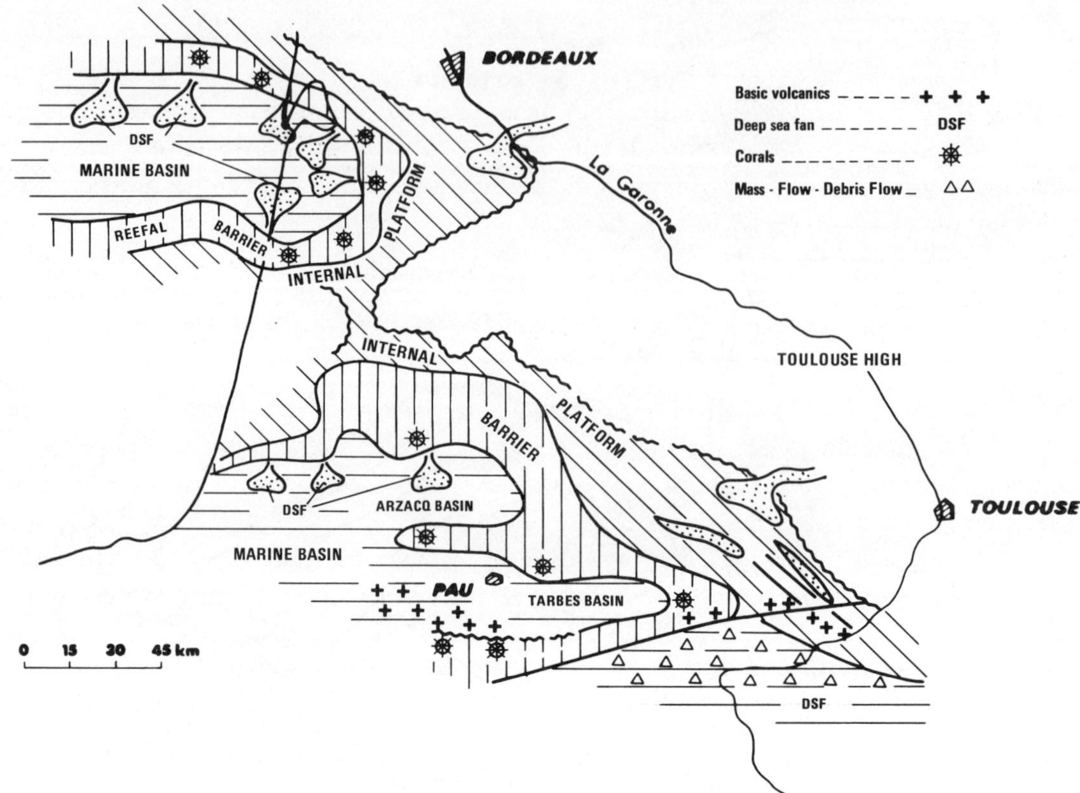

FIGURE 8. Aptian–Albian sedimentation zones (110–100 Ma).

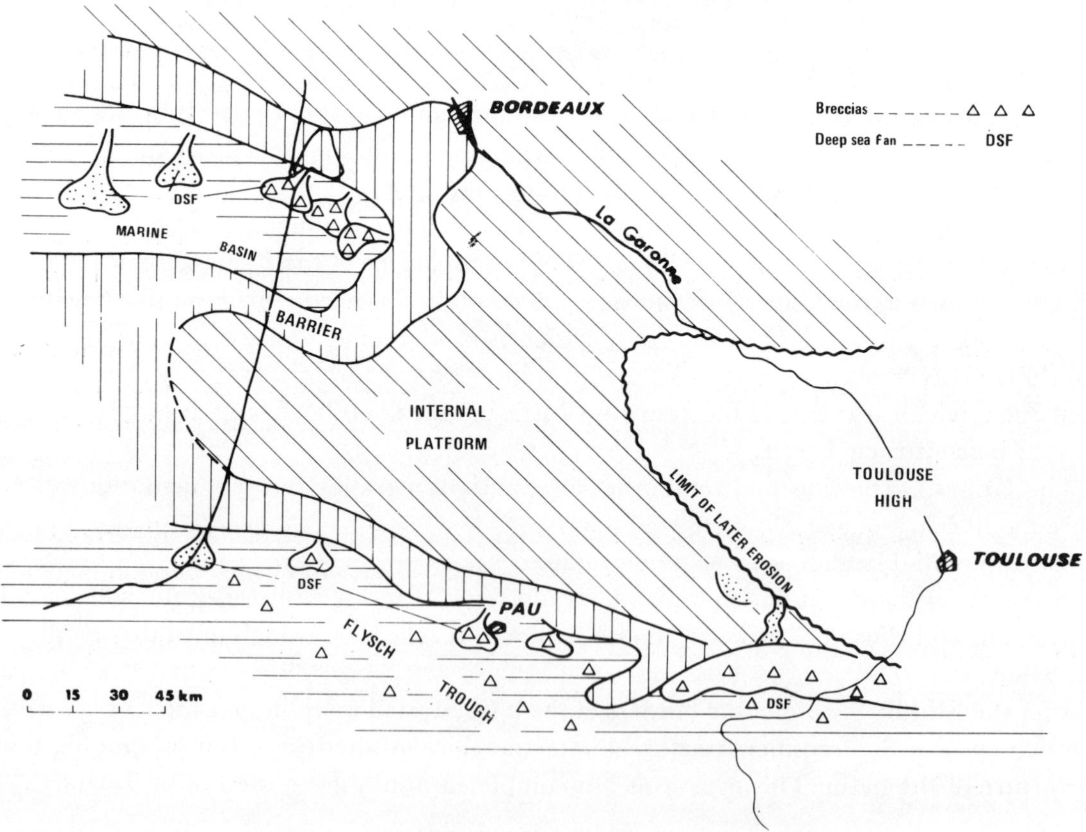

FIGURE 9. Lower Senonian sedimentation zones (85–80 Ma).

(ii) *The North Pyrenean Basin complex*

The North Pyrenean subsidence zone has a structural and sedimentary evolution very much more complex than that of the Parentis sub-basin. It is situated in a relatively narrow zone between the Central Aquitaine Platform and the Pyrenean axis. The latter remained a border high throughout the Aptian–Albian–Upper Cretaceous.

Several stages in the tectono-sedimentary evolution can be distinguished.

The Albo–Aptian taphrogenic phase of basin fragmentation seems to have developed either as a simple wide subsiding basin (Basin of Arzacq) or as narrow strongly subsiding troughs with deep palaeobathymetry (Comminges and Basque basins).

The Basin of Arzacq is filled with almost 3.5 km of open marine sediments, silty marl-with-spicules and limestones in thick coarsening-up regressive sequences. The basal unit, the Sainte-Suzanne Marl, dated Lower Aptian by ammonites, marks the first marine sedimentation in this region since the Kimmeridgian.

The narrow Comminges and Basque troughs of flysch are filled, in the south, with 2–3.5 km of fan deposits and to the north with shales and carbonates with many slumps, olistoliths and fan deposits.

The Albo–Aptian basins and troughs are separated by ridges on which the section is very thin and in shallow-water facies or the section is absent either by erosion or lack of sedimentation. Constructed reefs occur on the northern flanks towards the Central Aquitaine Platform, limiting the marine basins from the internal platform.

This taphrogenic phase is completed by basic and ultrabasic volcanic extrusions related to the ENE–WSW and WNW–ESE fault system.

Merging phase of the various Albian troughs (except the Basin of Arzacq). Gradually by Cenomanian–Turonian times the Albian troughs just to the north of the Pyrenean axis merged to form one simple trough, the Flysch Trough, which extends from the Ariège westwards to the Bay of Biscay. Its geometry is well known. This trough was strongly subsident, accumulating up to 4.5 km of turbidite – submarine fan deposits in very deep water. The fans were fed from canyons that cut into the northern flank of the Aquitaine basin and also the north Iberian margin. On the margins occur thick slumped or resedimented units that pass up-slope into external platform limestone, often with patch-reefs.

Migration phase of the Flysch Trough. From the Turonian to the end of the Maestrichian the Flysch Trough was displaced progressively, with the northern margin retreating about 30 km towards the north during this period. This movement was the result of compressive synsedimentary tectonics with associated folding and thrusting successively covered by the sedimentation.

Syn- and post-tectonic infilling of the Flysch Trough. The infilling was achieved by rapid lateral and frontal prograding units, which pass up into shallow-water detritics through lagoonal to continental facies. This regressive sequence is diachronous, taking place at the end of Cretaceous times in the east (Ariège–Comminges) and towards the end of the Eocene in the west (Arzacq–Basque region).

(iii) *The Central Aquitaine Platform*

Between the northern and southern sub-basins the high zone or Central Aquitaine Platform which is a vestige of the old Jurassic platform, has also (where it remains) a thin but relatively complete section, indicative of a low subsidence rate. Much of this platform was emergent at the Albo-Aptian notably along the 'Les Landes' Ridge – Toulouse High. This high suffered significant erosion with the removal of much of the Lower Cretaceous section and part of the Jurassic. On the northern side of this axis, during the Albo-Aptian, sand–shale littoral facies occur, whereas on the southern flank are supratidal to infratidal carbonates.

At Cenomanian–Turonian times the Central Aquitaine Platform was again drowned and covered as far as the edge of the Massif Central by marine facies shale and limestone and chalk, all very fossiliferous and thin (200–300 m). This episode represents the last great marine transgression in the Aquitaine Basin.

From the beginning of the Senonian onwards, the shorelines, often picked out by reefs, regressed towards the northern and southern marine basins. Behind them followed internal platform deposits, rich in Hippurites, then supratidal and continental shales and sands that progressively spread out from the Landes high, which again emerged at the end of the Cretaceous.

(iv) *Cretaceous salt tectonics*

The break-up of the Jurassic platform triggered the movement of Triassic salt over the whole of the Aquitaine Basin. This movement caused considerable syn-sedimentary effects in all the sedimentary regions, whether in the basins or on the central platform.

The results of the halokinesis are as follows.

Salt walls that grew during the Albo-Aptian created rim synclines with thick sediments. These synclines migrated laterally at the same time as the walls increased in amplitude. The summits of the ridges are often eroded and both local and generalized piercement took place; the best examples are in the Parentis sub-basin and in the Arzacq Basin.

Salt diapirs, more or less cylindrical and, by definition, piercing, are concentrated in the region of Dax and on the 'Les Landes' ridge.

Diapiric anticlines, either symmetrical or in the form of a factory roof, are related to the major fractures. The crests are often eroded and the rim synclines show syn-sedimentary thickening.

The main salt movement ended in the Albian, since a number of Triassic cap-rocks are covered by Albian series; but locally the movement continued into the Upper Cretaceous and often in the Tertiary there was intense reactivation.

The halokinetic phase had considerable impact on the Lower Cretaceous sedimentation. Also, the lithological and rheological changes as a result of the diapiric movement greatly influenced the geometry of the structures created or modified by the later Pyrenean tectonics. Thus the Cretaceous salt tectonics represented a very important stage in the geological history of Aquitaine.

(v) *Summary of the Mesozoic sedimentation*

The sedimentation throughout the Trias to the Upper Cretaceous, over the greater part of the Aquitaine Basin, was in limestone–shale and evaporitic facies typical of platform development and thus is analogous to all the great platform basins related to the Mesogean sea.

4. Mesozoic structural evolution of the Aquitaine Basin

The changes in palaeogeography throughout the Mesozoic often reflect the stages in the structural development during this period. Briefly one can summarize this development in four stages: (*a*) initiation of the basin in the Trias, (*b*) establishment of the Aquitaine Platform during the Jurassic, (*c*) break-up of this Platform at the end of the Jurassic and during the Lower Cretaceous, (*d*) development of the Pyrenean orogenic system during the Upper Cretaceous.

(*a*) *The initiation of the Triassic Aquitaine Basin* (*figure* 10)

Since the geometry and facies of the early stages are poorly known at present, the initiation of the Triassic phase of the Aquitaine Basin is little understood. The picture of the basin is confirmed only at the Upper Trias with a large rift that subsided between WNW–ESE-trending faults, and extended out into what is now the Bay of Biscay. This rift is infilled with evaporitic playa-type deposits. The presence of considerable volumes of ophitic volcanics confirms the crustal nature of the bounding faults.

This simple basin development was preceded by a series of poorly known narrow fault-controlled troughs trending NE–SW infilled mainly with detritics of uncertain Permian to lower Trias age. This zone extends as far as the Armorican and Central Massifs.

The relation between these two stages remains to be established, as is an understanding of the mechanism for the opening of the Upper Trias salt basin. The choice remains between a simple separation or rifting between the Iberian and European plates (if already differentiated) or alternatively an effect of stretching related to shear within the European plate.

(*b*) *The establishment of the Aquitaine Platform during the Jurassic* (*figure* 11)

The Jurassic palaeogeography, with its isopachs trending N–S, which replaced the generally E–W Triassic trends, indicates the establishment of a new structural régime.

The whole of Aquitaine was a more or less submerged platform that tilted towards the west away from the Occitan High, where, by the end of the Lower Lias, an open marine environment was established. The more subsident zones on this platform trend N–S, i.e. parallel to the facies zones.

This structural topographic arrangement seems to be the result of an E–W extension régime which can be ascribed to an early but feeble opening of the Atlantic, to the north of the fault separating the Iberian and European plates.

(*c*) *The break-up of the Aquitaine Platform at the end of the Jurassic and the Lower Cretaceous* (*figure* 12)

Between the Kimmeridgian and the Cenomanian a series of important events resulted in a change in sedimentary trends, from N–S ‘Atlantic’ to E–W ‘Pyrenean’. This change can be followed through three successive stages.

(i) *Kimmeridgian*

In an open marine environment, the isofacies lines remain N–S but the isopachs mark the appearance of E–W trending sub-basins, e.g. Parentis, Arzacq and Mirande.

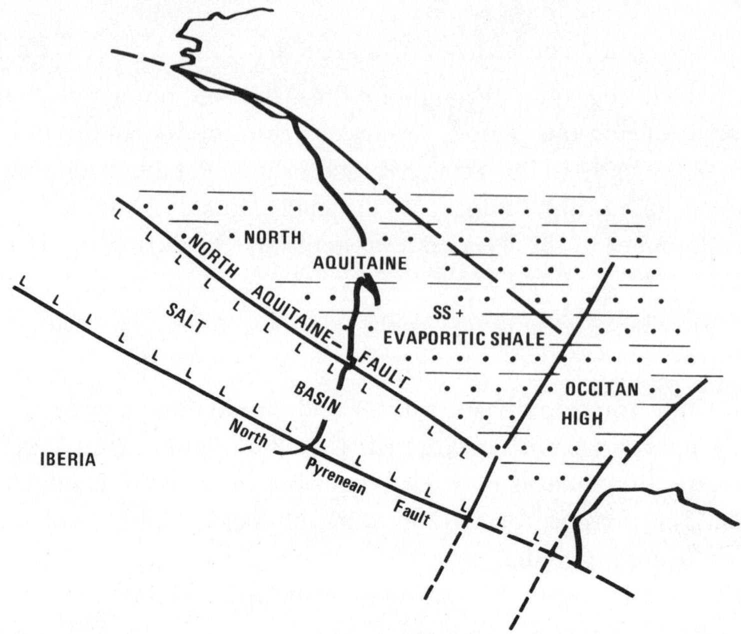

FIGURE 10. Tectonic and palaeogeographic sketch of Upper Triassic basin.

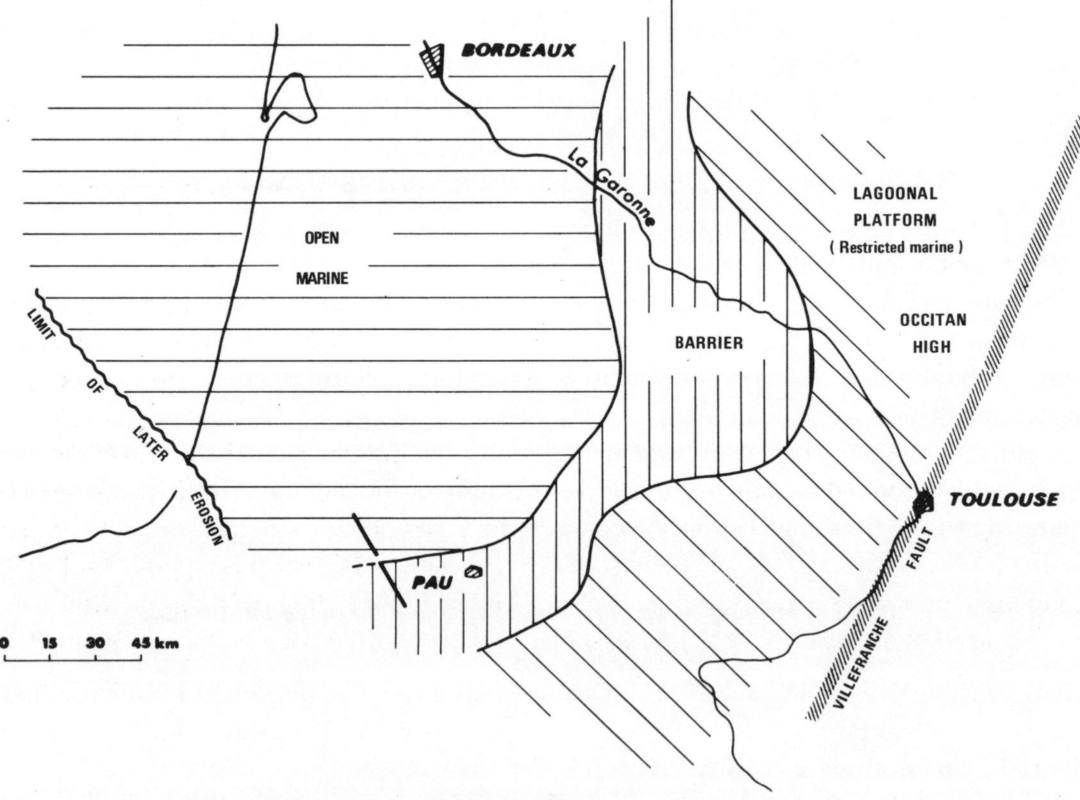

FIGURE 11. Geometry of basin in Middle Jurassic to Oxfordian.

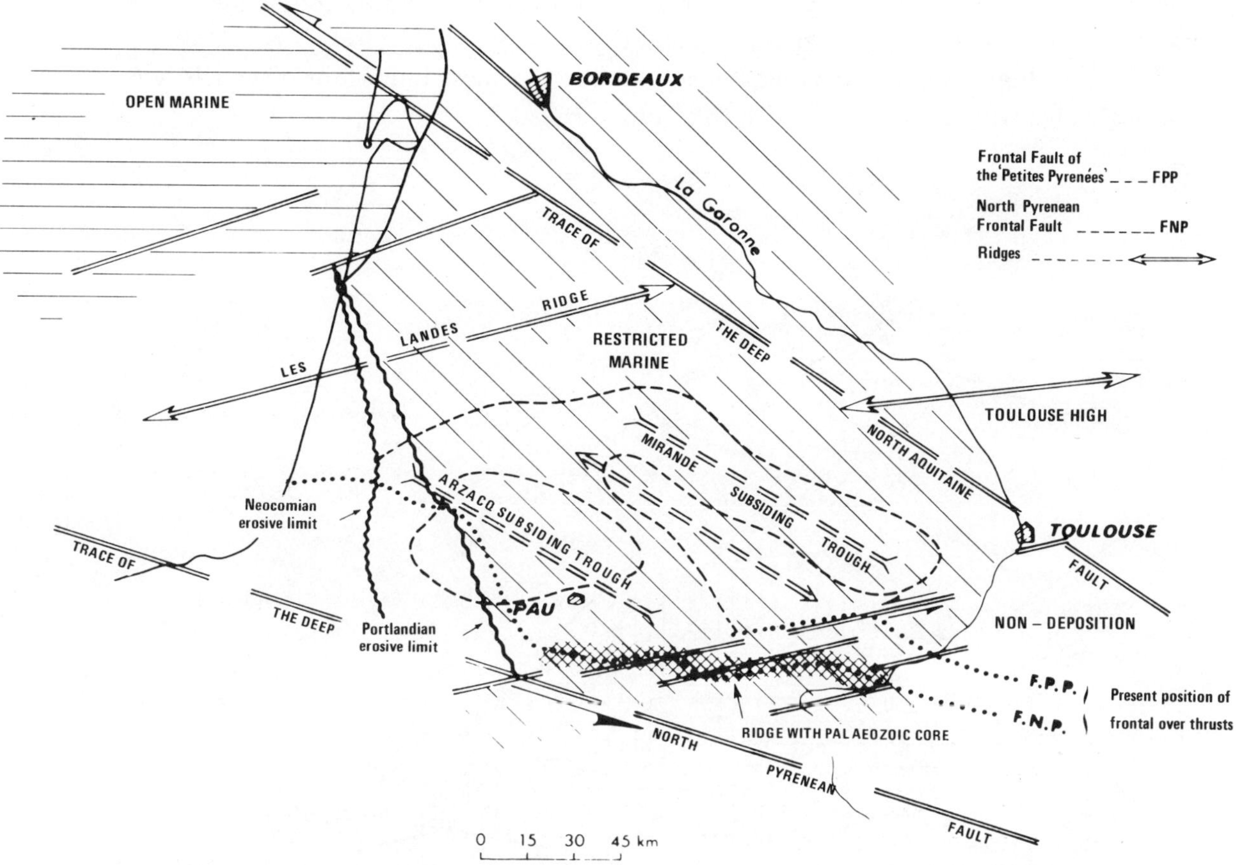

FIGURE 12. Tectonic and palaeogeographic sketch map of the Portlandian to Neocomian.

(ii) *Portlandian to Barremian*

The open marine environment was restricted to the Parentis sub-basin while the rest of Aquitaine was uplifted, with resultant emergence and erosion of the northern and southern flanks. However, towards the south of the region the Arzacq and Mirande sub-basins persisted, receiving either an evaporitic-sabkha or restricted marine sedimentation.

At the same time two structural systems progressively developed, firstly, ENE–WSW trending anticlines of which the main ones are the 'Les Landes' ridge, a wide high on which was triggered intense halokinesis of Triassic salt, and a series of high amplitude anticlines *en echelon* with the north Pyrenean border fault, which were eroded penecontamporaneously to their Hercynian basement cones; and secondly, WNW–ESE trending salt ridges with penecontemporaneous salt extrusion, e.g. Antin – Maubourguet and Audignon ridges (figure 1).

(iii) *Albo-Aptian* (*figure* 13)

The process of break-up reached its greatest mobility with structural amplitude reaching 4 km.

The palaeography was dominated by ENE–WSW trending structures:

a series of *en echelon* sub-basins – Parentis, Tarbes, Comminges – with a maximum rate of subsidence greater than 1 km Ma^{-1} and also a high rate of sedimentation, accumulated several kilometres of either flysch or silt–shale in a deep-sea environment, indicating considerable palaeobathymetry and also steep depositional dips;

a series of high ridges separating the sub-basins, the most important of which is the 'Les Landes' ridge which is *en echelon* with the Toulouse high.

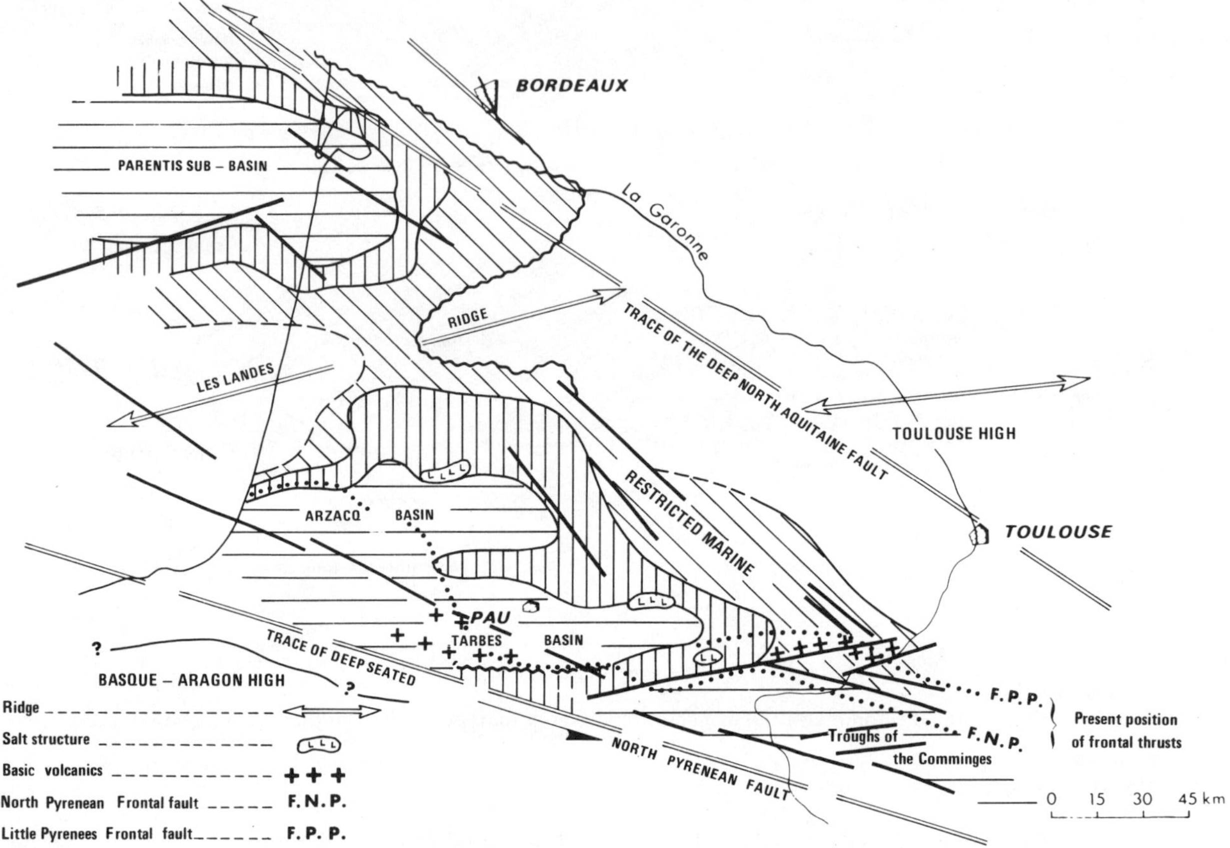

FIGURE 13. Tectonic and palaeogeographic sketch map of the Albian.

The dislocations that controlled these structures were of crustal magnitude, at least for the southern sub-basins, where ultrabasic volcanics occur and also where metamorphism with growth of diopside occurs in a narrow band the length of the Pyrenees.

All the above movements (Kimmeridgian to Albo-Aptian) have been grouped into a single tectonic phase, the pre-Cenomanian phase. This tectonism is mainly taphrogenic with tilted fault blocks and basins subsiding in regions of intense fracturing of the Aquitaine Platform. These movements again triggered the halokinesis of Triassic salt.

In the zone close to the eventual Pyrenees this phase is expressed, at least locally, by compression, which created *en echelon* anticlines and by development, again locally, of a penetrative schistosity.

Generally, the dislocation of the Aquitaine Platform seems to have followed a logical sequence that can be explained by a system of sinistral shears that affected the intracontinental segment of the European plate between the crustal Pyrenean fault zone and the deep north Aquitaine fault.

Within this segment the resultant ENE–WSW secondary shears controlled the Albo-Aptian basinal and anticlinal trends and also the halokinetic ridges. The opening of the sub-basins seems to have been in a system of pull-apart or rhomb-graben, which are either very large (Parentis and Arzacq) or very small (trenches of the Comminges). This shear system has been interpreted as a result of interactive shear between the Iberian and European plates. However, with the present knowledge of the palaeogeography (i.e. the position of isopachs on either side of the Pyrenees) one is unable to quantify the interplate movement.

(*d*) *The pre-Pyrenean evolution during the Upper Cretaceous*

This evolution has two phases: the development of the Flysch Trough with an E–W trend, and the start of the Pyrenean compression.

(i) *Development of the Flysch Trough*

After the Cenomanian–Turonian, the various Albian troughs along the edge of the Pyrenees (Basque region, Tarbes, Comminges) coalesced into a single E–W-trending unit, the Flysch Trough. As its name implies, this trough is filled with a very thick flysch sequence deposited in a deep marine environment.

The original geometry of the Flysch Trough, i.e. before the Pyrenean orogeny, has been obtained by analysis of the magnetic anomalies observed in the Atlantic near the Bay of Biscay. Anomalies 33/34 (figure 14) indicate that 80–85 Ma B.P., i.e. during the Upper Cretaceous, a triple junction developed with one of its arms trending E–W into the Bay of Biscay and in continuation with the Flysch Trough. It thus seems probable that the trough developed under the tensional régime associated with this arm. Superposition of the American and European anomalies 33 show that the Flysch Trough was considerably wider during the Upper Cretaceous than at present. The later reduction in width, of the order of between a third and a half, was the result of the rotational squeeze of the Pyrenean orogeny, the magnitude of which diminished from the east towards the Bay of Biscay.

(ii) *The start of the Pyrenean compression* (figure 15)

The first Pyrenean compressional phase started in the Middle Senonian and continued through to the end of the Upper Cretaceous. The average compressional direction was N 20° and the effects varied from one structural régime to another.

Aquitaine Platform, frontal to the Flysch Trough. The Pyrenean compression caused a progressive low-amplitude arching of the Aquitaine Platform, thus joining the Toulouse and 'Les Landes' highs to form a miogeanticline which separated the Parentis sub-basin from the Flysch Trough. Shorter wavelength compressive, often asymmetric, anticlines developed on the old salt ridges and on the Albo-Aptian fault-block highs. All these structures trend E–W and are cut by NW–SE and NE–SW trending dextral and sinistral strike-slip faults. During this phase the salt was reactivated and it injected into the cores of the anticlines and sometimes along the strike-slip faults.

The Flysch Trough. The deformation was increasingly intense towards the south where in the more extremely deformed zones an Upper Senonian penetrative schistosity developed. The tectonics were syn-sedimentary with the front part of thrust blocks being taken up in the Upper Cretaceous flysch, at the same time as the trough was being tightened and filled.

Another effect of the increased deformation taking place on the southern flank of the Flysch

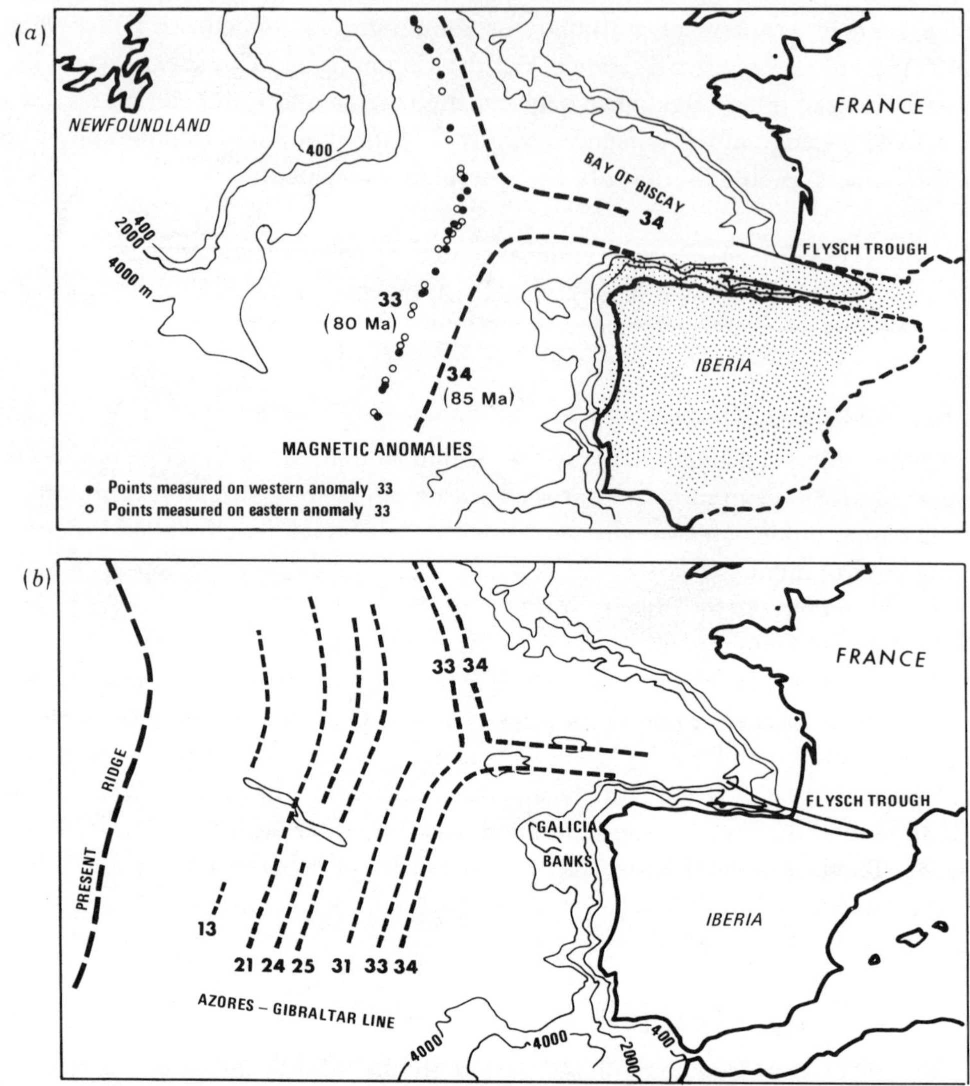

FIGURE 14. (a) Situation at Upper Cretaceous before the Pyrenean (modified from Boillot & Capdevila 1974); (b) present situation.

Trough was a migration throughout the Turonian–Maestrichtian of both the axis and the northern flank towards the north.

5. Tertiary and Quaternary structural evolution

The Pyrenean compression, which began in the Senonian, continued throughout the Lower Tertiary with final closure of the Flysch Trough migrating progressively east to west. This closure, with infilling facies changing from marine to continental fluvio-lacustrine, took place at the end of the Cretaceous in the eastern Pyrenees, at Middle to Upper Eocene in the Tarbes Basin and in the Oligocene in the Bay of Biscay (figure 15).

With the start of the uplift of the Pyrenees, during the Oligocene and Neogene, a thick series of fluvio-deltaic sediments (or Molasse) developed in the form of vast alluvial fans. Along the southern border of the Aquitaine Basin, where the Molasse oversteps onto eroded folded

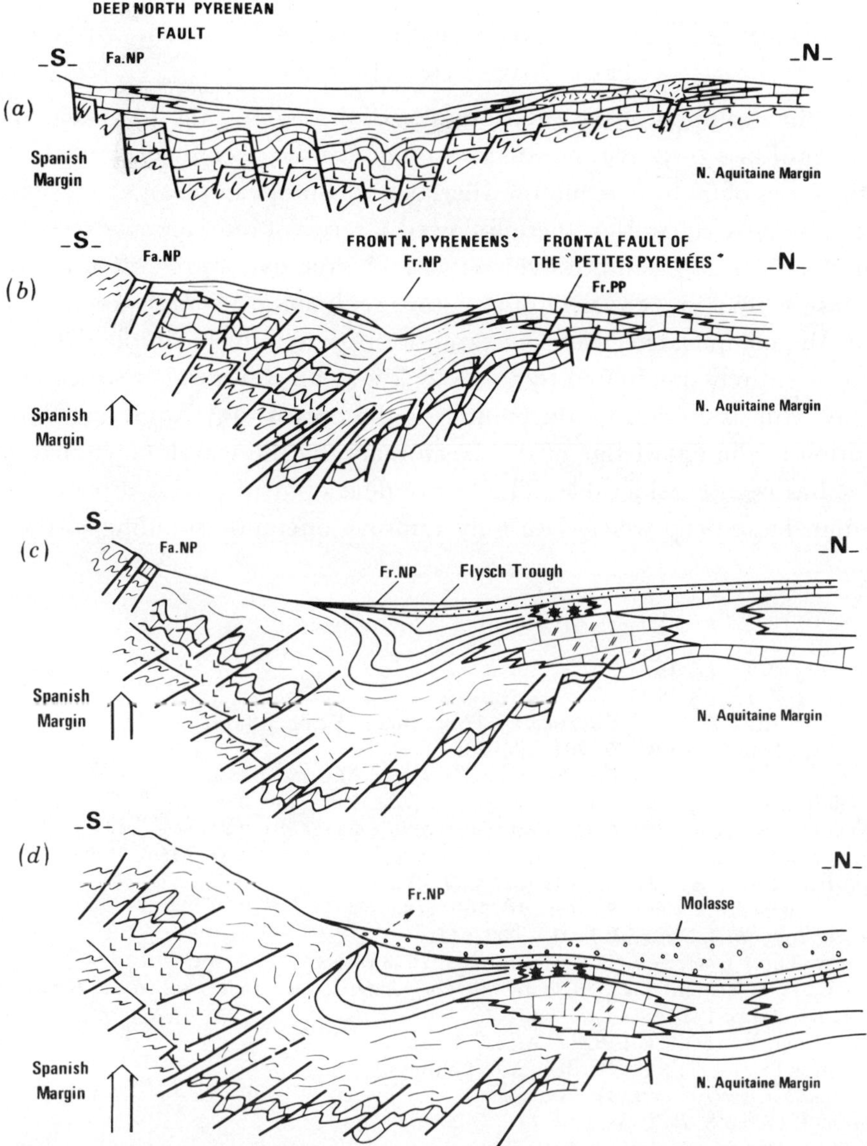

FIGURE 15. Evolution of the flysch troughs: (*a*) Albian–Cenomanian–Turonian (Comminges Zone); (*b*) Maestrichtian–Dano/Montian (Comminges Zone); (*c*) Eocene (Tarbes Basin); (*d*) Molasse Basin (Neogene of the Tarbes Basin).

Mesozoic rocks, the Molasse in its turn was gently flexed and folded, indicating waning N–S Pyrenean compression throughout this period.

The Plio-Quaternary saw the epeirogenic uplift of the entire Pyrenean chain, lifting Oligocene and Neogene erosion surfaces to an altitude of 2.5–3 km which were then dissected by deep-cutting rivers. These developed huge dejection cones of fan-glomerate (e.g. Lannemezan) on the northern flanks of the Pyrenees. The morphology of the Pyrenees was only slightly affected by the Pleistocene glaciation.

The Neogene to Recent post-orogenic sedimentation built out from the Pyrenees in a prograding system and caused a progressive regression of the coast from the east to the west and northwest. The continuation of this process formed the present-day continental shelf of the Atlantic.

6. Conclusions on the structural evolution of the Aquitaine Basin

The analysis of the evolution of the Aquitaine Basin throughout the Mesozoic and Tertiary shows that the overall structural régime was controlled by the opening of the Atlantic and the interaction of the series of faults limiting the Iberian and European plates.

After an initial Triassic separation there followed a series of interactions: separation, transform movements and variable compressive contact. The relative simplicity of this sequence of events is now masked by a great variety of structuro-sedimentary complexities.

In contrast to the complicated Mesozoic evolution, the relatively simple end Cretaceous – Tertiary phase was entirely dominated by the N–S Pyrenean compression which emplaced the Pyrenees of today with its erosion products infilling the present-day Aquitaine Basin.

The contribution to the knowledge of this basin made by the search for hydrocarbons over the last 40 years has been considerable. The future development in this direction – improved seismic techniques, more deep wells – can only improve our understanding of the geology of southwest Europe.

Bibliography (Curnelle *et al.*)

Anon. 1973 *Bull. Soc. geol. Fr.* (7) **15** (1).
Anon. 1978 *Bull. Soc. geol. Fr.* (7) **20** (5).
Arents, J. *et al.* 1975 In *IXe Congrès Int. Sédimentologie, Nice,* thème V, pp. 13–17.
Bertrand, L. 1940 *Bull. Servs Carte géol. Fr.* **204**, 205–282.
Boillot, G. & Capdevila, R. 1974 In *2è Réunion Sci. Terre, Pont-à-Mousson*, p. 59.
Bouroulec, J. & Deloffre, R. 1969 *Bull. Centre Rech. Pau – SNPA* **3** (2), 287–328.
Bouroulec, J. & Deloffre, R. 1970 *Bull. Centre Rech. Pau – SNPA* **4** (2), 381–429.
B.R.G.M., ELF r.e., ESSO-REP & S.N.P.A. 1974 *Géologie du Bassin d'Aquitaine.* Mém. B.R.G.M.
Casteras, M. 1933 *Bull. Carte – géol. Fr.* no. **189** (515 pages.)
Choukroune, P. *et al.* 1972 *Earth planet. Sci. Lett.* **18**, 109–118.
Choukroune, P. *et al.* 1973 *Bull. Soc. géol. Fr.* **15**, 600–611.
Choukroune, P. 1974 Thèse Doct. Sci. Nat., Montpellier. (276 pages.)
Curnelle, R., Dubois, P. & Seguin, J. C. 1980 In *26e Congrès Géol. Int., Bull. Cent. Rech. Explor. – Prod. ELF-Aquitaine, Mém.* no. 3, pp. 1–78.
Debroas, E. J. & Souquet, P. 1976 *Bull. B.R.G.M.* (2) **1** (4).
Delfaud, J. 1969 Thèse Doct. Sci. Nat., Bordeaux. (5 volumes.)
Dubar, G. 1925 *Mém. Soc. géol. Nord* **9** (1). (322 pages.)
Dubois, P. & Seguin, J. C. 1978 *Bull. Soc. géol. Fr.* (7) **20** (5).
Editions Technip 1971 *Histoire structurale du Golfe de Gascogne.*
Feuillee, P., Villanova, M. & Winnock, E. 1973 *Bull. Soc. géol. Fr.* (7) **15** (1).
Ferragne, A. & Vignueaux, M. 1978 *Bull. B.R.G.M.* (2), sect. 4, no. 2, pp. 35–142.
Le Pichon, X., Bonnin, J. & Sibuet, J. C. 1970 *C.r. hebd. Séanc. Acad. Sci., Paris* **271**, 1941–1944.
Mattauer, M. 1968 *Revue géogr. phys. géol. dyn.* (2) **10** (1).
Peybernes, B. & Souquet, P. 1975 *Bull. Soc. Hist. nat. Toulouse* **111**, 204–210.
Peybernes, B. 1976 Thèse Doct. Sci. Nat., Toulouse. (459 pages.)
Schoeffler, J. 1971 Thèse Doct. Sci. Nat., Bordeaux.
Souquet, P. 1971 In *96e Congr. Soc. Sav. Toulouse.*
Souquet, P. *et al.* 1977 *Géol. Alpine* **53** (2).
Williams, C. A. 1973 *Nature, Lond.* **244**, 86–88.
Williams, C. A. 1975 *Earth planet. Sci. Lett.* **24**, 440–456.
Ziegler, P. A. 1981 In *Petroleum geology of the continental shelf of north-west Europe*, pp. 3–39. London: Institute of Petroleum.

Discussion

D. Whitaker. How well explored are the Cretaceous submarine canyons and deep-water fans in the Parentis and southern basins? Could the canyons be clearly defined morphologically, and if so, do hydrocarbons occur in their sandy fills? How well are the fans delimited, and what is the nature of their reservoir rocks (e.g. sandy channel fills)?

R. Curnelle. There is a considerable amount of detrital material filling all the channels. Their base is eroded into the rocks below and has often exposed Barremian rocks, sometimes even rocks as old as Jurassic. In some areas the channels can be followed on the seismic sections. The channel sands often form reservoirs for hydrocarbons.

D. H. Matthews, F.R.S. What do we know about the relation between Pyrenean deformation in the south of the Aquitaine Basin and the formation of the North Spanish trough?

D. G. Roberts. The North Spanish Trough continues north of Gallicia bank and then stops. The seismic reflexion shows that the thrusting that formed the North Spanish Trough continues into the Atlantic, and presumably continues along the Azores Biscay Rise in a manner which is not yet clear. Spain must have moved westward, as well as northward, relative to France during this period of thrusting to produce the observed structures. These observations suggest that the Pyrenees were produced by a combination of shortening and right-handed strike-slip movement.

Sir Peter Kent, F.R.S. The North Pyrenean fault is generally believed to be a strike-slip fault, but is the North Aquitaine also a strike-slip structure?

R. Curnelle. The North Aquitaine is also a shear fault. Extension in the basin occurred during Albian times and continued during the Upper Cretaceous. The normal faulting is associated with widespread strike-slip movement on both right- and left-handed fault systems. The resulting disruption makes the geometry of the Albian and later movements difficult to reconstruct. This tectonic style is rather different from that further west on the northern margin of the Bay of Biscay, described earlier by Dr Roberts.

A. S. Laughton, F.R.S. The subsidence in the Aquitaine Basin was as rapid as 250 $\mathrm{m\,Ma^{-1}}$ during the Albo-Aptian. How does the timing and the rate of this subsidence compare with the results obtained from leg 48 of the D.S.D.P. further to the northwest?

D. G. Roberts. We believe that the history of the Aquitaine Basin is very similar to that of the region to the northwest where holes were drilled during Leg 48. The amount of subsidence is, however, difficult to determine because Hole 400A did not penetrate to the pre-Albian section.

M. M. Kholief. What is the oil potential of the Aquitaine Basin and is it confined to these submarine deltaic fans or are other depositional facies important? Also, what is the relation between the structural style of this basin and the hydrocarbon traps?

R. Curnelle. The total reserves in place are about 2.4×10^8 t of oil and about 3.5×10^{11} m^3 of gas. The estimated amounts that are recoverable are 6×10^7 t of oil and 2.8×10^{11} m^3 of gas.

There are two source rocks in the Aquitaine Basin. In the Parentis Basin the main source is a sapropel of Kimmeridgian age. Elsewhere it is a black shale formed during a period of restricted circulation during the Barremian, which is associated with evaporites. Most of the oil reservoirs were deposited at the same time as the source rocks.

Phil. Trans. R. Soc. Lond. A **305**, 85–99 (1982) [85]
Printed in Great Britain

Lithology and subsidence in the North Sea

BY G. G. LECKIE
British Petroleum Company Limited,
Britannic House, Moor Lane, London EC2Y 9BU, U.K.

The North Sea sedimentary basin has developed on the northwestern margin of the European tectonic plate and contains an almost continuous record of epeirogenic marine and deltaic sedimentation from Carboniferous to Recent times. The subsidence required to accommodate the pile of sediment deposited, which in places exceeds 12 km, has been brought about at various times and in various places by differing geodynamical processes. As a result the types of sedimentary rocks deposited vary widely both in time and space, but the nature of the mechanism is reflected in the sedimentary type deposited.

The following broad generalizations can be made. The late Carboniferous was a period of deltaic sedimentation during which eustatic changes in sea level or local variations in subsidence rates are reflected in the typical Coal Measures swamp deposits. Late Carboniferous – early Permian times saw the silting up of this basin, and in an arid climate aeolian sands were deposited grading laterally to sabkha shales and evaporites. The Permian culminated in a series of widespread marine incursions during which repetitive evaporites were deposited.

Triassic times were marked by a period of major rifting and the deposition of thick sequences of continental clastics in the north, while widespread marine sedimentation persisted in southern areas. Jurassic times saw the re-establishment of marine to deltaic deposition in a series of basins possibly controlled in their distribution by the Triassic fault systems. Late Jurassic deposits were laid down in a sea whose bathymetry reflected the structure of the underlying horsts and grabens inherited from Triassic times, and towards the close of the Jurassic the bottom waters at least of this sea become increasingly stagnant. Sands deposited during the late Jurassic were deposited as near-shore marine bars, beach sands, and proximal and distal submarine fans.

Triassic to early Cretaceous deposition was concentrated in the areas now occupied by the main grabens of the North Sea, i.e. the Viking, Central and Moray – Witch Ground grabens. Subsequent deposition in late Cretaceous to Tertiary times took place in a more widely subsiding area, resulting in progressive onlap onto the surrounding basin margins. Deposition within this broadly subsiding and relatively unfaulted basin is characterized by chalky limestones in southern areas, giving way laterally to shales and minor sands to the north. During early Tertiary times a large delta was formed in the area beneath the present Moray Firth, and from this delta a supply of sand was fed into submarine fans to the northeast and southeast of the delta front.

Late Tertiary deposition is largely represented by a monotonous sequence of marine shales.

INTRODUCTION

The configuration and geological history of the North Sea Basin and adjacent areas has been well documented by various authors in the past few years, in particular P. A. Ziegler (1981), W. N. Ziegler (1975) and Kent (1980). Very detailed reports on various aspects of the North Sea with particular emphasis on the petroleum geology of the area have been published in the proceedings of the Institute of Petroleum's Bloomsbury Conference (Woodland (ed.) 1975), the Norwegian Petroleum Society's Mesozoic Northern North Sea Symposium (Finstad &

Selley (eds) 1977), and most recently the Institute of Petroleum's Lancaster Gate Conference (Illing & Hobson (eds) 1981).

As a result of these many papers, based as they are to a large extent on the vast amount of data held in the files of petroleum exploration companies and other institutions, the tectonic and lithostratigraphical framework of the North Sea is relatively well known. However, there are still quite large areas of the North Sea yet to be tested by the drill in which interpretation of the geology still depends on the imaginative interpretation of reconnaissance seismic data.

While the nature of the region's geological history is seemingly well understood the mechanisms that brought it about are less well known. An explanation of this mechanism is beyond the scope of this paper, which will attempt to highlight in a short, simplified form, as a starting point for the papers to follow, the relation that seems to exist between the types of subsidence that have occurred in the North Sea and the lithologies of the sediments that have filled the depressions created.

Structural setting

The North Sea Basin in its broadest sense is located on the northwestern margin of the Eurasian tectonic plate (figure 1). Before the break-up of the North Atlantic the area was bounded to the northeast by Fennoscandia, to the northwest by the Caledonides and to the

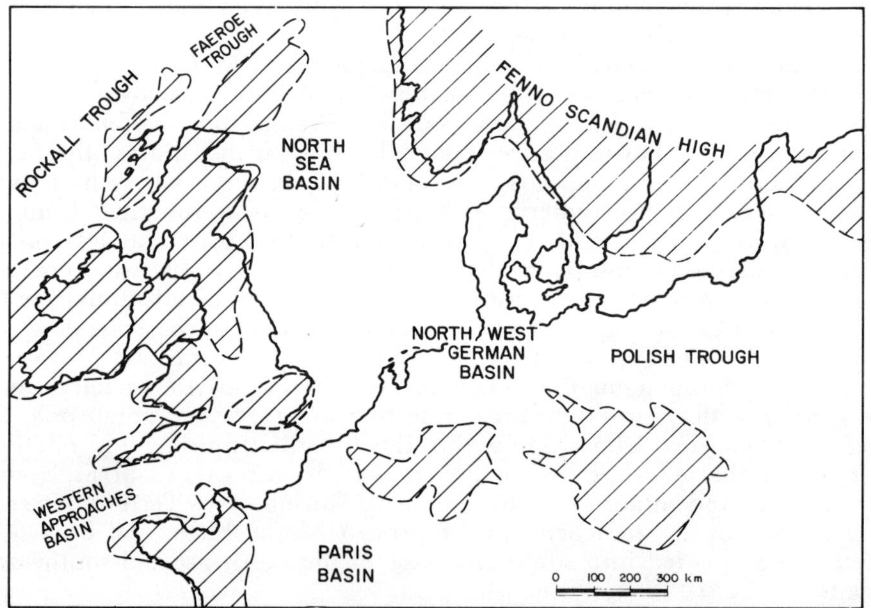

Figure 1. Location of the North Sea Basin showing the relative positions of major structural elements of northwest Europe.

south by various structural highs including the Brabant Massif, the Rhenish Massif and the Bohemian High. During the Carboniferous the area now underlying the southern North Sea subsided steadily in an epeirogenic downwarp. This relatively simple structural configuration appears to have persisted throughout the ensuing Permian period and to have created two east–west-oriented basins (figure 2).

During the Triassic, incipient opening of the North Atlantic was accompanied by the establishment within the northern North Sea region of a series of sharply defined rifts with a predominantly north–south orientation. Within the North Sea these comprise the Horn, Oslo, Central and Viking grabens (figure 2). Peripheral to this area a series of Triassic fault-controlled basins opened up to the northwest of Scotland (Steel 1977). These relatively narrow depocentres were thus established in a broad arc around the main southern North Sea Triassic basin, which continued the shallow synclinal pattern inherited from the Carboniferous to Permian periods.

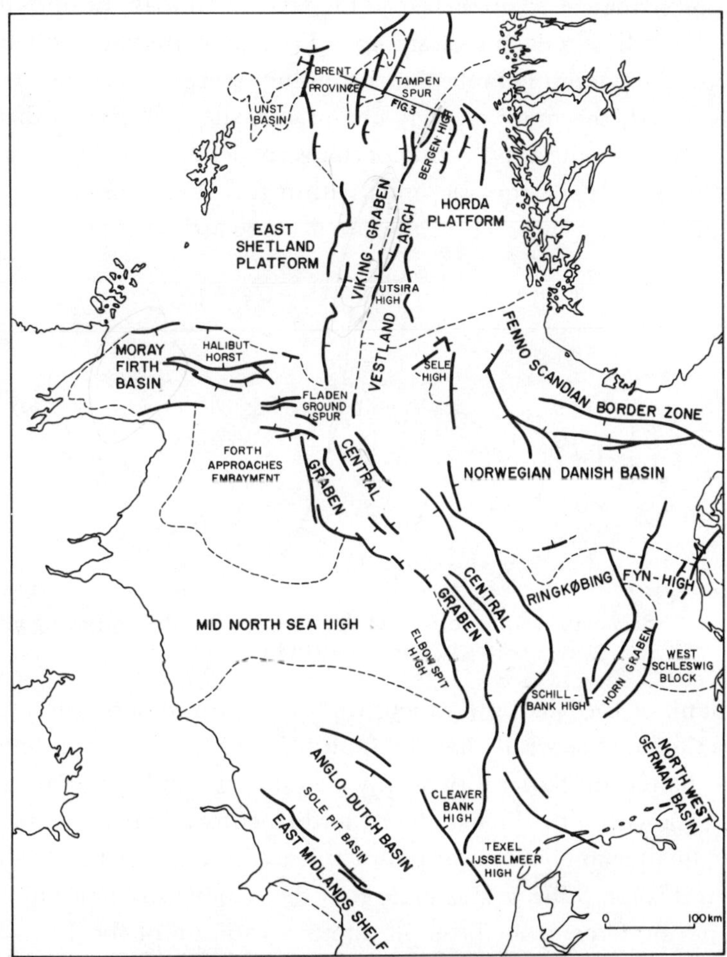

FIGURE 2. Tectonic elements map, North Sea, showing major fault trends and the location of the basin cross section.

In the Jurassic the rifting along the Viking, Central and Moray Firth grabens was firmly established and exerted a dominant control on the patterns of sedimentation in the area. Lower Jurassic sedimentation followed a similar pattern of the Triassic, i.e. a broad shallow synclinal area in the south contrasting with the narrow confined grabens existing to the north. A significant tectonic event occurred in the early Middle Jurassic when the area at the trilete junction of the North Sea grabens was domally uplifted, leading to the erosion of Lower Jurassic and some Triassic sediments (Ziegler 1981) and the outpouring of thick sequences of Jurassic volcanics (Howitt *et al.* 1975). The volcanism was concentrated in the area now underlying the Forties field, but volcanics are encountered as far west as the Piper field.

Subsidence continued during the Upper Jurassic and was concentrated along the old established rift system, but towards the end of the Jurassic, sea level changes had brought water levels sufficiently high to drown parts of the adjacent structural high areas.

Seismic sections across the northern part of the Viking Graben clearly show the nature of the horst and graben tectonics that have affected the region (figure 3). On the western side of the main graben the Jurassic sequence has been let down in a broad terrace that separates the western flank of the East Shetland Platform from the axial portion of the graben in which a very thick Cretaceous sequence is preserved. This terrace, the Brent Province, comprises at least three chains of horst blocks each containing a Triassic to Jurassic sequence which is tilted gently to the west with the bounding fault planes dipping steeply to the east. In contrast to what might have been expected, the more easterly chains, i.e. those closest to the axis of the main graben, are structurally the highest, reflecting perhaps earlier arching before rifting. Isopaching of the Lower, Middle and Upper Jurassic units within this Brent Province can be interpreted to show that the main Jurassic depocentre migrated westward with time.

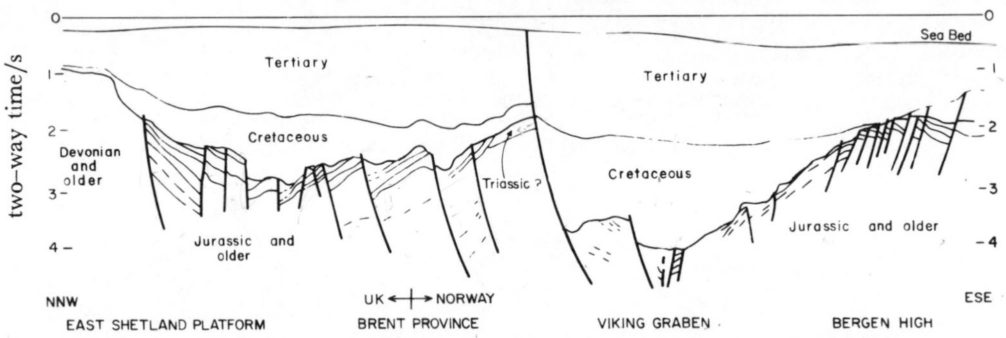

Figure 3. Schematic seismic profile of the northern North Sea (see figure 2 for location of section).

On the eastern flank of the Viking Graben, in Norwegian waters, the attitude of the pre-Cretaceous strata in the more centrally located fault blocks contrasts with that seen in the Brent Province. Here, the bounding faults dip steeply to the west and thus into the main graben, while the Jurassic strata within the blocks closest to the centre of the graben also dip westwards but at considerably smaller angle. The origin of this asymmetry in the Viking Graben is not fully understood, but it is tempting to speculate that the major subsidence along the edge of the Tampen Spur during the Cretaceous brought about a rotation of the Jurassic fault blocks on the eastern margin of the Viking Graben.

The Cretaceous was a period of widespread marine transgression, and the North Sea region is no exception. After filling in the irregularities still remaining on the Jurassic sea floor, Cretaceous deposition took place in a broad synclinal depression centred in the middle of the present North Sea. Few of the faults that controlled the Jurassic and earlier deposits were still active, and over a large part of the area the Cretaceous is unaffected by faulting. Salt tectonics in the southern North Sea locally had some effects on Cretaceous strata during their deposition.

The broad synclinal depositional trough established during the later Cretaceous continued throughout the Tertiary and is reflected in the present-day Tertiary isopachs. Seismic data clearly show that faulting, which had such important control of Triassic – Late Jurassic sedimentation, is not a feature of Late Cretaceous to Recent sedimentation and that subsidence since Early Cretaceous time has formed a broad synclinal flexure whose maximum points of subsidence have occurred in the areas overlying the older graben system.

Sclater & Christie (1980) published a very detailed analysis of the subsidence in the North Sea based on studies of well and seismic data from the Central Graben and incorporated in their study analysis of sediment backstripping, corrections for compaction, porosity–depth relations, thermal conductivities and palaeotemperatures. Their geological model for the post-Permian tectonic history of the Central Graben concluded, *inter alia*, that reinitiation of the Triassic graben system in the Middle Jurassic could have extended the underlying basement by 50–75 km. When the period of extension ended, thermal relaxation of the asthenosphere brought about general subsidence and the present-day synclinal downwarp was started. Seismic refraction studies in the Viking Graben and the Witch Ground Graben, together with gravity profiles (Donato & Tully 1981), suggest that crustal thinning has occurred beneath the main graben relative to the adjacent flanking areas.

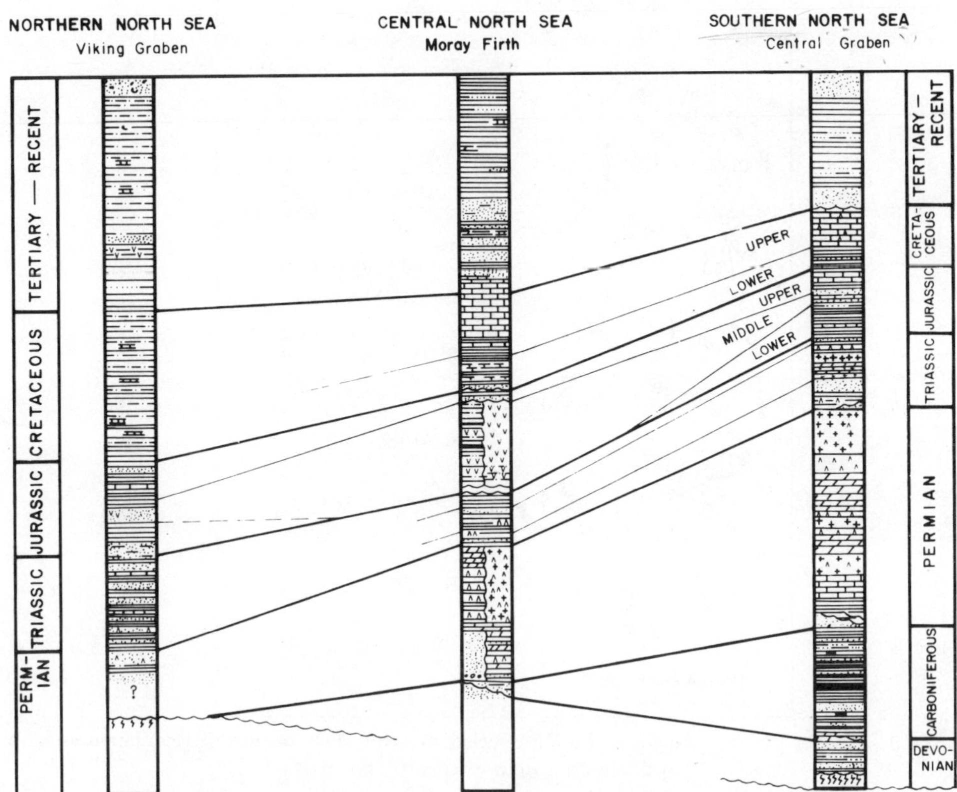

FIGURE 4. General stratigraphic summary of the North Sea Basin.

NORTH SEA LITHOLOGIES

Carboniferous

The generalized stratigraphic columns for the North Sea are summarized in figure 4. It will suffice here to say that the main Carboniferous depocentre of the North Sea basin was located under the southern North Sea as a continuation of the Carboniferous basins of northwestern Europe and the English Midlands. Coal swamps were developed on delta plain sediments and it is these coals that were the source of the gas found in the overlying Rotliegendes (Eames 1975). Sedimentation obviously matched subsidence during this period, and probably minor fluctuations in sea levels were sufficient to allow periodic marine incursions.

Permian

Towards the end of the Carboniferous, more arid climatic conditions set in and the Permian sequence (figure 5) opens with the deposition of a thick sand–shale sequence best known from the southern North Sea and northwestern Europe. The sandstones of the Rotliegendes were largely deposited in an aeolian environment but locally have been reworked by subsequent marine transgression. In addition to the aeolian sands, the Rotliegendes contains sediments representing other typical desert facies such as wadis and sabkhas (van Veen 1975). The dune

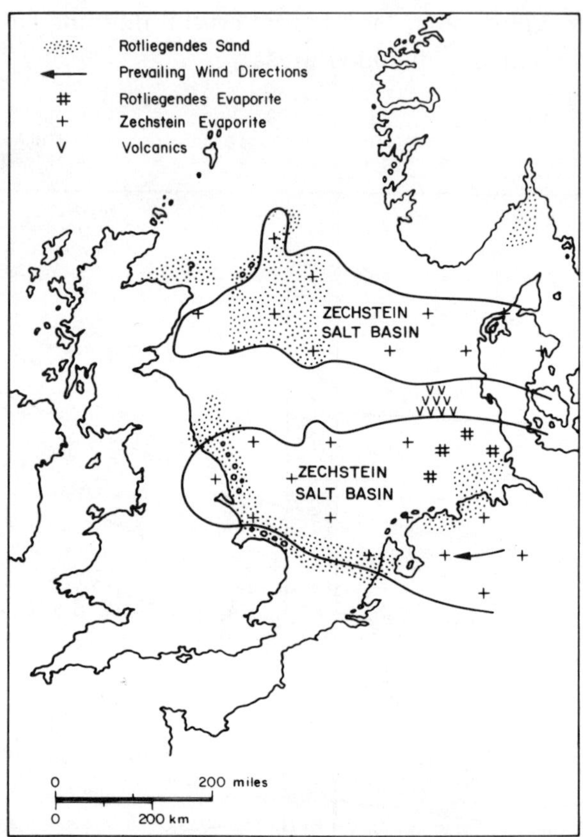

FIGURE 5. Map of Permian strata in the North Sea showing distribution of Rotliegendes sands and the Zechstein evaporitic basins.

sands, where preserved as such, show very good subaerial cross-bedding, and dipmeter studies of Rotliegendes wells have shown that the prevailing wind direction during the early Permian was from the east. An average of 300 m of sands were deposited along the southern margin of the Rotliegendes basin, while to the north of the aeolian sand belt, up to 1500 m of Rotliegendes was deposited in an evaporitic shaly facies adjacent to the Ringkobing–Fyn High. The concentration of coarse clastics along the southern flank of the basin and their virtual absence to the north indicates that the northern margin remained topographically subdued. Although the subsiding Rotliegendes basin appears to have been a relatively quiescent feature, some eruptive igneous activity did take place in the north Netherlands and German area.

Some coarse clastics of questionable Permian age have been recorded from beneath dated Zechstein strata in wells north of 56 °N, but there are insufficient well data to construct a satisfactory picture of Rotliegendes sedimentation in the more northerly Permian basin. The scant evidence indicates that aeolian sands similar to those of the southern basin were deposited (Jenkins & Twombley 1980).

Sedimentation later in the Permian produced a thick series of evaporites in two broad basins containing up to 2.5 km of sediments on either side of the Mid North Sea High. At times the high itself was submerged and was the locus of fringing carbonate sedimentation. Before the beginning of the Zechstein evaporitic cycles of sedimentation, the whole Permian basin was subjected to an anaerobic phase during which the Marl Slate or Kupferscheifer was deposited. Despite its thinness, often only 0.3–0.6 m thick, this laterally persistent marker unit can be found over a wide area of northwest Europe and the North Sea. It has a sharp boundary with the underlying sediments and in the north of England it can be seen to blanket the underlying Permian Yellow Sands even though they still preserve their original aeolian dune topography. These sands were probably not lithified before the deposition of the Marl Slate, which must have been deposited in an extremely quiet marine environment.

Taylor & Colter (1975) suggest that the first deposits of the Zechstein were laid down in a barred basin that was already well below contemporaneous world sea level before the invasion of sea waters. The interfingering and interbedding of the subsequent evaporitic and carbonate units of the Zechstein reflect the various stages of evaporite production on the margins of the basin as sea levels fluctuated and access to open waters opened and closed. The sediments thin both to the margin and to the basin centre, reflecting in one case a lack of space for deposition, and in the other a decrease in the amount of sedimentation because of increased water depth and lack of evaporation. Concentric to the basin, facies are uniform (Taylor & Colter 1975). This pattern and the uniformity of sedimentation within the Zechstein show that subsidence and sedimentation were balanced and that the basin subsided evenly across its entire area.

Triassic

The Triassic sediments of the North Sea (figure 6) mark the profound tectonic changes that accompanied the closing of the Palaeozoic in the region. In the southern North Sea sedimentation in the Triassic followed the broad synclinal downwarp pattern seen in the underlying Permo-Carboniferous and complete sequences of Triassic sediments, up to 2 km thick, were deposited and are locally preserved. Although subsequently affected by tectonism (halokinesis) and subdivided into several small basins, the pattern of sedimentation shows that the Triassic here comprised remarkably uniform sandstones, shales and evaporites. The whole sequence represents a broadly fining upward cycle (Brennand 1975). The distribution of evaporites in the basin shows a concentration towards the centre and is reminiscent of the pattern seen in the Rotliegendes.

In the central area of the North Sea only the Bunter Shale is preserved, since any younger sediments appear to have been stripped off during a period of Middle Jurassic erosion.

To the north of the southern North Sea basin a series of grabens were established during the Triassic. In contrast to the widely correlatable units of the gently flexured basin, the sedimentary infill of the Triassic grabens comprises very thick (up to 3.5 km) sequence of clastic sediments ranging from conglomerates to claystones (Brennand 1975). The Triassic basins described by Steel (1977) from northwest Scotland serve as a model for the totally unexposed North Sea grabens. In the Minch Basin, Steel showed that the basin fill consisted of thick fanglomerate wedges of sediment associated with fault scarps. As the fault scarps retreated – in this case westwards – up to 4 km of stratigraphical thickness of sediments were deposited. The basins were assymetrical half-grabens in which thick fanglomerates accumulated on the tectonically active side, while a thin condensed sequence was deposited on the opposite, flexured margin.

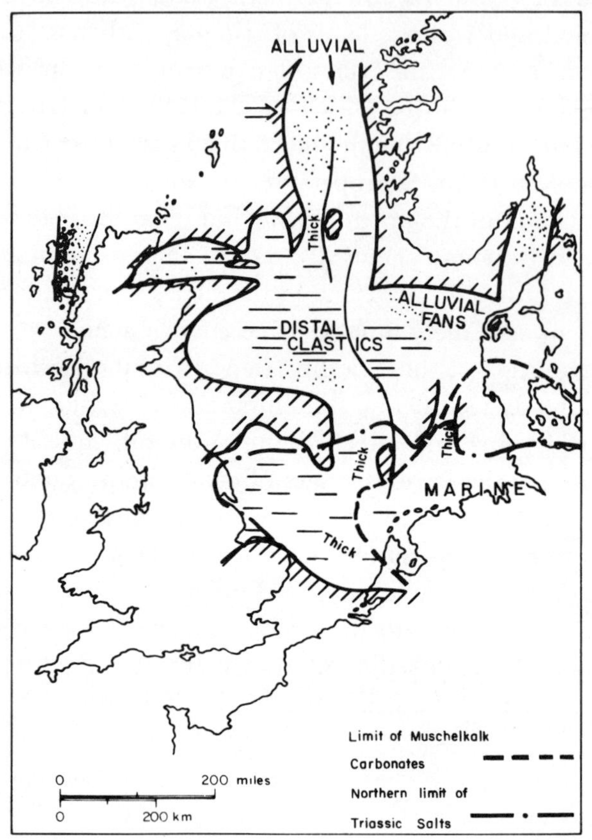

FIGURE 6. Distribution of Triassic strata in the North Sea and interpreted depositional environments.

Jurassic

Lower Jurassic deposition followed to some extent the pattern established during the Triassic, but part of the record has been lost in the central part of the North Sea through Middle Jurassic and later erosion (figure 7). In southern areas sedimentation continued in a quiet marine environment and locally gave rise to the formation of some 800 m of organic-rich shales in the Toarcian, which were subsequently to act as source rocks for small oil accumulations in the Netherlands (Bodenhausen & Ott 1981). In the northern North Sea, within the grabens formed earlier, the first Jurassic sandstones are conformable with and inseparable from the underlying Triassic.

The environment of deposition of the Statfjord Formation passed from continental alluvial to marginal marine as the graben continued to founder and as the seas encroached. The Dunlin Formation comprises prodeltaic mudstones deposited at the foot of the deltaic deposits of the Statfjord. The dominant source of sediments was still the active fault margins, but with the widening of the basin and the suppression of relief the overall clastic content is generally finer in a regional sense than in the Triassic section, and coarse fanglomerates are lacking.

As noted earlier, the close of the early Jurassic appears to have been accompanied by a significant tectonic event in the central North Sea which brought about emergence and active volcanism in the area now occupied by the trilete junction of the North Sea graben system. Middle Jurassic sedimentation is concentrated in the Viking Graben, the Norwegian Danish Basin and the Moray Firth Basin (figure 8) where 600–800 m of sediments were deposited in

FIGURE 7. Reconstructed map of the distribution of Lower Jurassic strata in the North Sea and interpreted environments: Dunlin, Fjerritslev, Statfjord and Gassum Formations, and Lias Group.

each area. In all three areas marginal marine to deltaic environment dominated and sediments were once more confined to the faulted troughs. However, the margins no longer seem to have been the main source of sediment supply and there is evidence that much of the sand could have been derived – at least in the Viking Graben – from the central domal high described above. Middle Jurassic sands in the Brent area appear to have been derived in part from the south (Bowen 1975). Variations in thickness of the Middle Jurassic sands across the fault blocks of the Brent Province suggests that the seabed had some topography that affected their deposition, but the main depocentre had moved somewhat west of that of the Lower Jurassic.

The close of the Middle Jurassic was accompanied by a widespread marine incursion and a rapid deepening of the marine basin overlying the North Sea graben system. Within the Moray Firth embayment and the Central Graben the Upper Jurassic is marked by marginal marine sedimentation (figure 9), where about 800 m of coastal sands and shales were deposited in the proximal areas of the Moray Firth Basin. Elsewhere in the North Sea a wider variety of depositional régimes were established. In the areas bordering the Vestland Arch, sediments appear to be related to the marginal marine conditions, which were inherited from the Middle Jurassic and comprise offshore bars and various shallow water intratidal deposits. The Viking Graben itself was a deep-water basin filled mostly with the shales of the Heather and Kimmeridge Clay Formations. The seabed appears to have had significant relief and submarine highs were progressively overlapped by 500 m and more of late Jurassic sediments.

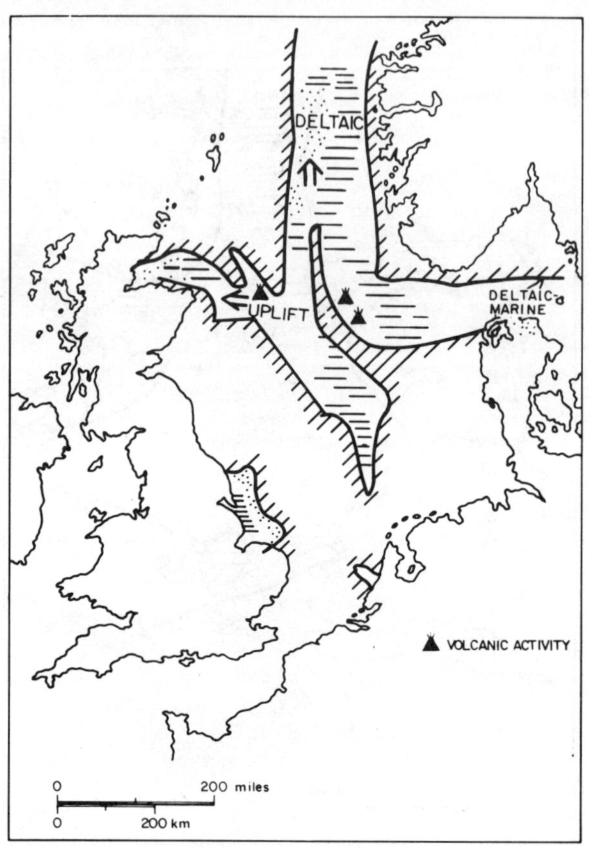

FIGURE 8. Middle Jurassic lithofacies map of the North Sea and interpreted depositional environments: Brent Group, West Sole Group, Haldager and Rattray Volcanics Formations.

Thicknesses, judged from seismic profiles, in the Upper Jurassic show that over 1 km was deposited in the Central Graben and in the axial portion of the Viking Graben. While the Upper Jurassic is predominantly a shale sequence, several commercially important sand units are developed in the Viking Graben. In the Magnus Field, Block 211/12, a sand unit within the Kimmeridge Clay formation, has been interpreted as a deep-water submarine fan derived from the East Shetland Platform to the west. With the considerable amount of core data available, De'ath & Schuyleman (1981) were able to show that the Magnus Sand represents the deposits of a middle fan environment and that Kimmeridge Clay was deposited contemporaneously in the outer fan régime. The inner fan deposits were subsequently eroded. The Magnus Sand can thus be interpreted as a sequence of submarine fan lobes derived from the East Shetland Platform, which lay some distance to the west.

The Brae Field, which also borders the East Shetland Platform and is located 280 km further south, where the Viking Graben was considerably narrower, is broadly contemporaneous with the Magnus Sand but presents an interesting contrast. The field is more closely associated with the edge of the graben and lateral changes within the Brae sand reservoir complex are more abrupt than in Magnus. Furthermore, a greater range of depositional environments is represented. The proximal, western portion of the complex is considered to be a subaerial fan delta that passes eastward into marine foresetted units (Harms *et al.* 1981). However, the subaerial origin of the Brae complex is not considered to be a unique explanation and a possible submarine origin not unlike that of Magnus is mentioned as a possibility by Harms *et al.* (1981).

FIGURE 9. Upper Jurassic lithofacies map of the North Sea and interpreted depositional environments: Kimmeridge Clay and Heather, Magnus, Brae and Piper Formations.

Cretaceous

The main period of faulting was completed by the beginning of the early Cretaceous. Some of the major bounding faults of the grabens persisted into and occasionally later than the Cretaceous but for the most part throughout the North Sea the Cretaceous was a time of prolonged synclinal basin subsidence with the maxima of subsidence located over the older graben system. The sedimentary environments blocked out in the late Jurassic persisted into the Cretaceous but water circulation and oxidation must have been more effective since the Cretaceous shales for the most part lack the high organic content of the Kimmeridge Clay formation.

Eustatic changes in sea level at the opening of the Cretaceous locally resulted in the spread of sedimentation beyond the limits of the earlier Jurassic. Despite evidence of periods of emergence in the North sea during the early Cretaceous, there is little development of sands associated with highs developed within the basin (figure 10). The greatest concentration of sands is in the Moray Firth embayment where a 1 km thick marginal marine to deltaic sequence exists that may pass laterally eastwards into pro-delta sands.

Some 0.8–1.0 km of the early Cretaceous deposits appear to have been laid down in a broadly synclinal depositional basin in which sandstones were concentrated along the basin edge but from which lobes of gravity-fed sand moved into the axial portions of the basin.

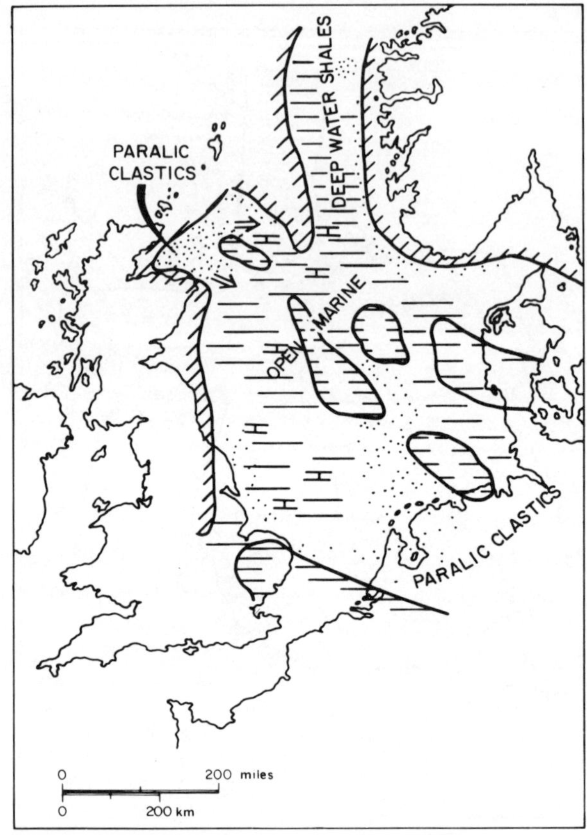

FIGURE 10. Lower Cretaceous lithofacies map of the North Sea and interpreted depositional environment: Valhall and Devils Hole Formations.

During the late Cretaceous (figure 11) widespread marine conditions were well established and resulted in the deposition of up to 1 km of chalky limestone in the central and southern North Sea. These pass laterally northwards into claystones and marls. Two very prominent troughs were created on either side of the Tampen Spur (figure 2) where very thick Cretaceous deposits (up to 2.5 km) are interpreted from seismic in the North Shetland Trough and the northern Viking Graben. Well evidence to date suggests that the Cretaceous in the northern areas is deficient in sand and it is difficult to explain the origin of the thick piles of monotonous claystones. The transition from chalky limestones in the south to claystones in the deeper waters to the north may be partly explained by subsidence in the northern areas taking the seabed below the carbonate compensation level.

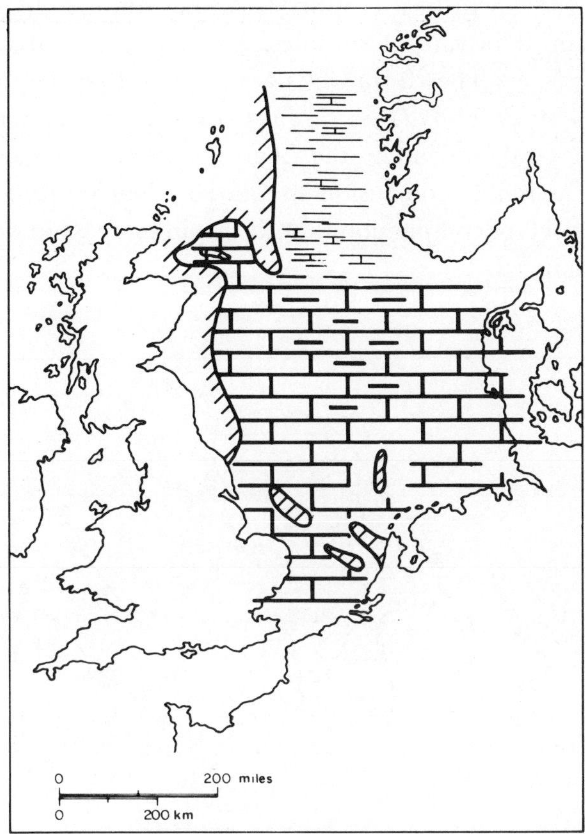

FIGURE 11. Upper Cretaceous – Danian lithofacies map of the North Sea: Shetland Group, Chalk Group, Ekofisk Formation.

Tertiary

During the Palaeogene, uplift of the area west of the northern North Sea Basin created a prominent easterly prograding in the Moray Firth delta (figure 12). This acted as a source of sediment supply to a series of deep-water turbiditic sands that were shed in a southeasterly direction into the axial portion of the North Sea Basin. Water depths ranged from near zero on the delta platform to several hundreds of metres in the axial portions (Parker 1975), and down these slopes a series of sand lobes were supplied with a constant supply of shallow-water sediments, including reworked marginal marine sands and terrestrial organic debris. These sands are best known from their occurrence in the Forties Field and adjacent areas.

A similar, but younger, sand sequence is to be found in the Frigg Field (Héritier *et al.* 1981) 250 km further north than Forties. Here an Eocene submarine fan was deposited on the sea floor of the central portion of the Viking Graben and the sands were derived from shelf sediments originally deposited on the edge of the Shetland Platform. In this aspect the two sand bodies differ, since there is no obvious delta build-up in the proximal portion of the Frigg complex. The Forties oil field is contained in a well defined anticlinal feature possibly caused by drape over the underlying Jurassic volcanic high. In contrast the relief at top sand level in Frigg, which gives the trapping mechanism, is very much an original bathymetric feature that has been subsequently enhanced by differential compaction of the surrounding marine shales (Héritier *et al.* 1981).

Since the Palaeocene the North Sea Tertiary basin has undergone a progressive subsidence in places of at least 2 km. The rate of subsidence is, however, subject to some controversy. According to Clarke (1973), evidence from well data suggests that the rate accelerated throughout the Tertiary, but Donato & Tully (1981) argue to the contrary that it decreased from Eocene to Miocene in the Viking Graben while in the Central Graben it remained constant. Sclater & Christie (1980) reached a similar conclusion to Clarke when plotting uncorrected data, but found on removing the effects of compaction that the rapid increase in deposition of the younger sediments was less pronounced.

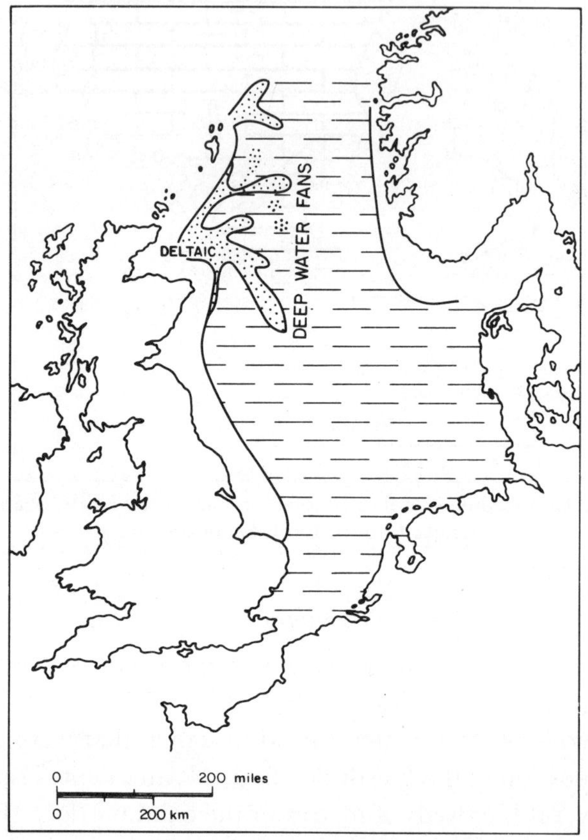

FIGURE 12. Palaeogene lithofacies of the North Sea and interpreted depositional environment of Forties and Frigg sands.

CONCLUSION

The North Sea Basin has subsided sufficiently since the pre-Carboniferous to accommodate a very thick, locally 12 km, sequence of sediments. The nature of the subsidence has varied in time and at differing places but essentially consisted of broad synclinal downwarping before the Triassic, a period of rifting in the Triassic to Jurassic, which may have been accompanied by a domal rise in the centre of the North Sea, and a final broad synclinal downwarp, which in effect has continued from late Jurassic to Recent. Within this structural régime the earlier synclinal phase was associated with deltaic clastics that gave way to an aeolian – evaporitic basin. There is some indication that initially the basin floor was substantially below existing worldwide sea levels.

Rifting in more northerly areas initiated a period (Triassic) of continental fanglomerate deposition in narrow, steep-sided grabens. Widening of these grabens both by erosion and more widespread faulting allowed the deposition of less restricted sediments in deltaic to marine facies and culminated in the deposition of thick organic marine shales in a deep-sea basin of restricted circulation. Into this deep trough a system of submarine fans were able to build up a series of constructive sand lobes.

A second stage of synclinal downwarping (Upper Cretaceous and Tertiary) was associated initially with carbonate and shale deposition, but the basin morphology developed sufficient bathymetric relief to allow the formation of deltas and shelf-edge sands along its margins. These in turn were able to supply sediment to gravity-flow sand lobes in the axial portions of the basin.

I thank the Chairman and Board of Directors of British Petroleum Co. Ltd for permission to publish this work and my colleagues in BP for their assistance and discussions during its preparation.

References (Leckie)

Bodenhausen, J. W. A. & Ott, W. F. 1981 In Illing & Hobson (eds), pp. 301–309.
Bowen, J. M. 1975 In Woodland (ed.), pp. 353–361.
Brennand, T. P. 1975 In Woodland (ed.), pp. 295–311.
Clarke, R. H. 1973 *Earth planet. Sci. Lett.* **18**, 329–332.
De'ath, N. G. & Schuyleman, S. F. 1981 In Illing & Hobson (eds), pp. 342–351.
Donato, J. A. & Tully, M. C. 1981 In Illing & Hobson (eds), pp. 65–75.
Eames, T. D. 1975 In Woodland (ed.), pp. 191–201.
Finstad, K. G. & Selley, R. C. (eds) 1977 *Mesozoic Northern North Sea Symposium Proceedings*. Oslo: Norwegian Petroleum Society (NPF).
Harms, J. C., Tackenberg, P., Pickles, E. & Pollock, R. E. 1981 In Illing & Hobson (eds), pp. 352–357.
Héritier, F. E., Lossel, P. & Wathne, E. 1981 In Illing & Hobson (eds), pp. 380–394.
Howitt, F., Aston, E. R. & Jacque, M. 1975 In Woodland (ed.), pp. 379–387.
Illing, L. V. & Hobson, G. D. (eds) 1981 *Petroleum geology of the continental shelf of northwest Europe*. London: Heyden & Son Ltd.
Jenkins, D. A. L. & Twombley, N.N. 1980 In *Petrol. Inst. Min. Metall. Trans. spec. Iss.*, pp. 6–24.
Kent, P. E. 1980 *Can. Soc. Petrol. Geol. Mem.* no. 6, pp. 633–652.
Parker, J. R. 1975 In Woodland (ed.), pp. 447–453.
Sclater, J. G. & Christie, P. A. F. 1980 *J. geophys. Res.* **85**, 3711–3739.
Steel, R. 1977 In Finstad & Selley (eds), pp. 7–17.
Taylor, J. C. M. & Colter, V. S. 1975 In Woodland (ed.), pp. 249–263.
van Veen, F. R. 1975 In Woodland (ed.), pp. 223–231.
Woodland, A. W. 1975 *Petroleum and the continental shelf of northwest Europe*, vol. 1 (*Geology*). London: Applied Science Publishers.
Ziegler, P. A. 1981 In Illing & Hobson (eds), pp. 3–39.
Ziegler, W. H. 1975 In Woodland (ed.), pp. 165–187.

Phil. Trans. R. Soc. Lond. A **305**, 101–112 (1982) [101]
Printed in Great Britain

Interpretation of refraction experiments in the North Sea

By P. A. F. Christie†

Department of Earth Sciences, Bullard Laboratories, Madingley Rise, Madingley Road, Cambridge CB3 0EZ, U.K.

Before 1977, only three long-range seismic refraction profiles had been shot in the North Sea Basin, with somewhat inconsistent results in terms of a description of the Mohorovičic discontinuity (Moho). In the summer of 1977, the Marine Group of Cambridge University fired three refraction lines with the use of pull-up, shallow-water seismometers, the longest reversed profile extending for 400 km along the 0.5° E meridian from east of the Shetland Islands to the latitude of Dundee.

Although records varied in quality, a time term interpretation of the results was made with the use of detailed velocity–depth information supplied by the oil industry. This interpretation indicates that there is an anticlinal form to the Moho beneath the deepest sediments of the Moray Firth Basin, as demanded by gravity data from the area. It is further suggested that the seismic evidence supports an extensional model to explain the origin of the North Sea Basin.

Introduction

The Cambridge North Sea Experiment was conceived against a background of intense activity in the field of long-range seismic refraction seismology. Much work had already been performed by the Groupe Grands Profils Sismiques and the Lithospheric Seismic Profile of Britain (Lispb) working party in an effort to study the deep crustal and uppermost mantle velocity structures beneath continental Europe (Sapin & Prodehl 1973) and mainland Britain (Bamford *et al.* 1978).

At the Bloomsbury meeting, of industrial and academic geologists and geophysicists, where so much proprietary data was released for the first time, Matthews (in Woodland (ed.) 1974, p. 453) announced that the Cambridge Marine Group intended to fire a long-range refracton profile in the North Sea during the summer of 1977. The intention of the experiment was to compare the compressional wave velocity structure beneath the subsiding North Sea Basin with the velocity structures beneath the stable areas of western Europe. Between the conception and execution of the experiment a great deal of interest arose in explaining the origins of sedimentary basins in general and the North Sea Basin in particular, with important contributions being made by Bott (1976), Collette (1968), Sleep & Snell (1976), Illies (1970), Artemjev & Artyushkov (1971), Ziegler (1975) and McKenzie (1978), to name but a few. The Cambridge experiment, in conjunction with the newly released data, should clearly provide important constraints in determining the types of model applicable to the North Sea.

The refraction data available in 1977 provided inconclusive information concerning the crustal structure of the North Sea. Although Collette's (1960) early studies of the North Sea

† Present address: Schlumberger Inland Services, 1 Kingsway, London WC2B 6XH, U.K.

gravity field eventually led him to propose an Airy type of compensation for the basin, his seismic investigations (Collette *et al.* 1970) found the Mohorovičic discontinuity (Moho) at a depth of about 31 km on his Dogger Bank Profile, which was shot by using the expanding spread technique from a fixed midpoint. Two lines by Sørnes (1971) and Solli (1976) were fired from the Shetlands across the northern North Sea Basin to Norway with conflicting results. Sørnes's profile was originally published as a time term interpretation but an inversion of the delay times to depths was published by Ziegler (1975) showing a deep crustal root below the Viking Graben with a Moho depth of some 38 km. Solli's profile, however, published in Ziegler (1977), shows a dramatic thinning of the crystalline crust to about 10 km beneath the Viking Graben with the shallowest Moho depth at about 18 km. There is thus a discrepancy of about 20 km in the estimated Moho depths from two effectively coincident refraction profiles, although the later profile had the benefit of a greater knowledge of the velocity structure of the sedimentary cover.

Experimental procedure

The recording equipment used in the North Sea experiment consisted of pull-up, shallow-water seismometers (Puss) designed and built at Cambridge (Smith & Christie 1976). Each instrument contained a three-component set of geophones mounted in 'sticky' gimbals and a pressure-sensitive hydrophone. The seismic signals were amplified by preset gain amplifiers and recorded in frequency modulated form on a standard cassette tape unit fitted with a four-track head. A clock and programmer allowed the preselection of shooting windows to avoid the requirement for continuous operation and also provided timing signals recorded with the shots.

The charges used were of 'Geophex' (I.C.I. registered trade mark) and were fired either as free-sinking single units varying in mass from 45 to 182 kg, or as dispersed charges comprising three simultaneously detonated charges of 204 kg per unit. Dispersed charges were chosen for their efficiency in long-range propagation (Jacob 1975) and because of the difficulty in firing a large single charge at its optimum depth (Burkhardt & Vees 1975) in the shallow waters of the North Sea.

The observation scheme in relation to the Tertiary isopach map from Ziegler (1978) is shown in figure 1. Three profiles in all were fired, the Firth of Forth profile, the Main Line and the Crustal Control Line. The first of these was fired by using seven free-sinking charges from the Firth of Forth towards an array of four Pusses placed near to the southern end of the Main Line. These receivers were then recovered and 13 were laid at about 30 km spacing on the 0.5° E meridian from a point east of the Shetlands to Devils Hole, a local deep at the latitude of Dundee. Reversed coverage was achieved by firing two dispersed charges at each end of the Main Line offset by about 15 km, being half the receiver spacing. Finally, six Pusses were laid in an extended array towards the northern end of the Main Line, and the Crustal Control profile was obtained by firing eight free-sinking charges to the south of this array. This procedure gave reversed coverage over the crystal refractors at the northern end of the Main Line.

Shot and receiver positions were obtained by using the standard Decca Navigator system. Shot times were computed by using a hull geophone to record the water arrival at the shooting ship simultaneously with the timing signals broadcast by M.S.F. Radio Rugby. A correction was made to allow for the travel time of the water wave arrival from the charge to the ship, which was typically 0.2 s for the free-sinking charges and 0.8 s for the dispersed charges. Upon return to Cambridge all records were digitized, allowing filtered, computer-drawn record sections to be prepared.

The quality of the data varied widely, with the hydrophone traces being more reliable on the whole than the geophone signals, owing to the uncertain ground coupling of the geophones. However, few recording points were totally lost since four channels of information

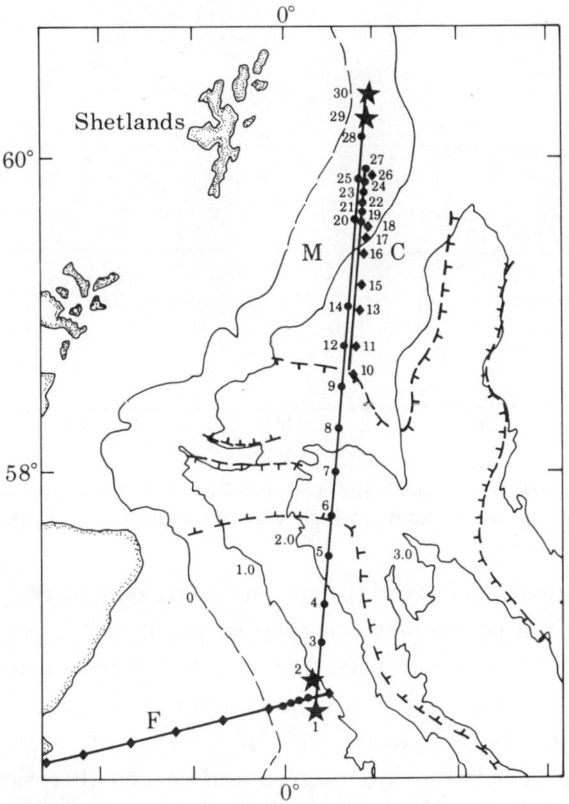

FIGURE 1. Seismic observation scheme in relation to the North Sea Tertiary isopach map from Ziegler (1978). The isopachs are marked in kilometres. Three refraction profiles were recorded as indicated by the letters C (Crustal Control Line), M (Main Line) and F (Firth of Forth Line). Free sinking shot points are denoted by diamonds, dispersed shot points by large stars and recording points by solid circles.

were recorded from each shot. Positioning errors were estimated by Decca to be ± 50 m at a 68 % confidence level. Timing precision was estimated at a few hundredths of a second while the probable error in picking arrival times was about one tenth of a second, a figure that was shot-dependent.

ARRIVAL TIME ANALYSIS

(a) *Crustal Control Line*

The travel time – distance curve for Crustal Control Line shots fired to the south of the PUSS array is shown in figure 2. The arrival times have been computed by using a reduction velocity of 6.7 km s^{-1} and the observations have been linked together for each shot. There is evidence for an upper crustal refractor out to about 25 km range, where there is a cross-over to a travel time branch from a higher velocity refractor. However, the linked observations have an *en échelon* pattern and are offset to increasing travel times with increasing range. Although the apparent velocity across the Puss array between 25 km and 115 km is 6.7 km s^{-1}, the apparent velocity across the shot array as observed at a single receiver is rather less than 6 km s^{-1}. The *en échelon* pattern is an indicator of refractor dip with a southerly direction consistent with the basinwards direction of shooting (figure 1).

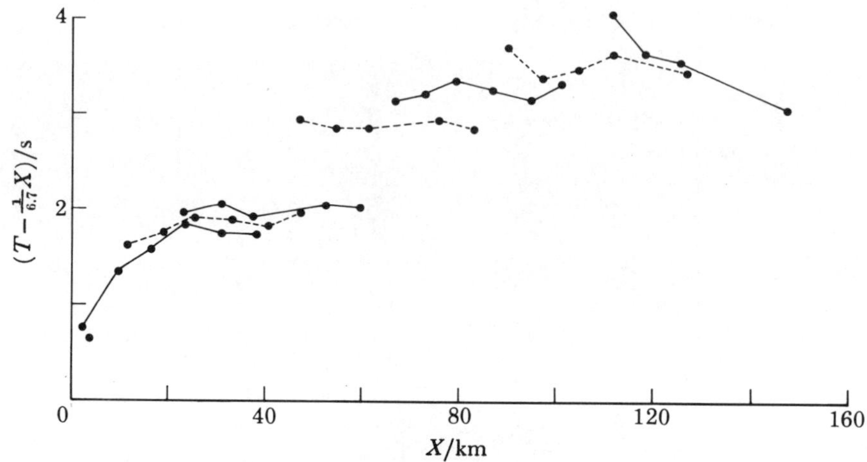

FIGURE 2. Crustal Control Line arrivals reduced at 6.7 km s^{-1}. Arrivals from each shot are linked revealing an *en echelon* pattern, which suggests a southward-dipping refractor.

Since the Crustal Control Line overlaps the northern end of the Main Line and there is a coincidence of observation points between the two profiles, it is possible to carry out a time term analysis by using the reversed coverage thus afforded. The time term method (Scheidegger & Willmore 1957; Berry & West 1966) is usually applied to shot and receiver networks of random geometry, but the method works well in the case of reversed linear profiles where there are points in common between the forward and reverse directions.

In fact, time term analysis is necessary when, as in this case, the refractors are not dipping planes, as can be seen from the jump in reduced time from the linked arrivals terminating at 60 km to the group starting at 47 km (figure 2). From an examination of the curves of travel times, there is a sharp break from an upper crustal refractor of relatively low velocity to a deeper refractor of significantly higher velocity. The assignment of travel time – distance observations to a given refractor for the time term analysis was made by using the travel time curves, phase correlation and iterative examination of the distribution of residuals with range from successive solutions of the time term equations. The final solution indicated the presence of two crustal refractors, over which there was reversed control, with true velocities of 5.66 km s^{-1} and 6.18 km s^{-1}. The corresponding delay times are displayed in table 1.

By using the first arrival times out to 3.5 km and after correcting for the thin Tertiary cover by using the Tertiary isopach data published by Ziegler (1978), it was possible to arrive at an unreversed estimated of 4.1 km s^{-1} as being the velocity of the Devonian Old Red Sandstone of this part of the East Shetland Platform. This compares with a velocity of 4.0 km s^{-1} as found from borehole measurements in a nearby well.

TABLE 1. TIME TERMS FOR THE TWO CRUSTAL REFRACTORS COVERED BY THE NORTHERN PART OF THE MAIN LINE AND THE CRUSTAL CONTROL LINE

site	term time / s	s.d. / s	number of observations	term time / s	s.d. / s	number of observations
27	0.26	—	1	0.41	0.03	5
26	0.21	0.01	4	—	—	0
24	—	—	0	0.50	0.02	7
23	0.50	0.01	3	0.64	0.05	3
22	0.44	0.01	4	0.65	0.03	3
21	0.42	0.01	3	0.80	0.05	4
19	0.51	0.00	2	0.87	0.02	4
18	0.71	0.01	3	0.88	0.03	2
17	0.79	0.01	3	0.88	0.04	3
16	0.82	—	1	0.84	0.02	4
15	—	—	0	1.44	0.02	5
13	—	—	0	1.51	0.03	5
11	—	—	0	1.50	0.05	3
refractor velocity/(km s^{-1})	5.66	0.06	13	6.18	0.06	24

(*b*) *Firth of Forth profile*

The reduced-time record section of the observations made in the Firth of Forth refraction line is shown in figure 3. Many data were also recorded by the LOWNET and Eskdalemuir arrays, and by temporary stations installed in the Southern Uplands by Cambridge personnel. However, the recording geometry of the lands stations was such that split profile observations, incorporating both land and sea data, could not be made over any one refractor. Consequently, the analysis of the land data set is not treated in this contribution, and the data from the PUSS recording positions must be interpreted essentially as a single-ended profile, although, as figure 3 demonstrates, there is a reversed coverage out to 25 km provided by shots at each end of the PUSS array. From the absence of an *en échelon* pattern as seen in the Crustal Control Line, there is apparently little structural dip below the profile. This is further supported by the reversed coverage on the upper crustal refractor, which is the first arrival out to 25 km and has apparent velocities of 5.63 km s^{-1} shooting west–east and 5.69 km s^{-1} shooting east–west. The least-square time term solution gives 5.66 km s^{-1}, which agrees very well with the value from the Crustal Control Line.

From 25 km range out to about 100 km, the first arrival is that due to the main crustal refractor with an apparent velocity greater than 6 km s^{-1}. Since little dip is apparent from these arrivals a linear least-squares fit can be applied, yielding a velocity of 6.2 km s^{-1} with an intercept time of 1.54 s. At ranges greater than 70 km, the dominant phase is a large-amplitude ringing arrival, which is interpreted as being the post-critical reflexion from the Moho. This phase persists out to 200 km, even though the charge mass decreases with increasing range since the profile was fired principally for land-based recorders.

By using the asymptote of the post-critical Moho reflexion as a guide, and by making a phase correlation of second arrivals in the interval 60–90 km and at about 2.7 s reduced time, it is possible to fit a straight line indicating the presence of a lower crustal refractor of velocity 7.2 km s^{-1} with an intercept time of 4.55 s. Although the existence of a fast crustal refractor is somewhat tentatively postulated, it is consistent with the position of the Moho critical point at about 70 km and is also suggested by an analysis of the Moho reflexion arrival times, which results in an average crustal velocity of 6.4 km s^{-1}. With such a strong wide-angle Moho reflexion, suggesting a sharp transition between crust and upper mantle, the Moho refraction is weak and is not observed as a first arrival. The best indication of where the travel time curve should lie is given by assuming the value of 8.16 km s^{-1} for the Moho velocity (from the Main Line analysis) and fitting the line through the critical point, which is at about 70 km in range and 3.2 s reduced time. This gives an estimate of 6.6 s for the Moho intercept time, which is in agreement with the time term for shot point 2 (figure 1) of the Main Line determined from Puss observations augmented with Eskdalemuir and Lownet data.

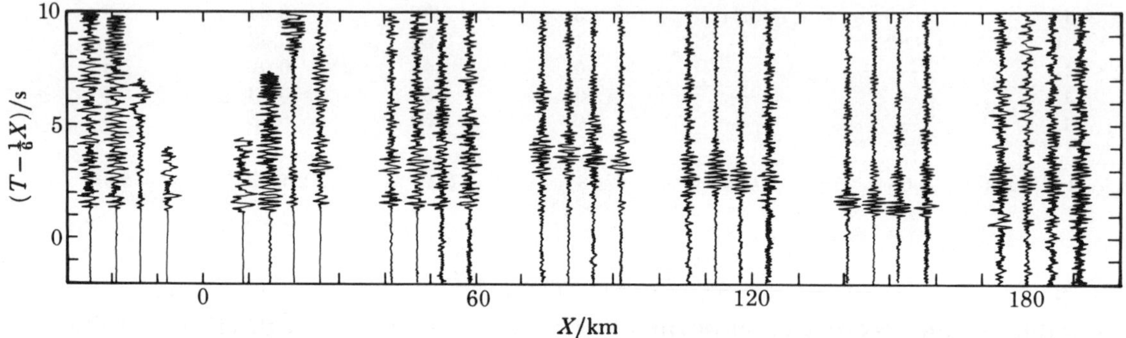

Figure 3. Firth of Forth hydrophone record section reduced at 6 km s^{-1}. The arrivals have been low pass filtered at 15 Hz. Both directions of cover are displayed.

Using data from a nearby borehole and an interpreted seismic reflexion profile, the intercept times can be inverted to a velocity–depth profile beneath the Puss array. A comparison of this structure with that obtained by the Lispb profile beneath the Southern Uplands of Scotland is shown in figure 4, where there is a broad agreement. Apart from the superficial sedimentary cover at the seaward end of the section, the main difference between the two structures lies in the velocity–depth function of the upper 20 km of crust. The Southern Uplands structure displays a velocity gradient from 5.8 to 6.0 km s^{-1} at a depth of about 14 km, where there is a step to a velocity of 6.3 km s^{-1}. Over approximately the same interval, the plane refractor interpretation at Devil's Hole results in a two-layer upper crust with velocities of 5.7 and 6.2 km s^{-1}. On both interpretations, there is a lower crustal refractor of slightly differing velocities, apparently thinning towards the basin. There is a significant difference in Moho velocities, although the depth to the Moho is similar on both structures, shallowing from about 32 km beneath the Southern Uplands to 30 km near Devil's Hole. It should be borne in mind that the Lispb profile has a much greater density of observations and has been tested by ray tracing, whereas the Puss profile has only had an arrival-time analysis made upon it.

(c) *Main Line*

The first arrival time – distance curves for the Main Line observations are shown in figure 5, where the first arrival points are plotted by using a reduction velocity of 8 km s^{-1} and are linked by shot. The break to the upper mantle refractor is clearly seen in both forward and reverse directions, but the apparent velocity of the crustal arrivals in both directions is about 5.8 km s^{-1} over ranges from 25 to 120 km. From the reversed cover at the northern end of the Main Line, the main crustal refractor has a velocity of 6.2 km s^{-1} (figure 6), which agrees with the unreversed estimate from the Firth of Forth profile. By adopting this value it appears that the refractor is also dipping into the basin at the southern end of the Main Line. The intercepts of the travel time curves give an indication of the relative thicknesses of sediments beneath the northern and southern shot points, with a considerably greater delay at the southern end being consistent with its position in relation to the Tertiary isopachs (figure 1).

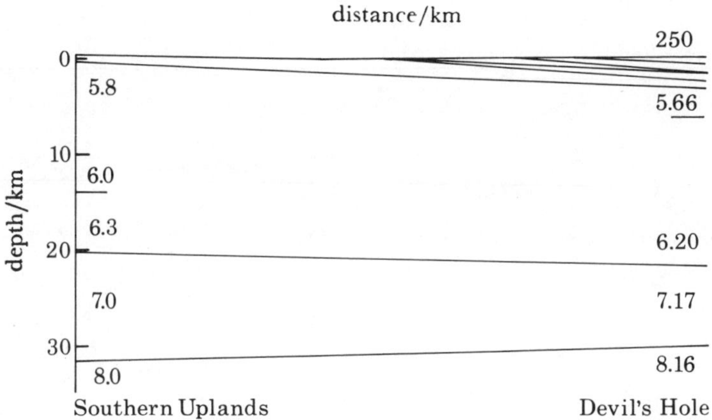

FIGURE 4. Velocity–depth structures beneath the Southern Uplands as revealed by the LISPB experiment and beneath Devil's Hole as determined by the PUSSES in the Firth of Forth Profile. Vertical exaggeration × 4. Refractor velocities are shown in kilometres per second.

A time term analysis of the upper mantle arrivals gives a well constrained value for the Moho velocity of 8.16 km s^{-1}, which is significantly higher than the value obtained beneath mainland Britain (8.0 km s^{-1}). However, the velocity is in good agreement with the estimate from Collette's Dogger Bank profile (8.15 km s^{-1}) and Sørnes's Norway Shetlands profile (8.12 km s^{-1}), although somewhat less than Solli's value of 8.3 km s^{-1}. The Moho velocity beneath France as reported by Sapin & Prodehl (1973) is 8.1 km s^{-1}. There is thus consistent evidence that the Moho velocity beneath the North Sea Basin is higher than that beneath the stable areas of western Europe.

An attempt was made to solve for a horizontal velocity gradient beneath the Moho by modifying the travel time equations to allow for the possibility of a velocity function varying linearly with range. With this model, an apparent horizontal velocity gradient can provide an estimate of the change of velocity with depth by inverting the function with the Weichert–Herglotz relation. However, a least-square solution of the time–distance data predicted a decrease in velocity with increasing range with a standard deviation greater than the absolute value of the horizontal gradient. The standard deviation of the solution as a whole was greater than that for the uniform refractor model and since the physical implications of a negative velocity gradient were inconsistent with the observed strong Moho arrivals, it was concluded that the data set could resolve neither the magnitude nor the sign of a horizontal velocity gradient. The uniform refractor solution was adopted, and the values of the Moho time terms are listed in table 2.

There is a discrepancy between the time terms of the two southern shot points. An investigation was made into possible sources of shot timing errors, since the magnitude of the dispersed charge time-of-flight correction was about 0.8 s, without finding any correlation between errors and wind or sea conditions. Since the discrepancy is visible on both crustal and Moho branches of the travel time curves for shots 1 and 2 in figure 5 and is also observed on LOWNET and Eskdalemuir records, the source of the discrepancy probably lies in near-surface structure since there are abrupt changes in the thickness of the high-velocity Zechstein salt in the vicinity of Devil's Hole, but because of the uncertainty it was decided to adopt the time terms for the PUSS array from the Firth of Forth profile.

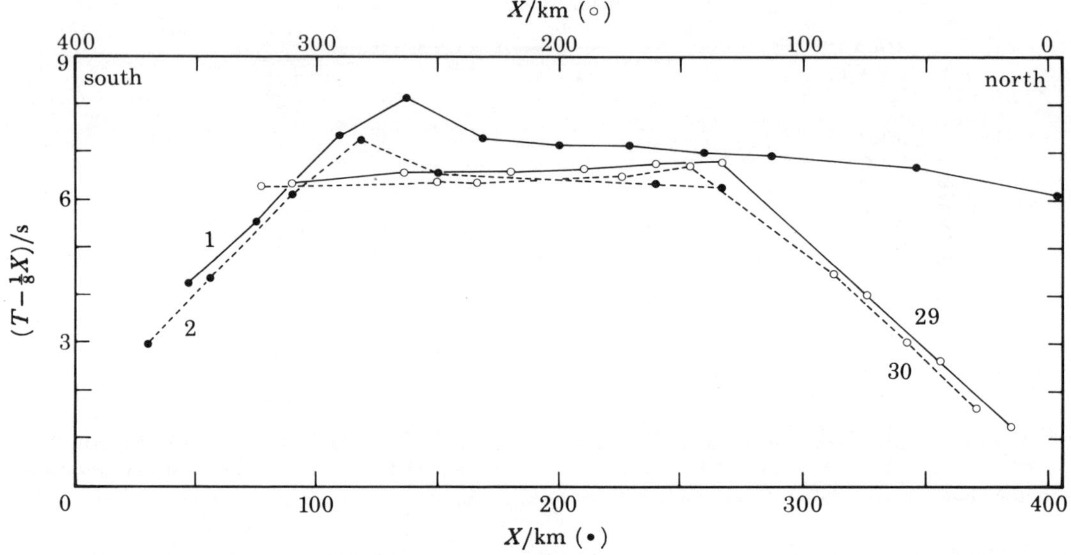

FIGURE 5. Reduced time – distance curves for all shots fired in the Main Line. Solid circles are observations from southern shots, open circles from northern shots.

Because of the good agreement between the Firth of Forth profile and the reversed crustal control line in the estimate of 6.2 km s^{-1} for the main crustal refractor, crustal time terms at the southern end of the Main Line were obtained by constraining the refractor velocity to 6.2 km s^{-1}, and are displayed in table 2.

TIME TERM INVERSION

In order to strip off the effects of the sediments from the time terms, recourse was made to seismic, borehole and velocity survey data kindly provided by many oil companies, which allowed the delay due to the upper 4 km of sediment to be removed. This had the effect of reversing the trend of increasing Moho time terms towards the centre of the basin. The inversion was continued by using the refraction information from all three profiles, and by making certain assumptions in regions of missing coverage the velocity–depth profile in figure 6 was obtained. Control points are indicated by a solid circle. The assumptions made in the course of the inversion were as follows:

(i) missing time terms on the Moho and 6.2 km s^{-1} refractor were obtained by interpolation;

(ii) the lower crustal refractor was assumed to exist over all the profile and was constrained to be half the thickness of the 6.2 km s^{-1} refractor, as observed on the Firth of Forth profile;

(iii) material below the depth of well control was assumed to have a velocity of 4.8 km s^{-1};

Table 2. Time terms to the main crustal refractor and to the Moho for the area covered by the Main Line

site	time term (s)	s.d. (s)	number of observations	time term (s)	s.d. (s)	number of observations
30	0.39	0.01	3	3.21	0.04	5
29	0.45	0.01	3	3.30	0.03	6
28	0.18	0.01	2	3.26	—	1
25	0.50	0.02	7	—	—	0
20	0.87	0.02	4	3.63	—	1
14	—	—	0	3.78	0.02	4
12	—	—	0	3.77	0.04	4
9	—	—	0	3.85	0.04	2
8	—	—	0	3.79	0.04	3
7	—	—	0	3.81	0.04	4
6	2.28	—	1	—	—	0
5	2.16	0.01	2	3.77	0.02	3
4	1.67	0.05	2	—	—	0
3	1.33	0.05	2	—	—	0
2	0.55	0.03	4	3.13	0.02	3
1	1.13	0.04	3	3.83	0.01	8
refractory velocity/(km s^{-1})	6.20			8.16	0.02	22

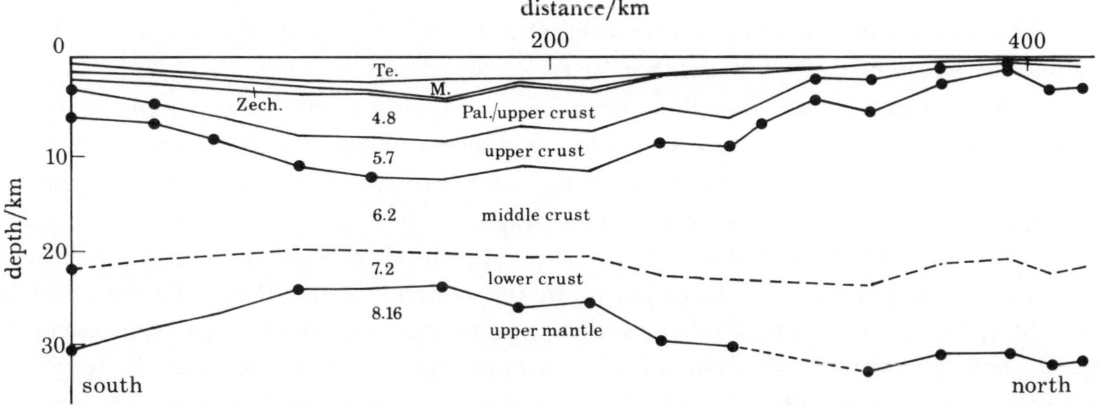

Figure 6. Velocity–depth section along the Main Line profile after inverting the time terms according to the procedure outlined in the text. Upper sedimentary sequence given by borehole information. Velocities are shown in kilometres per second. Time term control points are indicated by solid circles. Vertical exaggeration × 4. Te., Tertiary; M., Mesozoic; Zech., Zechstein; Pal., Palaeozoic.

(iv) although there is little control over the behaviour of the 5.7 km s^{-1} refractor, the interval between the base of the well control and the top of the 6.2 km s^{-1} refractor was assumed to be approximately equally divided between 4.8 km s^{-1} material and 5.7 kms^{-1} material. A velocity gradient without such an arbitrary division may be more appealing but there is good evidence for a distinct 5.7 km s^{-1} refractor from reversed short-range observations on the two crustal profiles.

Discussion

The most important feature of the velocity–depth profile is the anticlinal form to the Moho, which reaches its shallowest point, 24 km, beneath the deepest sediments. Although some broad assumptions have been made in the delay time inversion, they are not unreasonable. The average velocity of the crystalline crust is over 6.2 km s^{-1} and is somewhat greater than Solli's estimate of 6.09 km s^{-1}. The higher crustal velocity, in a time term inversion, will lead to a greater estimate of the Moho depth at all points along the profile. Although the contrast in terms of depth between the deepest and shallowest Moho soundings along the profile increases with average crustal velocity, the ratio between the two is independent of average crustal velocity if this is constant along the profile. It is the ratio of unstretched and stretched crustal thicknesses that is the attenuation or 'β' factor of the McKenzie model. If the same value of 32 km for the unstretched crustal thickness is taken for both Solli's line and the Cambridge profile, then the Cambridge profile results in a more conservative estimate of the attenuation factor. Furthermore, the Moho depths along the profile are in good agreement with those obtained by modelling gravity data over the same area (Smythe *et al.* 1980), where the shallowest Moho depth is also estimated at 24 km.

The anticlinal form to the Moho imposes considerable constraints upon the type of model used to explain North Sea Basin formation (Christie 1979). If it is assumed that the North Sea is an epicontinental basin and that the original crustal thickness in the region of the present North Sea was about 30 km, comparable with the crustal thickness beneath the Southern Uplands or the East Shetlands Platform (32 km), then a large class of models must be excluded. These include the classical graben formation mechanism of Vening Meinesz (1950), the viscoelastic loading model of Beaumont (1978) and mechanisms involving an increase in density of the lower crust such as that due to Haxby *et al.* (1976), all of which predict a deeper Moho beneath the centre of the basin than at the basin edges.

The gabbro–eclogite phase transition proposed by Collette (1968) is appealing in that it permits compensated subsidence to take place, but thermal balance is not maintained by the process, which predicts a smaller heat flow than normal in the centre of the basin, contrary to what has been measured (Evans & Coleman 1974).

The seismic evidence from all North Sea profiles appears to rule out the existence of a rift pillow of the form proposed by Illies (1970) in the context of the Rhine Graben and by Ziegler (1975) for the North Sea Graben, and indeed it should be noted that in the case of the Rhine Graben the existence of a rift pillow is strongly questioned by Edel *et al.* (1975) and other workers. According to this model, the rift pillow of fractionated upper mantle material is created during the extensional stage of graben formation and results in an uplifted rift dome. Supra-crustal erosion then creates the depression required for the deposition of material once the rift pillow starts to be absorbed back into the upper mantle. The absence of rift pillow velocities from the refraction data, and the difficulty of thinning the crust by the required amount to produce space for the subsequent sedimentary basin, effectively eliminates this model and other models appealing to subaerial erosion as a mechanism for crustal attenuation when they are considered in the context of the North Sea.

The most appealing mechanism to explain North Sea Basin formation is that due to McKenzie (1978), which relates the graben generation stage with lithospheric extension, to which the lower crust and upper mantle respond by ductile thinning, and the upper crust by brittle failure. Thinning the lower crust and upper mantle raises the isotherms and creates a thermal anomaly, which decays by conduction through the crust producing gently uniform subsidence with an exponentially decaying rate. The amount of extension required to produce the observed subsidence was estimated by Sclater & Christie (1980) from well log data to be 50–60% in the centre of the Main Line profile, which is consistent with the seismic evidence if an original crustal thickness of 32 km is assumed. However, recent work by R. Wood (personal communication) suggests that a value of 35% extension may be more appropriate. A local crustal attenuation factor of 1.3–1.5, especially if achieved by areal as opposed to simple linear extension, goes far towards meeting the amount of extension estimated by backtracking the faults interpreted from reflexion profiles (Smythe *et al.* 1980).

The temporal correlation between North Sea activity and the evolution of the North Atlantic (Ziegler 1978; Christie 1979) allows the McKenzie model as applied to the North Sea to be set against a background of extensional tectonics without the necessity of appealing to a catastrophic initiation event. With the possible exception of an as yet unresolved space problem in the fault pattern of the upper crust, the model satisfies the gross requirements of the observed North Sea data in terms of its deep velocity structure, subsidence history and present-day heat flow (Christie & Sclater 1980; Sclater & Christie 1980). Further refraction work currently in progress by the Cambridge Marine Group may help to define in more detail the velocity depth structure across the Central Graben, where the model predicts the greatest crustal attenuation.

While there is but one author ascribed to this paper, the work presented would not have been possible without the help of the Marine Group at Cambridge University. In particular, the guidance of Dr D. H. Matthews is gratefully acknowledged. During the course of this work, I was supported by a research studentship awarded by the Natural Environmental Research Council.

References (Christie)

Artemjev, M. E. & Artyushkov, E. V. 1971 *J. geophys. Res.* **76**, 1197–1211.

Bamford, D., Nunn, K., Prodehl, C. & Jacob, B. 1978 *Geophys. Jl R. astr. Soc.* **54**, 43–60.

Beaumont, C. 1978 *Geophys. Jl R. astr. Soc.* **55**, 471–497.

Berry, M. J. & West, G. F. 1966 *Bull. seism. Soc. Am.* **56**, 141–171.

Bott, M. H. P. 1976 *Tectonophysics* **36**, 77–86.

Burkhardt, H. & Vees, R. 1975 *J. Geophys.* **41**, 463–474.

Christie, P. A. F. 1979 Ph.D. thesis, Cambridge University.

Christie, P. A. F. & Sclater, J. G. 1980 *Nature, Lond.* **283**, 729–732.

Collette, B. J. 1960 In *Gravity expeditions*, vol. 5 (ed. G. J. Bruins). Delft: Netherlands Geodetic Commission.

Collette, B. J. 1968 In *Geology of shelf seas* (ed. D. T. Donovan), pp. 15–30. Edinburgh: Oliver & Boyd.

Collette, B. J., Laagay, R. A., Ritzema, A. R. & Schouten, J. A. 1970 *Geophys. Jl R. astr. Soc.* **19**, 183–199.

Edel, J. B., Fuchs, K., Gelbke, C. & Prodehl, C. 1975 *J. Geophys.* **41**, 333–356.

Evans, T. R. & Coleman, N. L. 1974 *Nature, Lond.* **247**, 28–30.

Haxby, W. F., Turcotte, D. L. & Bird, J. M. 1976 *Tectonophysics* **36**, 57–75.

Illies, J. H. 1970 In *Graben problems* (ed. J. H. Illies & S. Mueller) (*International Upper Mantle Project Scientific Report* no. 27), pp. 4–27.

Jacob, A. W. B. 1975 *J. Geophys.* **41**, 63–70.

McKenzie, D. P. 1978 *Earth planet. Sci. Lett.* **40**, 25–32.

Sapin, M. & Prodehl, C. 1973 *Ann. Geophys.* **29**, 127–145.

Scheidegger, A. E. & Willmore, P. L. 1957 *Geophysics* **22**, 9–22.

Sclater, J. G. & Christie, P. A. F. 1980 *J. geophys. Res.* **85**, 3711–3739.

Sleep, N. H. & Snell, N. S. 1976 *Geophys. Jl R. astr. Soc.* **45**, 125–154.

Smith, W. A. & Christie, P. A. F. 1976 *Mar. geophys. Res.* **3**, 235–250.

Smythe, D. K., Skuce, A. G. & Donato, J. A. 1980 *Nature, Lond.* **287**, 467–468.

Solli, M. 1976 M.Sc. thesis, Bergen University.
Sørnes, A. 1971 Ph.D. thesis, Edinburgh University.
Vening Meinesz, F. A. 1950 *Bull. K. Belg. kolon. Inst.* **21**, 539–552.
Woodland, A. 1974 *Petroleum and the continental shelf of north-west Europe.* London: Applied Science Publishers.
Ziegler, P. A. 1975 *Bull. Am. Ass. Petrol. Geol.* **59**, 1073–1097.
Ziegler, P. A. 1977 *GeoJournal* **1**, 7–31.
Ziegler, P. A. 1978 In *Tectonics and geophysics of continental rifts* (ed. I. B. Ramberg & E. R. Newman), pp. 249–277. Dordrecht: D. Reidel.

Discussion

A. S. Laughton, F.R.S. Dr Christie started by showing two completely different interpretations of the crustal structure beneath the North Sea. His own preferred model closely resembles one of them. Could he comment on why the interpretations were so different, and why he believes his own section is better constrained?

P. A. F. Christie. The first profile that I showed was obtained by Sørnes by using data from a number of different experiments carried out at different times. There may therefore, be internal inconsistencies in the data. Furthermore, Sørnes used his data only to produce time terms to the Moho and did not convert these into a crustal model. As I remember it was Willmore who made this conversion to Moho depths, about which he had certain reservations. Probably the principal difference between this profile and the second one that I showed, which was obtained by Solli, is caused by the values of the delay times introduced by the superficial sediments. When Sørnes made his analysis these delays were not well known. The more recent analysis by Solli made use of very detailed values of the delay times that Ziegler supplied, obtained from reflexion studies. Solli also used a velocity of about 6 km s^{-1} for the whole of the crystalline crust. His low crustal velocity and high sub-Moho velocity of 8.3 km s^{-1} combine to produce a much thinner crust beneath the graben than Sørnes obtained. The profile that I proposed had a two-layer crust beneath the sediments. The existence of the lower layer, which depends on the interpretation of the Firth of Forth line, increases the average velocity of the crystalline crust from 6 km s^{-1} to about 6.25 km s^{-1}. This change increased the crustal thickness. Hence my interpretation lies between that of Sørnes and Solli. Any increase in the mean crustal velocity would result in greater crustal thickness everywhere. Hence the change in crustal thickness going from the platforms on either side into the graben would also be increased, though the extension required to produce this change would not.

D. G. Roberts. Could Dr Christie comment on how much extension is required to account for his profiles, and how this compares with Ziegler's estimates for the central North Sea?

P. A. F. Christie. The minimum thickness of crystalline crust on the section is about 20 km The crustal thickness beneath both the Southern Uplands and the East Shetland Platform is about 32 km. If this is representative of the original crustal thickness beneath the graben, the maximum stretching factor required would be about 1.6. To produce the observed variation along the profile would then require a displacement of about 65 km if all of the stretching resulted from movement on faults striking east–west. This estimate is probably too large. If the extension is caused by movement of faults with a variety of strikes, as it may well do where the Moray Firth basin joins the Central Graben, then the resulting areal extension may require considerably smaller linear extension in any direction. To estimate the extension produced by areal extension requires detailed seismic coverage with lines run in several directions, which is not available to me.

Phil. Trans. R. Soc. Lond. A **305**, **113–143** (1982) [**113**]
Printed in Great Britain

Faulting and graben formation in western and central Europe

By P. A. Ziegler

Shell Internationale Petroleum Mij. B.V., Carel van Bylandtlaan 30,
The Hague, The Netherlands

Rifts and rift-related basins play a pre-eminent role among the sedimentary basins of western and central Europe. Through time, grabens developed in a number of different megatectonic settings whereby the principal mechanisms governing their subsidence was crustal stretching and 'subcrustal erosion'. The level of volcanic activity associated with rifting is highly variable and can change significantly during the development history of a rift. Some rifts are totally non-volcanic. The development of rift domes is generally associated with volcanic activity. Uplifting of a rift dome can induce a reversal in the subsidence pattern of a rift. Intracontinental rifts, with their thinned crust, are prone to inversion when the respective craton is subjected to tangential stresses. In the process of inversion the crust of rifts is mechanically thickened again.

1. Introduction

Rifts play a pre-eminent role among the sedimentary basins of western and central Europe. Several more or less distinct rifting cycles can be distinguished whereby the formation of grabens took place in a number of different megatectonic settings.

Back-arc rifting played a significant role during the Devonian and early Carboniferous development of the Variscan geosynclinal system. Essentially wrench-induced pull-apart features developed during the Devonian and early Carboniferous sinistral translation between Laurentia–Greenland and Fennoscandia–Baltica as well as during the late Variscan dextral translation between Europe and Africa. Regional crustal extension, preceding continental splitting and the onset of seafloor spreading, was the dominant mechanism that governed the development of the Carboniferous and Mesozoic Arctic – North Atlantic rift and of the Mesozoic rifts of western and central Europe. Mechanisms controlling the Neogene collapse of the Mediterranean basins and the development of Oligocene and younger rifts in the northern Alpine foreland are still the subject of dispute.

During the Variscan and Alpine orogenic cycles a number of the previously formed graben became deformed and inverted in response to compressional stresses transmitted through the foreland crust of the respective fold belts.

In the following a summary is given of the development of the various graben systems of western and central Europe. For a more detailed discussion, including comprehensive literature indexes, the reader is referred to Ziegler (1978*a*, *b*, 1980, 1981) and to the *Geological atlas of western and central Europe* that is currently in preparation (Ziegler 1982).

2. Variscan Geosynclinal System, a case of back-arc rifting?

The late Caledonian framework of the North Atlantic domain is summarized in figure 1, while figures 2 and 3 provide an overview of its mid-Devonian and early Carboniferous setting.

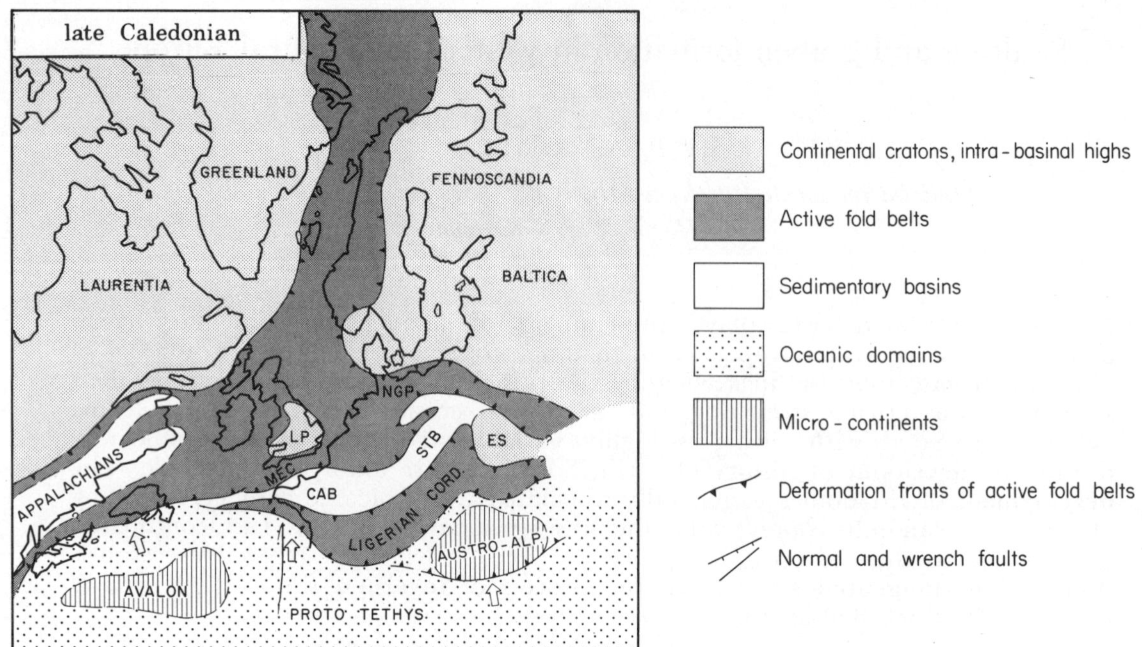

FIGURE 1. Schematic late Caledonian tectonic map of the North Atlantic area.

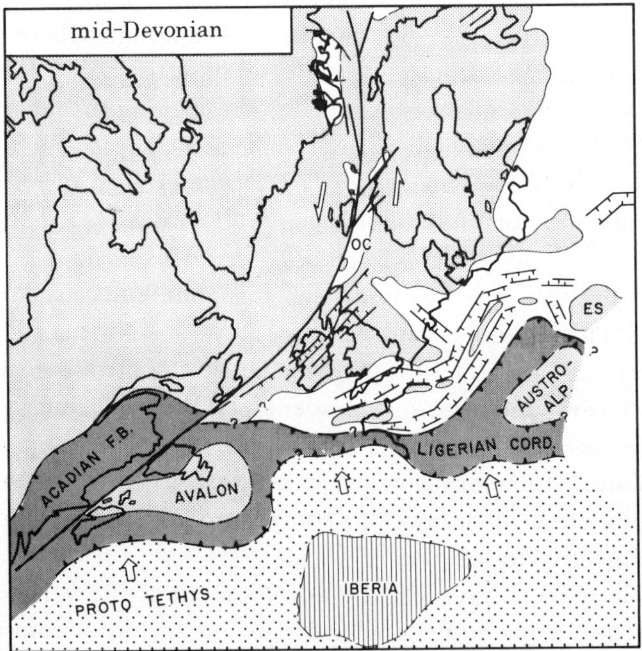

FIGURE 2. Schematic mid-Devonian tectonic map of the North Atlantic area. For key see figure 1.

The Precambrian and Caledonian micro-cratons of the London Platform, Central Armorica, Moldanubia and East Silesia, which were apparently rifted off the northern margin of Gondwana during Cambrian and Ordovician time, were accreted in the course the late Ordovician and Silurian to the southern margin of Fennoscandia–Baltica. By late Silurian – early Devonian time much of western and central Europe was occupied by Caledonian fold belts. Only in the

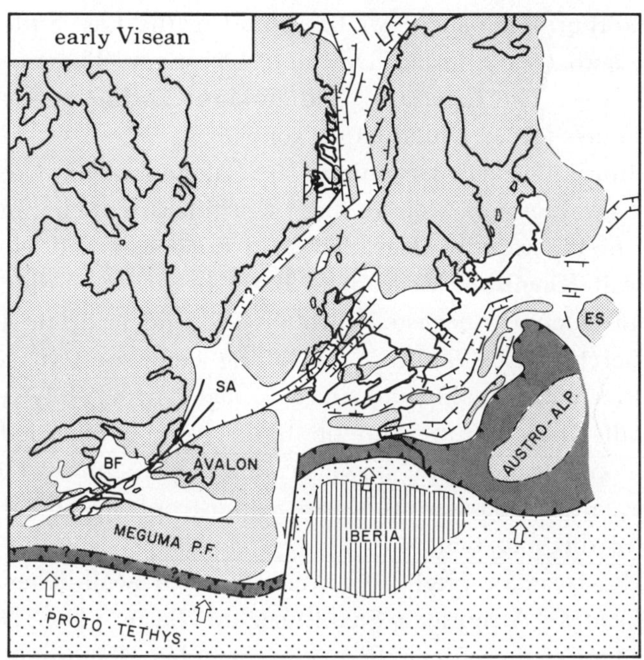

FIGURE 3. Schematic early Carboniferous tectonic map of the North Atlantic area. For key see figure 1.

TABLE 1. ABBREVIATIONS USED ON THE FIGURES

AA	Austro-Alpine Block	MEC	Mid European Caledonides
AB	Altmark–Brandenburg Basin	MF	Moray Firth Basin
ALH	Alemanic High	NGP	North German–Polish Caledonides
BC	Bristol Channel Trough	MM	Morton-in-the March Axis
BF	Bay of Fundy Basin	MNH	Mid-North Sea High
BU	Brugundy Trough	MW	Market Weighton Axis
CA	Cardigan Bay Trough	MX	Manx-Furness Basin
CAB	Central Armorican Basin	NF	Normandy Fault
CB	Channel Basin	PB	Paris Basin
CG	Central Graben	PC	Porcupine Trough
CN	Central Netherlands Basin	PD	Pays-de-Bray Fault
CS	Celtic Sea Basin	PS	Pompeckj Swell
DB	Dublin Trough	PT	Polish Trough
DT	Danish Trough	RFH	Ringkøbing–Fyn High
EB	Egersund Basin	RM	Rhenish Massif
EC	East Carpathian Gate	SA	Saint Antony Basin
ES	East Silesian	SE	Slyne–Erris Trough
ET	Emsland Trough	SH	Sub-Hercynian Basin
GG	Glückstadt Graben	SM	Silesian-Moravian Gate
GRLD	Greenland	SP	Sole Pit Basin
HB	Horda Basin	SS	Saxonian Strait
HD	Hessian Depression	ST	Sticklepath Fault
HF	Haig–Fras Depression	STB	Saxothuringan Basin
HG	Horn Graben	TB	Trier Bay
HP	Hampshire Basin	UB	Ulster Basin
KB	Kish Bank Basin	VG	Viking Graben
LS	Lower Saxony Basin	WA	Western Approaches Trough
LP	London Platform	WG	Worcester Graben
MA	Mendip Axis	WM	Welsh Massif
MB	Minch Basins	WN	West Netherlands Basin
MC	Massif Central		

Central Armorican–Saxothuringian successor basin and in the East Sudetic area did marine sedimentation continue across the Silurian–Devonian boundary. These basins were underlain by thinned continental crust. The Ligerian–Moldanubian Cordillera was associated with the north-plunging Proto-Tethys (Rheic) subduction zone.

Following the late Silurian – early Devonian late Caledonian diastrophism the evolution of areas to the north of the Ligerian-Moldanubian Cordillera was characterized by tensional tectonics. In the area of the rapidly degraded Mid-European and North German–Polish Caledonides the Cornwall–Rhenish–East Sudetic Basin began to subside rapidly during the early Devonian. Crustal extension persisted through the mid and late Devonian into early Carboniferous time whereby block faulting controlling facies patterns was associated with a distinctly bimodal felsic–mafic alkaline volcanism (Ziegler 1978*b*; Sawkins & Burke 1980).

As such the Cornwall–Rhenish–East Sudetic Basin can be regarded as a complex rift system. Details of its configuration are, however, difficult to unravel owing to its strong deformation during the late Carboniferous Variscan diastrophism and the lack of a reliable palispastic restoration of this fold belt.

The Normannian–Mid-German High separated the Cornwall–Rhenish Basin from the southward adjacent Central Armorican–Saxothurigian Basin. The Devonian and early Carboniferous development of the latter reflects the interplay of tensional and compressional tectonics. Periods of extension induced basin subsidence and the repeated outpouring of alkaline bimodal volcanics. During periods of compression, rift-related volcanic activity was interrupted.

The early Middle Devonian Acadian–Ligerian diastrophism, which marked the collision of the Avalon and Austro-Alpine micro-cratons with Laurasia, is strongly expressed in the Ligerian–Moldanubian Cordillera but affected only the southern margin of the Central Armorican–Saxothuringian Basin (Autran & Cogné 1980). This orogenic event was followed by a renewed tensional accentuation of the Central Armorican–Saxothuringian successor basin and correspondent volcanic activity. At the transition from the Devonian to the Carboniferous the Bretonian diastrophism, which strongly affected the Ligerian–Moldanubian Cordillera, caused minor folding and a temporary interruption of the rift volcanism in the Central Armorican–Saxothuringian Basin. During the early Visean this basin became once more accentuated by tensional tectonics. With the late Visean onset of the Sudetic diastrophism, the Central Armorican–Saxothuringian Basin became progressively filled with the synorogenic Culm flysch deposits. By latest Visean time this successor basin became folded, uplifted and partly destroyed.

In the Cornwall–Rhenish–East Sudetic Basin the last pulse of rift-related volcanism occurred during the early Visean. During the late Visean underthrusting began along the northern margin of the Normannian–Mid-German High and initiated the shedding of coarse clastics into the Cornwall–Rhenish–East Sudetic Basin where they were deposited as the synorogenic Culm flysch. At the same time the axis of this basin was shifted to the south, presumably in response to tectonic loading of the foreland by the advancing thrust sheets. With this the Cornwall–Rhenish–East Sudetic Basin assumed the asymmetric geometry of a classical foredeep. During the late Carboniferous the axis of this foredeep migrated northward as a consequence of the advancement of the Variscan deformation front. By late Namurian time sedimentation exceeded subsidence rates and paralic conditions were established along the southern basin margins as topsets of large clastic fans that prograded northward into its deeper parts. With the onset of the Westphalian a paralic, molasse-type, depositional régime was established throughout the Variscan foredeep basin (Paproth & Teichmüller 1961; Teichmüller 1973). Last compressional deformations are dated as late Westphalian – early Stephanian (Asturian phase).

The Devonian and early Carboniferous evolution of the Variscan geosynclinal system can be explained in terms of back-arc rifting. Back-arc rifting and sea-floor spreading is thought to occur if the subducting and overriding plates diverge or if their convergence rates decrease below the level at which partial coupling occurs between them at the Benioff zone (Uyeda 1981; Hsui & Toksöz 1981).

After the late Caledonian diastrophism these conditions were apparently fulfilled between Laurasia and the subducting Proto-Tethys plate. Back-arc rifting set in during the early Devonian and persisted till early Visean time. It affected wide areas to the north of the Ligerian–Moldanubian Cordillera but apparently never proceeded to the point of crustal separation and the opening of back-arc oceanic basin. Back-arc rifting was partly contemporaneous with a major mid-Devonian to early Carboniferous sinistral displacement between Laurentia–Greenland and Fennoscandia–Baltica (see below).

Temporary partial coupling at the Proto-Tethys B-subduction zone during the Acadian and Bretonian orogenies was apparently strong enough to interrupt rifting in the Central Armorican–Saxothuringian Basin but did not affect the Cornwall–Rhenish–East Sudetic Basin.

The onset of the late Visean Variscan main orogeny presumably reflects an increase in the convergence rate between the Laurasian and Proto-Tethys plates. This was followed by the collision of Iberia and later also of Gondwana with Laurasia. Resultant regional compressive stresses caused the sharp termination of back-arc rifting and the scooping-out of the back-arc basins by thrust sheets and nappes.

3. Devonian and Carboniferous Arctic – North Atlantic wrench and rift system

The early Devonian development of the Arctic – North Atlantic Caledonides was characterized by their rapid erosion, isostatic uplifting and a post-orogenic calc-alkaline plutonism. From mid-Devonian until early Dinantian time wrench-tectonics related to a sinistral displacement between Laurentia–Greenland and Fennoscandia–Baltica amounting to 1500–2500 km dominated the evolution of the Arctic – North Atlantic area (Morris 1976; Kent & Opdyke 1979; Van der Voo & Channel 1980).

This is illustrated by the evolution of the partly fault-bounded intramontane Old Red 'Molasse' basins of Svalbard, Eastern Greenland, Western Norway and the British Isles. Subsidence of these basins was accompanied by syndepositional wrench-related compressional deformations. Earliest compressive movements are dated as mid-Devonian and are thus contemporaneous with the Acadian-Ligerian diastrophism, while the last compressional deformations are dated as early Dinantian (Haller 1971; Harland 1978). From Visean time onward the development of the Arctic – North Atlantic realm was governed by tensional tectonics. This is illustrated by the evolution of eastern Greenland and the Svalbard–Barents Sea area where thick Carboniferous and younger strata accumulated in complex graben systems (Vischer 1943; Haller 1971; Harland *et al.* 1974; Rønnevik 1981).

In the area of the future Norwegian–Greenland Sea, crustal extension persisted from late early Carboniferous time over some 270 Ma until crustal separation was achieved during the late Palaeocene. In this rift system the level of volcanic activity was extremely low during late Palaeozoic and Mesozoic times but increased sharply immediately before crustal separation.

4. Devonian and Carboniferous rifts of the British Isles

In the British Isles the Orcadian Basin, the Midland Valley Graben and the Northumberland–Solway–Dublin Trough started to subside rapidly during the early Devonian. Lower Old Red clastics reach thicknesses of 2.5 km in the Orcadian Basins and 6.6 km in the Midland Valley Graben (House *et al.* 1977). Subsidence of these basins was accompanied by widespread calc-alkaline intrusions and an intense felsic to mafic volcanism. The mid-Devonian deformation of these basins is thought to reflect the onset of transpression along the Arctic – North Atlantic transform system. In the Midland Valley Graben and the Northumberland–Solway–Dublin Trough the mid-Devonian corresponds to a regional unconformity, while in the Orcadian Basin Middle Old Red clastics are up to 5 km thick. Rapid subsidence of the Orcadian Basin as well as the syndepositional deformation of its sedimentary fill reflects the interplay between tensional and compressional stresses associated with sinistral translation of major proportions along the Great Glen fault (Van der Voo & Scotese 1981).

In the course of the late Devonian sedimentation was resumed in most of the Old Red Basins of northern and central Britain and Ireland. Volcanic activity was at a low level and subsidence rates were generally relatively low.

Sedimentation was continuous across the Devonian–Carboniferous boundary (George *et al.* 1976; Francis 1978*a*). During the Dinantian the Midland Valley Graben and the Northumberland–Solway–Dublin Trough continued to subside differentially, although the Orcadian Basin became largely inactive. At the same time the Craven Basin began to subside rapidly. Sharp lateral facies and thickness changes are indicative of syndepositional tensional faulting (Leeder 1974; George 1958). Locally this was accompanied by an early Dinantian alkaline, mafic–felsic bimodal volcanism. In the course of the late Namurian and early Westphalian the individual grabens gradually ceased to subside differentially and rift-related volcanic activity abated (Francis 1978*b*). During the late Westphalian the sedimentary fill of these tensional basins became folded and uplifted to various degrees (inversion). Owing to the ensuing erosion of the Carboniferous strata it is not clear whether these rifts had become inactive altogether before their inversion. In spite of an incomplete Permian and younger stratigraphic record these rifts appear to have reached thermal and isostatic equilibrium through their inversion.

The Devonian and Carboniferous rifts of the central and northern British Isles occupy an intermediate position between the back-arc rifts of the Variscan geosynclinal system and the Arctic – North Atlantic wrench–rift system.

The rapid early Devonian subsidence of the Midland Valley and the Northumberland–Solway–Dublin grabens parallels the development of the Cornwall–Rhenish–Lower Silesian and the Central Armorican–Saxothuringian basins and may also be related to back-arc extension. During the mid and late Devonian the development of the rifts of the central and northern British Isles was dominated by the Arctic – North Atlantic translation. Crustal extension resumed, however, at the onset of the Carboniferous and persisted into late Carboniferous times, while rifting in the Variscan geosynclinal system ceased with the late Visean onset of regional compression. In contrast, the development of the Barents Sea and Norwegian–Greenland Sea area was governed from late Visean time onward by crustal extension. In this respect the Carboniferous rifts of Ireland and the United Kingdom as well as the St Anthony and the Bay of Fundy basins of the Canadian Maritime Provinces (Howie & Barss 1975) could be regarded as forming part of the Arctic – North Atlantic rift system.

While rifting persisted in the Norwegian–Greenland and Barents Sea area throughout the late Carboniferous, crustal extension abated in the British Isles. During the late Westphalian terminal phases of the Variscan orogeny tangential stresses were transmitted through the foreland crust and caused the transpressional deformation and partial inversion of these grabens and troughs at considerable distances to the north of the Variscan thrust-front.

5. Late Variscan wrench and rift tectonics

Following the late Westphalian – early Stephanian consolidation of the Variscan fold belt convergence between Gondwana and Laurasia apparently changed from an essentially north–south-directed collision to an east–west-oriented one. Orogenic movements continued during Stephanian and Autunian time in the Appalachian–Mauretanides and in the Uralides, whereas the Variscides remained inactive. This was accompanied by the development of a right lateral transform system that linked the southern Uralides with the northern Appalachians via Europe where it caused the development of a complex pattern of conjugate shear faults and related pull-apart structures (Arthaud & Matte 1977). This fault system remained active until the late early Permian consolidation of the Appalachians–Mauretanides and the Uralides

Main elements of the late Variscan fracture system of northern Africa and Europe are summarized in figure 4. Displacements along the Kelvyn–Agadir and the Chedabucto–Gibraltar faults were compensated for by wrench and compressional deformations in the northern Appalachians and by the inversion of the Carboniferous troughs in the Canadian Maritime provinces (Rast & Grant 1977; Schenk 1978). Displacements along the Bay of Biscay fault system were partly taken up in the Arctic – North Atlantic rift in which crustal distension persisted through late Carboniferous and Permian time. Right lateral wrench movements along the Tornquist–Teisseyre lineament, at its northern termination, induced the development of the Oslo–Bamble–Horn Rift.

Areas located between the Agadir fault zone and the Tornquist–Teisseyre lineament became transected during the Stephanian and Autunian by a network of subsidiary wrench and tensional faults (figure 5). This was accompanied by the rapid subsidence of narrow wrench-related basins and the extrusion of often thick volcanics. Examples of such wrench basins, in which several thousand metres of coal-bearing clastics were deposited, are the St Étienne, Lodève and Cévennes basins in the Massif Central. Extensive late Carboniferous to early Permian volcanics occur in the subsurface of northern Germany and Poland and reach thicknesses up to 2 km. Volcanic centres appear to coincide with the intersection of fault systems and are probably related to pull-apart structures at the termination of subsidiary wrench faults paralleling the Tornquist–Teisseyre line. The latter is not represented by a single major fracture but corresponds to a broad essentially northwest–southeast striking fault zone. In the western Baltic and in the Kattegat, where it is referred to as the Fennoscandian Border Zone, the pattern of this fault system is strongly suggestive of strike-slip displacement. Associated dyke swarms permit dating of faulting as late Carboniferous to early Permian.

In this area the early Palaeozoic shelf sediments of the Caledonian foreland were strongly dissected whereby up to 4 km thick Cambro-Silurian sediments are preserved in narrow down-faulted blocks, while in adjacent upthrown blocks the entire section is missing owing to late Palaeozoic and Mesozoic erosion (see Ziegler 1978*a*, encl. 1).

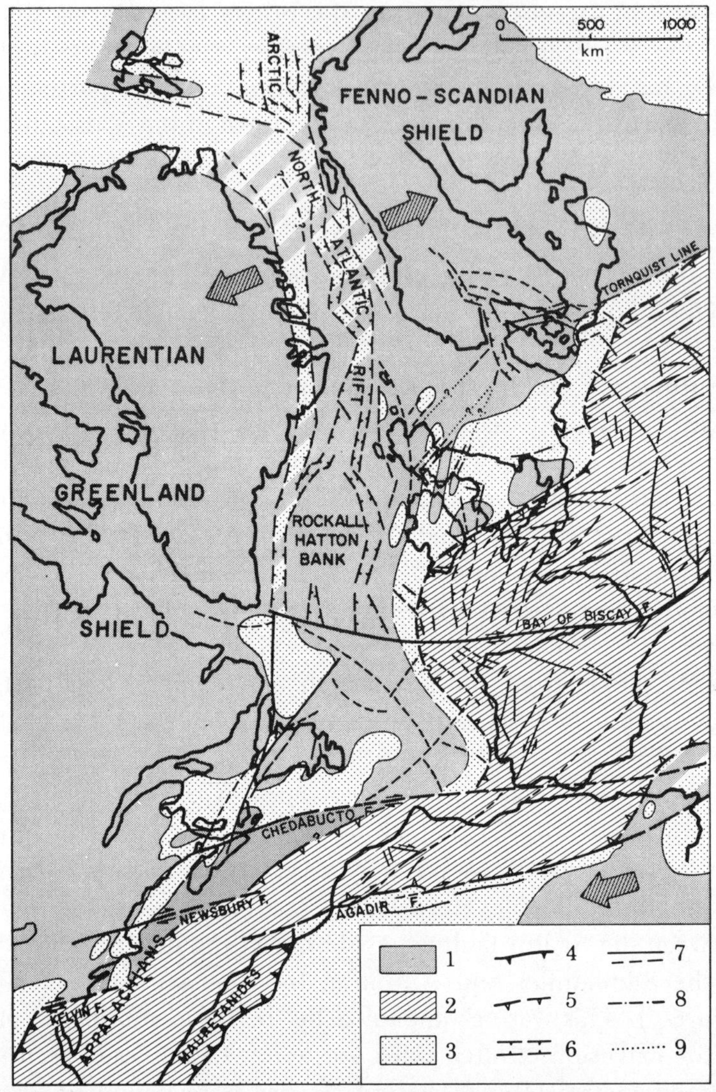

FIGURE 4. Schematic late Variscan tectonic map of the North Atlantic area: 1, pre-Variscan stable elements; 2, Variscan fold-belt; 3, Carboniferous basins in Variscan foreland; 4, Alleghenian deformation front; 5, Asturian deformation front; 6, grabens, rifts; 7, faults; 8, dyke swarms; 9, alignments.

The Oslo–Bamble–Horn Graben, which marks the northwestern termination of the Tornquist–Teisseyre line, came into evidence during the latest Westphalian to Stephanian. Igneous activity started in the southern Oslo Graben and gradually spread northward and presumably also southward. This probably reflects rift development by fracture propagation. In view of its relation to the Tornquist–Teisseyre line, the Oslo–Bamble–Horn Graben can be considered as a pull-apart feature at the termination of a transcurrent fault, rather than

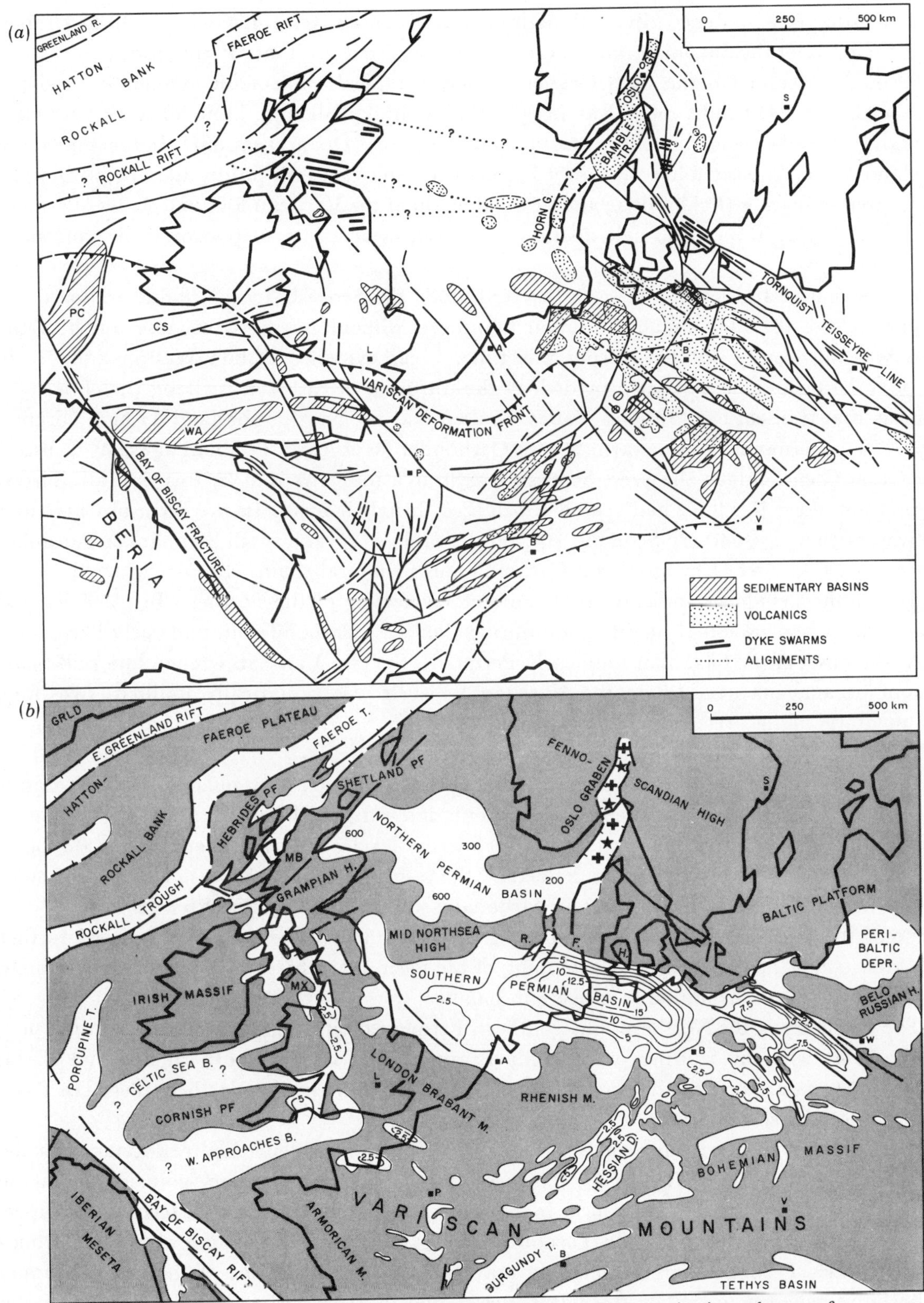

FIGURE 5. (*a*) Stephanian–Autunian fault patterns. (*b*) Tentative isopach map of Rotliegend sediments, contour values in hundreds of metres; thickness values in metres.

as a mantle plume-induced rift. Although crustal distension terminated in the Oslo Graben with the onset of the Saxonian, magmatic activity persisted until late Permian time.

Permian volcanics in the Oslo Graben display a typical mafic–felsic bimodality and are highly alkaline (Oftedahl 1968; Ramberg 1976). A similar alkaline bimodality characterizes the Lower Permian volcanics of the central North Sea area (Dixon *et al.* 1981). In the immediate foreland of the Variscan fold belt, Lower Permian volcanics are only mildly alkaline, with mafic rocks predominating (Eckhardt 1979). In the domain of the Variscan fold belt, Lower Permian volcanics display the typical calc-alkaline composition of a late to post-orogenic volcanism (Kramer 1977).

In the central and northern North Sea and along the Scottish–Irish Atlantic seaboard the importance of the Stephanian–Autunian faulting is difficult to assess because much of the area was apparently uplifted during this time. Clear evidence of late Westphalian – early Stephanian tectonic activity is provided by the intrusion of east–west striking tholeiitic dyke swarms that transect the Midland Valley Graben and the Northumberland Trough (Francis 1978*b*). Furthermore a dyke swarm of late Carboniferous to early Permian age that transects the Great Glen Fault in the West Scottish Argyll area indicates that by this time all sinistral translation along this fault had ceased (Speight & Mitchell 1978). In the Firth of Clyde and Solway Firth area (southern Scotland) the border faults of several small Stephanian–Autunian basins clearly transect Devonian and Carboniferous structural trends (Brookfield 1978).

This limited evidence indicates that the tectonic setting of the northern British Isles had also undergone a profound modification during the latest Carboniferous and early Permian.

At the onset of the Saxonian (late early Permian), the late Variscan wrench and pull-apart system became inactive. While the Variscan fold-belt was isostatically uplifted, two large sedimentary basins, referred to as the Northern and Southern Permian basins, began to subside in its foreland.

During the Saxonian the continental Rotliegendes redbeds accumulated in these basins (figure 5). Subsidence rates exceeded sedimentation rates and caused the development of topographic depressions that were located below the level of the world oceans. At the onset of the Thuringian these basins were flooded by the transgressing Arctic seas. This was followed by the accumulation of thick Zechstein carbonate, sulphate and halite series. Isostatic adjustment of the crust, initially to water loading and later to sediment loading, caused further subsidence and the progressive overstepping of the basin margins. At the end of the Permian, subsidence and sedimentation rates were in balance.

The geometry of the Northern and Southern Permian basins reflects regional downwarping of the crust whereby faulting played only a minor role. In the Southern Permian Basin, Permian strata attain a maximum thickness of 3.5 km, while in the Northern Permian Basin they do not exceed 1.5–2 km.

Isopach maps of the Rotliegendes series show that two subsidence centres occurred in the Southern Permian Basin (figure 5). The western subsidence maximum coincides with the area of extensive Autunian volcanism in northern Germany, while the eastern one is superimposed on the Polish part of the Tornquist–Teisseyre lineament. Cooling of a late Carboniferous – early Permian thermal anomaly may explain the subsidence of the north German depocentre, whereas to explain the subsidence pattern of the Polish sub-basin, limited crustal extension, possibly related to early rifting phases in the Tethys, may have to be invoked. On the other hand, lateral ductile flow in the upper mantle, possibly induced by the isostatic uplift of the Variscan fold belt and its thickened lithosphere, may also have played a role in the subsidence of this basin. In this respect a parallel can be drawn between the megatectonic setting of the Southern Permian Basin relative to the Variscan fold belt and the Black Sea relative to the Alpine fold belt; both basins have a similar areal extent and both can be considered as post-orogenic foreland collapse basins.

The geometry of the Northern Permian Basin is difficult to decipher owing to its strong overprinting by Mesozoic rift tectonics. Mechanisms governing its subsidence remain enigmatic. While the Oslo Graben, in which magmatism persisted until late Permian time, did not show any evidence of differential subsidence, the Horn Graben and possibly also the Bamble Trough did begin to subside during the Thuringian.

6. Mesozoic rift systems

After its temporary Permian consolidation the Pangaean supercontinent was again showing signs of inherent instability during the late Permian and even more so during the Triassic. This was manifested initially by the reactivation of the pre-existing peripheral Arctic – North Atlantic rift system and by the inception of the Gondwana rifts but later also by the development of new, interior rift systems. In the course of the Triassic Pangaea became transected in a north–south direction by the Arctic – Central Atlantic rift and in an east–west direction by the Tethys – Central Atlantic – Gulf of Mexico rift–wrench system (Dewey *et al.* 1973; Laubscher & Bernoulli 1977; Biju-Duval *et al.* 1977).

Both of these mega-rifts flanked the already deeply fractured metastable platform of western and central Europe which became affected as a whole by regional tensional stresses. This gave rise to the subsidence of a multidirectional system of grabens and troughs that partly reflects the reactivation of late Carboniferous – early Permian faults.

This rift system remained active until crustal separation was achieved in the Iceland and Norwegian–Greenland Sea during the late Palaeocene. However, the progression of crustal extension and the onset of sea-floor spreading in the central Atlantic and in the western Neo-Tethys during the mid-Jurassic, and in the North Atlantic and the Bay of Biscay during the early Cretaceous, in time caused a reorientation of the stress systems that affected western and central Europe (figure 6). This resulted in a polarization of its rift systems in the course of which a number of Triassic and Jurassic grabens became inactive. Major tectonic events affecting the Atlantic and Tethys rift systems are also reflected in the stratigraphic record of the Mesozoic basins of western and central Europe.

(*a*) *Triassic rifts*

In western and central Europe, Triassic series consist of continental redbeds, evaporites and shallow marine sediments. As sedimentation rates generally kept pace with subsidence rates, isopachs of their depositional thickness give a fair image of the overall subsidence pattern of the Triassic basins. From figure 7 it is evident that the Northern and Southern Permian Basin continued to subside during the Triassic along lines established during the Permian. However, their framework became modified by the development of the North Sea rift system, the Polish–Danish Trough and a number of subsidiary grabens such as the Horn and Glückstadt Graben, the Emsland Trough and the Hessian Depression. Triassic series reach a thickness of over 4 km in the Polish–Danish Trough and in the Glückstadt Graben and some 3 km in the northern Viking Graben.

The Viking and Central Graben as well as the Horda–Egersund half-graben began to subside during the early Scythian. These rifts, developed presumably by fracture propagation of the East Greenland – Western Norway rift in response to accelerated crustal extension in the Arctic – North Atlantic domain (figure 6).

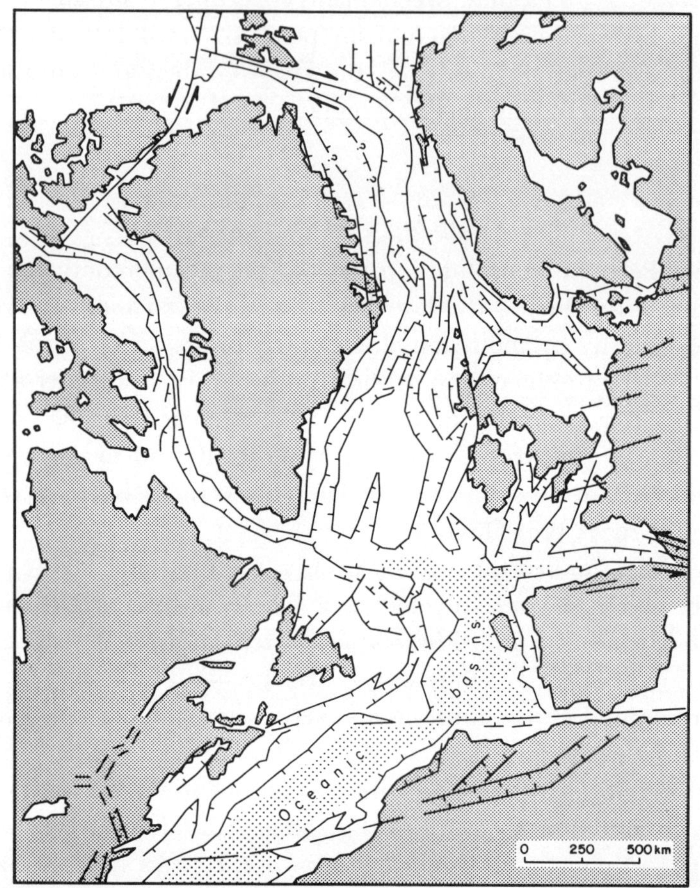

FIGURE 6. Schematic pattern of Mesozoic Arctic – North Atlantic rift system (continental fit approximately early Cretaceous).

Differential subsidence of the Minches, West Hebrides and West Shetland half grabens and probably also of the Rockall – Faeroe Rift, which had set in during the late Permian, persisted through Triassic time. In the Minches Graben, poorly dated Permo-Triassic redbeds reach a thickness of some 4 km.

The Bay of Biscay Rift subsided relatively slowly during the early and mid-Triassic but considerably faster during the late Triassic. Accelerated subsidence was accompanied by the widespread extrusion of basaltic volcanics. This volcanic activity was probably induced by contemporaneous left-lateral wrench movements that compensated for differential crustal stretching in the Newfoundland–Lusitania Rift and in the rifts on the Irish and French shelves. The latter include the Porcupine, Celtic Sea, Bristol Channel and Western Approaches troughs. Differential subsidence of these graben probably started during the Permian but became pronounced during the Triassic. In the Celtic Sea and Western Approaches troughs,

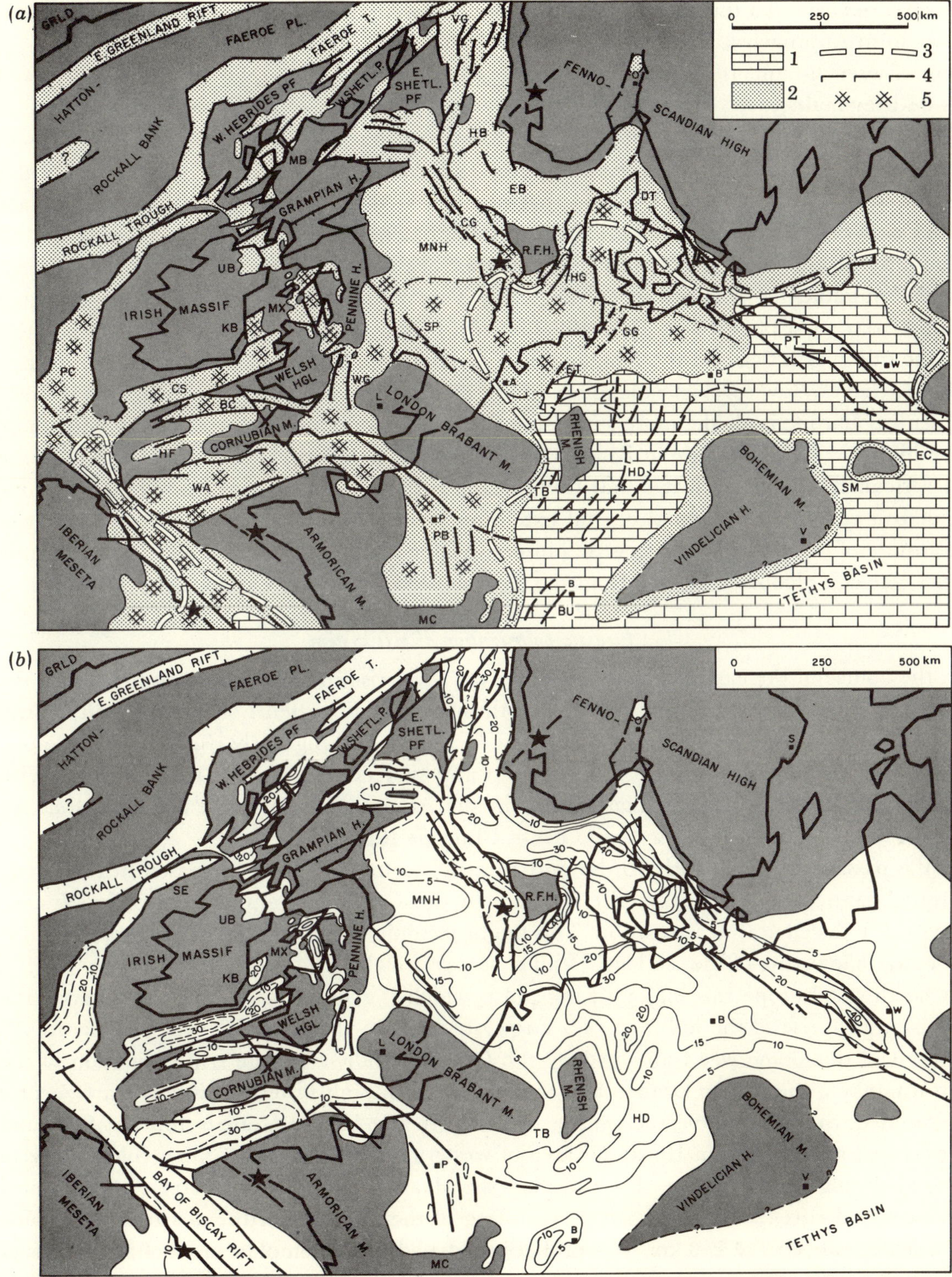

FIGURE 7. (*a*) 1, Distribution of massive Muschelkalk; 2, maximum distribution of Triassic strata; 3, depositional edge of Muschelkalk; 4, distribution of Röt salt; 5, Keuper salts. (*b*) Tentative isopach map of depositional thickness of Triassic sediments. Contour values in hundreds of metres.

Triassic series exceed a thickness of 3 km and in the Porcupine and Bristol Channel some 2 km. The western Approaches, Bristol Channel and Celtic Sea troughs are cut off to the east by a system of wrench-faults that strike into the Paris Basin (Sticklepath, Normandie–Sennely and Pays-de-Bray faults).

The Worcester Graben, The Cheshire and Manx-Furness, the Kish Bank, Solway-Vale-of-Eden and the Ulster Basin may be interpreted as forming the northern continuation of this complex wrench and pull-apart system that compensated for crustal dilatation taking place in the Celtic Sea, Bristol Channel and Western Approaches grabens.

The Triassic rifts of western and central Europe are essentially non-volcanic; the wrench-related Aquitaine Basin is an exception. Only very minor Triassic volcanics are reported from the west coast of Norway (Faerseth *et al.* 1976) from the Central Graben where it intersects the mid North Sea and the Ringkøbing-Fyn High and from the south coast of Brittany.

The extremely low level of volcanic activity associated with the Triassic rifts of western and central Europe is all the more surprising as their dimensions compare readily with those of the highly volcanic Cainozoic Rhine Graben system, or for that matter with parts of the East African Rift. This indicates that the Triassic rift system of western and central Europe subsided in response to regional crustal extension rather than owing to the development of a multitude of local hotspots or mantle plumes. Major rift domes, the telltale signs of mantle plumes, are conspicuously lacking from the Triassic scene of western and central Europe.

(*b*) *Jurassic polarization of rift system*

In the course of the early Jurassic a relative sea level rise induced the transgression of the Arctic and Tethys seas and their linking up through the rift systems of western and central Europe. These graben and troughs continued to subside during early Jurassic times along patterns established during the Triassic without any evidence for discrete rifting pulses. With the exception of minor Rhaeto-Liassic tuffs in the Aquitaine Basin there is no evidence for early Jurassic volcanism in western and central Europe (figure 8).

At the passage from the early to the mid-Jurassic, tectonic activity increased in the Arctic Central Atlantic and the Tethys rift systems. In the Central Atlantic and in the Western Tethys this induced crustal separation. This major tectonic event was apparently accompanied by a eustatic lowering of the sea level (Vail *et al.* 1977; Hallam 1978).

In northwest Europe this major rifting pulse, which is referred to as the Mid-Cimmerian tectonism, gave rise to the uplifting of a large rift dome in the central North Sea (figure 8). The crest of this dome was transected by the Central Graben. At the triple junction of the Central Graben, the Viking Graben and the Moray Firth fault system a large volcanic centre was initiated. Subsidiary volcanic centres occurred in the Viking Graben and the Egersund Basin as well as in the Sunn Hordland area of western Norway. Most of these volcanics were extruded during the Bajorcian. They display the alkaline bimodal chemistry that is typical for continental rifts (Dixon *et al.* 1981). Uplifting of the central North Sea rift dome, which had a structural relief of 2–3 km, was probably induced by the emplacement of a low density asthenolith at the crust–mantle interface.

Uparching of this rift dome induced drastic palaeogeographic changes in northwest Europe. While early Jurassic and older strata were deeply truncated over the crest of this dome, erosion products were shed northward into the continuously subsiding Viking Graben and southward into the incipient Sole Pit, West and Central Netherlands and Lower Saxony basins.

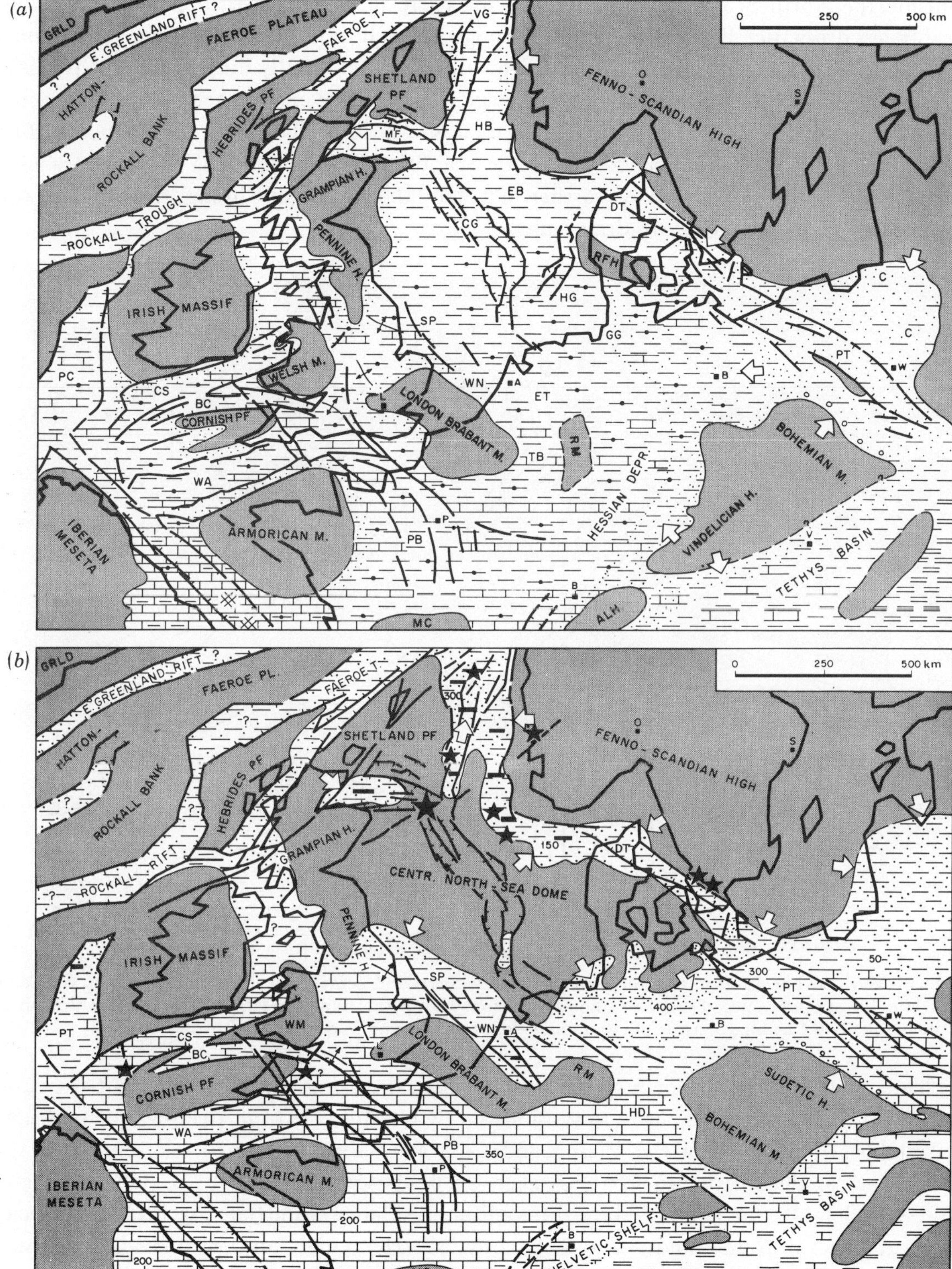

FIGURE 8. (*a*) Early Jurassic palaeogeography; (*b*) Bajocian–Bathonian palaeogeography. For key see figure 14.

Outside the North Sea area, the effects of the early mid-Jurassic rifting pulse were considerably less dramatic. The Polish–Danish Trough, as well as the Celtic Sea, Bristol Channel and Western Approaches grabens continued to subside differentially during the mid-Jurassic. There is no evidence for up-doming of these large rifts. Minor volcanic activity was restricted to the southwestern parts of the Celtic Sea Graben, southern England and the Fennoscandian Border Zone. Crustal extension across the Celtic Sea, Bristol Channel and Western Approaches grabens was compensated for by further left lateral movements along the Sticklepath, Normandie–Sennely and Pays-de-Bray faults.

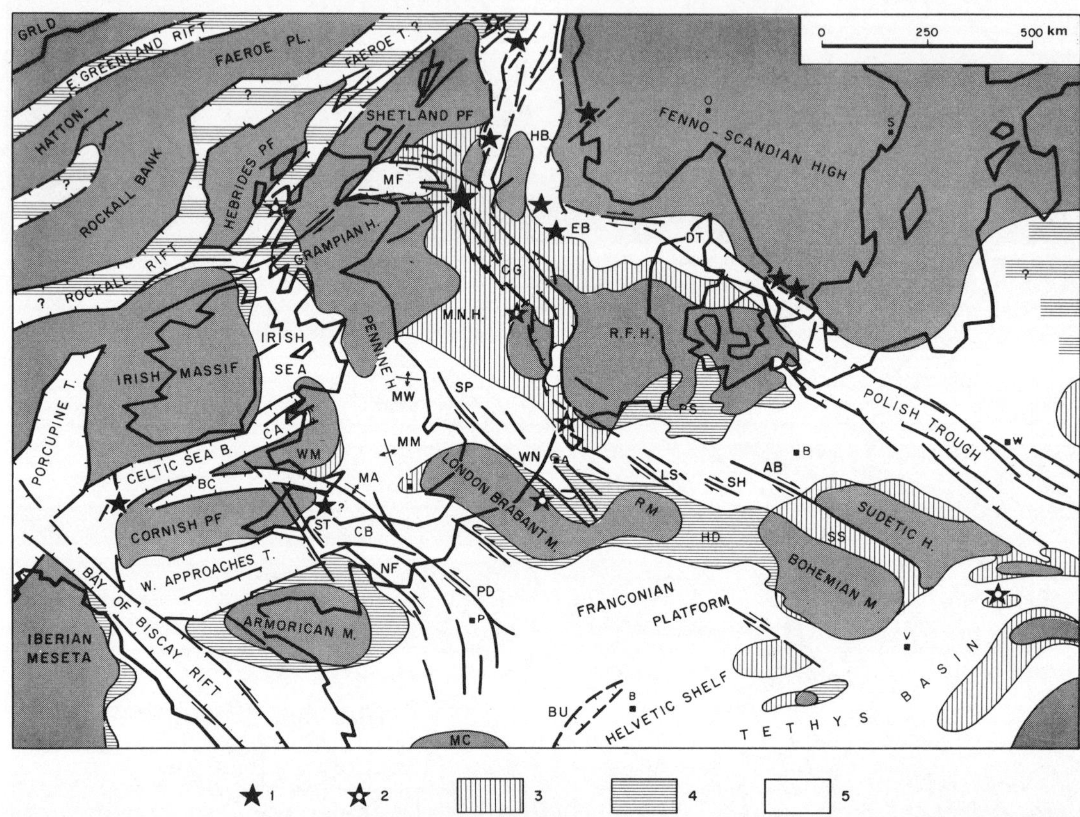

FIGURE 9. Mid and late Jurassic tectonic framework of western and central europe: 1, mid-Jurassic volcanics; 2, late Jurassic volcanics; 3, areas of non-deposition during the mid-Jurassic; 4, areas of non-deposition during the late Jurassic; 5, areas of continuous deposition during mid and late Jurassic.

Volcanic activity in the North Sea essentially ceased during the Bathonian. At the same time the North Sea rift dome began to subside and by late Jurassic time, deep-water conditions were established in much of the Viking and Central Graben (figure 9). Continued crustal extension across the North Sea rift was taken up at its southern end by a system of right-lateral wrench faults that controlled the differential subsidence of the Sole Pit, Broad Fourteens, West and Central Netherlands, Lower Saxony, sub-Hercynian and Altmark-Brandenburg basins ('marginal troughs' of Voigt (1962)). The progressive accentuation of these wrench-induced basins was accompanied by minor volcanic activity and the uplift of the Rhenish and Bohemian Massif whereby the latter became transected by the Saxonian Strait. At the same time the longstanding Hessian Depression was closed and the northeast–southwest-striking Horn Graben, Glückstadt Graben and the Emsland Trough became inactive.

The late mid-Jurassic and late Jurassic polarization of the European rift system can be related to a change in the overall stress pattern that was induced by crustal separation in the western Tethys. With this the northern shelves of the Tethys became tectonically quiescent and the evolution of northwest Europe became increasingly dominated by the Arctic – North Atlantic rift system in which crustal extension apparently accelerated during the late Jurassic, as reflected by the updoming of the triple junction between the Viking Graben and the West Norway – West Shetland/Faeroe Rift.

Late Jurassic crustal extension across the North Sea Rift and the rifts on the Celtic Sea – Western Approaches shelf was accompanied by a westward movement of the Anglo-Saxon Block (Ireland and the United Kingdom, including the London–Brabant Massif) relative to the Armorican Massif – Massif Central Block to its south and the Fennoscandian–Ringkøbing–Fyn block to its north. The resulting tectonic picture can be compared to a tooth being extracted from a jaw. Cracks that started to gape along the sides of its crown, the Anglo-Saxon Block, are represented by the North Sea Rift and the Celtic Sea – Western Approaches graben system, while slip planes developing along the roots of this 'tooth' correspond to the wrench faults and tension-gash basins that developed along the margins of the London–Brabant–Rhenish–Bohemian Block (figure 9).

(*c*) *Early Cretaceous rifting phases*

The early Cretaceous tectonic evolution of western and central Europe continued to be dominated by crustal stretching in the Arctic and North Atlantic domain whereby relatively little change occurred in its framework. Although the Bay of Biscay and the Rockall–Faeroe Graben were the principal rifting axes during the early Cretaceous continued crustal extension across the North Sea Rift and the Celtic Sea – Western Approaches graben system resulted in a sharp accentuation of the wrench-induced basins flanking the London–Brabant, Rhenish and Bohemian massifs. On the other hand the Horda–Egersund Basin and the Danish–Polish Trough subsided only mildly during early Cretaceous times (figure 10).

During the earliest Cretaceous a major rifting pulse, referred to as the Late Cimmerian tectonism, affected the entire northwest European and Arctic – North Atlantic rift system. It preceded the Neocomian onset of sea-floor spreading between the Azores and Charlie Gibbs fracture zones and was accompanied by a significant eustatic sea-level drop that is expressed in western and central Europe by a regional regression. During Valanginian to early Aptian times sea levels rose again, but this trend was temporarily reversed during the mid-Aptian. This regression coincides with the Austrian tectonic pulse that preceded the onset of sea-floor spreading in the Bay of Biscay and in the South Atlantic in areas north of the Walvis Ridge. A further regionally correlative, albeit minor, regression preceded the late Albian onset of sea-floor spreading in the Rockall Trough (Roberts *et al.* 1981*a*).

In the Arctic – North Atlantic rift zone the Late Cimmerian pulse gave rise to the rapid subsidence of large rotational fault blocks in eastern Greenland (Surlyk 1975), the western Barents Sea (Rønnevik 1981) and on the West Norway shelf (Jorgensen & Navrestad 1981) and to the development of a significant submarine relief.

Also in the North Sea Rift the Late Cimmerian tectonism gave rise to the rapid subsidence of rotational fault-blocks along listric normal faults and a regional, largely submarine unconformity (figure 11). The resulting sea-floor relief was of the order of 1 km. In the Viking Graben, the tectonically undisturbed marine onlap of the early Cretaceous deep-water shales

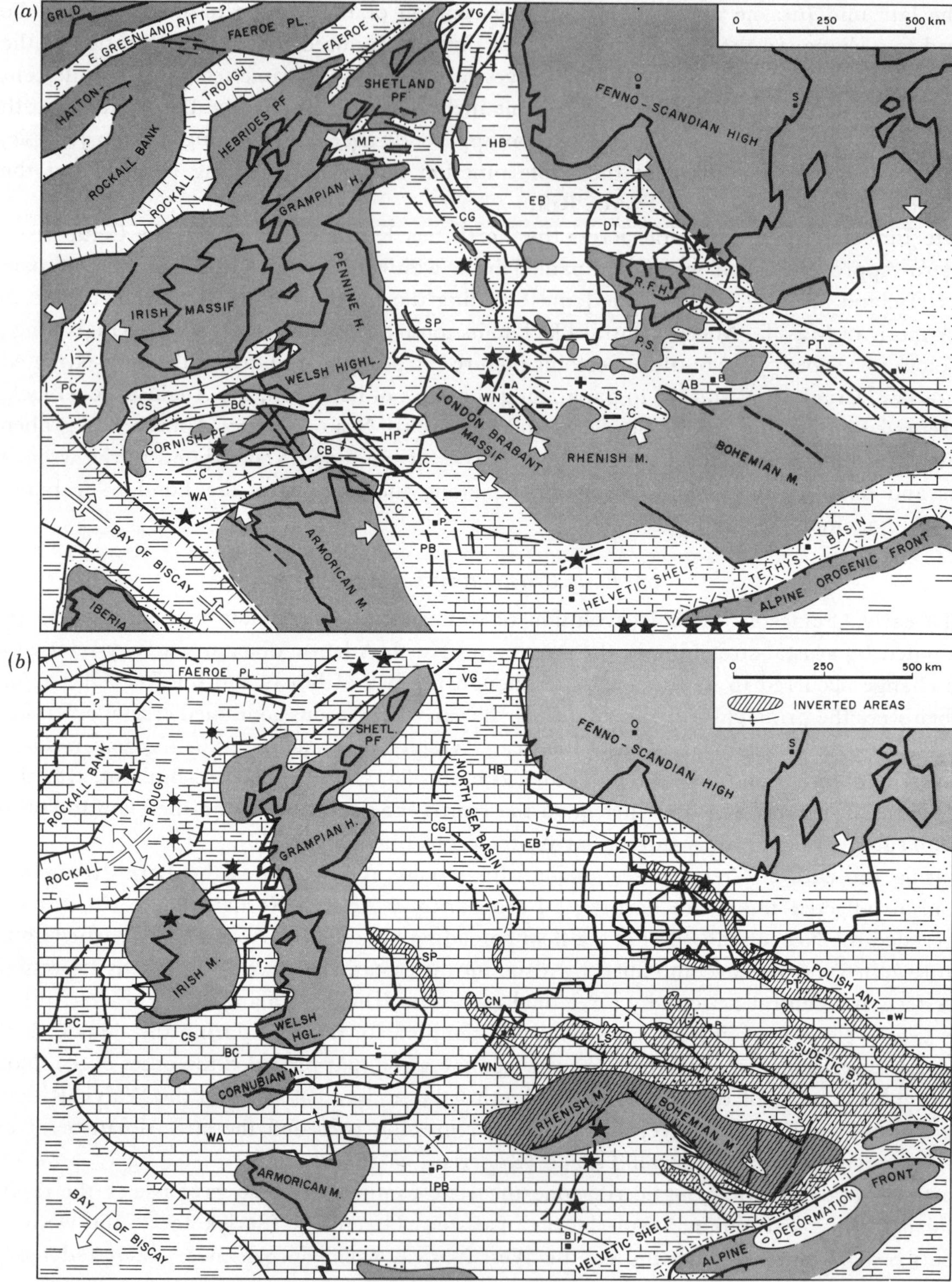

FIGURE 10. (*a*) Early Cretaceous palaeogeography; (*b*) late Cretaceous palaeogeography. For key see figure 14.

against the Late Cimmerian unconformity clearly illustrates that this rifting phase was very short-lived. In the central North Sea, early Cretaceous rift tectonics are less obvious owing to the overprinting effects of the intense diapirism of the Zechstein salt. The Austrian tectonic pulse finds little expression within the Viking and Central Graben, which continued to subside differentially during the Aptian and Albian. Early Cretaceous crustal extension across the Viking and Central Graben amounted to some 10 km but did not induce any major volcanic activity. Contemporaneous crustal stretching across the Moray Firth Rift gave rise to minor wrench movements along the Great Glen Fault.

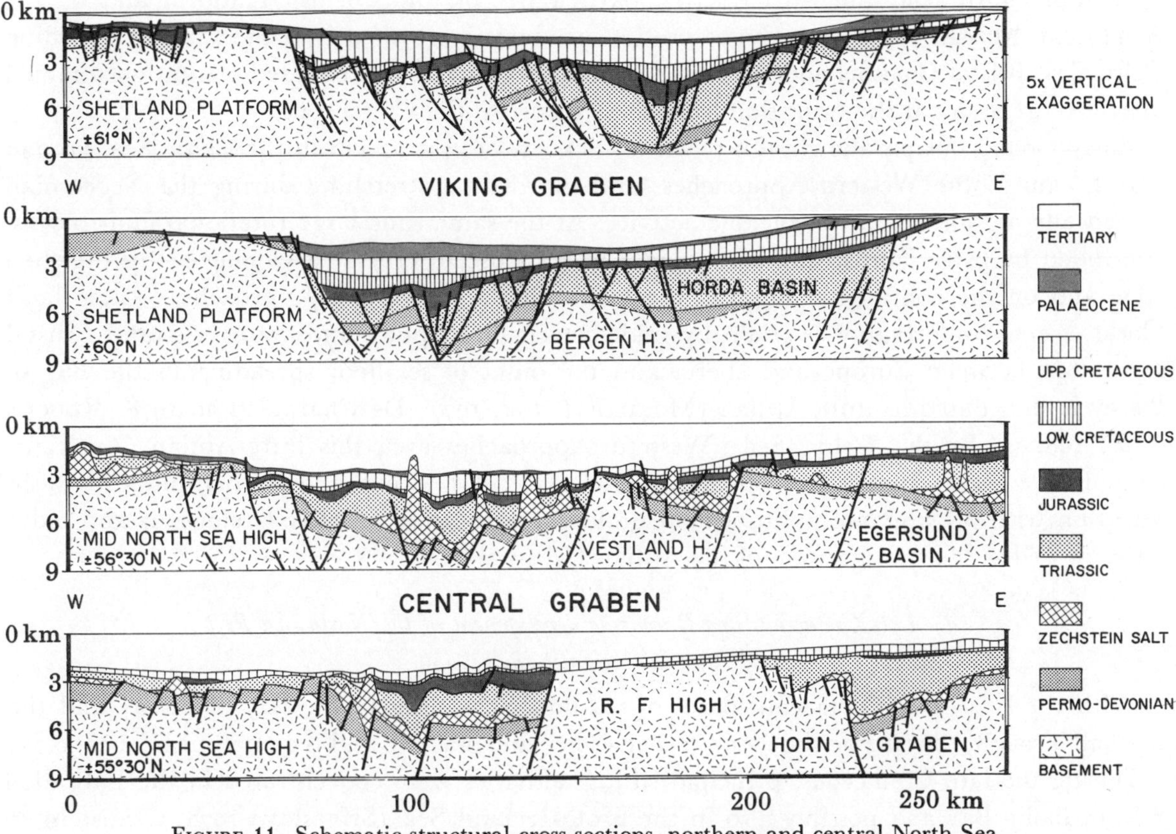

FIGURE 11. Schematic structural cross sections, northern and central North Sea. Approximate location of each section given by latitudes in western corner.

Early Cretaceous crustal stretching across the North Sea Rift was paralleled by the rapid subsidence of the wrench-induced basins flanking the London–Brabant, Rhenish and Bohemian massifs in which early Cretaceous series are developed in the clastic 'Wealden' facies (figure 10). Deep crustal fracturing associated with the Austrian tectonic pulse induced a number of short-lived volcanic centres in the Dutch offshore and the intrusion of batholiths in the Lower Saxony Basin.

In areas bordering the Bay of Biscay Rift the Late Cimmerian rifting and its associated sea-level drop caused the termination of the carbonate-dominated late Jurassic depositional régime and the onset of widespread Wealden-type clastic sedimentation.

Accelerated crustal extension preceding the onset of sea-floor spreading between Iberia and the Grand Banks was accompanied by uplifting and eastward tilting of the Iberian Meseta. The northward continuation of the Newfoundland–Lusitania rift is represented by the

Porcupine Trough, in the southern parts of which a geophysically defined major volcanic centre came into evidence during early Cretaceous time (figures 6 and 10). Crustal extension in the Lusitania–Newfoundland rift induced further left-lateral wrench movements in the Bay of Biscay Rift where they caused the rapid subsidence of the Parentis and Ardour sub-basin in the Aquitaine area and of the Duero Basin in northern Spain. In these relatively small basins early Cretaceous strata attain thicknesses of 4 km and more.

Associated with rift–wrench movements along the Biscay Rift, the Armorican Massif and the Western Shelves became uparched and tilted northeastward. This caused widespread truncation of Jurassic and earlier series, particularly on the Cornish Platform and on the Armorican Massif. From these highs erosion products were shed into the rapidly subsiding Celtic Sea, Bristol Channel and Western Approaches troughs, in which sedimentation resumed after a short break at the transition from the Jurassic to the Cretaceous.

Early Cretaceous paralic clastics reach a thickness of up to 2 km in the Celtic Sea Trough and 1.5 km in the Western Approaches Graben. Crustal stretching during the Neocomian was locally accompanied by volcanic activity. At the same time large rotational fault blocks, controlled by listric faults, subsided rapidly in response to crustal stretching along the shelf edge and on the continental slope of the Celtic Sea and Western Approaches. These fault blocks are unconformably overlain by undeformed late Aptian and younger strata. Crustal separation between Europe and Iberia and the onset of sea-floor spreading in the Bay of Biscay is thus dated as intra-Aptian (Montadert *et al.* 1977; De Charpal *et al.* 1978; Roberts *et al.* 1981*b*). In the Celtic Sea – Western Approaches area this intra-Aptian (Austrian) tectonic event corresponds to a distinct phase of block-faulting and wrench-induced basin inversion whereby regional uplifting and profound truncation of early Cretaceous and older series preceded the transgression of the Aptian Greensands.

(*d*) *Late Cretaceous and Cainozoic development of the North Sea Rift and the Atlantic shelves*

With the late early Cretaceous onset of sea-floor spreading in the Bay of Biscay and the Rockall Trough, the rifts on the western shelves became inactive and regional subsidence set in.

During the late Cretaceous, principal rifting activities were concentrated in the Labrador Sea, in Baffin Bay and possibly also in the Proto-Iceland Sea (Srivastava 1978; Gradstein & Srivastava 1980). At the same time rifting activities in northwestern Europe decreased substantially and regional subsidence of Atlantic shelves, the North Sea and Polish rift commenced, presumably in response to lithospheric cooling and sedimentary loading of the crust. A last tensional pulse affecting the North Sea area, but even more so the Atlantic shelves of Scotland and Ireland, occurred during the late Palaeocene and preceded the onset of sea-floor spreading in areas north of the Charlie Gibbs fracture zone (Kristoffersen 1977; Eldholm & Thiede 1980). This was associated with a significant sea-level drop and a major volcanic outburst that affected a broad belt extending from the Porcupine Trough over a distance of 1800 km to Kristiansand (western Norway) (figure 12).

After crustal separation in the Iceland and Norwegian–Greenland Sea, the evolution of the Atlantic shelves and the North Sea area was characterized by tectonic quiescence and regional subsidence.

In the North Sea late Cretaceous chalks and marls, attaining thicknesses up to 2 km, progressively infilled the topography of the Viking and Central Graben and onlapped against

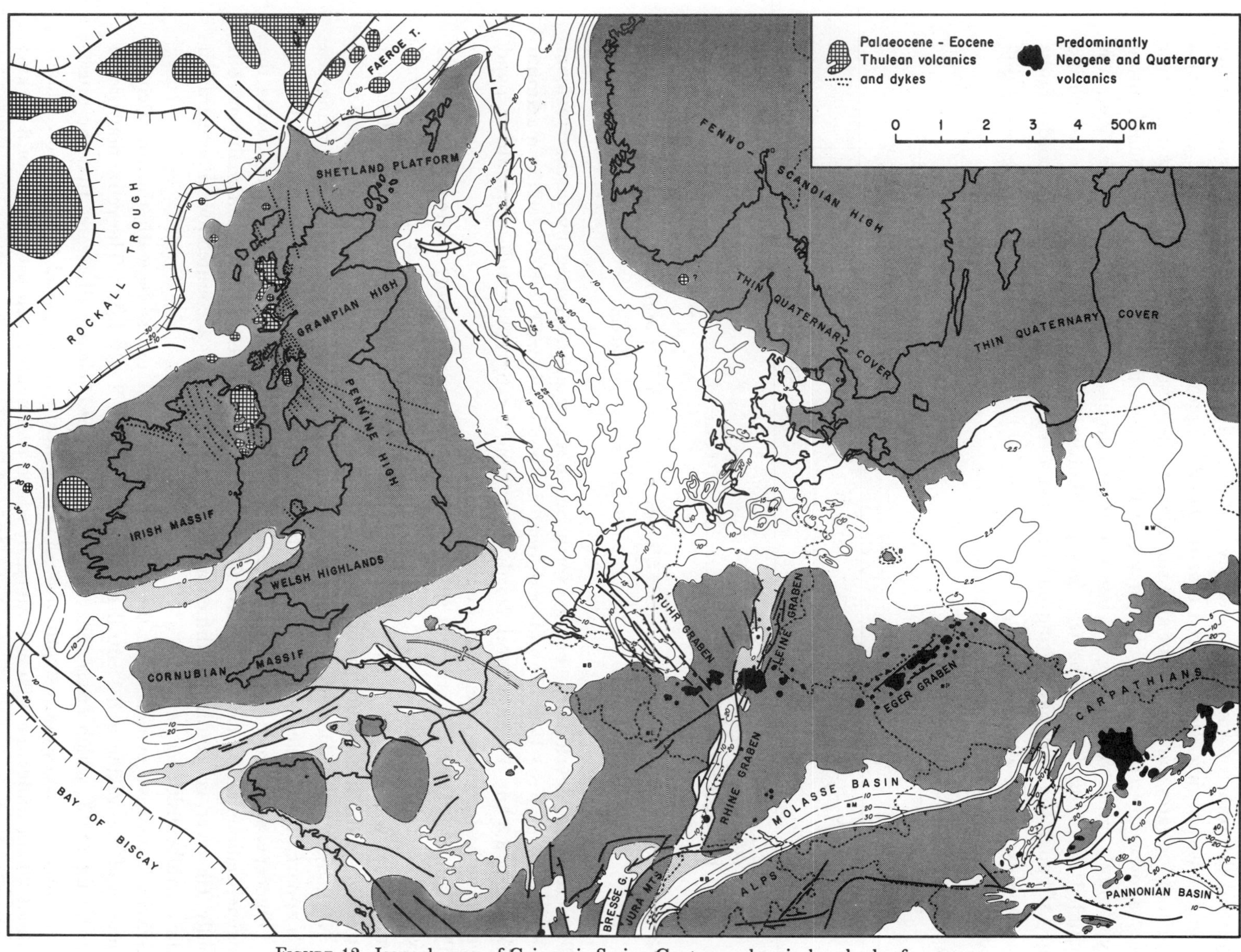

FIGURE 12. Isopach map of Cainozoic Series. Contour values in hundreds of metres.

highs and the graben flanks. This probably had a loading effect on the crust, causing further regional downwarping. It is inferred that in the Viking and Central Graben, late Cretaceous sedimentation rates somewhat exceeded subsidence rates and eustatic sea-level rises; however true shallow-water conditions were not yet established in them. As syndepositional faulting played only a minor role during the late Cretaceous development of the North Sea Rift, it is concluded that crustal stretching was only of subordinate importance during its late rifting stage. Although the Laramide rifting pulse caused a reactivation of some of the border faults of the Viking and Central Graben, it is unlikely that this induced a significant reversal of the lithospheric cooling processes.

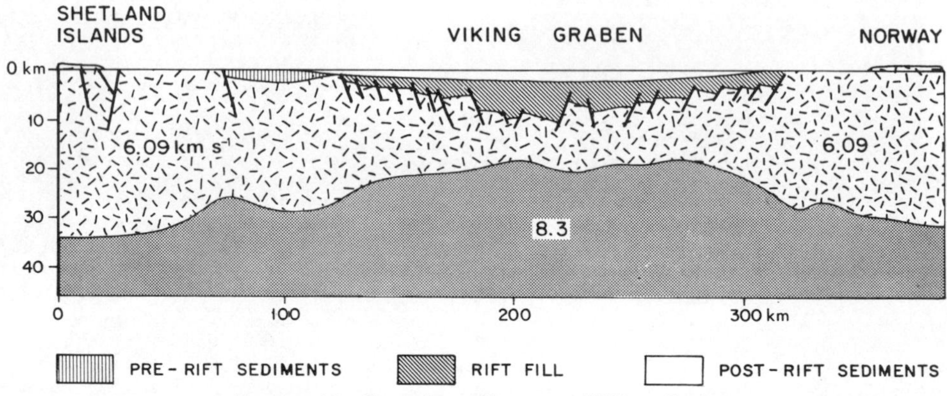

FIGURE 13. Crustal profile across Viking Graben.

The Cainozoic development of the North Sea Basin is characterized by further regional downwarping and the lack of syndepositional faulting. Cainozoic series reach a maximum thickness of 3.5 km in the central North Sea. Time stratigraphic units generally expand towards the basin centre (Ziegler & Louwerens 1979).

The Cainozoic Central North Sea subsidence maximum is probably related to an area of maximum crustal thinning (Donato & Tully 1981) and to the decay of a maximum thermal anomaly that was induced during its Jurassic and early Cretaceous rifting stage. Thermal contraction of the lithosphere probably controlled the Cainozoic post-rifting development of the North Sea Basin. During the Palaeocene and Eocene, subsidence rates and rising sea levels apparently exceeded sedimentation rates and caused the deepening of the basin. During the Oligocene, subsidence and sedimentation rates kept more or less in balance. During the Neogene, sedimentation rates exceeded the tectonic subsidence and the basin shallowed out rapidly. However, concomitant sedimentary loading of the crust accounted for an acceleration of the overall subsidence rates.

Gravity data give evidence for the presence of a mass excess beneath the low density sediments of the Viking and Central grabens (Donato & Tully 1981). The refraction profile given in figure 13, which crosses the northern North Sea, clearly illustrates that the crust–mantle interface rises from a depth of 30–35 km beneath Norway and the Shetland Islands to about 20 km under the Viking Graben in which sediments reach a thickness of 8–10 km; correspondingly, its continental crust is 60–70 % thinner than beneath the Shetland Platform and Fennoscandia. Throughout this refraction profile the upper mantle displays a normal velocity of 8.1–8.3 km s^{-1}.

Based on these data and on a quantitative analysis of the late Cretaceous and Cainozoic subsidence of the North Sea Basin, Sclater & Christie (1980) propose that thinning of the crust under the Viking Graben was accomplished during the Mesozoic by stretching it by a factor 1.8–2.0. This would correspond to a crustal extension by 75–100 km. Multichannel reflexion data show, however, that extension of the sediments by faulting is at the Jurassic level only of the order of 10–15 km. As the base of the Triassic series can only be mapped locally in the Viking Graben, the amount of crustal extension at the base of the syn-rift sediments cannot be readily determined. However, judging by the geometry of the individual fault blocks and the lack of significant divergence of Triassic reflectors, it is very unlikely that the total amount of Mesozoic crustal extension exceeds 20–25 km; this would correspond to a stretching factor of only 1.2 to a maximum of 1.3. Similarly the amount of crustal extension postulated by the stretching model for the Witchground and Buchan Graben (Christie & Sclater 1980; Christie, this symposium) is at variance with the amount of extension observed on reflexion seismic data.

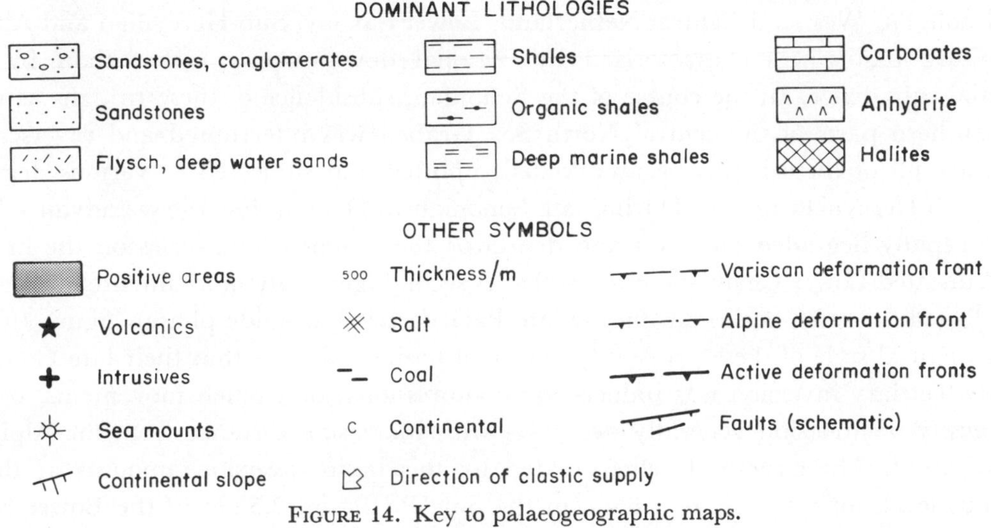

FIGURE 14. Key to palaeogeographic maps.

For the Danish part of the Central Graben and the Horn Graben, reflexion data indicate that at the base Zechstein level the combined amount of crustal stretching is about 10 km, to a maximum of 15 km. This is compatible with the values determined for the Viking Graben. A considerably greater amount of crustal extension for the central North Sea, for which gravity data indicate a wide zone of thinned crust, is, however, difficult to visualize in view of the overall geometry of the North Sea Rift. Yet the deepest parts of the Cainozoic North Sea Basin do coincide with the axis of the Central Graben and the largest gravity anomaly (Donato & Tully 1981).

Results of these preliminary investigations indicate that crustal thinning across the North Sea Graben system was achieved during its Mesozoic rifting stage not only by mechanical stretching, but partly also by thermally induced physico-chemical processes affecting the lower crust. These processes, which are vaguely circumscribed as 'subcrustal erosion', are apparently irreversible. Moreover they appear to have been more effective in the area of the mid-Jurassic central North Sea rift dome than beneath the Viking Graben.

The late Cretaceous and Cainozoic development of the Atlantic shelves of Ireland and Scotland differs from that of the North Sea in that lithospheric cooling processes were temporarily reversed by the Palaeocene – early Eocene Thulean volcanism or only set in after its extinction. Moreover, as the thickness of the crust under the different parts of the Atlantic shelves varies considerably and as sediment availability to these shelves was far from uniform, their Cainozoic subsidence patterns show great variations. For instance a Cainozoic clastic wedge, up to 2 km thick, progrades from the West Shetland shelf into the Faeroe Trough. Also, the Porcupine Trough with its thinned crust subsided rapidly during the Cainozoic; however, owing to a limited clastic supply Neogene subsidence rates outpaced sedimentation rates.

On the other hand, the Hebrides and Northwest Irish shelves, which are characterized by only limited Mesozoic extension, were largely bypassed by Cainozoic clastics.

7. Late Cretaceous and Cainozoic foreland compression

The early late Cretaceous development of the Polish–Danish Trough and of the wrench-induced Sole Pit, West and Central Netherland, Lower Saxony, Sub-Hercynian and Altmark-Brandenburg basins was characterized by regional downwarping rather than by sharp differential subsidence. In the course of the Senonian, subsidence of these troughs as well as of the southern parts of the central North Sea Graben was interrupted and reversed. The sedimentary fill of these basins became folded, uplifted and subjected to various degrees of erosion (Sub-Hercynian phase). During late Senonian to Danian time the sea advanced again over the rapidly degraded fold axes and deposited the Upper Chalk series on the in places deeply truncated Lower Chalk and older series. A second, generally more intense, deformation phase affected the same basins during the late Palaeocene (Laramide phase) (figure 10).

The structural style of these previously tensional basins indicates that their late Cretaceous and early Tertiary inversion was induced by compressional or wrench movements, or both. The intensity of inversion generally decreases with increasing distance from the Alpine deformation front. The structural relief created by this basin inversion amounts at the late Cretaceous level, for instance, to 1.5 km in the Sole Pit Basin, 2.5 km in the Lower Saxony Basin and 2–3 km in the Polish Trough.

Concomitant with the inversion of these basins, the Rhenish and the Bohemian Massifs became dissected and in part uplifted along a set of wrench and steep reverse faults whereby the reactivation of Stephanian–Autunian faults probably played a major role.

The subcrop pattern of the Mesozoic series below the Tertiary sediments of the Carpathian foredeep and the Eastern Molasse Basin illustrates that also these areas were strongly deformed during late Cretaceous and early Tertiary time and that major structural elements extend well under the Carpathian and Alpine nappes.

Late Cretaceous and early Tertiary compressional and wrench-induced deformations are restricted to the Carpathian and northern Alpine foreland and fade out to the west (figure 10). For instance the Channel and Hampshire Basins were only very mildly deformed during the Palaeocene compressive phase. Their main inversion, similar to the Western Approaches, Celtic Sea and Bristol Channel Troughs, is dated as Oligocene and Miocene (figure 12).

The onset of the late Cretaceous and early Tertiary compressional foreland deformation coincides with major orogenic events in the Carpathians and the Eastern Alps and probably reflects the collision between the Central Carpathian Massif and the Italo-Dinarid promontory with the fractured European craton. Compressional stresses exerted on the latter gave rise to the tilting and uplifting of the Variscan massifs and the inversion of major Mesozoic troughs at distances up to 1000 km to the north of the present Alpine deformation front. The crustal configuration of the strongly inverted Polish Trough (Guterch *et al.* 1976) shows that the thinned crust of this previously tensional basin was mechanically thickened during its inversion to the degree that it reached thermal and isostatic equilibrium.

From this it is inferred that compressional foreland deformation causing the inversion of rifts and the uplifting of basement blocks along reverse and wrench faults involves the entire crust and therefore requires a decoupling between it and the upper mantle.

In the northern Alpine and Carpathian foreland no further compressional deformations occurred during the post-Palaeocene orogenic phases of the Alpine fold belt. With the onset of the Eocene, elastic down-flexing of the foreland crust under the load of the advancing nappes (onset of underplating?) gave rise to the development of the Carpathian and Alpine foredeeps. This was accompanied by intense basin-parallel block-faulting of the down-bent foreland crust. It is reasoned that the lack of further compressional deformation of the northern Alpine and Carpathian foreland reflects a decoupling between the European craton and the overriding thrust masses at the Carpathian and the East Alpine A-subduction zones. On the other hand, strong coupling at the Central and West Alpine A-subduction zones induced the late Oligocene and Miocene inversion of the rifts of the Celtic Sea – Western Approaches shelf.

8. Cainozoic rifts of the Alpine foreland

During the late Eocene and early Oligocene, the volcanic rift system of the Limagne, Bresse, Rhine, Ruhr, Leine and Eger grabens came into evidence as entirely new tectonic features on the face of western and central Europe (figure 12). The Limagne and Rhine–Leine grabens and possibly also the Eger Graben are superimposed on late Carboniferous – early Permian fractures. The Ruhr Graben is partly superimposed on the inverted West Netherlands Basin.

This complex graben system remained intermittently active until the present. Its evolution is contemporaneous with the Alpine late orogenic phases, and parallels the Neogene collapse of the Mediterranean and the Pannonian Basin and the inversion of the Channel area and of the Western Approaches, Bristol Channel and Celtic Sea troughs.

Volcanic activity set in the Rhine Graben simultaneously with its initial subsidence during the late Eocene, in the Massif Central in the course of the Oligocene and in the Eger Graben during the late Oligocene. These volcanics display an alkaline, mafic–felsic bimodal composition.

Upwarping of the Rhenish Massif at the triple junction between the Rhine and Leine and Ruhr Graben began during the early Miocene and was followed by the mid and late Miocene extrusion of extensive trap basalts. Uplifting of the Vosges – Black Forest rift dome at the southern end of the Rhine Graben started somewhat earlier but apparently also postdates the onset of graben subsidence. The Massif Central became updomed during the late Miocene and Pliocene.

The different parts of the Bresse–Rhine–Ruhr–Leine rift subsided only intermittently during the Neogene and Quaternary; this reflects changes in the regional stress pattern. During the Miocene, the Bresse Graben and the southern parts of the Rhine Graben ceased to subside differentially (Rat 1978; Illies 1978); this coincides with the main inversion phase of the Mesozoic troughs in the Celtic Sea – Western Approaches and Channel areas. During the Pliocene, however, the Bresse Graben once more began to subside, while the evolution of the Rhine Graben was controlled by sinistral shear motions. During the Pliocene the southern margins of the Rhine Graben and the eastern rim of the Bresse Graben were overridden by the frontal thrust sheets of the Jura Mountains. At present the Rhine Graben is being deformed by sinistral strike-slip movements. In contrast, the Ruhr Graben is currently subsiding actively in response to tensional stresses (Illies & Greiner 1978; Ahorner 1978). This graben feathers out to the northwest in the Dutch onshore areas. Similarly, the currently inactive Leine Graben dies out to the north at the edge of the North German lowlands.

Cainozoic series reach maximum thicknesses of some 3 km in the Rhine Graben and 2 km in the Bresse Graben.

Geotectonic processes that governed the development of the Cainozoic rifts of western and central Europe are the subject of much speculation.

Regional crustal extension apparently preceded the development of discrete hot-spots that are marked by upper mantle anomalies as observed under the Massif Central, the Vosges – Black Forest dome and possibly the Rhenish Massif (Perrier & Ruegg 1973; Edel *et al.* 1975; Giese 1978). For example, crustal extension acrosst he Rhine Graben is estimated to amount to some 5 km and is considerably larger than could result solely from the uparching of a rift dome in response to the emplacement of an asthenolith (Laubscher 1970). Yet the early onset of volcanic activity distinguishes the Cainozoic rifts from the Mesozoic grabens of Europe.

Viewed in a broader framework, the Cainozoic rifts in the Alpine foreland may be considered as forming part of the Neogene Mediterranean collapse system that presumably developed in response to a reorientation of the convergence direction between Africa and Europe (Laubscher 1974; Biju-Duval *et al.* 1977). It is conceivable that this reorientation of the regional stress field was also responsible for the interplay of tensional and compressional tectonics in the Alpine foreland.

Alternatively, the volcanic rifts of western and central Europe could be regarded as forming the northern extension of the East African – Red Sea and Libyan–Tyrrhenian rift systems, which became active during the early Oligocene and the Miocene (Richter-Bernburg 1974; Illies 1974).

A third model, which is based on the Himalayan setting, visualizes the Rhine–Bresse Graben as so-called 'impactogens' (Sengör *et al.* 1978; see also Molnar & Tapponier 1975). This model is, however, not fully compatible with the evolution of the Rhine and Bresse Grabens, in which periods of active subsidence appear to alternate with the main compressional events in the Alpine fold belt.

None of the above geotectonic processes is able to explain fully the development of the Cainozoic rift in the Alpine foreland. Thus some combination of these processes may have to be envisaged whereby one or the other processes was variously, in time and space, the dominant mechanism controlling basin subsidence and the initiation of volcanic activity.

Overall, the Neogene and Quaternary development of the Alpine domain and its foreland may be interpreted as heralding the break-up of the current continent assembly.

9. Conclusions

In Europe, rifts have developed through time in a number of megatectonic settings.

Rifting, leading to the break-up of continents and the opening of major oceanic basins, for instance on the scale of the Atlantic, is the most important process of graben formation. In this setting the rifting stage preceding crustal separation can be relatively short, as with the central Atlantic (40 Ma) or can be long-lived, as in the Norwegian Greenland Sea (270 Ma).

Wrench faulting associated with the translation of either rift systems or orogenic belts, dependent on fault geometries, can induce the rapid subsidence of pull-apart or tension gash basins, or both. Pull-apart grabens at the termination of wrench faults, as for instance the Oslo Rift, can be highly volcanic. Subsidence of tension-gash basins (e.g. Lower Saxony and Parentis Basin) may or may not be associated with short-lived igneous activity.

Back-arc rifting is thought to have played a significant role in the development of the Variscan geosynclinal system. Mechanisms governing back-arc rifting and sea-floor spreading are still the subject of much debate. In the same way that compressional and extensional phases appear to alternate in an island arc setting, back-arc rifts are prone to destruction, particularly under the impact of continent–arc collision.

Wrench faulting and rifting of the Himalayan type (Molnar & Tapponier 1975), which are thought to be the consequence of continent–continent collision, do not appear to play a major role in Europe. However, intraplate compressional deformation involving the inversion of rifts and the uplifting of basement blocks was of major importance during the Variscan and Alpine diastrophism.

Elastic down-flexing of the foreland plate under the weight of the advancing nappes controlled the development of the Variscan and Alpine foredeep basins. The subsidence of such basins can be associated with tensional deformation of the upper crust whereby synthetic faults predominate, as for instance in the Molasse Basin.

Regardless of which megatectonic setting rifts formed under, their subsidence is governed by lithospheric thinning and sedimentary loading of the crust. During rifting phases the continental crust is stretched and thinned in response to regional extension that is presumably induced by convection currents in the asthenosphere, which drive the lithospheric plates apart. Necking of the continental crust is attained at shallow levels by listric normal faulting and at deeper levels by ductile flow. However, the amount of crustal thinning observed, for instance across the North Sea rift and the Celtic Sea – Western Approaches shelf (Avedik *et al.*, this symposium), cannot be fully accounted for by the crustal stretching model. It is therefore inferred that during periods of active crustal extension, thermally induced physico-chemical processes affect the lower crust and contributes significantly to crustal thinning; these processes appear to be irreversible.

Many rifts are totally non-volcanic or show only a very low level of volcanism. Other rifts display a high level of volcanic activity right from the onset of crustal extension or become temporarily volcanic after an initial stage of non-volcanic subsidence. A high level of volcanism is generally associated with a high rate of crustal extension and the uplifting of a wide-radius rift dome that is centred over the axis of the rift. Uplifting of a rift dome can cause a substantial reversal in the subsidence pattern of a rift and generally induces extensive erosion mainly over the rift flanks; on a restricted scale this can contribute to crustal thinning. Crustal extension resulting from the uplifting of a rift dome is, however, rather small and amounts for instance to some 200 m for a dome with a width of 200 km and a height of 3 to 4 km (Artemjev & Artyushkov 1971).

Upwarping of such a rift dome is caused by the emplacement of a low-density, low-velocity upper mantle anomaly at the crust–mantle interface; the development of such an anomaly is presumably caused during periods of intensified crustal extension by failure of the lithosphere or by thermally induced lower mantle diapirism, or both (Osmaston 1971, 1973, 1977; Bott 1976).

The physical properties of such upper mantle anomalies, also referred to as 'asthenoliths', 'rift pillows' or 'rift cushions', can be explained by melting processes. Magmas intruding the crust from an asthenolith may eventually reach the surface where they display at first the chemical characteristics of a typical alkaline initial rift volcanism. In time and with persisting crustal extension this volcanism will gradually change over to a tholeitic basalt type. This gives rise to the bimodal felsic–mafic alkaline suites that characterize the volcanism of continental rifts (Martin & Piwinski 1972).

Triple junctions, where crustal extension is most intense, are likely places for an early manifestation of rift volcanism (e.g. Rhine Graben and North Sea Rift). However, volcanic activity is not necessarily restricted to the actual rift zone; yet the occurrence of lateral volcanic centres appears to be limited to the area of the rift dome and, with this, to the confines of the correspondent asthenolith.

Asthenoliths are thermally unstable upper mantle anomalies and are resorbed in the latter upon cooling once crustal extension has fallen below a certain rate or has ceased altogether. Correspondingly, beneath thermally stabilized palaeorifts the upper mantle displays a normal density of about 3.3 and velocities of 8.1–8.3 km s^{-1}.

Following crustal separation the subsidence of the newly formed 'passive continental margins' is controlled by lithospheric cooling and sedimentary loading of the crust. Similar mechanisms govern the subsidence of rifts that have become inactive (Sleep 1973, 1976; McKenzie 1978; Sclater & Tapscott 1979; Watts & Steckler 1981; Jarvis & McKenzie 1980; Royden *et al.* 1980).

The amount of subsidence caused by lithospheric contraction is controlled by the magnitude of the thermal anomaly that was induced during the lithospheric stretching – crustal separation stage. It is inferred that maximum thermal anomalies are induced during continental separation and that thermal anomalies induced by rifting are comparatively smaller. Correspondingly, the post-separation development of a passive continental margin probably reflects the decay of a maximum thermal anomaly, while the subsidence of inactive rifts is probably governed by the decay of smaller thermal anomalies. On the other hand, the post-rifting development of highly volcanic rifts that were underlain by an asthenolith during their rifting phase is probably associated with the decay of a larger thermal anomaly than of non-volcanic rifts. In the post-rifting subsidence pattern of volcanic rifts the resorption of the asthenolith into the mantle by cooling processes presumably plays a significant role. Moreover, erosion of upper crustal rocks over the crest of a rift dome can contribute to crustal thinning and thus will be reflected in the subsidence of an inactive rift.

A further aspect that has to be considered in quantitative subsidence models of rifts is the fact that crustal extension and concomitant sub-crustal thinning can take place intermittently over very long periods. In long-lived rifts, thermal anomalies induced by, and associated with, crustal extension can already start to decay during periods of decreased rate of crustal stretching. Thus the thermal anomaly associated with a rift may not be at its maximum when crustal extension terminates altogether and the respective rift becomes inactive. Similarly, late rifting pulses may interrupt and even reverse the normal lithospheric cooling processes. This is illustrated by the evolution of, for instance, the North Sea Rift in which the maximum thermal anomaly was presumably induced during the early Bajocian, while significant crustal extension persisted into early Cretaceous times. On the other hand, the evolution of the West Shetland–Faeroe Rift is characterized by a late Jurassic thermal surge and a second late Palaeocene – early Eocene one.

References (Ziegler)

Ahorner, L. 1978 In *Alps, Apennines, Hellenides* (Inter-Union Commission on Geodynamics, Scientific Report no. 38) (ed. H. Closs, D. Roeder & K. Schmidt), pp. 17–19.

Artemjev, M. E. & Artyushkov, E. V. 1971 *J. geophys. Res.* **76**, 1197–1211.

Arthaud, F. & Matte, P. 1977 *Bull. geol. Soc. Am.* **88**, 1305–1320.

Autran, A. & Cogné, J. 1980 In *Géologie de l'Europe du Précambrien aux bassins sédimentaires post-Hercyniens* (*Mém. B.R.G.M.* no. 108) (ed. J. Cogné & M. Slansky), pp. 90–111.

Biju-Duval, B., Dercourt, J. & Le Pichon, X. 1977 In *Structural History of the Mediterranean Basin* (ed. B. Biju-Duval & L. Montadert), pp. 143–164. Paris: Editions Technip.

Bott, M. H. P. 1976 *Tectonophysics* **36**, 1–4.

Brookfield, M. E. 1978 *Geol. Rdsch.* **67**, 110–149.

de Charpal, O., Guennoc, P., Montadert, L. & Roberts, D. G. 1978 *Nature, Lond.* **275**, 706–711.

Christie, P. A. & Sclater, J. C. 1980 *Nature, Lond.* **283**, 729–732.

Coisy, P. & Nicolas, A. 1978 *Nature, Lond.* **274**, 429–432.

Dewey, J. F., Pitmann, W. G., Ryan, W. B. F. & Bonnin, J. 1973 *Bull. geol. Soc. Am.* **84**, 3137–3180.

Dixon, J. E., Fitton, J. G. & Frost, R. T. C. 1981 In *Petroleum geology of the continental shelf of northwest Europe* (ed. L. V. Illing & G. D. Hobson), pp. 121–137. London: Institute of Petroleum.

Donato, J. A. & Tully, M. C. 1981 In *Petroleum geology of the continental shelf of northwest Europe* (ed. L. V. Illing & G. D. Hobson), pp. 65–75. London: Institute of Petroleum.

Eckhardt, F. J. 1979 *Geol. Jb.* D **35**, 1–84.

Edel, J. B., Fuchs, K., Gelbke, C. & Prodehl, C. 1975 *J. Geophys.* **41**, 333–356.

Eldholm, O. & Thiede, J. 1980 *Palaeogeogr. Palaeoclim. Palaeoecol.* **30**, 243–259.

Faerseth, R. B., McIntyre, R. M. & Naterstad, J. 1976 *Lithos* **9**, 331–345.

Francis, E. H. 1978*a* In *Tectonics and geophysics of continental rifts* (ed. I. B. Ramberg & E.-R. Neumann) (N.A.T.O. Advanced Study Institute, Series C), pp. 133–148. Dordrecht: D. Reidel.

Francis, E. H. 1978*b* In *Crustal evolution in NW Britain and adjacent regions* (*Geol. J. Spec. Iss.* no. 10) (ed. D. R. Bowes & B. E. Leake), pp. 279–296.

George, T. N. 1958 *Proc. Yorks. geol. Soc.* **31**, 1–227.

George, T. N., Johnson, G. A. L., Mitchell, M., Prentice, J. E., Ramsbottom, W. H. C., Sevastopulo, G. D. & Wilson, R. B. 1976 *Geol. Soc. Lond. spec. Rep.* no. 7. (87 pages.)

Giese, P. 1978 *Z. dt. geol. Ges.* **129**, 513–520.

Gradstein, F. M. & Srivastava, S. P. 1980 *Palaeogeogr. Palaeoclim. Palaeoecol.* **30**, 261–295.

Guterch, A., Kowalski, T., Materzok, R. & Toporkiewicz, S. 1976 *Publs Inst. geophys. Pol. Acad. Sci.* A **2**, 15–23.

Hallam, A. 1978 *Palaeogeogr. Palaeoclim. Palaeoecol.* **23**, 1–32.

Haller, J. 1971 *Geology of the East Greenland Caledonides.* (413 pages.) London: Interscience.

Harland, W. B. 1978 In *IGCP Project no. 27, Caledonian–Appalachian Orogen of the North Atlantic Region* (*Geol. Surv. Can. Pap.* no. 78–13), pp. 3–11.

Harland, W. B., Cutbill, J. S., Friend, P. F., Gobbett, D. J., Holliday, D. W., Maton, P. I., Parker, J. R. & Wallis, R. H. 1974 *Spitsbergen. Skr. norsk Polarinst.* no. 161. (72 pages.)

House, M. R., Richardson, J. B., Chaloner, W. G., Allen, J. R. L., Holland, C. H. & Westoll, T. S. 1977 *Geol Soc. Lond. spec. Rep.* no. 8. (110 pages.)

Howie, R. D. & Barss, M. S. 1975 *Geol. Surv. Can. Pap.* no. 74–30, pp. 35–50.

Hsui, A. T. & Toksöz, M. N. 1981 *Tectonophysics* **74**, 89–98.

Illies, J. H. 1978 In *Tectonics and geophysics of continental rifts* (ed. I. B. Ramberg & E.-R. Neumann) (N.A.T.O. Advanced Study Institute, Series C), pp. 63–72. Dordrecht: D. Reidel.

Illies, J. H. & Greiner, G. 1978 *Bull. geol. Soc. Am.* **89**, 770–782.

Jarvis, G. T. & McKenzie, D. P. 1980 *Earth planet. Sci. Lett.* **48**, 42–52.

Jorgensen, F. & Navrestad, T. 1981 In *Petroleum geology of the continental shelf of northwest Europe* (ed. L. V. Illing. & G. D. Hobson), pp. 407–413. London: Institute of Petroleum.

Kent, D. V. & Opdyke, N. D. 1979 *Earth planet. Sci. Lett.* **44**, 365–372.

Kramer, W. 1977 *Z. geol. Wiss., Berl.* **5**, 7–20.

Kristoffersen, Y. 1977 In *N.P.F. Mesozoic Northern North Sea Symp., Oslo*, 17–18 October, 1977 (*Norw. Petrol. Soc. Publ.* no. MNNSS/5), pp. 1–25.
Laubscher, H. P. 1970 In *Graben problems* (*Int. Up. Mantle Project, Sci. Rep.* no. 27) (ed. J. H. Illies & S. Müller), pp. 79–87.
Laubscher, H. P. 1974 *Scienze* **72**, 48–59.
Laubscher, H. P. & Bernoulli, D. 1977 In *Structural history of the Mediterranean basins* (ed. B. Biju-Duval & L. Montadert), pp. 129–132. Paris: Éditions Technip.
Leeder, M. R. 1974 *Scott. J. Geol.* **10**, 283–296.
Martin, R. F. & Piwinski, A. J. 1972 *J. geophys. Res.* **77**, 4966–4975.
McKenzie, D. P. 1978 *Earth Planet. Sci. Lett.* **40**, 25–32.
Molnar, P. & Tapponnier, P. 1975 *Science, N.Y.* **189**, 419–426.
Montadert, L., Roberts, D. G., Auffre, G. A., Bock, W., Peuble, P. A. du, Hailwood, E. A., Harrison, W., Kagani, H., Lumsden, D. N., Müller, C., Schnitke, D., Thompson, R. W., Thompson, T. L. & Timotev, P. P. 1977 *Nature, Lond.* **268**, 305–309.
Morris, W. A. 1976 *Can. J. Earth Sci.* **13**, 1236–1243.
Oftedahl, C. 1968 *Geol. Rdsch.* **57**, 203–218.
Osmaston, M. F. 1971 *Tectonophysics* **11**, 385–405.
Osmaston, M. F. 1973 In *Implications of continental drift to earth sciences* (ed. D. H. Tarling & S. K. Runcorn), vol. 2, pt 6, pp. 649–674. London: Academic Press.
Osmaston, M. F. 1977 In *Developments in petroleum geology* (ed. G. D. Hobson), vol. 1, pp. 1–52. London: Applied Science Publishers.
Paproth, E. & Teichmüller, R. 1961 In *C.r. 4e géol. Carbonif., Heerlen*, 1958, vol. 2, pp. 471–490.
Perrier, G. & Ruegg, J. C. 1973 *Annls Géophys.* **29**, 435–502.
Ramberg, I. B. 1976 *Norg. geol. Unders.* **325**, 1–194.
Rast, N. & Grant, R. 1977 In *La chaîne varisque d'Europe moyenne et occidentale* (*Colloques int. C.N.R.S. Rennes*, no. 243), pp. 583–586.
Rat, P. 1978 *Cent. Rech. somm. Soc. Géol. Fr.* (5), pp. 231–234.
Richter-Bernburg, G. 1974 In *Approaches to taphrogenesis* (*Inter-Union Commission on Geodynamics, Sci. Rep.* no. 8) (ed. J. H. Illies & K. Fuchs), pp. 13–43. Stuttgart: Schweizerbart.
Roberts, D. G., Masson, D. G. & Miles, P. R. 1981*a* *Earth planet. Sci. Lett.* (In the press.)
Roberts, D. G., Masson, D. G. & Montadert, L. 1981*b* In *Petroleum geology of the continental shelf of northwest Europe* (ed. L. V. Illing & G. D. Hobson), pp. 455–473. London: Institute of Petroleum.
Ronnevik, H. C. 1981 In *Petroleum geology of the continental shelf of northwest Europe* (ed. L. V. Illing & G. D. Hobson), pp. 395–406. London: Institute of Petroleum.
Royden, L., Sclater, J. G. & Herzen, R. P. 1980 *Bull. Am. Ass. Petrol. Geol.* **64**, 173–187.
Sawkins, F. J. & Burke, K. 1980 *Geol. Rdsch.* **69**, 349–360.
Schenk, P. E. 1978 In *IGCP Project 27, Caledonian–Appalachian Orogen of the North Atlantic Region* (*Geol. Surv. Canada, Paper* no. 78–13), pp. 111–136.
Sclater, J. G. & Christie, P. A. F. 1980 *J. geophys. Res.* **85**, 3711–3739.
Sclater, J. G. & Tapscott, C. 1979 *Scient. Am.* **240**, 120–133.
Sengör, A. M. C., Burke, K. & Dewey, J. F. 1978 *Am. J. Sci.* **278**, 24–40.
Sleep, N. H. 1973 In *Implications of continental drift to earth sciences* (ed. D. H. Tarling & S. K. Runcorn), vol. 2, pt 6, pp. 685–692. London: Academic Press.
Sleep, N. H. 1976 *Tectonophysics* **36**, 45–56.
Speight, J. M. & Mitchell, J. G. 1978 *J. geol. Soc. Lond.* **136**, 3–11.
Srivastava, S. P. 1978 *Geophys. Jl R. astr. Soc.* **52**, 313–357.
Surlyk, F. 1975 In *N.P.F. Jurassic Northern North Sea Symposium, Stavanger*, 28–30 September, pp. 7-1–7-31.
Teichmüller, R. 1973 *Z. dt. geol. Ges.* **124**, 149–165.
Uyeda, S. 1981 *Geol. Rdsch.* **70**. (In the press.)
Vail, P. R., Mitchum, R. M., Todd, R. G., Widmier, J. M., Thompson, S., Sangree, J. B., Bubb, J. N. & Hatfield, W. G. 1977 *Seismic stratigraphy application to hydrocarbon exploration* (*Am. Ass. Petrol. Geol. Mem.* no. 26) (ed. C. E. Payton), pp. 49–212.
Vischer, A. 1943 *Meddr Grønland* **133**, 1–195.
Voigt, W. A. 1962 *Z. dt. geol. Ges.* **114**, 378–418.
Voo, R. van der & Channel, J. E. T. 1980 *Rev. Geophys. Space Phys.* **18**, 455–481.
Voo, R. van der & Scotese, C. 1981 *Geology* **18**. (In the press.)
Watts, A. B. & Steckler, M. S. 1981 In *Implication of deep drilling results in the Atlantic Ocean* (Maurice Ewing Series, vol. 2). Washington, D.C.: American Geophysical Union.
Ziegler, P. A. 1978*a* In *Tectonics and geophysics of continental rifts* (ed. I. B. Ramberg & E.-R. Neumann) (N.A.T.O. Advanced Study Institute, Series C), pp. 249–277. Dordrecht: D. Reidel.
Ziegler, P. A. 1978*b* *Geologie Mijnb.* **57**, 589–626.
Ziegler, P. A. 1980 In *Géologie de l'Europe du Precambrian aux bassins sédimentaires post-Hercyniens* (*Mém. B.R.G.M.* no. 108) (ed. J. Cogné & M. Slansky), pp. 249–280.

Ziegler, P. A. 1981 In *Petroleum geology of the continental shelf of northwest Europe* (ed. L. V. Illing & G. D. Hobson), pp. 3–39. London: Institute of Petroleum.

Ziegler, P. A. 1982 *Geological atlas of western and central Europe*. Shell Publication. (In the press.)

Ziegler, P. A. & Louwerens, C. J. 1979 In *The Quaternary history of the North Sea* (*Acta Univ. Upps. Symp. Univ. Upps. Annum Quingentesimum Celebrantis*, vol. 2) (ed. E. Oele, R. T. E. Schüttenhelm & A. J. Wiggers), pp. 7–22.

Znosko, J. 1979 *Z. angew. Geol.* **25**, 447–458.

Discussion

Sir Peter Kent, F.R.S. Ziegler has suggested that the Bajocian–Bathonian mid North Sea dome, extending southwards from the Forties region across the mid North Sea High into the southern North Sea, had an amplitude of 2 km. This amplitude is difficult to reconcile with the sections across the oil fields, which are very well constrained. The missing sediments have a thickness of only 60 m at the most, and there is no evidence for the formation of a huge bulk of detritus of this age, which would have been produced by levelling such a structure.

P. A. Ziegler. Evidence in support of the postulated Bajocian–Bathonian rift dome in the central North Sea is provided by (*a*) the facies development of the Liassic series and their subcrop pattern against the base Bajocian (Mid-Cimmerian) unconformity (Ziegler 1980, 1981) (*b*) the distribution and facies development of the Bajocian and Bathonian series (see figure 8), and (*c*) in the central North Sea the progressive onlap of the transgressive Callovian, Oxfordian and Kimmeridgian series against the Mid-Cimmerian unconformity.

The importance of the Mid-Cimmerian unconformity is illustrated by the third cross section from the top on figure 11. This cross section extends from 56° N 1° E on the mid North Sea High via the Auk and Ekofisk fields to 57° N 6° E in the Egersund Basin. Throughout this cross section the base of the Jurassic series corresponds to the Mid-Cimmerian unconformity. While Bajocian and Bathonian series occur in the Egersund Basin, Callovian, Oxfordian and locally even Kimmeridgian strata transgress in the Central Graben, on the Vestland High and the Mid North Sea High over Triassic and older sediments. The angular relation between the series underlaying and overlaying the mid-Cimmerian unconformity is clearly evident on reflexion seismic data and indicates that the Central North Sea rift dome near its culmination along the Central Graben had a structural relief of some 2–3 km (see particularly the western part of Egersund Basin and Vestland High).

From this dome vast amounts of sediments were removed, particularly during Bajocian and Bathonian times. While paralic and deltaic series accumulated along the margins of this high, finer clastics were dispersed in the Arctic – North Atlantic rift, in the southern North Sea and in northern Germany.

No attempt has been made to establish a material balance between the volumes of sediments removed from the postulated rift dome and the volumes of sediments deposited in adjacent areas during times of its emergence. Uncertainties of such volumetric estimates are likely to be very large and presumably will not provide an overriding constraint to the palaeogeographic and seismostratigraphic observations that support the existence of an early mid-Jurassic rift dome in the central North Sea.

Phil. Trans. R. Soc. Lond. A **305**, 145–148 (1982) [145]
Printed in Great Britain

Late Palaeozoic basins of the southern U.S. continental interior [abstract]

J. F. Dewey†[1] and W. C. Pitman III‡
† *Department of Geological Sciences, State University of New York at Albany, Albany, New York* 12222, *U.S.A.*
‡ *Lamont-Doherty Geological Observatory of Columbia University, Palisades, New York* 10964, *U.S.A.*

The timing and geometry of late Palaeozoic inhomogenous deformation in the southern U.S. continental interior from the Appalachian foreland of Kentucky–Tennessee through Oklahoma and Texas to the Ancestral Rockies of Colorado–Wyoming–Utah can be definitively linked with a discrete sequence of collisional events in the Appalachian–Ouachita–Marathon orogenic belt to the south (Permian coordinates). A progressive staged collisional sequence beginning in late Mississippian times in the southern Appalachians and culminating in early Permian times in the Marathons led to a progressive deformation of the adjacent craton in a wide swath dominated by right-lateral shear (e.g. Rough Creek Fault zone) by which a Mexican promontory of North America was displaced towards a Pacific 'free face'. While the deformation in the Appalachian–Marathon belt was dominated by vertical plane strain leading to crustal thickening, the associated continental interior deformation can be averaged as a horizontal plane strain at or near sea level with localized deep extensional (Delaware) and flexural (Arkoma) basins and compressional uplifts (Amarillo–Wichita) giving local source areas.

Several distinct basin types can be recognized on the basis of position, geometry and subsidence history.

1. *Foreland basins immediately adjacent to the Appalachian–Marathon orogenic belt.* These, such as the Appalachian, Black Warrior and Arkoma Basins are asymmetric with cratonward onlap, clearly have a flexural origin with peripheral linear bulges (Cincinatti Arch, Llano Uplift), and resulted from the migrating thrust load of adjacent orogens. These basins show gradually increasing, then gradually decreasing, subsidence rates.

2. *Localized deep extensional basins linking offstepping strike-slip faults.* These basins (e.g. Magdalene Basin in the Canadian Maritimes and the Delaware Basin) are bounded by both steep and listrict faults, have stratigraphic sequences up to 10 km thick, and sometimes show basin margin thrusting of 'flower-structure' type. Their subsidence history follows a two-phase pattern predicted by the McKenzie (1978) model with a very rapid initial phase (crustal stretching) followed by a phase of exponential decline (thermal subsidence). For basins of this type, values for β of between 1.6 and 2.0 are usual, but in the Canadian Maritimes these are as high as 6.0 (Magdalene Basin) suggesting extremely attenuated crust, with local stretching of *ca.* 6 km. That marginal faults of such basins are growth faults is indicated by the eastern margin of the Delaware Basin where over 3 km of Wolfcampian black shale decreases to zero across the bounding fault system. These extensional basins have a distinctive cross-sectional shape that may be termed the 'Steer's head' or 'Texas Longhorn' geometry resulting from a rapid initial phase of fault-controlled subsidence in a stretching lithosphere with local isostatic compensation entirely within the basin. This is followed by a thermal

[1] Present address: Department of Geological Sciences, University of Durham, South Road, Durham DH1 3LE, U.K.

phase with a thickening and strengthening of the lithosphere inducing compensation over an ever-widening area with basin margin onlap to produce the 'horns'. The complete absence of late Palaeozoic volcanism in extensional basins from Oklahoma to Colorado suggests that McKenzie's (1978) crustal attenuation model is more appropriate than the Royden *et al.* (1980) multiple asthenosphere dyke model, at least for these basins.

3. *Complex intracratonic basins that appear to show neither a simple flexural or clear extensional origin.* The Anadarko basin, for example, is asymmetric, bounded by a steep, possibly listric, thrust system to the south and a gradually thinning transition to the Kansas Shelf to the north. The likely origin of the Anadarko Basin is transpressional, that is the Amarillo–Wichita uplift overthrusts the southern margin of the basin with a component of left-lateral strike-slip. The localization of the Anadarko Basin seems to be due to the pre-existence of an early Palaeozoic rift, the South Oklahoma Aulacogen. The Val Verde Basin appears to be dominantly a flexural foreland basin adjacent to the Marathons but with a strong right-lateral component. The Paradox Basin, with its adjacent Uncompaghre Uplift, shows a bimodal subsidence history that suggests a stretching origin. Later thrusting on a flower structure above a right-lateral transform may explain the Pennsylvanian clastic flood from the Uncompaghre Uplift.

Hydrocarbon maturation calculations for the Delaware and Anadarko Basins indicate early maturation and long-distance migration. Most of the Midland Basin oil and gas probably originated 200 km away in the Delaware Basin and the Hugoton Field gas migrated some 300 km from the deep Anadarko Basin.

The McKenzie (1978) stretching model has a number of fundamental implications for the evolution of sedimentary basins, their deformation and their crustal structure, as follows.

(i) All transitions exist from $\beta = 1$ (no stretching) to $\beta = \infty$ (continental crust attenuated to zero) giving total conformable sediment thicknesses ranging from zero to 16 km and crustal thicknesses from 35 to 10 km (where basins are filled to sea level). Such transitions are seen both in intracratonic and rifted continental margin basins.

(ii) Consequently, all transitions exist from slightly stretched, through severely attenuated continental crust (with disorganized separation by ultramafic–gabbroic diapiric intrusion) and eventual virtual exposure of subcontinental mantle where $\beta \rightarrow \infty$ (para-oceans) to organized plate accretion and the growth of new, rather than the stretching of existing, surface area.

(iii) Because the post-stretching thermal re-equilibration of the lithosphere and the restoration of pre-stretching lithosphere thickness has a time constant of over 150 Na, the stretched lithosphere forms a zone of structural weakness for long periods that nucleates inversion zones or zones of subsequent compressional deformation. Obviously, the sooner after stretching that compression occurs, the easier it will be to invert a sedimentary basin because the lithosphere is still thin. As a basin ages and the lithosphere thickens, it becomes progressively more difficult to deform the sub-basin lithosphere. An analogous situation appears to be the obduction of oceanic lithosphere. Most obducted ophiolite nappes are emplaced as thin (10 km) flakes with hot subjacent aureoles less than 30 Ma after their generation by plate accretion, suggesting the increasing difficulty of detachment in older, thicker, colder lithosphere. The Laramide basement deformation of Colorado and Wyoming coincides remarkably with the Ancestral Rockies deformation, which latter deformation during the Pennsylvanian may have preconditioned the lithosphere for late Cretaceous – early Tertiary deformations.

(iv) Stretching of the continental crust yields sections in which mantle, high-grade metamorphic lower crustal rocks and low-grade upper crustal rocks and supracrustal stratigraphic sequences become more closely vertically spaced, hence enabling subsequent thrust flaking to emplace thin nappe sheets containing granulitic and mantle rocks over low grade sequences, a common relation in many orogenic belts.

References (Dewey & Pitman)

McKenzie, D. P. 1978 Some remarks on the development of sedimentary basins. *Earth planet. Sci. Lett.* **40**, 25–32.

Royden, L., Sclater, J. G. & Von Herzen, R. P. 1980 Continental margin subsidence and heat flow: important parameters in formation of petroleum hydrocarbons. *Bull. Am. Ass. Petrol. Geol.* **64**, 173–187.

Discussion

A. S. MacKenzie. Do the authors have any direct measurements of present-day levels of organic metamorphism and if so, how do they fit their predicted values? Does the geometry of the traps require the organic matter to be oil or gas prone by the end of the Permian?

J. F. Dewey. All the ideas that I talked about have come from following the logical consequences of the stretching model. We have not examined whether measurements of maturation within these basins agree in detail with the model. The nature of the traps does not require the oil and gas to migrate earlier than the Permian. We have examined the maturity obtained from vitrinite reflectance of the Middle and Upper Permian rocks where we did not believe the maturity was sufficient to produce the hydrocarbons found nearby, and we found that the maturity was indeed insufficient to generate the oil and gas, which must therefore have migrated, in some cases over considerable distances.

M. F. Ridd. When calculating the maturation of these rocks did the authors take into account the effect on the maturation of the Permo-Carboniferous rocks of perhaps 3 km of Cretaceous and Tertiary rocks that are found on the flanks rather than the crest of the Hugoton Field?

J. F. Dewey. Much of the oil and gas in the Hugoton gas field could have come from Palaeozoic sediments in the Denver basin, which is a foreland basin. The section of Cretaceous and younger sediments above the Hugoton field is not sufficient to account for the observed level of maturation, though we did take them into account. The thick Cretaceous and younger section lies further to the west.

A. W. Bally. There is a great variety of source rocks, both lipid and humic, in west Texas and Oklahoma. For example, Pennsylvanian source rocks tend to be humic, but some of the Permian source rocks in the Delaware Basin are lipid. These source beds require different timespans to mature for oil and also for gas generation. Hood *et al.* (*Bull. Am. Ass. Petrol. Geol.* **59**, 986 (1975)) used a hole in the Anadarko Basin to calibrate their maturation scale, where there is no evidence for an upper Palaeozoic thermal event. In the Arkoma Basin, however, there appears to be a late Palaeozoic thermal event that is discordantly superposed on folded beds.

Before a stretching model is applied to the great variety of basins discussed in this paper, we should need a considerable amount of published information about their deep structure. Such information is generally difficult to obtain, and many of the seismic reflexion data are not very good. In the Anadarko Basin there appear to be some minor normal faults, on the north side, but if I remember correctly, these are synthetic faults, not antithetic (rotational) faults. Similar normal faults occur in the Arkoma Basin, and in the St Lawrence Lowlands of Quebec. These normal faults are formed during the same time brackets as the folding in the adjacent folded belts. I doubt that in the basins mentioned there is good evidence for an extensional event. The principal reason why the stretching model is unlikely to apply to this area is that the deformation that is most clearly observable was produced by compression. This is particularly well displayed in the Ardmore Basin, where there are many tight folds. Indications of reverse faults also occur on the Central Basin Platform. Thus there is hardly any evidence of late Palaeozoic rifting in west Texas and Oklahoma, and until such evidence is published, I am doubtful whether the stretching model can be applied there.

J. F. Dewey. I agree that the Arkoma and Ardmore basins were not produced by extension. The Arkoma basin is a foreland basin, with evidence of crustal thickening. The Ardmore basin is more complicated, and fits neither the loading nor the stretching model.

The faulting in the Delaware and Valverde basins of west Texas is not listric, since there is no evidence of rotation. The dip on the faults may even increase with depth.

D. H. Matthews, F.R.S. Present-day thermal gradients in the North Sea are thought to be profoundly affected by water flowing in aquifers. Might this effect disrupt the modelling of the maturation history of the sediments?

D. L. Turcotte. The first interpretation of the Cocorp line across the Anadarko basin suggests that it was produced entirely by compression.

Phil. Trans. R. Soc. Lond. A **305**, 149–168 (1982) [149]
Printed in Great Britain

Subsidence history of the Middle East Zagros Basin, Permian to Recent

By W. J. Koop† and R. Stoneley‡

† *Aramco Overseas Company, Canterbury House, Sydenham Road, Croydon, Surrey CR9 2LS, U.K.*

‡ *Petroleum Geology Section, Department of Geology, Royal School of Mines, Imperial College of Science and Technology, London SW7 2BP, U.K.*

[Overlay]

The Zagros Basin is broadly defined as the palaeodepositional wedge of sediments along the present belt of the Zagros Mountains. A series of ten regional isopach maps trace the development of a portion of this basin from Permian to Recent. From the Permian to the middle Cretaceous the area occupied a position along the stable northeastern Atlantic-type shelf margin of the Afro-Arabian continent, bounded by a rift zone that evolved into a southern Tethys ocean. Late Cretaceous to Recent subsidence patterns are influenced by plate margin tectonics, obduction, and eventual continental collision along the Zagros Suture as this ocean closed. The late Alpine Zagros folding and faulting took place from the Miocene onwards.

1. Introduction

In this paper regional time–isopach and facies maps are used to demonstrate the post-Carboniferous evolution of the Zagros Basin. These maps represent a compilation of surface and subsurface studies made in Iran in 1977, supplemented by published regional information from adjacent areas.

Figure 1 shows the present regional tectonic and geographical setting of the area under consideration. The NE limit of the area is shown on this map as the Zagros Suture, which is interpreted as a former plate boundary separating the Zagros fold belt from the complex Hamadan–Sirjan zone of Central Iran. The NW–SE-striking Zagros mountain chain exposes the entire sequence down to the Palaeozoic. The narrow interior fold belt plunges SW along the 'mountain-front flexure' into a folded 'foothills' belt, which is the setting for the major oilfields of Iran and Iraq in the Dezful and Kirkuk embayments, respectively. Further SW the stable Arabian platform rises gently to the Arabian Shield and the Huqf–Dhofar of south Oman.

The chrono-stratigraphic intervals covered by the ten isopach–facies maps are shown on the table of formations (figure 2), which includes most of the more commonly used formation names of the region.

All thicknesses shown on the maps are preserved thicknesses and no attempt has been made to estimate how much might have been removed by erosion at unconformities, nor to allow for differential compaction. It has also not been possible to make a reliable palinspastic restoration of the localities to their original relative geographic positions in the Zagros orogenic belt. Because of repeated transgressions and regressions, the facies boundaries drawn on the maps should be regarded as average positions for the period.

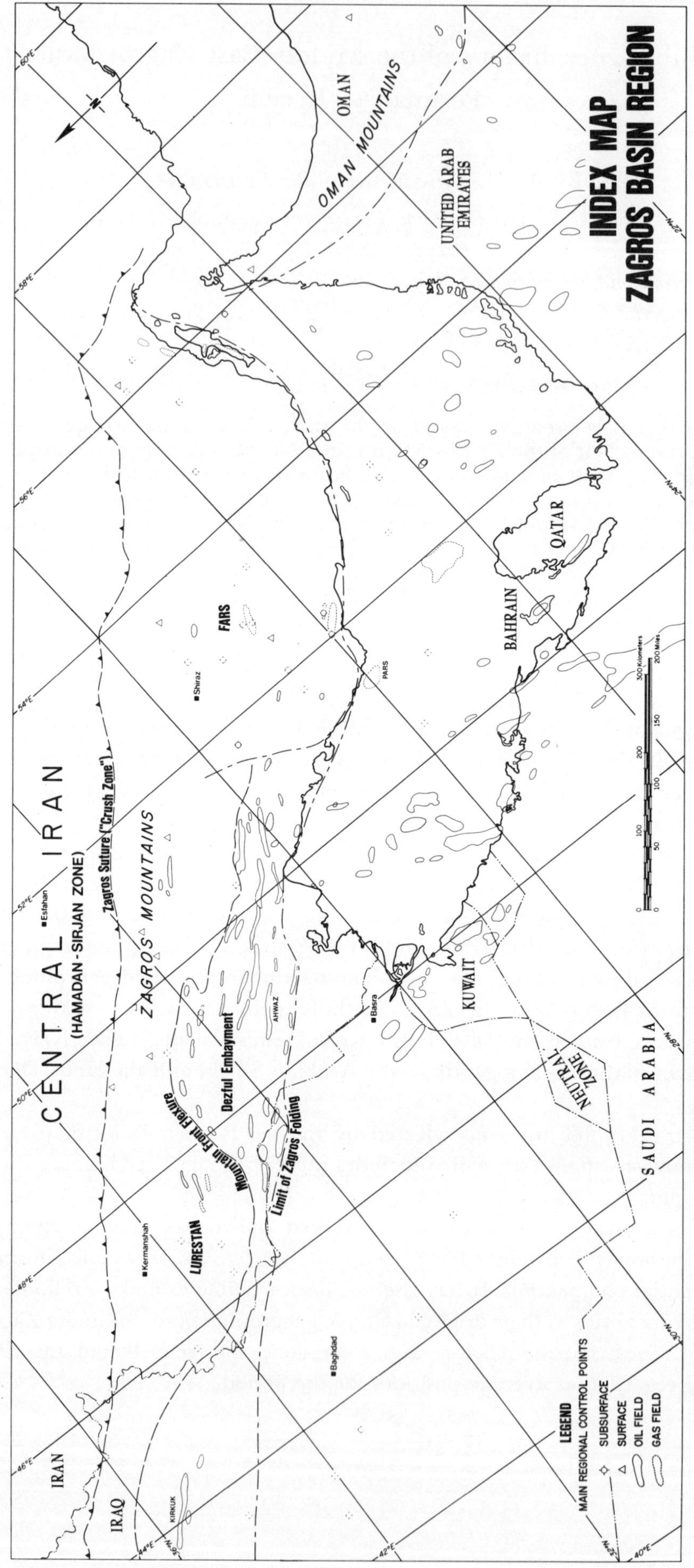

FIGURE 1. Index map. Regional geographic and tectonic setting of the Zagros Basin.

The isopach and facies maps are based on the work of a large number of geologists who have worked in Iran and adjacent countries, and on the results of many years of oil company drilling. Publications by Dunnington (1958, 1967), James & Wynd (1965), Glennie *et al.* (1973), Clarke (1975), Hassan *et al.* (1975), Szabo & Kheradpir (1978), Setudehnia (1978) and others have been particularly useful in extending the regional patterns into adjacent areas. Some of the maps were originally prepared by G. Orbell, who participated extensively in the study. Our dependence on him and other workers, and on the various exploration organizations, is gratefully acknowledged. R. Green drafted the illustrations.

AGE	ISOPACH MAPS	S.ARABIA-KUWAIT	IRAQ	IRAN LURESTAN-FARS	S.GULF	FIG. NO.
0–10	POST L.MIOCENE	HOFUF-FARS GP	FARS GP	BAKHTYARI; AGHA JARI/MISHAN/GACHSARAN		12
10–37.5	OLIGOC.-L.MIOCENE	GHAR	JERIBE/DHIBAN/EUPHRATES; KIRKUK GP	ASMARI		11
37.5–65	PALAEOC.-EOCENE	DAMMAM; RUS; UMM ER RADHUMA	JADDALA; AALIJI; PILA SPI; AVANAH; KHURMALA; SINJAR; REDBEDS; KOLOSH	SHAHBAZAN; KASHKAN; TALEH ZANG; AMIRAN; PABDEH; JAHRUM; SACHUN	DAMMAM; RUS; U.E.R.	10
65–88	U.CRETACEOUS (Coniacian-Maastrichtian)	ARUMA GP	PILSENER; SHIRANISH; TANJERO	AMIRAN; GURPI; ILAM; TARBUR; SURGAH; LAFFAN	SIMSIMA; FIQA-SHARGI; HALUL	9
88–107	M.CRETACEOUS (Albian-Turonian)	WASIA GP; MISHRIF/RUMAILA; AHMADI/MAUDDUD; BURGAN-NAHR UMR	KOMETAN; JAWAN; QAMCHUQA; BALAMBO	GARAU; SARVAK; KAZHDUMI	MISHRIF/SHILAIF; MAUDDUD; NAHR UMR	8
107–141	L.CRETACEOUS (Berriasian-Aptian)	SHU'AIBA; BIYADH; BUWAIB; YAMAMA; SULAIY; ZUBAIR; RATAWI; MINAGISH; MAKHUL	GHIA GARA; QAMCHUQA; SARMORD; GARAGU; ZANGURA; BALAMBO	GARAU; DARIYAN; GADVAN; FAHLIYAN	SHU'AIBA; HAWAR; KHARAIB; YAMAMA; SULAIY; THAMAMA	7
141–160	U.JURASSIC	HITH/ARAB; JUBAILA/HANIFA; TUWAIQ MTN.; GOTNIA; NAJMAH	BARSARIN; NAOKELEKAN	GOTNIA; NAJMAH; HITH; SURMEH	HITH/QATAR; DARB/DIYAB	6
160–195	L.&M. JURASSIC	DHRUMA; MARRAT; U.MINJUR	SARGELU; ALAN/MUS; ADAIYAJI/BUTMAH	(MAND MBR); SURMEH; NEYRIZ	ARAEJ; IZHARA; MARRAT	5
195–230	TRIASSIC	JILH; SUDAIR	BALUTI; KURRA CHINE; GELI KHANA; BEDUH/MIRGA MIR	DASHTAK; KANGAN	GULAILAH; SUWEI; KHAIL; SUDAIR	4
230–280	PERMIAN	KHUFF	CHIA ZAIRI	DALAN; FARAGHAN	KHUFF	3
MaBP						

FIGURE 2. Simplified time–stratigraphic table of formation names, Zagros Basin and adjacent areas.

2. PRE-PERMIAN PATTERNS

Regional stratigraphic comparisons and palaeogeographic patterns suggest that during late Precambrian (Infra-Cambrian) and Palaeozoic time the entire area on both sides of the later Zagros Suture was an extension of the Afro-Arabian continental shelf and was the site of widespread platform, largely marine, sedimentation flanking a continental block (Gondwanaland) to the SW. Distinct intrashelf salt basins lay, during the Infra-Cambrian, on both sides of the Zagros Suture (Stöcklin 1974; Murris 1980). These 'Hormuz' salt layers play an important halokinetic role in subsequent basin evolution, both in growth and piercement structures in the shelf areas, as well as forming a detachment horizon during the Neogene Zagros folding and thrusting. This initial Infra-Cambrian carbonate–evaporite deposition was followed by a predominantly clastic régime that prevailed until the Permian.

The distribution of Cambrian to Carboniferous sediments is affected by several intervening regional unconformities and in particular by the late Carboniferous Hercynian unconformity, which cuts down to the Precambrian basement in some areas. In the Zagros Basin the sparse control points show that the mid(?)-Permian carbonates, with a thin basal sandstone, rest on a truncated Silurian to Infra-Cambrian surface (Szabo & Kheradpir 1978). The pattern of the pre-Permian subcrop shows a probable series of highs on both sides of the Zagros Suture (e.g. Zagros High, figure 3). This pattern is interpreted as suggesting thermal expansion of the lithosphere along an incipient rift within the Iranian portion of the Afro-Arabian continent along the later Zagros Suture.

3. Zagros Basin evolution

The Zagros Basin provides an excellent example of all the stages of evolution of a basin, from a passive continental shelf to a rift and drift phase and, finally, through various stages of deformation associated with plate collision. Figures 3–12 show the shifting axes of depocentres and the facies patterns resulting from these plate interactions. The post-Carboniferous subsidence patterns illustrated by the ten maps fall logically into three phases.

(*a*) *Rifted continental shelf phase: Permian and Triassic*

Traces of late Carboniferous to early Permian glacial deposits are reported from Oman and Saudi Arabia, but in the Zagros Mountains the first sedimentary rocks deposited on the Hercynian unconformity surface were thin basal sands overlain by fossiliferous Middle Permian limestone. During the middle and late Permian a facies pattern of inner shelf carbonates and evaporites bounded to the NE by partly reefal organic carbonates became established (figure 3). The earlier Zagros High (pre-Permian arching) remained positive, resulting in a local source of redbed terrigenous clastics. Thiele *et al.* (1968) described a clastic–basic volcanic Permian sequence at Ab-e Barik immediately NE of the Zagros Suture, which supports the suggestion of early rifting along this line. Other areas of adjacent Central Iran show a very thick open marine Permian carbonate section. The oceanic realm of the main Tethys may have lain much further to the NE at this time, in the vicinity of the present Alborz Range (Stöcklin 1974).

The pattern of inner shelf evaporitic facies rimmed by massive shelf carbonates in the High Zagros was repeated in the Triassic (figure 4). Evaporites became the dominant facies across the entire Gulf region. In contrast to the Permian, the entire High Zagros area was now affected by thinning due to uplift and pre-Jurassic truncation. This continued arching along the line of the Zagros Suture has not resulted in any recognized source of terrigenous clastics to the NE. A strong late Triassic sand source was developed at the Arabian Shield margin to the SW. A basin-wide late Triassic to early Jurassic unconformity extending across Central Iran is recognized as marking the onset of drift separation of the Arabian plate and the Central Iranian plate(s) along the present Zagros Suture, so that the post-late Triassic history of the area NE of the Zagros Suture was totally different from that of the Zagros Basin. The new ocean was referred to as the Neo-Tethys by Stöcklin (1974).

(*b*) *Passive miogeoclinal phase: Jurassic to end Middle Cretaceous*

After drift separation along the Zagros Suture, a significant change of isopach and facies patterns is seen on the Lower–Middle Jurassic map (figure 5). The Fars Platform appears as a positive area of thin carbonate deposition. Local domal uplifts on this platform are attributed to Infra-Cambrian salt movement (see below). A complementary negative element, the deeper water silled evaporite–carbonate–shale basin of the Basrah–Lurestan area, was initiated during this same interval. High energy carbonate grainstones on the Surmeh–Dhruma shelf extend

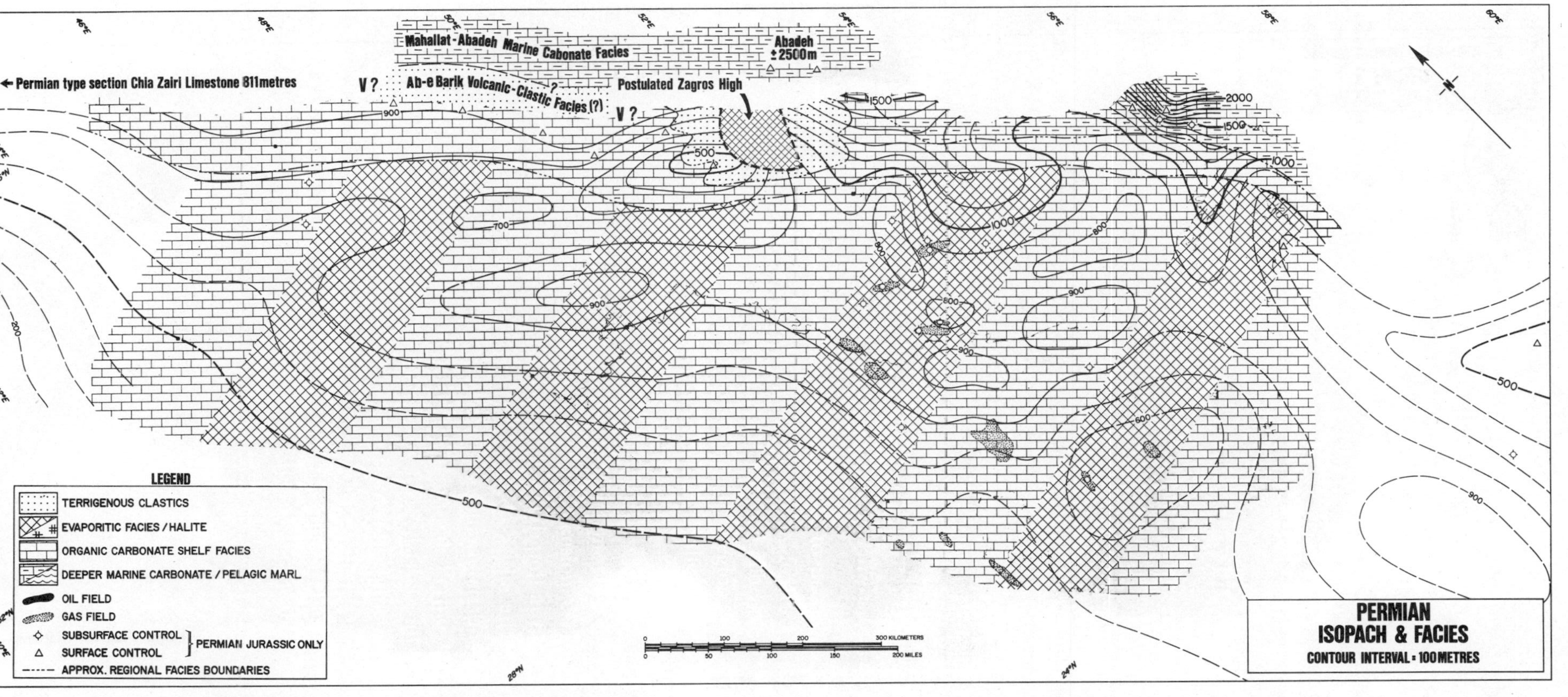

FIGURE 3. Permian isopach and facies distribution map. Facies boundaries on this and subsequent maps are extremely generalized and emphasize selected facies that are considered to be significant indicators of basin evolution. The legend applies to figures 3–12.

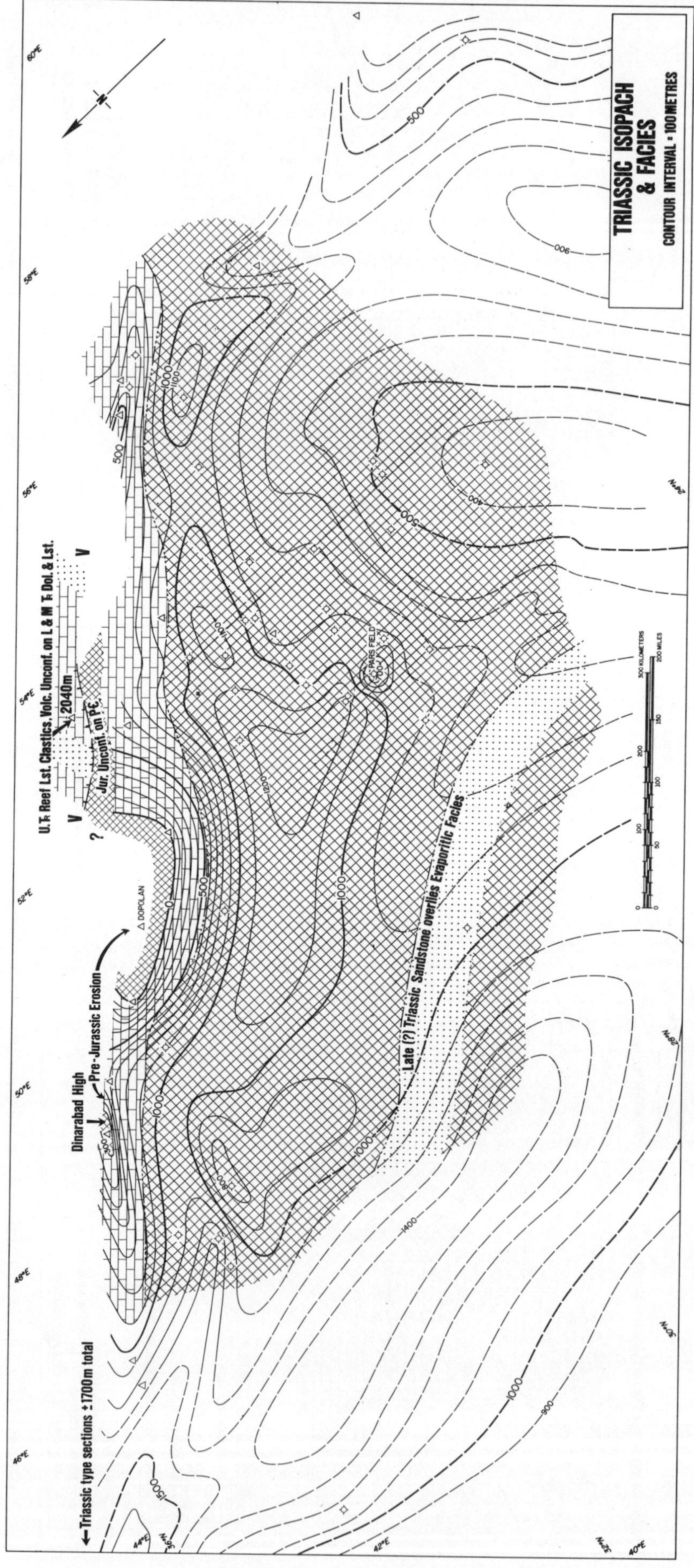

FIGURE 4. Triassic isopach and facies distribution map.

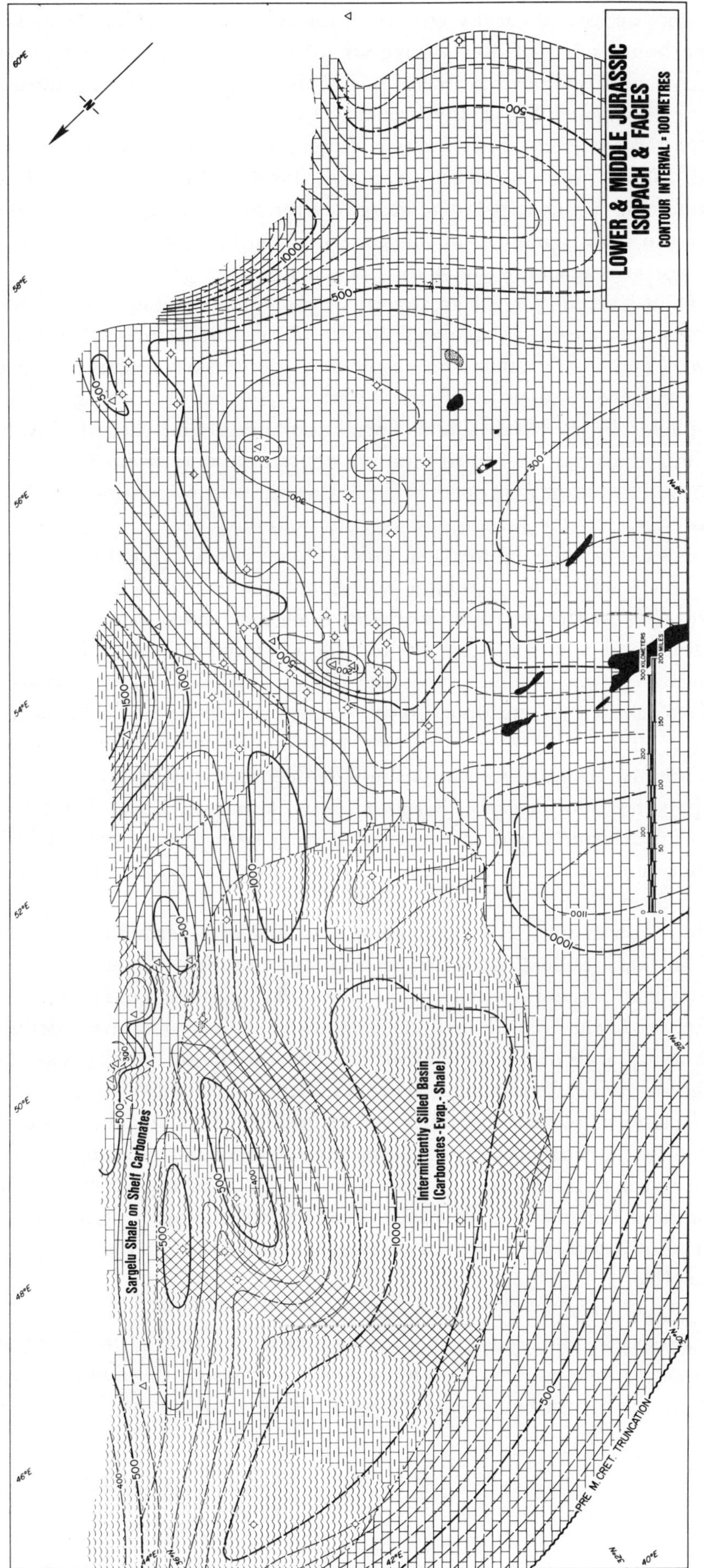

FIGURE 5. Lower and Middle Jurassic isopach and facies distribution map.

as sheets over large areas and locally constitute oil and gas reservoirs. A thickening section of outer shelf carbonates NE of Shiraz suggests proximity to the continental slope, but in general the inner Zagros remained a zone of thin sediment preservation separating the Zagros Basin from the newly opened Neo-Tethys ocean.

The overall patterns of the Upper Jurassic (figure 6) remained very similar to those of the Lower and Middle Jurassic. Basin tectonics, facies differentiation and arid climate conditions became more pronounced. The margin of the Surmeh carbonate platform rimming the Basrah–Lurestan basin remained in the same position but the basin became increasingly restricted with portions reaching the halite evaporite stage (Kuwait), while others were apparently sediment-starved (Dezful area) with a pronounced basal unconformity or depositional hiatus. The southern Gulf area was the site of intrashelf marls alternating with high-energy grainstones and finally anhydrite as the shelf depressions became infilled and more restricted. This sequence resulted in the ideal source–reservoir–seal juxtaposition responsible for the giant oilfields in the Arab–Qatar Formation underlying the Hith Anhydrite. The NE margin of the Zagros Basin received minimal deposition in Iraq–Lurestan, while the shelf margins of interior Fars were depressed and became the site of continental slope tintinnid-bearing carbonate facies at the Jurassic–Cretaceous boundary.

During the early Cretaceous (figure 7) subsidence patterns continued to reflect the Jurassic basin elements, but uplift of the Arabian Shield margin provided an abundant source of terrigenous clastics; evaporitic conditions ceased. Shelf carbonates persisted in the Fars and southern Gulf areas and now constitute hydrocarbon reservoirs. Along the southern margin of the former Basrah (Gotnia) Basin, a Neocomian oolitic facies and the overlying Barremian–Aptian sandstones resulted in multiple Lower Cretaceous reservoirs. The Basrah–Lurestan Basin was largely infilled and euxinic pelagic conditions retreated to a more distal shelf position in Lurestan–N. Iraq, the Garau Basin. The depressed shelf margin with open marine tintinnid facies persisted into the early Cretaceous in N. Fars and Oman. The early Cretaceous terminated with an extremely widespread Aptian carbonate transgression, which blanketed most of the earlier deposits. In the southern Gulf area local intrashelf basins were flanked by rudistid reefs that form the Shuaiba oil reservoirs.

The basin-wide Aptian marine transgression was abruptly terminated by a major pre-Albian regression that resulted in some erosion and another major clastic régime – the Burgan–Safaniya delta system – which pushed the carbonate–shale facies of the Dezful–Lurestan area further to the NE. The Albian Kazhdumi shales, which were deposited beyond the limits of the delta sands on the Aptian surface, are the major source rocks of the Iranian oilfilelds, and the overlying Cenomanian–Turonian limestones constitute important reservoir rocks. The Fars Platform remained a fairly stable positive element. The Middle Cretaceous isopachs (figure 8) are strongly affected by pre-late Cretaceous tectonism and erosion along conspicuous N–S linear trends of the northern Gulf as well as by more localized salt pillowing in the Fars – southern Gulf region. This tectonism marked the onset of collision phenomena along the outer continental shelf margin of the Arabian plate as the Neo-Tethys oceanic seaway began to close.

(*c*) *Collision phase: Upper Cretaceous to Recent*

The Upper Cretaceous map (figure 9) shows a significant change in thickness and facies patterns in response to the complex plate margin tectonics along the Zagros Suture line near the end of the Turonian. Intrashelf elements became more pronounced. Truncation and onlap

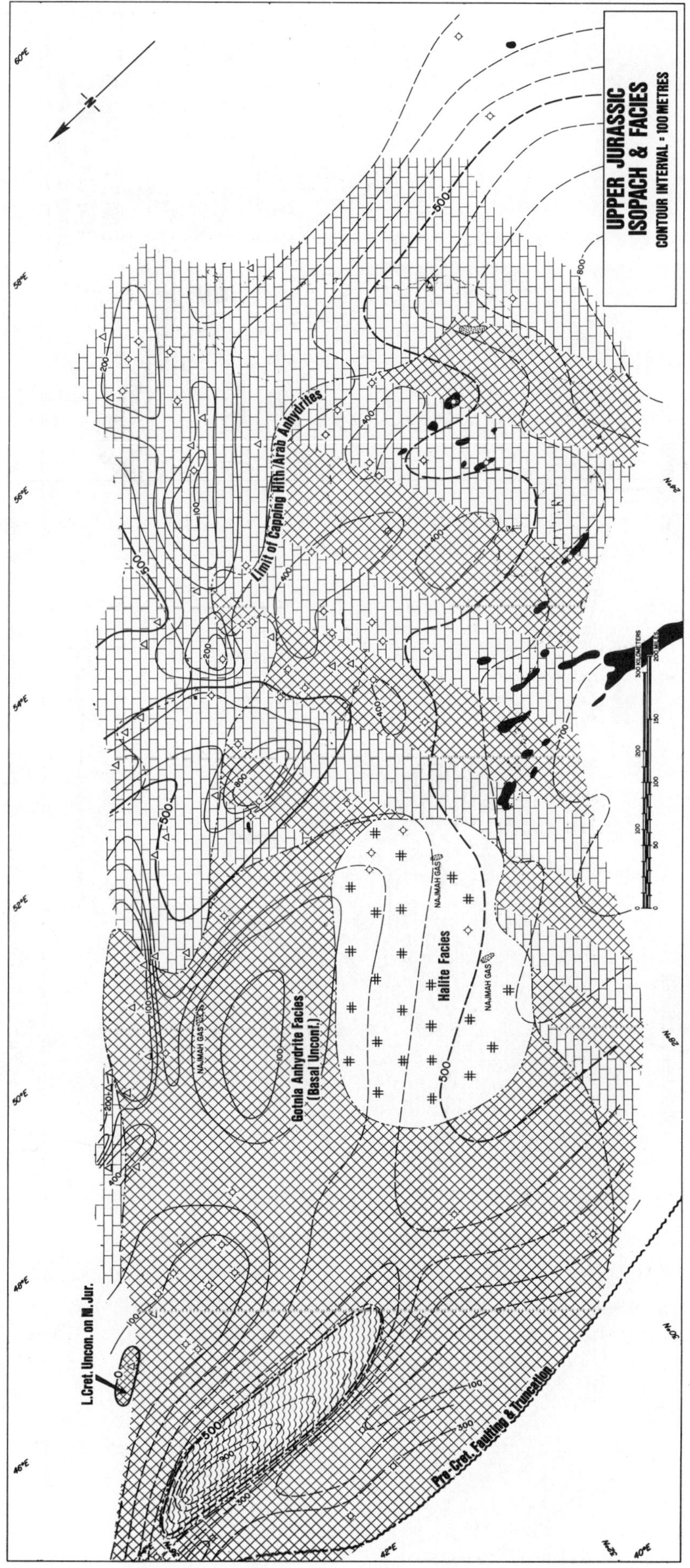

Figure 6. Upper Jurassic isopach and facies distribution map.

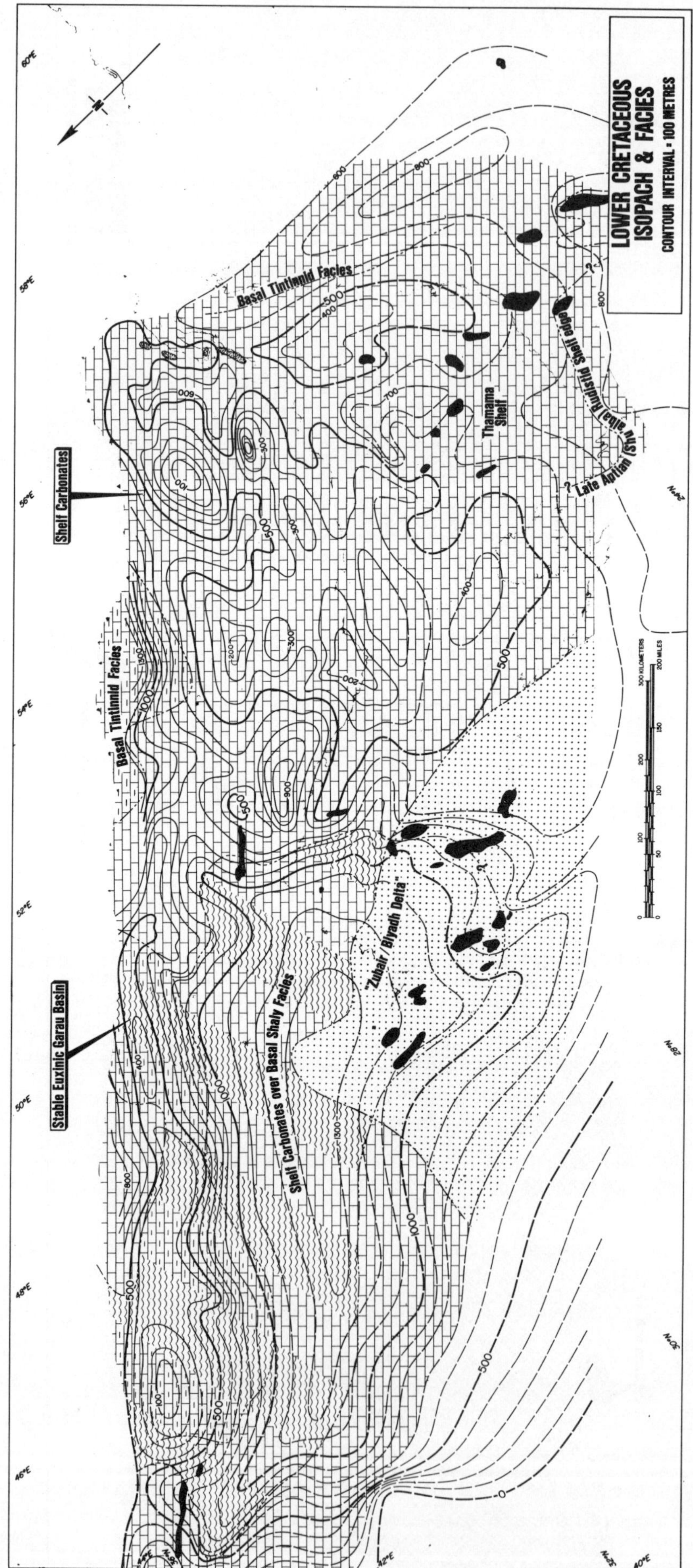

FIGURE 7. Lower Cretaceous isopach and facies distribution map.

are strongly in evidence at this pre-late Cretaceous unconformity. Structural growth at this time formed the main elements of closure on many of the giant oilfields on the Arabian side of the Gulf and these ancient N–S trends have been traced through the Iranian oilfields area into the Zagros Mountains. Oceanic sediments and ophiolites were emplaced onto the continent margin during the late Cretaceous (probably Campanian). A series of linear foredeeps, developed to the SW of this orogene, were filled with detritus derived from the oceanic rocks. SW of this 'flysch' foredeep, deeper water marine conditions prevailed in the Gurpi basin, where a thin pelagic shale was deposited NE of the carbonate platform fringing the Arabian Shield to the SW. These linear subsidence and facies trends, paralleling the collision zone along the Zagros Suture, persisted into the Tertiary.

Regional uplift late in the Cretaceous resulted in an unconformity that affected most of the elongate basin, except in parts of Lurestan where pelagic sedimentation was continuous from late Cretaceous to early Miocene. The Palaeocene–Eocene isopach–facies map (figure 10) shows a thin veneer of pelagic marls along the central axis of the basin between the carbonate margin and the mountain front foredeep which had persisted from the Cretaceous. Shallow-water marine carbonates fringed the 'flysch' trough. A carbonate–evaporite régime was developed in interior Fars, where there is considerable evidence that movement of the deep Infra-Cambrian salt was influencing sedimentation. Terrigenous clastic supply diminished towards the end of the Eocene and shallow marine limestone frequently blanketed the 'flysch' deposits suggesting that the southern Tethys seaway was still not completely closed. Evidence for this is provided also by an oceanic mélange, with a matrix age of Maastrichtian to early Tertiary, adjacent to the suture (Stoneley 1981).

Before the Oligocene a minor phase of uplift appears to have rejuvenated the ancestral Inner Zagros Mountains, shifting the axis of the depocentre to the SW and then confining marine sedimentation to a shallow elongate basin in the Oligocene – early Miocene (figure 11). Oligocene marl deposition continued in the central part of the basin from Iraq to N. Fars (upper Pabdeh Formation) with neritic carbonates (Asmari Formation) at the margins. A delta centred on Kuwait extended towards the Dezful Embayment (Ahwaz Sandstone Member). In the early Miocene the shallow water carbonates, and locally evaporites, spread over the deeper marine marls and eventually over the Ahwaz delta sands. The Asmari Formation is the most prolific oil reservoir rock in the Zagros fold belt. (It is emphasized that, in this instance, the top of the isopach interval (figure 11) is not a precise time line owing to lateral facies changes between the upper Asmari and the overlying evaporitic Fars Group: in the Oman foredeep and the southern Gulf the thick lower Fars salt sequence may be in part equivalent to the Asmari, but scarcity of information has necessitated its inclusion in the overlying interval.)

Figure 12 is a highly generalized map of the total 'post-Asmari' sedimentary thicknesses, embracing the entire Fars Group and also in some synclinal areas the Bakhtyari syn-orogenic conglomerates. The Zagros folds were already forming so that a true isopach map should properly indicate some thinning over every rising fold. The map, however, emphasizes the major subsidence of the pre-Zagros foredeep in the Dezful Embayment, where in excess of 5 km of post-Asmari sediments accumulated. Salt deposited during the initial stages of this interval constitutes the all-important seal over the Asmari oil accumulations, and is also the incompetent layer that forms the upper detachment zone above the buried folds of the Dezful oil province. The Miocene evaporites and carbonates were succeeded by Upper Miocene–

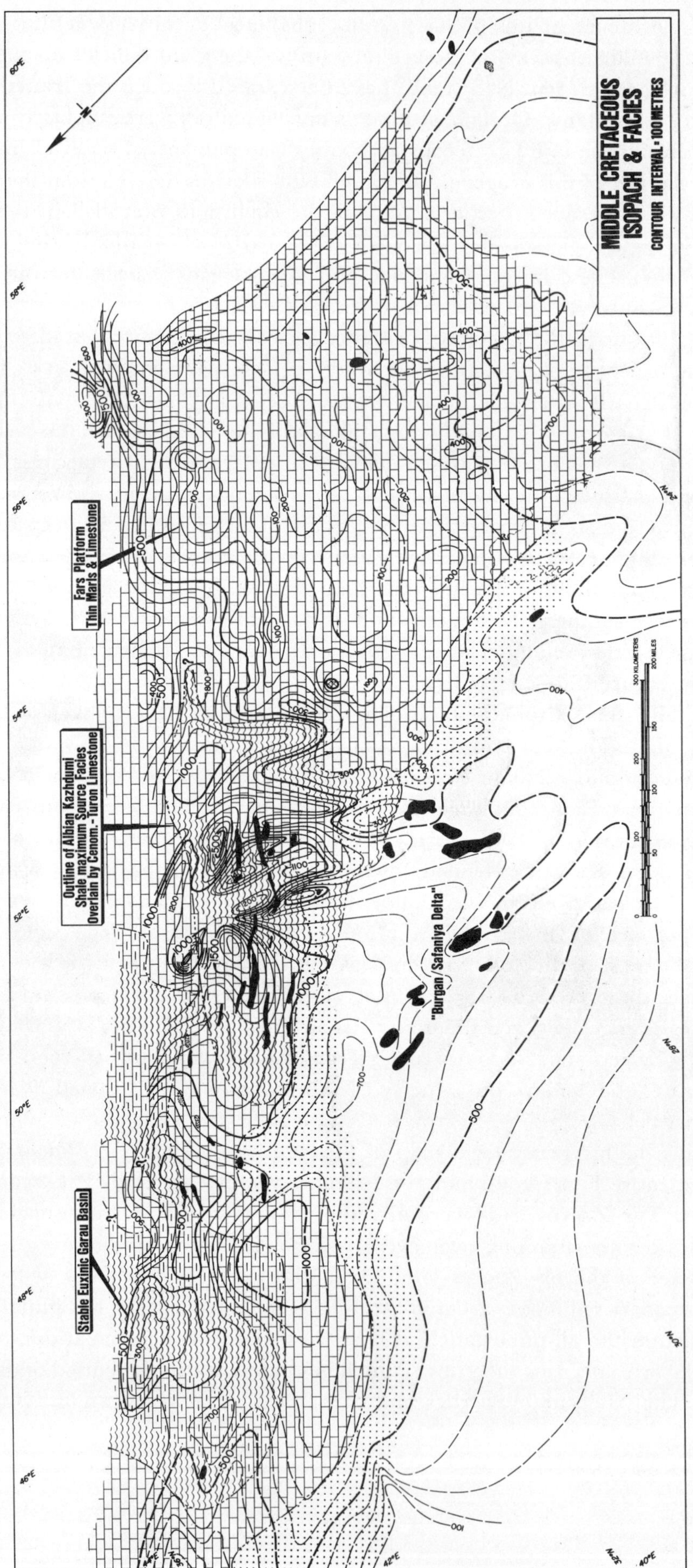

FIGURE 8. Middle Cretaceous isopach and facies distribution map.

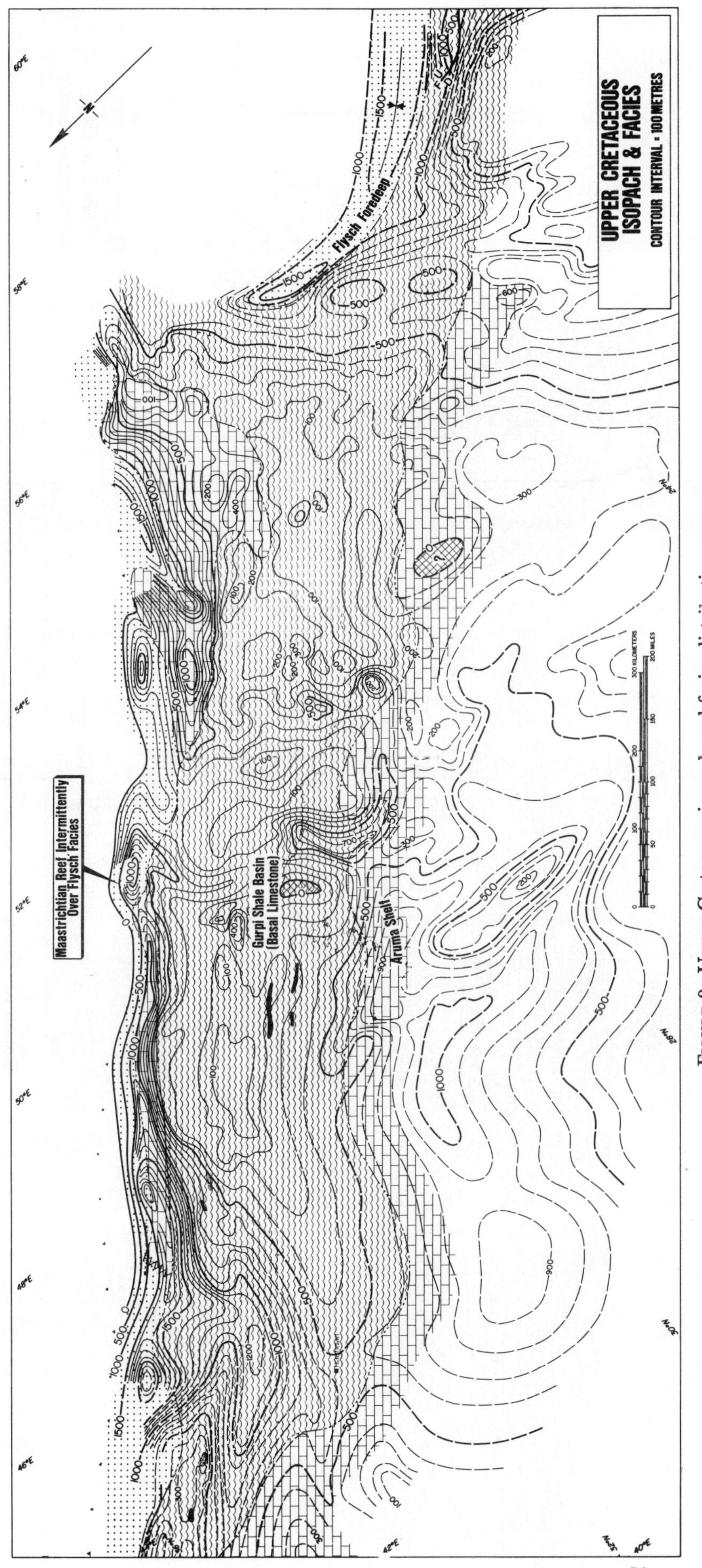

Figure 9. Upper Cretaceous isopach and facies distribution map.

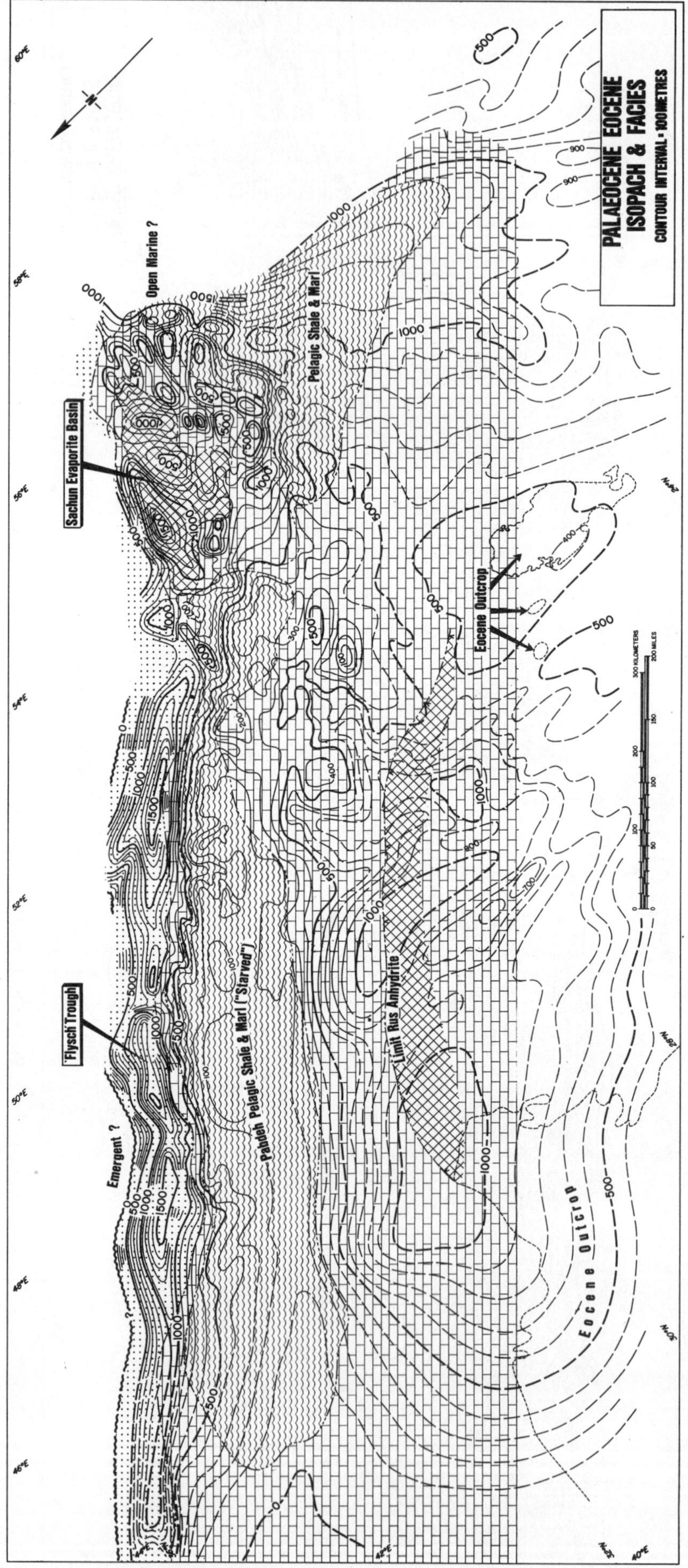

FIGURE 10. Palaeocene–Eocene isopach and facies distribution map.

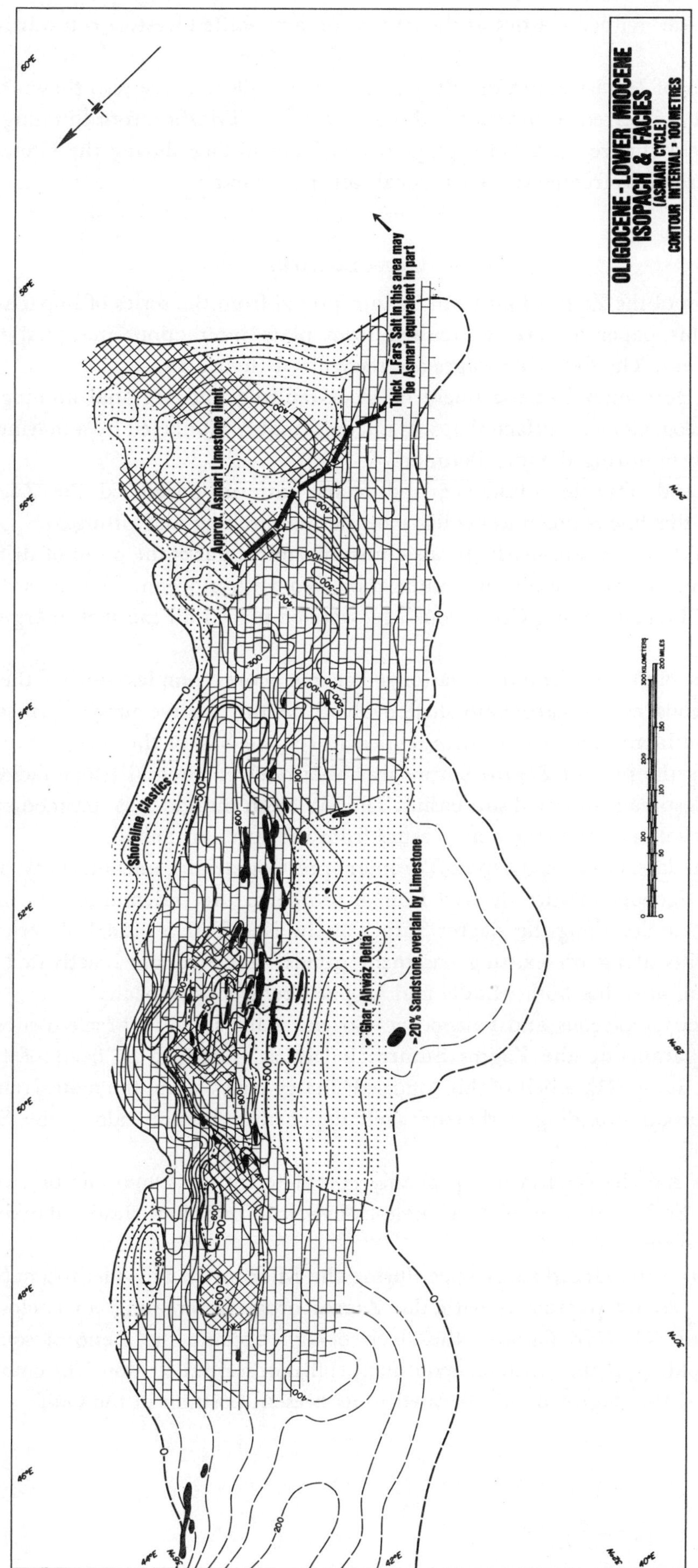

Figure 11. Oligocene – Lower Miocene isopach and facies distribution map: Asmari Cycle.

Pliocene 'molassic' redbed clastics as the deposition axis shifted westward towards the present Gulf.

The locations of Infra-Cambrian salt domes in the Gulf area are also shown on figure 12. Salt movement is inferred to have started as early as the Triassic, from thinning observed in the Pars gas field (figure 4). Several plugs reached the surface during the Cretaceous (Kent 1979) and the isopachs frequently reflect local salt movement.

4. Conclusions

The evolution of the Zagros Basin can be interpreted from the series of isopach–facies maps presented in this paper as having resulted from plate interactions marginal to the Afro-Arabian continent. The following stages are recognized (figure 13).

1. A major Hercynian tectonic phase of block-faulting and regional arching resulted in a complex palaeogeological surface that was unconformably blanketed by a marine carbonate–evaporite platform during the mid-Permian.

2. Permian and Triassic subsidence and facies patterns paralleled the Zagros Suture; thinning along this line is taken as evidence of thermal arching and rifting.

3. A regional late Triassic–early Jurassic unconformity marks the onset of drift separation along the Zagros Suture forming the southern or Neo-Tethys ocean. Oceanic sediments from this seaway, of Triassic to mid-Cretaceous age, were emplaced on the shelf margin in the late Cretaceous.

4. Subsidence and sedimentation patterns became more complex during the Jurassic to middle Cretaceous as the carbonate shelf evolved on the passive newly formed margin of the Arabian continent. The depocentres tended to remain SW of the present Gulf, while the shelf edge along the present Zagros Suture was alternately depressed (slope facies) or raised, resulting in widespread restricted sub-basins. The SW basin margin was spasmodically uplifted to supply large volumes of terrigenous clastics from the Arabian Shield.

5. A regional late Cretaceous (post-Turonian–pre-Coniacian) unconformity and the first evidence of terrigenous clastics derived from a source to the NE marked the beginning of plate margin tectonics along the Zagros Suture as the Neo-Tethys closed. A series of roughly N–S linear uplifts along pre-existing basement trends were strongly reactivated in response to this tectonism, affecting both Middle and Upper Cretaceous isopachs.

6. The Upper Cretaceous and Palaeocene–Eocene maps show a threefold subsidence and facies pattern paralleling the Zagros Suture: a 'flysch' foredeep in front of the complex 'collision' zone to the NE, a belt of thin 'starved' marly facies along the main Iranian oilfields trend, and a broad subsiding carbonate shelf depocentre roughly along the SW coast of the present Gulf.

7. Closure of the Neo-Tethys was probably complete by the Oligocene or early Miocene and the Zagros Basin was reduced to a single trough with marginal clastics flanking a central carbonate–evaporite seaway.

8. A final very pronounced subsidence during the Miocene to Plio-Pleistocene(?) preceded and was partly contemporaneous with the Zagros orogeny, creating an enclosed redbed–evaporite trough. The NE flank of this deep trough became the scene of southwestward thrusting and folding as the continent–continent collision continued along the complex Zagros Suture Zone and the Zagros Basin migrated to its present position in the Gulf.

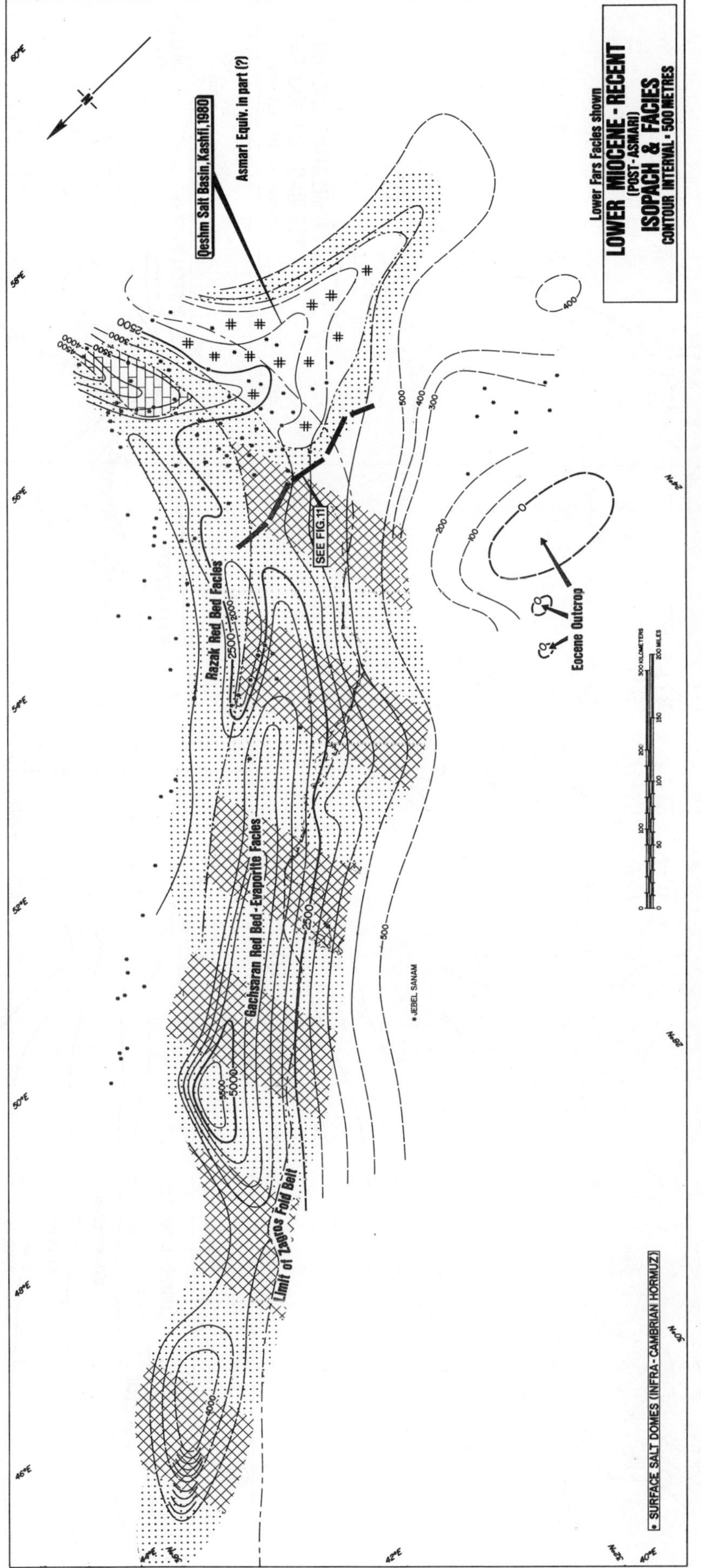

Figure 12. Miocene–Recent (Post-Asmari) isopach and facies distribution map, showing Infra-Cambrian surface salt plugs. Thicknesses in the fold-belt are highly interpretive as this stage was partly synchronous with Zagros folding.

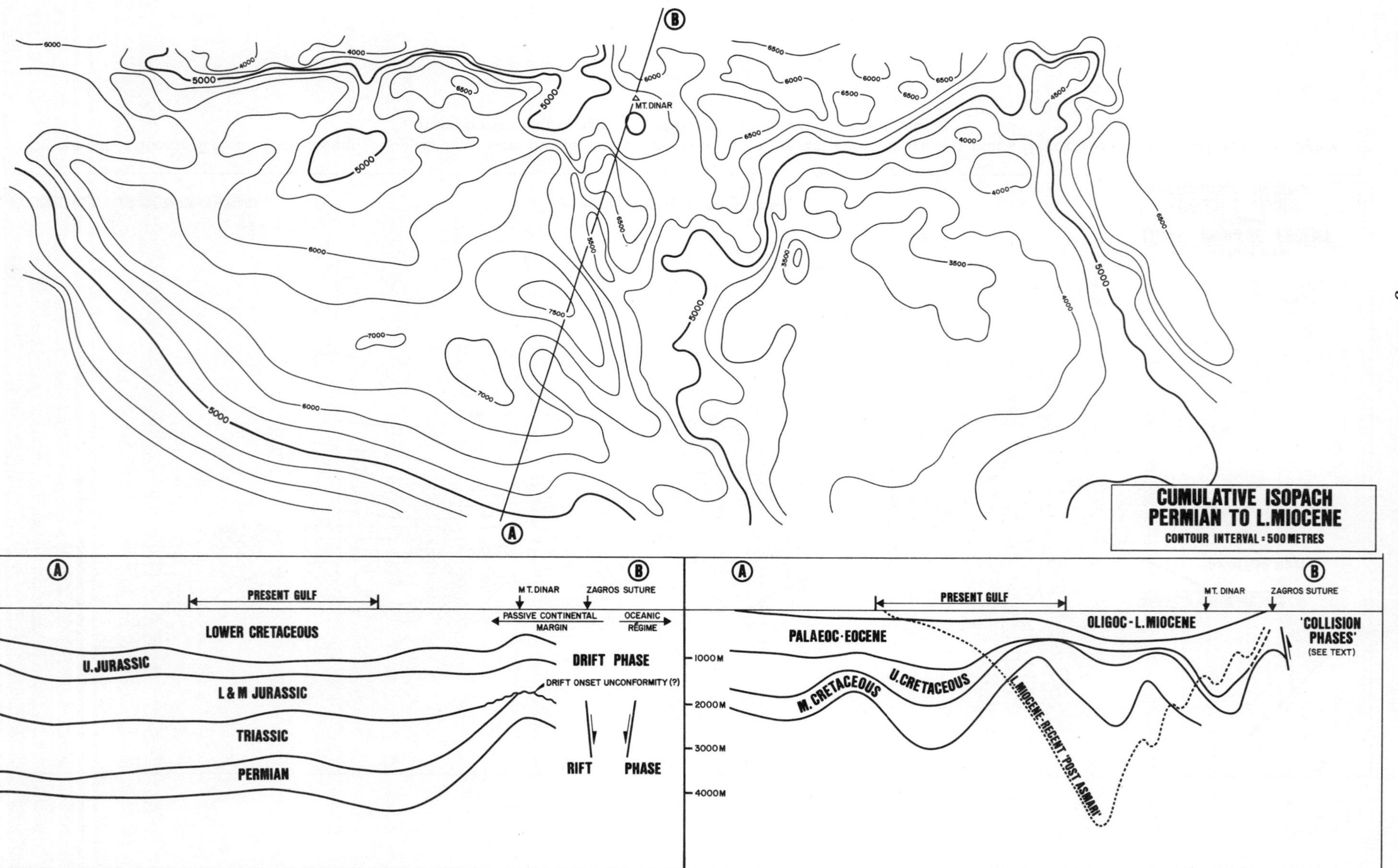

FIGURE 13. Isopach showing the cumulative thickness of post-Hercynian sediments in the Zagros Basin at early Miocene time (top Asmari Formation). Profiles A–B illustrate the nine time rock unit layers included in the isopach: left, Permian to Lower Cretaceous (figures 3–7); right, Middle Cretaceous to Lower Miocene (figures 8–11) with 'post-Asmari' (figure 12) superimposed. The final subsidence and tectonism (Zagros Orogeny) illustrated by the dotted profile (from figure 12) should be added to the isopach to obtain the total post-Hercynian basin subsidence.

References (Koop & Stoneley)

Clarke, R. H. 1975 In *Proc. 9th Arab Petrol. Cong., Dubai*, paper no. 120 (B-3), pp. 1–20.
Dunnington, H. V. 1958 In *Habitat of oil, a Symposium* (ed. L. G. Weeks), pp. 1194–1251. Tulsa: American Association of Petroleum Geologists.
Dunnington, H. V. 1967 *Inst. Petrol. J.* **53**, 129–161.
Glennie, K. W., Boeuf, M. G. A., Hughes Clarke, M. W., Moody-Stuart, M., Pilaar, W. F. H. & Reinhardt, B. M. 1973 *Bull. Am. Ass. Petrol. Geol.* **57**, 5–27.
Hassan, T. H., Mudd, G. C. & Twombley, B. N. 1975 In *Proc. 9th Arab Petrol. Cong., Dubai*, paper no. 107 (B-3), pp. 1–11.
James, G. A. & Wynd, J. G. 1965 *Bull. Am. Ass. Petrol. Geol.* **49**, 2182–2245.
Kashfi, M. S. 1980 *Bull. Am. Ass. Petrol. Geol.* **64**, 2095–2107.
Kent, P. E. 1979 *J. Petrol. Geol.* **2**, 117–144.
Murris, R. J. 1980 *Bull. Am. Ass. Petrol. Geol.* **64**, 597–618.
Setudehnia, A. 1978 *J. Petrol. Geol.* **1**, 3–42.
Stöcklin, J. 1974 In *The geology of continental margins* (ed. C. A. Burk & C. L. Drake), pp. 873–888. New York: Springer-Verlag.
Stoneley, R. 1981 *J. geol. Soc. Lond.* **138**, 509–526.
Szabo, F. & Kheradpir, A. 1978 *J. Petrol. Geol.* **1**, 57–82.
Thiele, O., Alavi, M., Assefi, R., Hushmand-Zadeh, A., Seyed-Emami, K. & Zahedi, M. 1968 *Explanatory text of the Golpaygan Quadrangle map, 1:250,000*, pp. 1–24. Geological Survey of Iran.

Discussion

M. F. Ridd. Normally on opposite sides of a suture the pre-collision rocks are of different facies, since they were deposited in geographically different regions; only after the suturing has occurred do the facies on either side become similar. Yet in the Zagros the rock types that the authors describe on either side of the present suture were very similar even before the Triassic collision. How did this come about?

W. J. Koop. I believe that the pre-Triassic rocks that are now next to each other on either side of the present suture were originally deposited on the same continental margin. Part of this margin was then separated by rifting from the remainder of the Arabian plate during the Triassic. Thereafter the sediments deposited on the two parts of the margin were very different, since they lay far apart. On the Iranian side, there is a section of Lower Jurassic coals, suggesting that even the climates on the two parts were different. When the suture closed it did so by bringing back together the two pieces of the margin that had been separated by the rifting.

R. W. Murphy. The timing of events along the Alpine–Himalayan chain is of great interest in unravelling the tectonic history. Stöcklin (1977, 1980) has shown the importance of a late Triassic compressive phase along the southern margin of Eurasia. This suture, which has been dated locally as post-Carnian, pre-Norian, can be traced eastwards into the mountains of Yunnan and northern Vietnam. This appears to be the same age as the rifting along the axis of what later became the Zagros suture. It would be instructive to discover (*a*) if the late Triassic suturing can be traced into Iran, and (*b*) if the Zagros rifting precedes the suturing or postdates it.

References

Stöcklin, J. 1977 *Mém. Soc. géol. Fr.* no. 8, pp. 333–353.
Stöcklin, J. 1980 *J. geol. Soc. Lond.* **137**, 1–34.

W. J. Koop. The Tethys was presumably near the Alborz mountains during Triassic and Permian times. Stöcklin (1974) suggests that the ophiolite slices at Rasht and Mashad may be evidence of late Triassic closing of this northern Palaeo-Tethys ocean at the same time that rifting and drifting opened up the southern Neo-Tethys along the Zagros zone.

R. Stoneley. A possible explanation is that the pre-Triassic Tethys lay entirely to the North of the present Iranian land mass. This is the ocean to which Murphy is referring, and which extended from Crimea through the Himalayan region. Then towards the end of the Triassic the piece of Gondwanaland which now forms central Iran broke away from the Arabian plate, and a new ocean opened up. The part of Tethys north of Iran closed, probably towards the end of the Jurassic or during the early Cretaceous. This disappearance of the Northern Tethys occurred as the new Southern Tethys was produced, along what is now the Zagros Suture. Finally the Southern Tethys closed during the mid-Tertiary.

R. W. Murphy. How does the timing of the closing of the Northern Tethys compare with that of the opening of the Southern ocean?

R. Stoneley. As far as the Iranian sector is concerned, both the location and the timing of the closure of Northern Tethys are uncertain. There is some evidence from northeastern Iran that it may finally have closed in the early Cretaceous.

M. F. Osmaston. What happened in the Zagros between the Maastrictian, when the Oman ophiolite was emplaced, and the middle Eocene when the first accretionary slice known in the Makran, containing Lower Eocene ooze, was emplaced. For instance, what was the Palaeocene palaeogeography?

R. Stoneley. Along the suture there is a mélange of oceanic sediments from which Maastrictian, Palaeocene and early Eocene ages have been obtained. The mélange also includes blocks of neritic shelf carbonates with ages from Maastrictian to early Miocene; however, the compressional folding in the Zagros itself did not start until the end of the Oligocene or the early Miocene.

M. M. Kholief. In the Zagros most of the oil production is confined to the Mesozoic sedimentary sequence of Jurassic or Cretaceous age. Is there also oil in the Tertiary sediments, especially those of Miocene age? If so, is there any resemblance between the structure and sedimentology of these rocks and those in the Gulf of Suez?

W. J. Koop. The bulk of the oil in the Zagros fold belt does in fact occur in Oligo-Miocene, mainly carbonate, reservoirs (the Asmari Formation and the Kirkuk Group), even though it is believed to have migrated upward from Cretaceous source rocks. These Tertiary limestones were deposited in a shallow marine trough between the Arabian platform and the Zagros Suture immediately preceding the main Zagros downwarp and folding. The oil-bearing section in the Gulf of Suez is in part the same age and has some lithological similarities, but the pull-apart régime is, I believe, completely different from the linear compressional setting of the Tertiary Zagros Basin.

Phil. Trans. R. Soc. Lond. A **305**, 169–192 (1982) [169]
Printed in Great Britain

The Western Canada Sedimentary Basin

By J. W. Porter†, R. A. Price‡[1] and R. G. McCrossan§

† *Canadian Superior Oil Limited*, 355 *Fourth Avenue S.W.*, *Calgary*, *Alberta T2P* 0*J*3, *Canada*
‡ *Department of Geological Sciences*, *Queen's University*, *Kingston*, *Ontario K*7*L* 3*N*6, *Canada*
§ *Esso Resources Canada Limited*, 500 6*th Avenue S.W.*, *Calgary*, *Alberta T2P* 0*S*1, *Canada*

The Western Canada Sedimentary Basin, a simple northeasterly tapering wedge of sedimentary rocks more than 6 km thick, extends southwest from the Canadian Shield into the Cordilleran foreland thrust belt. Its internal structure and the lateral variations in its shape reflect a long and complex history of development involving a foreland basin that was superimposed on a cratonic platform and continental terrace wedge. This history, which is inextricably linked to the evolution of the Canadian Cordillera, can be outlined succinctly with reference to the unconformity-bounded transgressive–regressive stratigraphic sequences established by Sloss (*Bull. geol. Soc. Am.* **74**, 93 (1963)), each of which has a distinctive character in Western Canada.

The continental terrace wedge was established with the deposition of the Proterozoic Purcell (1500–1350 Ma) and Windermere (850–600 Ma) sequences, but the first record of the platformal phase is the early Palaeozoic transgressive onlap of the early Proterozoic (> 1750 Ma) crystalline basement by the Sauk sequence. Early Palaeozoic subsidence of the margin of the craton may have been due to cooling of the lithosphere after renewed stretching at the ancient rifted western margin of the Precambrian craton, and to isostatic flexure of the lithosphere under the weight of the sediment that had accumulated at the margin in the oceanward prograding continental terrace wedge.

During a subsequent Middle Ordovician to Middle Jurassic phase, the cratonic platform became differentiated into an intersecting network of epeirogenic arches with intervening basins. Development of the basins was as much a result of erosion and uplift of the arches between transgressive–regressive cycles as it was a result of differential subsidence of the basins during the cycles. The cause of the long (> 300 Ma) episode of intermittent epeirogenic movements that produced the basins and arches is a major unsolved problem.

The foreland basin developed in two stages, in Middle Jurassic to early Cretaceous and late Cretaceous to Palaeocene time, as a result of collisions between North America and two pieces of a tectonic collage of oceanic terranes that were accreted to its western margin. During these two collisions, the continental terrace wedge, which had accumulated outboard from the rifted margin of the continental craton, was compressed and displaced over the western margin of the craton. Part of the supracrustal cover was scraped off the craton and accreted to the overriding mass to form a wedge of imbricate thrust fault slices that was tectonically prograded over the margin of the continental craton. Isostatic flexure of the continental lithosphere in response to the tectonic loading imposed on it by the displaced continental terrace wedge and the accretionary wedge of thrust slices produced the migrating moat in which the outwash of clastic detritus from the evolving thrust belt was trapped to form the foreland basin.

Introduction

The Western Canada Sedimentary Basin (figure 1) is basically a simple northeasterly tapering wedge of sedimentary rocks, more than 6 km thick, that extends southwards from the Canadian

[1] Present address: Geological Survey of Canada, 601 Booth Street, Ottawa, Ontario, K1A 0E8, Canada.

Shield into the Cordilleran foreland thrust belt (figures 1 and 2). However, its internal structure and the lateral variations in its shape reflect a long and complex history of development that involved superposition of contrasting patterns of differential subsidence and uplift, the causes of which have yet to be explained adequately.

Two fundamentally different phases in the development of the basin coincide with the two different stages in the evolution of the eastern part of the Cordillera, which is the deformed counterpart of the sediments that fill the undeformed basin. An initial platformal phase involved transgressive onlap of the Precambrian crystalline basement of the North American Craton, and the development of a series of epeirogenic arches and basins on the cratonic platform. It coincided with the Proterozoic to Jurassic stage of oceanward progradation of a pre-Cordilleran continental terrace wedge along the ancient rifted western margin of the North American Precambrian craton. A subsequent foreland basin phase, involving cratonward progradation of synorogenic clastic detritus shed by the evolving Cordillera, coincided with the Jurassic to Palaeocene interval of collisional orogeny during which foreign terranes were

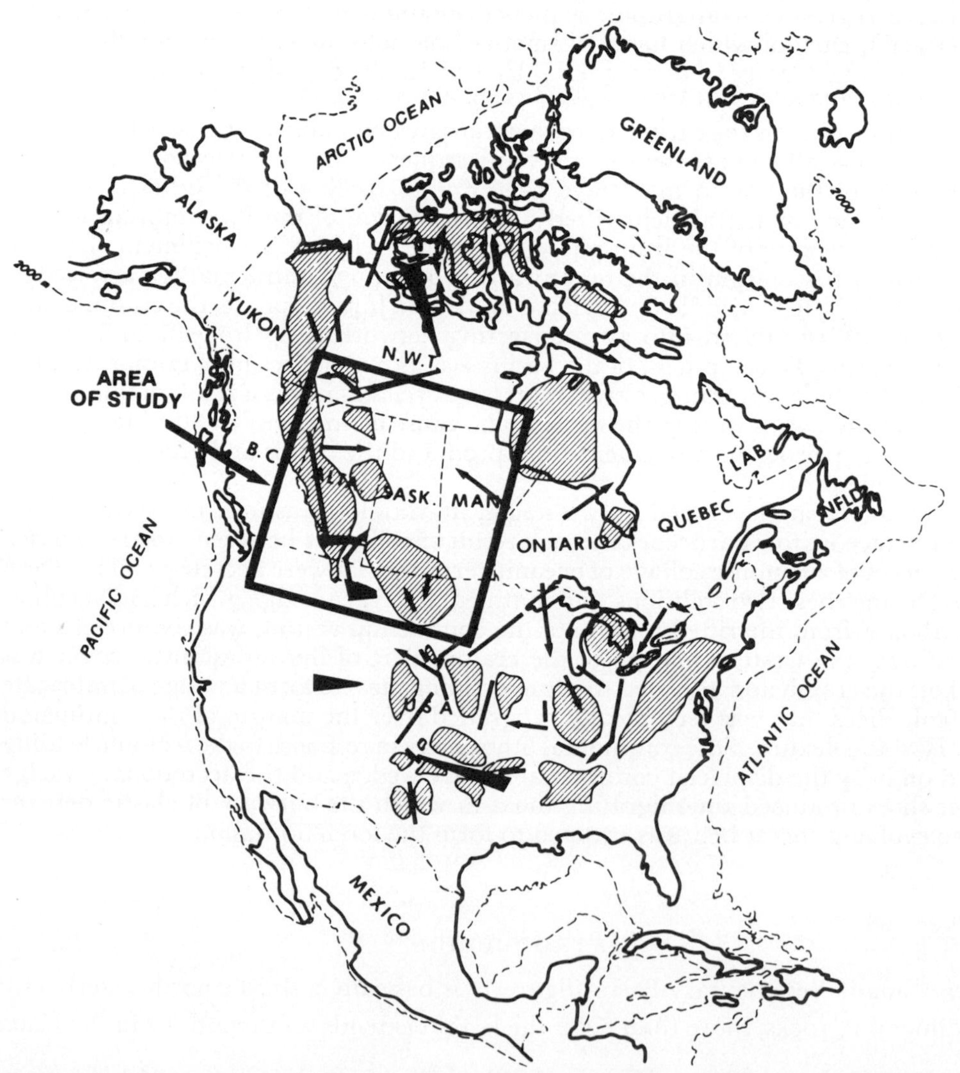

Figure 1. The Western Canadian Sedimentary Basin (area of study) in relation to the principal Palaeozoic cratonic basins (diagonal ruling) and arches (arrows) and Proterozoic aulacogens of North America.

accreted to the western margin of North America and the continental terrace wedge was compressed, detached from its basement, and displaced over the flank of the craton. The evolution of the Western Canada Sedimentary Basin is inextricably linked to the evolution of the Cordillera; to understand the one it is necessary to consider carefully the other.

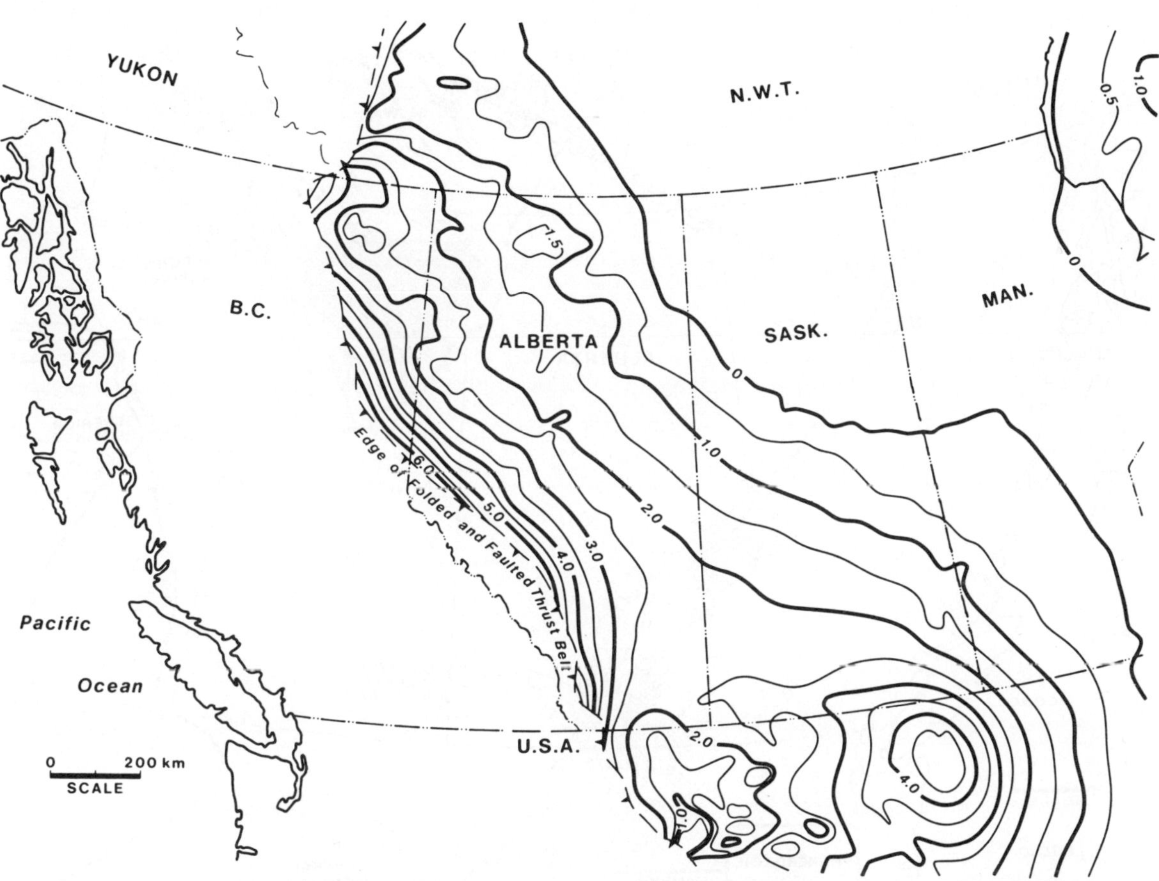

FIGURE 2. The Western Canada Sedimentary Basin. Total preserved thickness of Phanerozoic rocks east of the folded and faulted rocks of the Cordilleran foreland thrust belt. Contours in this and other figures are in kilometres.

This analysis of the stratigraphic record in the Western Canada Sedimentary Basin aims at elucidating the evolution of the basin, its relation to the evolution of the Cordillera, and its significance for models of basin development. The analysis involves a data base of 45000 exploratory wells within an area of 3×10^6 km^2 between the Cordilleran foreland thrust and fold belt and the Canadian shield, and published information from surface and subsurface studies of both the Cordillera and the cratonic platform. Porter and McCrossan are responsible for assembling and synthesizing the data from the undeformed basin fill; Price is responsible for the information on the Cordillera; the interpretations are a joint responsibility.

The stratigraphic record of the evolution of the basin can be outlined succinctly with reference to six stratigraphic sequences, each representing a major transgressive–regressive megacycle, and bounded by inter-regional unconformities (Sloss 1963). The nature of each sequence, and the relations between sequences in the undeformed cratonic platform, can be illustrated with regional stratigraphic maps and sections based directly on data from exploratory wells and bedrock maps (figures 5–14), but to portray relations in the deformed rocks of the Cordilleran foreland thrust belt it is necessary to use palinspastic reconstructions of balanced structural cross sections (figures 4, 14 and 15).

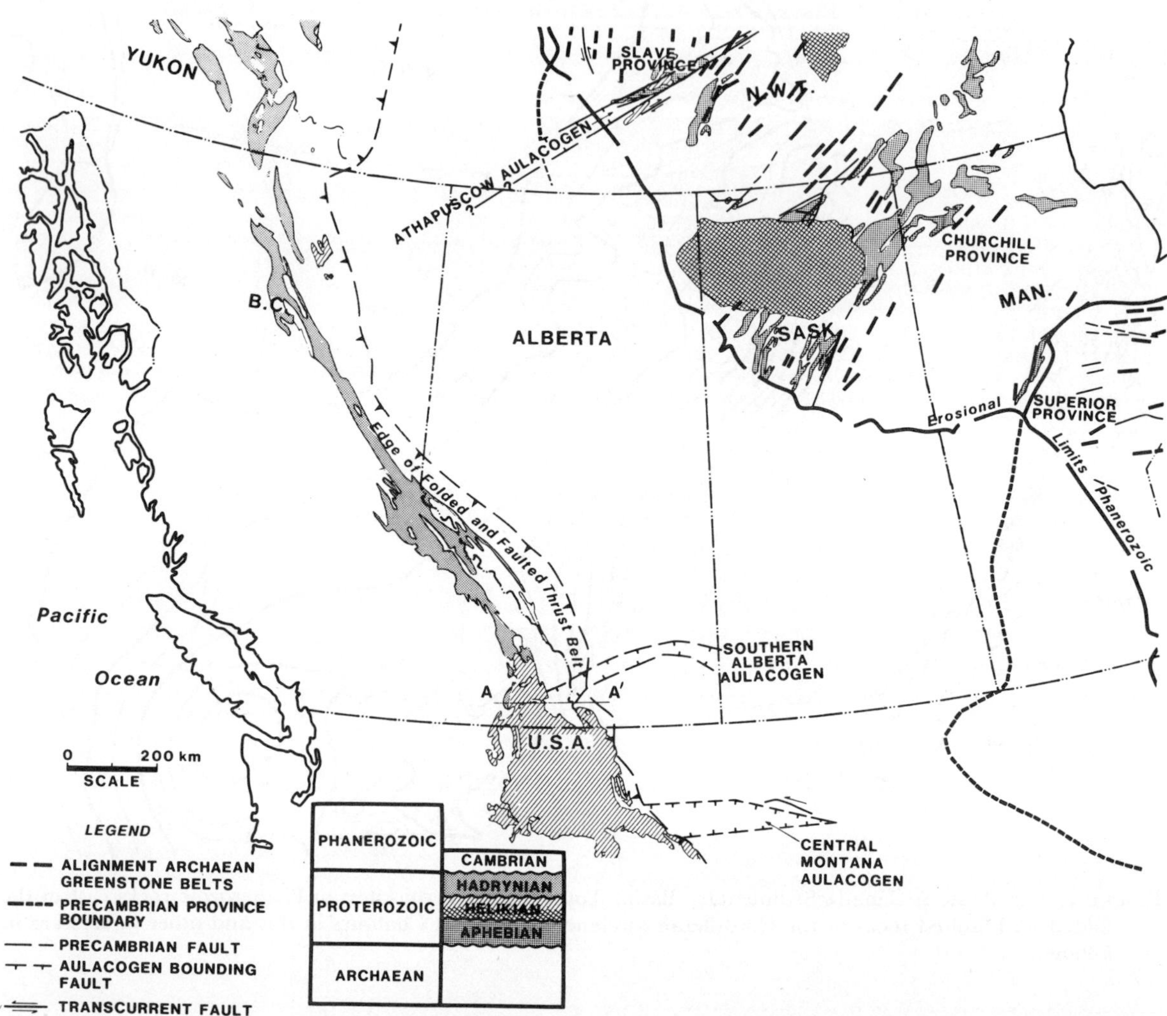

FIGURE 3. Precambrian geotectonic provinces and Proterozoic aulacogens (after Douglas (1969), Kanasevich *et al.*, (1969), Fraser *et al.* (1970) and Hoffman (1973).) Line A–A′ is the location of the section in figure 4.

THE BASEMENT

The Precambrian continental craton, which forms the basement of the Western Canada Sedimentary Basin, consists mainly of Archaean crystalline rocks and Aphebian (2500–1750 Ma) supracrustal rocks that were modified by deformation, metamorphism and magmatism during the early Proterozoic (1750 Ma) Hudsonian orogeny, and are exposed in the Churchill Province of the Canadian Shield. It also includes large, unmodified, remnant Archaean cratons in the north (Slave Province) and the east (Superior Province), as well as smaller Archaean crustal remnants within the Churchill Province and its buried counterparts. Flat-lying, unmetamorphosed, early Helikian (1750–1500 Ma) platformal cover deposits that are preserved locally in the Churchill Province (figure 3) define an upper limit for the time at which the Churchill, Slave and Superior Provinces became established as a single stable continental craton (Fraser *et al.* 1970). The structures in Churchill Province trend northeastward, and can be traced on the basis of their distinctive magnetic anomaly signature, under the deformed Phanerozoic rocks of the foreland thrust belt (Price 1981). These northeast-trending structures were overlapped discordantly by the northwesterly trending prism of late Helikian (1550–1350 Ma) Belt–Purcell Supergroup sedimentary rocks that forms the lowest part of the continental terrace wedge.

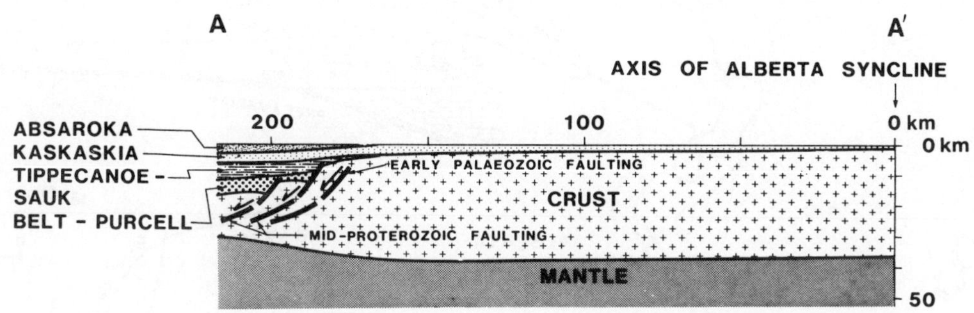

FIGURE 4. The continental terrace wedge: a palinspastically restored section through the Rocky Mountains along 49° 45′ N; line A–A′ of figure 3 (after Price 1981, Figure 2). No vertical exaggeration.

THE CONTINENTAL TERRACE WEDGE AND THE CRATONIC PLATFORM

Belt–Purcell sequence

The Purcell Supergroup in Canada and the Belt Supergroup of the United States consists mainly of fine-grained terrigenous clastic sediment that appears to be exclusively of North American provenance. In southern Canada the Belt–Purcell sequence (figures 3 and 4) forms a northeasterly tapering wedge up to 15 km thick (Price 1964). It appears to represent a continental terrace wedge that was prograded westward into the ocean basin formed by the rifting that extablished the western margin of the North American Precambrian craton (Sears & Price 1978). Thickness and facies variations indicate that rifting persisted during deposition of the Purcell Supergroup, and was influenced, at least locally, by the older northeasterly trending structures that characterize the Churchill Province of the Canadian Shield (McMechan 1979, 1980). Aulacogen-type rifts that project into the craton from this ancient rifted margin in southern Alberta and central Montana (figure 3) also may have been controlled by these older structures in the Precambrian basement. Deposition of the Belt–Purcell sequence appears to have terminated about 1350 Ma ago, with an episode of deformation that involved regional metamorphism, folding and granitic intrusion locally (McMechan 1979; McMechan & Price 1982), but only broad epeirogenic uplift on a regional scale.

Windermere sequence

A second major episode of faulting (rifting?) and oceanward progradation of the continental terrace wedge occurred during the deposition of the late Proterozoic (850–600 Ma) Windermere Supergroup, after a hiatus of about 500 Ma. It is recorded by the erosion of up to 4 km of the Purcell Supergroup and the deposition of up to 9 km of Windermere strata, before the early Cambrian regional transgression that initiated the overlying Sauk sequence (Lis & Price 1976).

The Windermere sequence is characterized by coarse clastic deposits ('grits') comprising detritus eroded from the granitic basement of the North American Craton, and by the occurrence of conglomeratic mudstones that may represent glacial deposits. It appears to comprise part of a suite of similar deposits of the same general age, marking an episode of rifting that extended almost continuously around the North American craton (Stewart 1976).

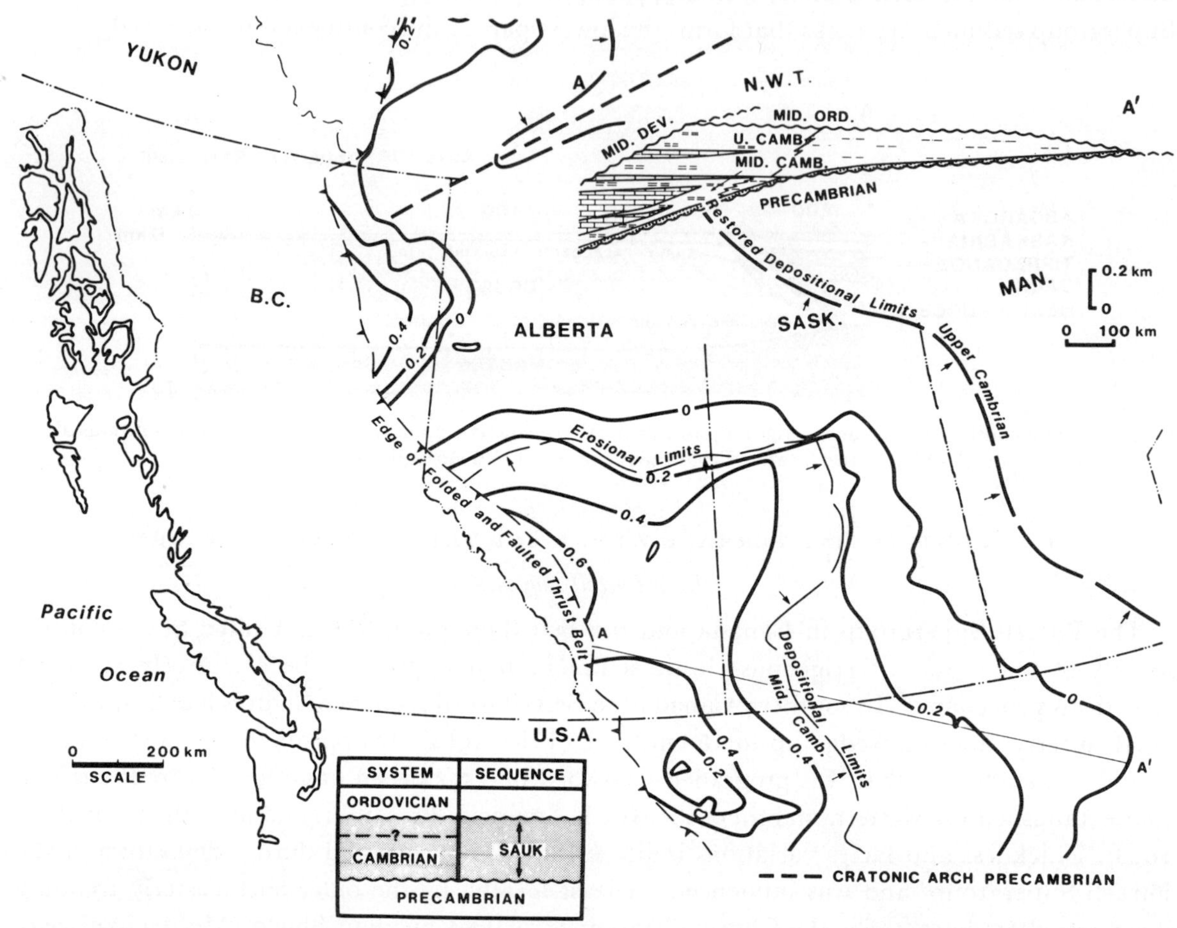

FIGURE 5. Sauk sequence, regional map and stratigraphic section A–A′ (modified after Fuller & Porter 1962, Figure 3).

Sauk sequence

The Sauk sequence (figure 5), which marks the beginning of the record of deposition on the cratonic platform, is a simple northeasterly tapering pericratonic wedge. It shows no signs of the influence of the cratonic arches and basins that dominated patterns of erosion and deposition during the subsequent evolution of the Western Canada Sedimentary Basin. The Sauk sequence began with a basal diachronous quartz sandstone that transgressed the entire western flank of the craton. The sandstone ranges in age from early Cambrian in the continental terrace wedge, where it overlaps the Helikian and Hadrynian (mid and late Proterozoic) rocks of the Purcell and Windermere sequences, to late early Ordovician (post-Tremadoc to pre-Arenig) in the east on the cratonic platform, where it overlies the deformed, Aphebian (early Proterozoic) and

Archaean rocks of the crystalline basement complex (figure 11). In the outboard, western part of the continental terrace wedge, the early Cambrian sandstone succession is more than 2 km thick, contains major intercalations of basic volcanic rocks (Wheeler 1963, 1965), and may have been deposited during an episode of crustal stretching. An overlying succession of younger onlapping units, showing progressively less eastward attenuation, advanced toward the centre of the craton until late Croixan time; and a regional regression began with the earliest Tremadocian deposits. Much of the centre of the craton remained emergent during the Sauk cycle; no Cambrian sediments are known to have been deposited in the Hudson Bay Basin and Moose River Basin, which lie east of the Western Canada Sedimentary Basin, in the centre of the Canadian shield.

The mid and late Cambrian and early Ordovician strata comprise a succession of grand sedimentary cycles, each of which begins with shales and grades upward into carbonate rocks. These cycles are superimposed upon, and reflect lateral shifts in the positions of, three broad facies belts (Aitken 1966). An inner detrital belt extends over the craton. It contains a higher proportion of shale, siltstone and sandstone, representing detritus eroded from the Precambrian Craton. A middle carbonate belt is characterized by thick carbonate bank deposits that are exposed in the thrust sheets of the Main Ranges of the Rocky Mountains. It separates the inner detrital belt from an outer detrital belt that is characterized by thick sequences of calcareous shales, most of which were deposited in relatively shallow water. Further outboard from the continent, in westernmost exposures of the continental terrace wedge deposits, homotaxial strata are black graphitic phyllites that contain intercalations of basic volcanic rocks, and may represent deposition in a continental slope and rise environment.

The carbonate bank margin that separates the middle carbonate and outer detrital facies belts persisted at essentially the same location from Middle Cambrian to Middle Ordovician time (Aitken 1971). Palinspastic reconstructions of the overthrust belt show that it was situated over the ancient rifted margin of the Precambrian continental craton (Price 1981), and suggest that there was intermittent 'down-to-the-basin' faulting along this zone from early Cambrian to Middle Devonian time (figure 4). In southern British Columbia and Alberta the southeastern edge of the basin in which the outer detrital facies accumulated coincided with the northwestern margin of a 'Montania', a large high-standing block that extended into the United States in central Montana (figure 5).

The northwestern margin of Montania was an important northeast-trending hinge zone from early Cambrian to late Devonian time (Norris & Price 1966). In southeastern British Columbia, within the continental terrace wedge, the net stratigraphic separation across this zone, along the unconformity at the base of the Upper Devonian rocks, is 8 km. This hinge zone was aligned with older northeast-trending structures that had controlled thickness and facies variations in underlying rocks of the Middle Proterozoic Purcell sequence (McMechan 1980) and in the late Proterozoic Windermere sequence (Lis & Price 1976); and these structures, in turn, were aligned with still older structures in the early Proterozoic crystalline basement complex of the craton (Kanasevich *et al.* 1969). The magnitude, nature and location of the differential movements across the hinge zone and the carbonate bank margin suggest that they were controlled by faulting in the basement beneath the continental terrace wedge (figure 3), and that this faulting followed the ancient rifted margin of the Precambrian craton southward to where it was deflected by some of the younger northeasterly trending structures that had formed when the Purcell and Windermere sequences were being deposited.

The pattern of variation of thickness of the Sauk sequence (figure 5) is a result of two episodes of erosional bevelling. A pre-Middle Ordovician (pre-Tippecanoe) episode affected the Sauk sequence over the entire Western Canada Sedimentary Basin. Within the Williston Basin, the pre-Tippecanoe erosional truncation of the Sauk sequence is approximately parallel with its depositional strike. This shows that the Williston Basin did not become established until later. The regional eastward bevelling marked a relatively uniform westward tilt of the craton that was associated with the westward progradation of the continental terrace wedge. A pre-Middle Devonian (pre-Kaskaskia) episode of erosion was more pronounced, but affected only those parts of the Sauk sequence not covered by Middle Ordovician strata. This second phase of erosion marked the inception of major cratonic arches, including the Central Montana Uplift, Peace River Arch, Western Alberta Arch and Tathlina Arch. Montania stood relatively high during deposition of the Sauk sequence; its northwestern margin was bevelled by pre-Kaskaskia erosion (Norris & Price 1966; Benvenuto & Price 1979). Along the axes of the Peace River Arch and Tathlina Arch, the entire Sauk sequence was eroded and the basement exposed before the Middle Devonian transgression (figure 7).

The only evidence of a cratonic arch during deposition of the Sauk sequence occurs across the locus of the northeast-trending, early Proterozoic Athapuscow Aulocogen (figures 3 and 5). Upper Cambrian rocks to the northwest comprise a red bed assemblage with numerous halite beds; but Upper Cambrian rocks to the southeast are marine clastics.

Tippecanoe sequence

The Tippecanoe sequence (figure 6) began with a transgressive basal sandstone that overlapped the eastward bevelled upper surface of the Sauk sequence from the part of the continental terrace wedge that is now exposed in the Western Ranges of the Rocky Mountains to the eastern limit of the Sauk sequence on the cratonic platform. The Williston Basin was established during the early stages of the deposition of the Tippecanoe sequence. It was circumscribed by the embryonic Western Alberta, Peace River, Severn, Transcontinental and Sioux cratonic arches. Montania may also have been relatively emergent and linked to the Western Alberta Arch during the deposition of the Tippecanoe sequence.

The pattern of variation in thickness of the Tippecanoe sequence in the Williston Basin is a result of both pre-Middle Devonian uplift and erosional truncation and the deposition of thinner condensed equivalents of the Tippecanoe sequence on the surrounding cratonic arches. The Middle Ordovician sandstone, which is thickest in the central and eastern parts of the Williston Basin (figure 6), was derived in part from Cambrian sandstones on the adjacent arches, as well as from Proterozoic sandstones on the Canadian Shield. A succeeding shallow-water carbonate shelf environment persisted until late Silurian time. Thinning of the carbonate deposits over the adjacent arches was a result of slower deposition rather than onlap. The carbonate platform deposits of the Tippecanoe sequence are the most widespread and of the longest duration of any of the Phanerozoic sequences. The maximum transgression probably occurred during late Ordovician time when most of the North American craton was inundated. Pre-Middle Devonian erosion has removed most of this sequence, leaving remnant small outliers of stratigraphically similar deposits at various sites on the interior of the craton, as well as the extensive deposits in cratonic basins such as the Williston, Hudson Bay, Moose River and Michigan basins (figure 7).

The carbonate facies that dominates the Ordovician in the Williston Basin is interrupted by some five sabkha evaporite cycles that are limited to the central and north-central part of the basin (Porter & Fuller 1959). Their development appears to be related to the initial growth of Central Montana and Western Alberta arches because the locus of the maximum sedimentary thickness during the interval in which they were deposited shifted some 60 km to the southeast. Subsequent shallow-water biogenetic carbonate deposits extended across the entire Western Canada Basin, and included biostromal reefs that were replaced in late Silurian time by micro-grained dolomites and anhydrite-bearing dolomites that are the counterparts of the Cayugan evaporites of the Michigan Basin.

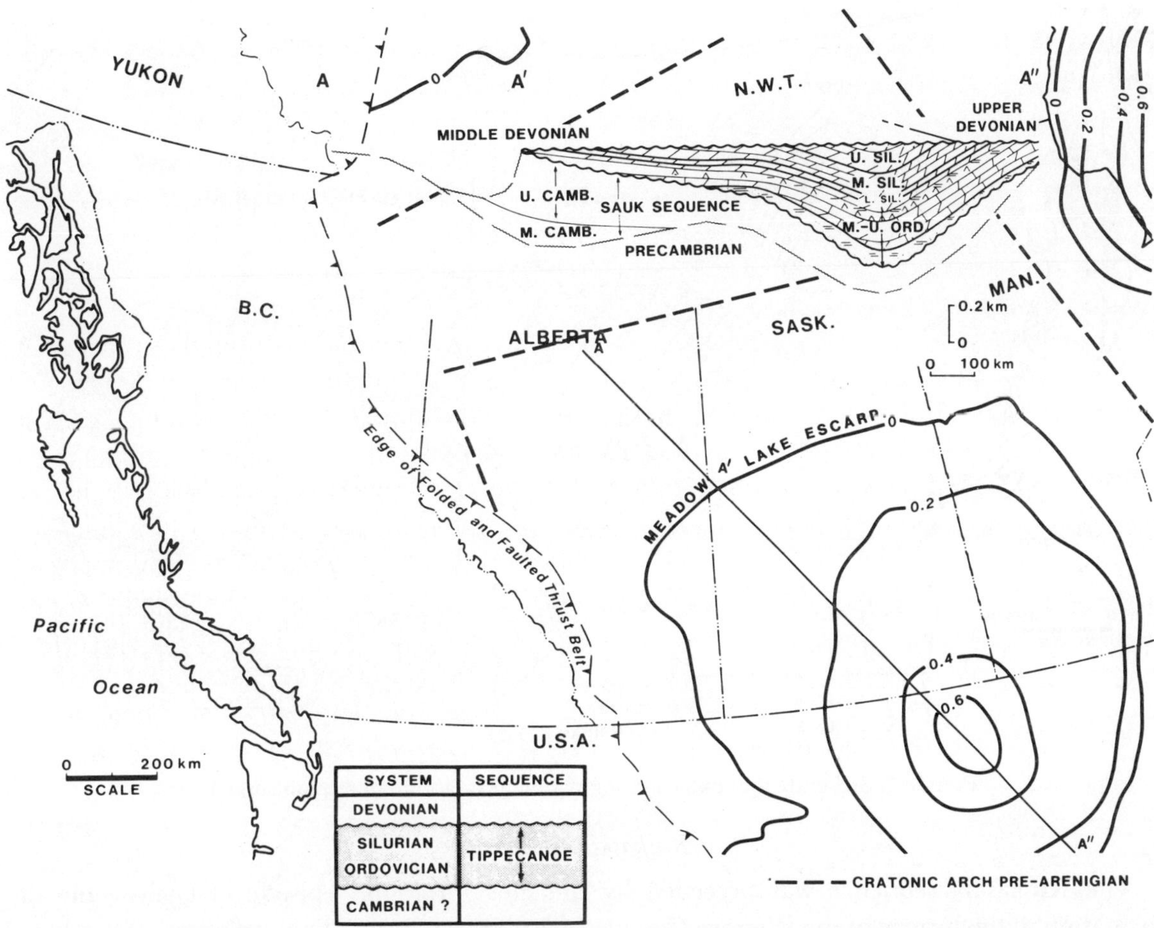

SYSTEM	SEQUENCE
DEVONIAN	
SILURIAN	TIPPECANOE
ORDOVICIAN	
CAMBRIAN ?	

Figure 6. Tippecanoe sequence, regional map and stratigraphic section A–A′ (modified after Fuller & Porter 1962, Figure 7).

The restricted circulation reflected by these evaporite deposits can be attributed to the emergence of the cratonic arches that circumscribe the basin. A number of very thin, discrete synchronous sheets of sandstone that interrupt the massive carbonate strata of the Silurian system (Porter & Fuller 1959) probably result from periodic uplift of the central craton, which was the source of the sands.

The erosion that removed much of the Silurian and older strata of the Tippecanoe sequence before the deposition of the Middle Devonian was concentrated along the axial regions of the cratonic arches, which became high-standing topographic features and strongly influenced the depositional environment of the succeeding Kaskaskia sequence. The asymmetrical depositional profile of the Williston Basin was accentuated by deeper erosion along the east and south sides of the Severn, Transcontinental and Sioux arches. This shape and the lack of any deep water facies suggests that the geometry of the basin was controlled primarily by the growth of the arches.

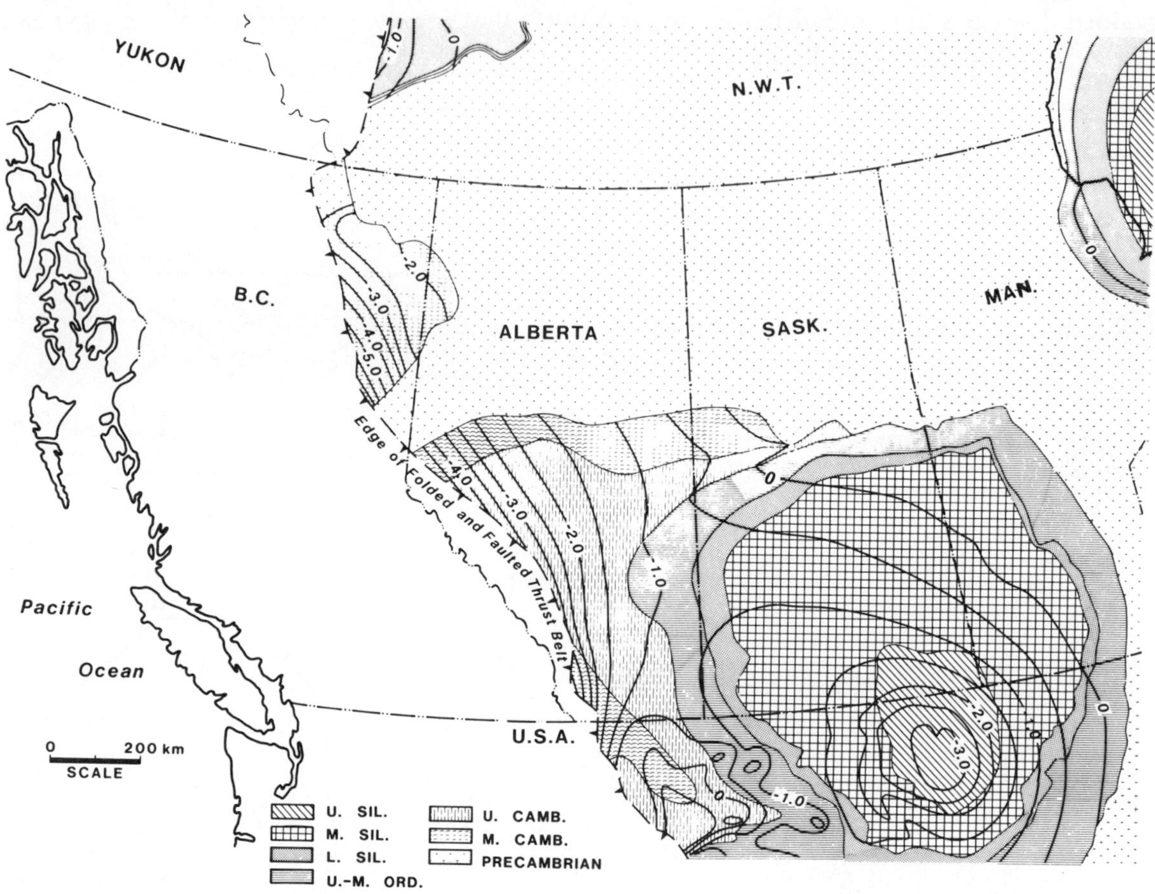

FIGURE 7. Structure and palaeogeology of the pre-Devonian erosion surface.

Kaskaskia sequence

The Kaskaskia sequence was preceded by the most profound episode of epeirogenic deformation in the history of the Western Canada Sedimentary Basin. The surface relief produced by early Devonian epeirogenic uplift of the cratonic arches and by the ensuing erosion of the Tippecanoe and Sauk sequences controlled the distribution, thickness and character of the initial deposits of the Kaskaskia sequence. This surface relief forced epeiric seas to transgress from northwest to southeast instead of from the previous westerly direction. The Tathlina, Western Alberta and Peace River arches were strongly emergent and shed sands into the transgressive basal Devonian deposits. The Williston Basin, where the Silurian and Ordovician strata were preserved, persisted as a regional depression with little topographic relief, bounded on the east by the Seven Arch and on the south by the Transcontinental and Sioux arches. In central Alberta and Saskatchewan, where the protective cover of resistant Ordovician carbonate rocks was breached, deep erosion of the less resistant clastic rocks of the underlying Sauk sequence created the northward-facing Meadow Lake Escarpment, which was about 250 m high (figure

6). The initial Middle Devonian (Eifelian) deposits, which consisted of massive halite beds, associated red mudstones and a basal sandstone, completely filled the topographic depressions that were enclosed on the west by a combination of cratonic arches (Western Alberta, Peace River and Tathlina), and on the north and south by inward facing Cambro-Ordovician escarpments (figure 10*a*). The three pods of Eifelian salt that remain are relics of a much more extensive evaporite deposit, the eastern parts of which have been removed by subsequent dissolution, and by erosion during Mesozoic time. The early Middle Devonian (Eifelian) northwest margin of the evaporite basin was controlled by a massive reef barrier (Manetoe) which was situated

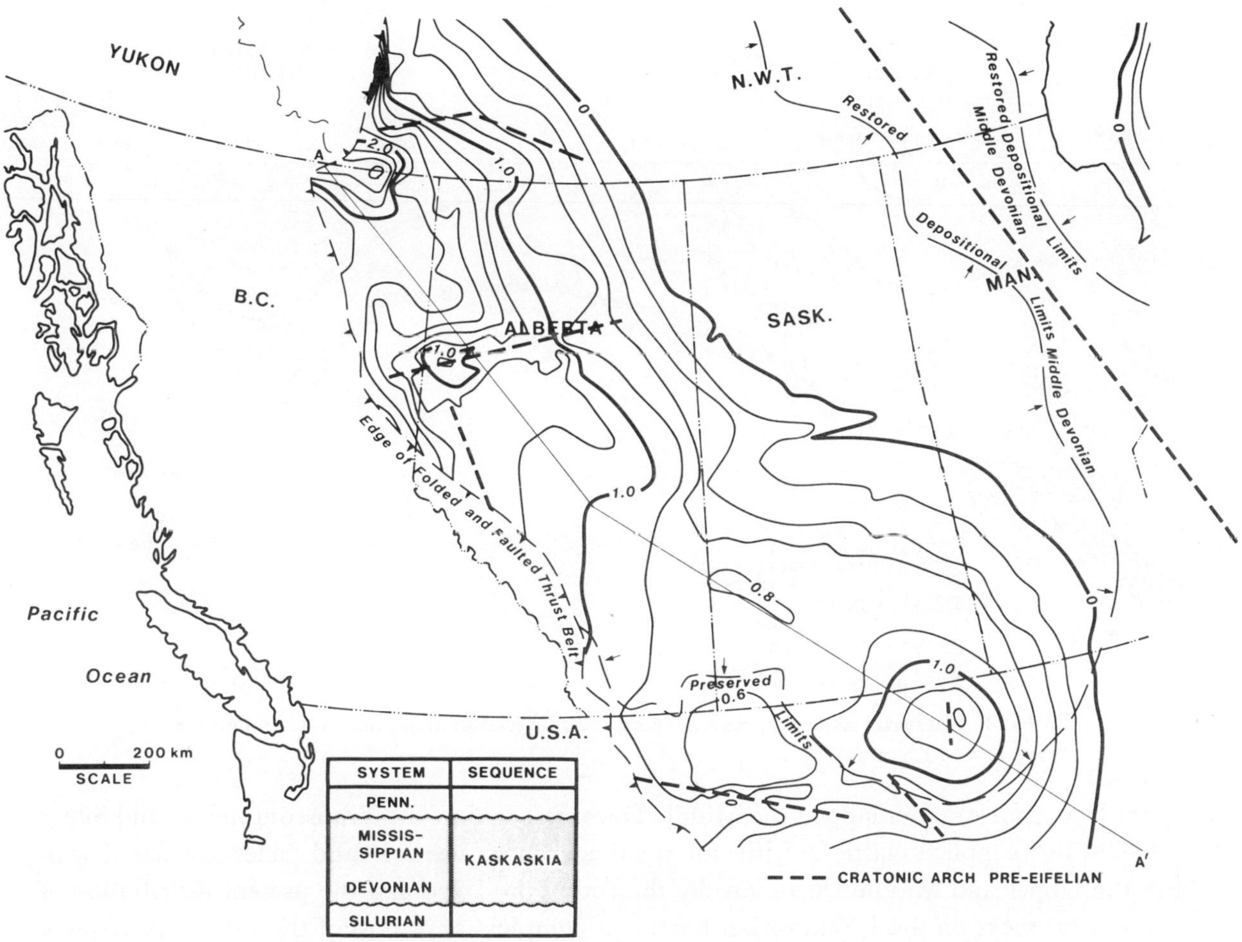

FIGURE 8. Kaskaskia sequence, regional map, showing line of section A–A′ of figure 9.

north of the map area, and which limited the southern encroachment of unrestricted marine deposition (figure 10*a*). Subsequent dissolution of thick halite deposits by meteoric water resulted in complex adiastrophic structures within the Western Canada Basin. The dissolution of the salt affected both the preservation and the facies and thickness of overlying beds.

The changing patterns of distribution of successively younger Devonian salt deposits (figure 10) involved a progressive southward shift into the central part of the Williston Basin, and was part of a regional pattern involving southward transgression of barrier reefs and unrestricted marine facies (figures 9 and 10). The Peace River Arch and Western Alberta Arch remained emergent and supplied sand to transgressive Givetian deposits. A massive reef complex (Presqu'ile) that developed across the south flank of the Tathlina Arch formed the barrier behind which was formed an extensive (600 × 1600 km) Givetian evaporite basin, bounded by the Western Alberta, Severn, Transcontinental and Sioux arches (figure 10 *b*). Successively younger (Frasnian and Fammennian) barrier reefs enclosed progressively smaller salt basins (figure 10*c*,*d*). Frasnian deposits transgressed the Western Alberta Arch, reestablishing the marine connection between the cratonic platform and the adjacent continental shelf, and Famennian deposits transgressed the last inliers of the Precambrian basement complex on the crest of the Peace River Arch (figures 7 and 11). At the south end of the Williston Basin, the

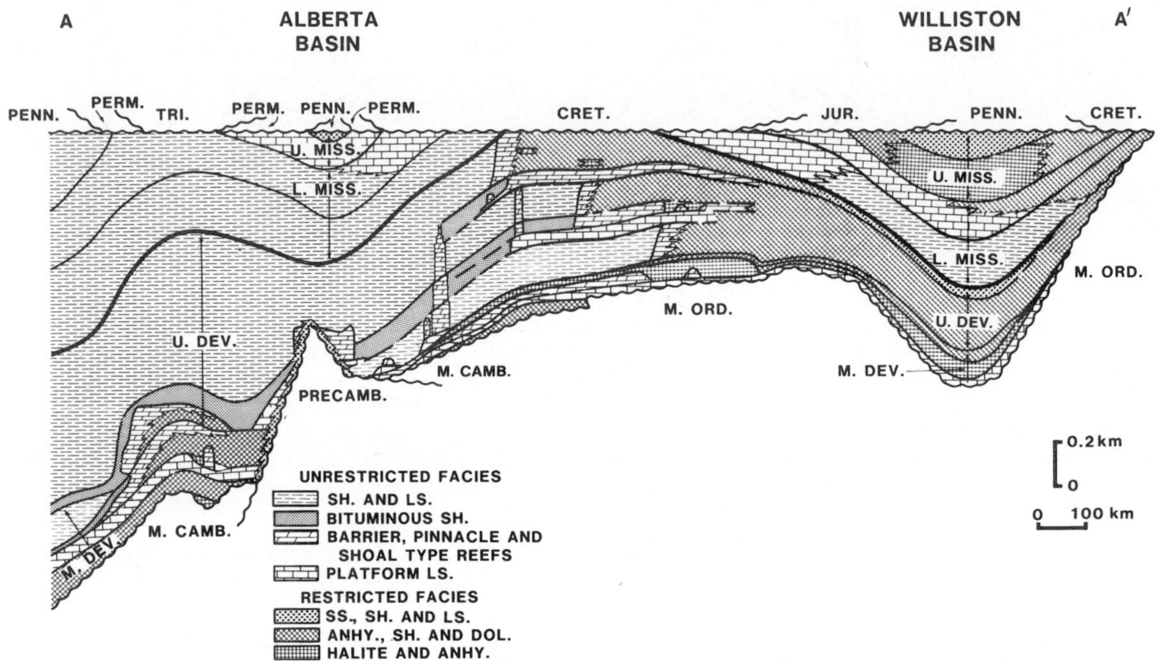

FIGURE 9. Kaskaskia sequence, regional stratigraphic section along line A–A′ of figure 8.

Upper Devonian strata onlapped the Middle Devonian against the Transcontinental and Sioux arches, which supplied clastic detritus for the diachronous inshore sand facies associated with both the Upper and Middle Devonian. By the end of the Devonian the present distribution of supracrustal rocks on the Precambrian basement complex, and most of the major epeirogenic structures on the cratonic platform, had been established (figure 11). Progressively younger deposits had onlapped eastwards over the Precambrian basement complex in the southern part of the area; but in the northern part the Upper Devonian had onlapped the Middle Devonian westward over the Peace River Arch, which had been stripped of its cover of Sauk and Tippecanoe sequences by erosion during early Devonian uplift.

On the cratonic platform much of the stratigraphic record of the Absaroka sequence and the Mississippian and Upper Devonian parts of the Kaskaskia sequence was destroyed by the Pennsylvanian to Middle Jurassic emergence and erosion of the Sweetgrass Arch, and by the regional northeastward erosional truncation that preceded the deposition of the foreland basin (Zuni) sequence (figure 12).

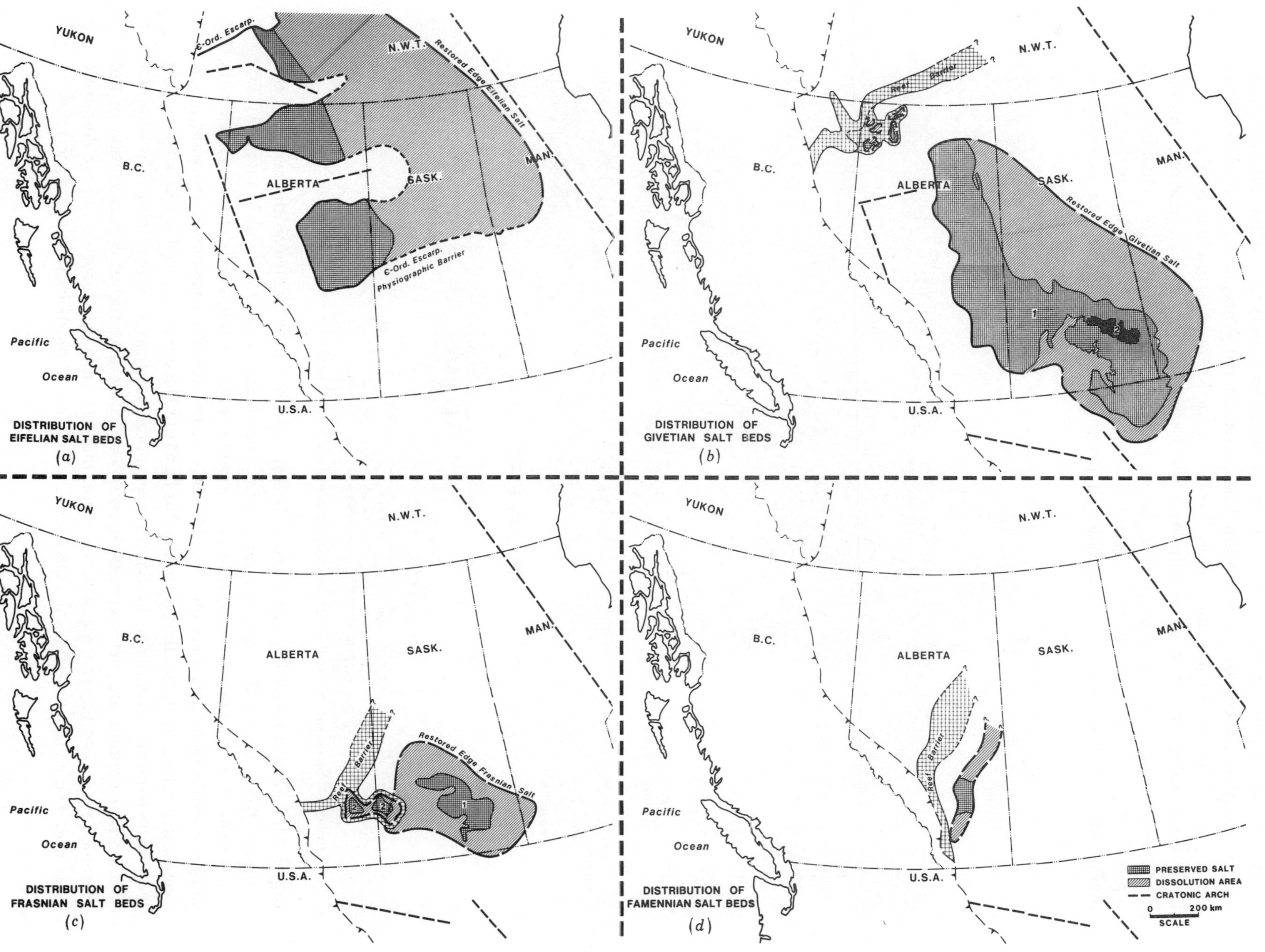

FIGURE 10. Devonian salt deposits.

Mississippian deposition began with an abrupt transgression marked by a thin but widespread veneer of dark organic shale. The associated fine clastic detritus may have been eroded from the Severn, Transcontinental and Sioux arches. There followed a gradual transition from Kinderhookean restricted euxinic conditions to an Osagean transgressive stage of open marine calcareous shales and limestones, and to lower Meramecian deposition of crinoidal sands representing the culmination of the transgression. The latter are gradational, on the eastern margin of the Williston Basin, into sabkha anhydrites and dolomites. They are succeeded by

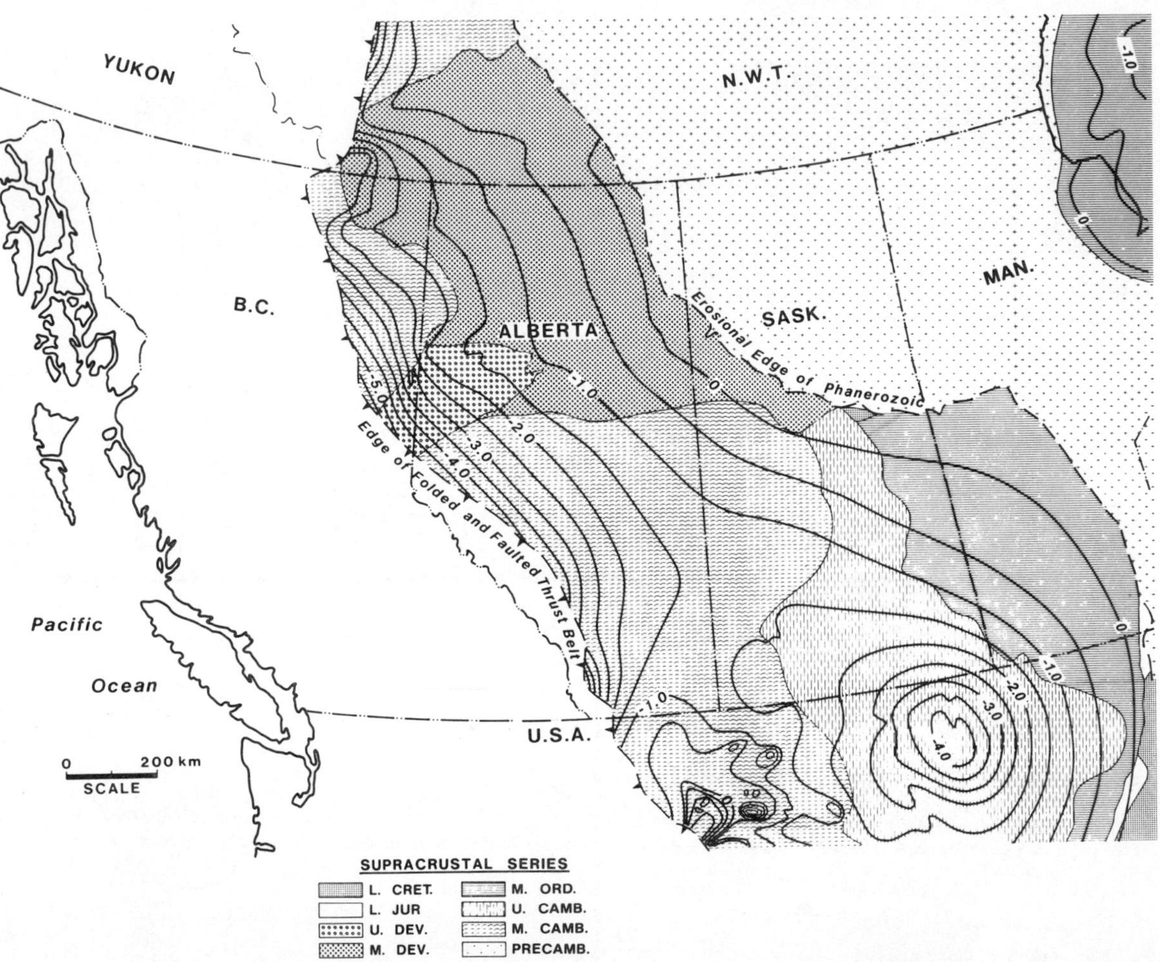

FIGURE 11. Structure of the Precambrian basement surface, and ages of overlying supracrustal rocks.

upper Meramecian regressive halite and anhydrite cycles, which formed in the centre of the Williston Basin, behind the barrier created by the emergence of the Central Montana Uplift. These general conditions also obtained in Alberta but were much less restricted and evaporitic since the arches in Alberta were inactive. Here the carbonate cycles pass northwestward into a deep-water fine clastic facies that is continuous from the Devonian (figure 9). The whole of the Kaskaskia sequence eventually becomes shale to the northwest. Uppermost Mississippian (Chesterian) deposits are characterized by silts and sands that record a reactivation of the cratonic arches, and signal a regional regression that preceded the deposition of the Absaroka sequence. The conspicuous lateral variations in thickness of the Kaskaskia sequence reflect a combined influence of pre-Kaskaskia erosional relief, lateral variation in the amount of subsidence during deposition of the Kaskaskia sequence, and differential preservation due to lateral variations in the amount of uplift and erosion after the deposition of the Kaskaskia sequence.

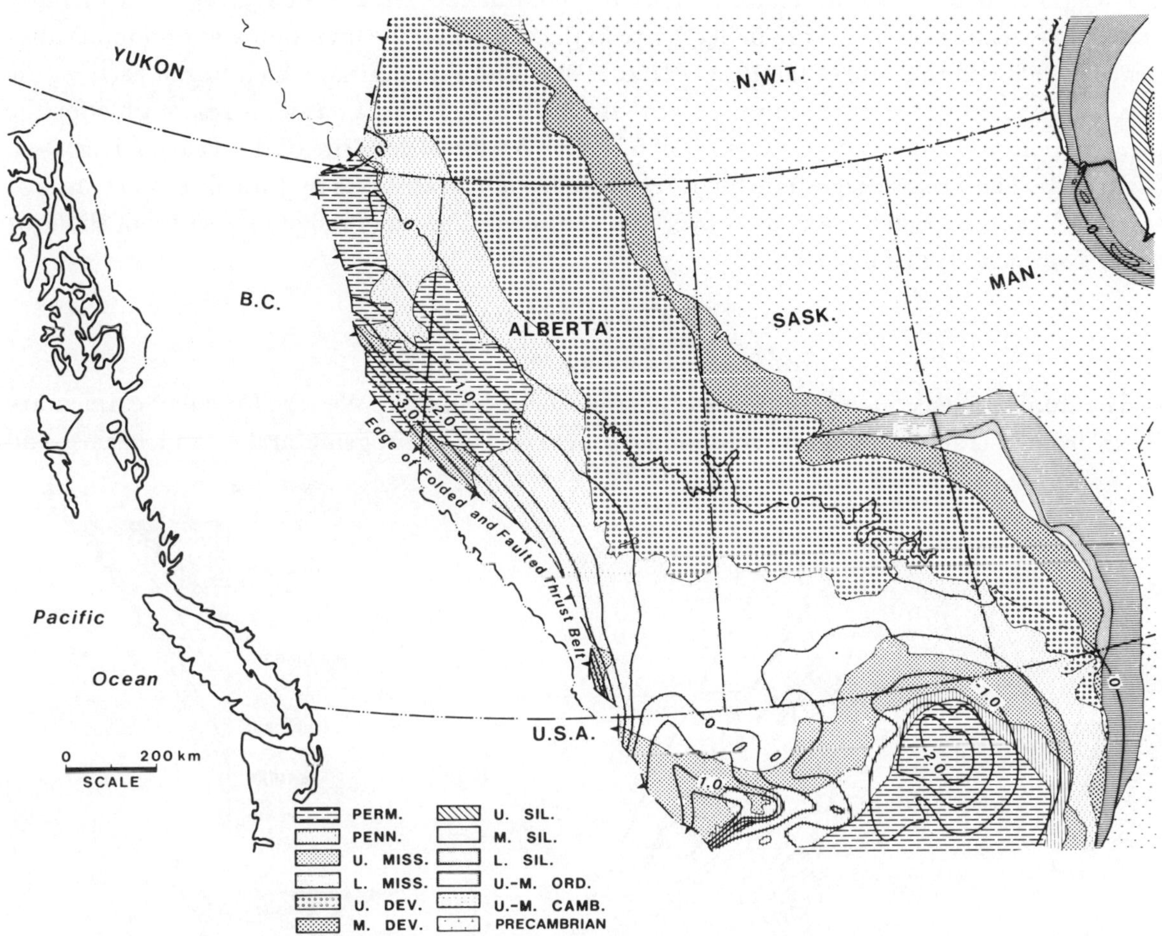

FIGURE 12. Structure and palaeogeology of the pre-Mesozoic erosional surface.

Absaroka sequence

The Absaroka sequence (figure 13) is characterized by a conspicuous westward shift in the limits of sediment accumulation, which records an emergence and westward tilt of the craton. This regional long-term regression is a continuation of a pattern established during deposition of the upper part of the Kaskaskia sequence, and is accompanied by an upward increase in the amount of terrigenous clastic detritus. The abrupt eastward thinning and termination of the Pennsylvanian, Permian and Triassic strata is a combined result of depositional attenuation and intermittent erosional bevelling.

The Absaroka sequence (figure 13) was separated into two contrasting facies domains by the early Pennsylvanian (Morrowan) to Middle Jurassic emergence of the northeast-trending Sweetgrass Arch. Southeast of the Sweetgrass Arch, in the Williston Basin, non-marine deposition was important, and during the Pennsylvanian and Permian deposition was influenced by the northward dispersal of sands and silts from the uplift of the ancestral Rocky Mountains (Mallory

1972; Rascoe & Baars 1972). During the Permian (Leonardian), Triassic and early Middle Jurassic, uplift of the bordering cratonic arches isolated the Williston Basin and it become a locus of evaporite (halite) and red bed deposition. This is in sharp contrast with the unrestricted marine deposits that characterize almost all of the Absaroka sequence northwest of the Sweetgrass Arch, on the cratonic platform and the continental terrace wedge. One particularly noteworthy feature of the Absaroka Sequence in this domain is the inversion and profound subsidence over the former site of the Peace River Arch. This foundering, which began early in the Mississippian and continued into the Permian, but not into the Triassic, is responsible for the preservation of Pennsylvanian and Permian rocks over the former site of the Peace River Arch (figure 12). The Triassic deposits, which were bevelled northeastwards by Jurassic erosion, appear to have been prograded westwards over the continental terrace wedge as a series of shallow-water continental shelf deposits.

THE FORELAND BASIN

Zuni sequence

The Zuni sequence (figure 14) records the transformation of the Western Canada Sedimentary Basin from a cratonic platform containing a network of epeirogenic arches and basins, and

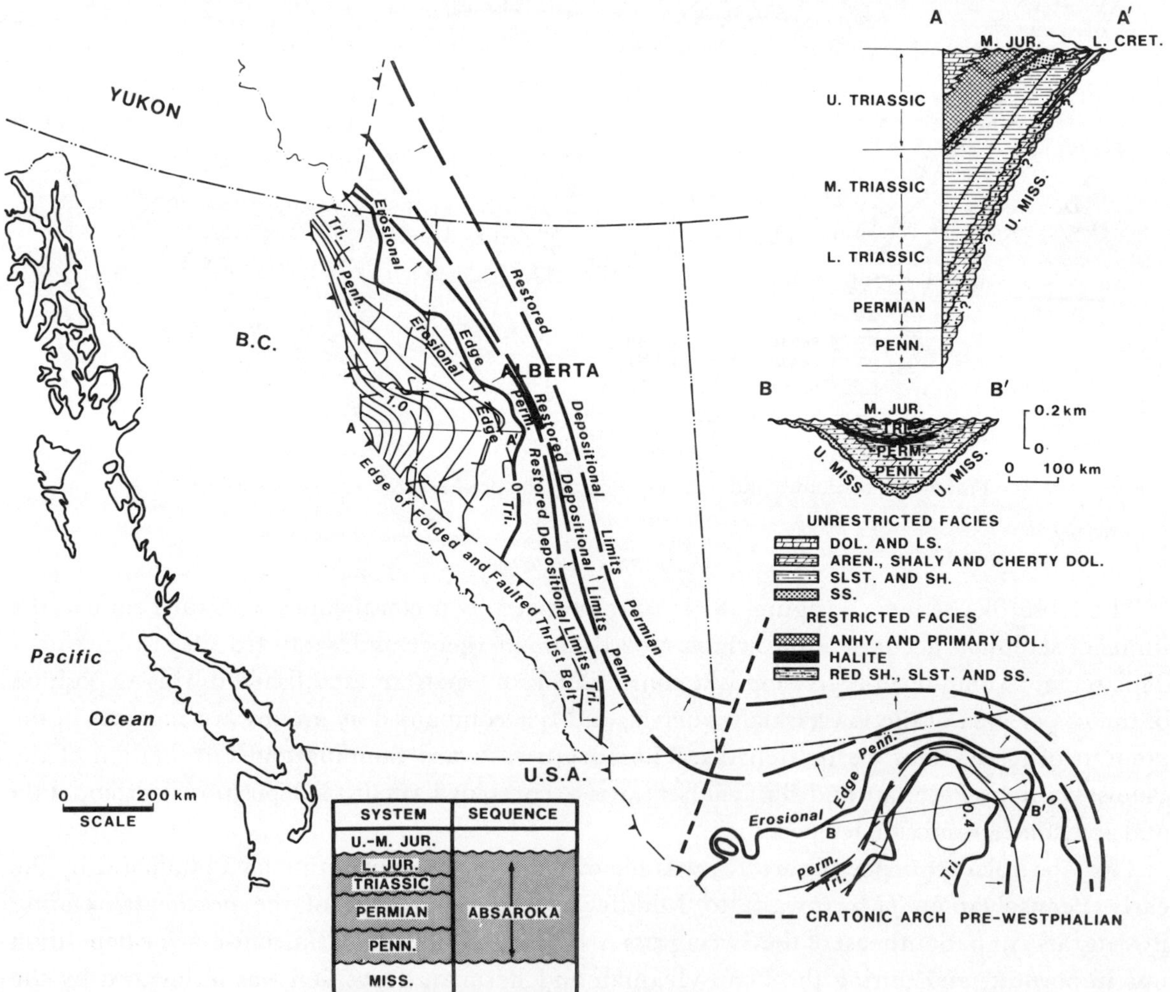

FIGURE 13. Absaroka sequence, regional map and stratigraphic sections A–A′ and B–B′.

bordered by a continental terrace wedge that was prograded into an adjacent ocean basin, into a foreland basin containing stacked northeastward tapering wedges of synorogenic clastic detritus that were prograded toward the interior of the craton. The Middle Jurassic deposits that initiated the Zuni sequence reflect the same pattern of tectonic controls on erosion and deposition that had characterized the Absaroka Sequence: deposition of a restricted facies within the lowermost Bajocian southeast of the Sweetgrass Arch in the Williston Basin, and of normal shallow-water marine facies on the craton and the continental shelf northwest of the Sweetgrass Arch. However, during the late Jurassic (Kimmeridgian and Portlandian) there

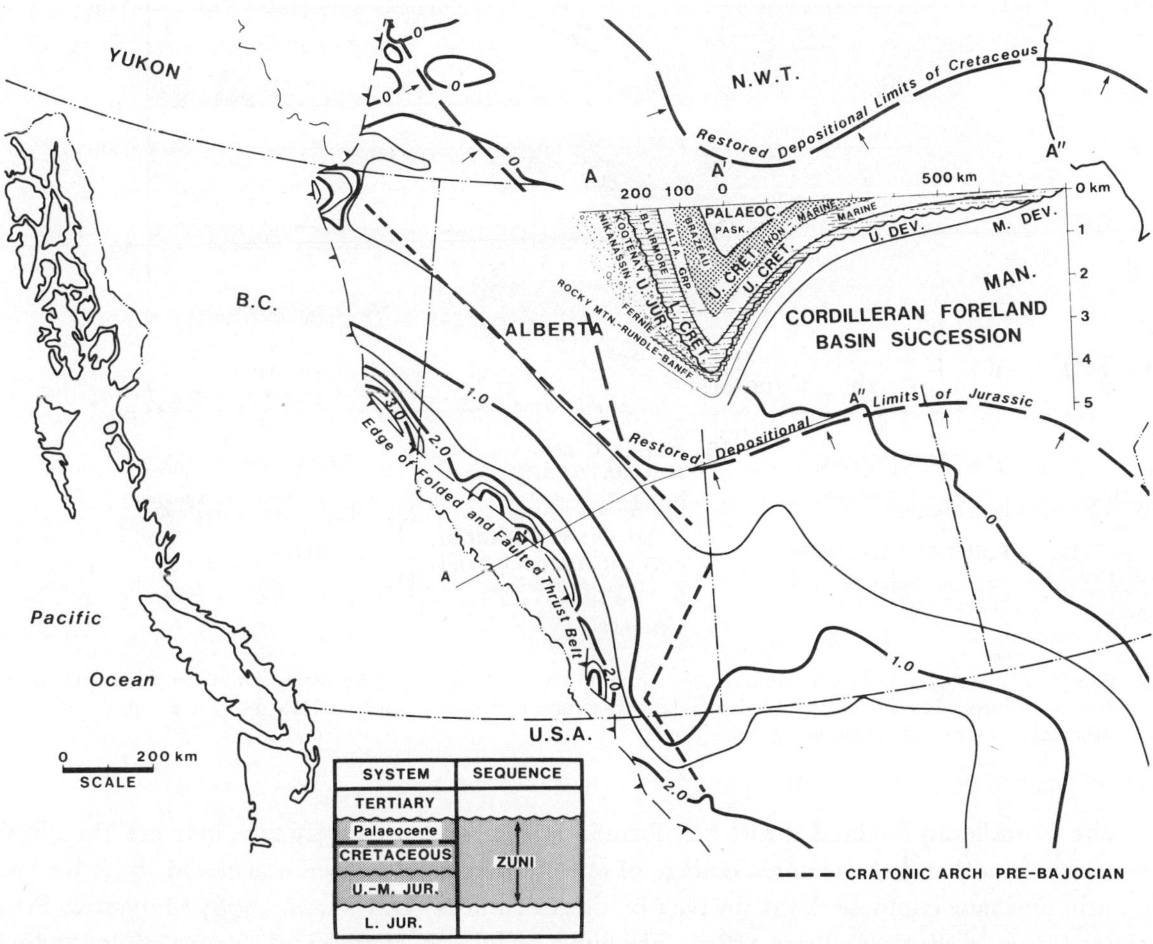

FIGURE 14. Zuni sequence, regional map and structure section. The segment A–A′ of the stratigraphic section is a palinspastic reconstruction through the foreland thrust belt, based on a structure section by Price & Mountjoy (1970, Figures 2 and 3), and the segment A′–A″ on a regional stratigraphic section edited by Gussow (1962).

was a reversal in the direction of sediment transport and an abrupt influx of clastic detritus from the emerging Cordilleran foreland thrust belt. Within the thrust belt, the latest Jurassic and earliest Cretaceous synorogenic clastic wedge, comprising the non-marine Kootenay and Nikannassin Formations, consists of sedimentary detritus eroded from the continental terrace wedge, and locally is more than 1 km thick. The succeeding early Cretaceous synorogenic clastic wedge, comprising the Blairmore Group, is mainly non-marine, contains metamorphic and igneous, as well as sedimentary detritus, and is locally more than 2 km thick. It spread out across the basin unconformably overlapping the Palaeozoic and older Mesozoic deposits, finally onlapping the crystalline basement (figure 12). An extensive marine transgression, which had been encroaching during the earlier Cretaceous from both the north and south, covered by Albian to Cenomanian time most of the Western Canada Basin and appears to coincide with a lull in the development of the Cordilleran orogenic belt. The late Cretaceous and Palaeocene synorogenic clastic wedge deposits, which are also of mixed provenance, comprise a major regressive cycle, up to 4 km thick, and record the final stages of the evolution of the foreland basin.

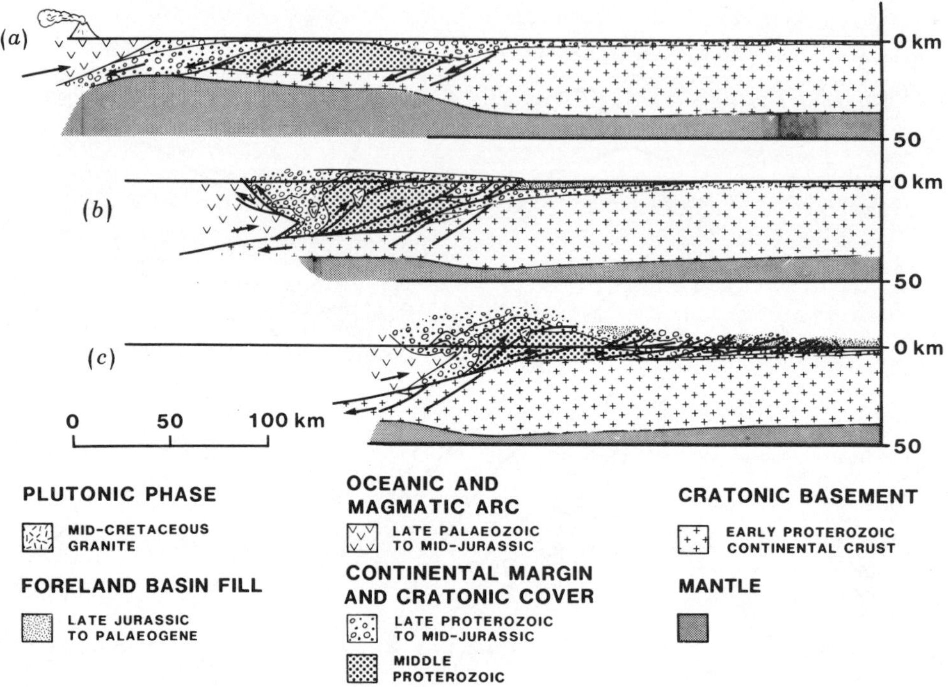

FIGURE 15. Three stages in the evolution of the foreland thrust belt in the southeastern Canadian Cordillera, based on a palinspastic reconstruction of a balanced structure section along 49° 45′ N (Price 1981). (*a*) Mid-Jurassic; (*b*) mid-Cretaceous; (*c*) mid-Tertiary.

The Cordilleran foreland thrust belt formed in the zone of convergence between the North America craton and the tectonic collage of oceanic terranes that were accreted to its western margin and now compose the main part of the Cordillera (Davis *et al.* 1978; Monger & Price 1979; Coney *et al.* 1980; Price 1981). The pre-Cordilleran continental terrace wedge, which had been deposited outboard from the Proterozoic rifted margin of the North American craton on a basement of oceanic or tectonically attenuated continental crust, was compressed, tectonically thickened, detached from its basement, and juxtaposed over the western margin of the continental craton (figure 15). Part of the supracrustal cover of platformal and foreland basin deposits was scraped off the craton and accreted to the overriding mass to form a wedge of imbricate thrust fault slices. This accretionary wedge was tectonically prograded over the margin of the continental craton. The foreland basin formed as a cratonward migrating moat because of isostatic flexure of the lithosphere in response to the tectonic loads imposed on it by the overriding continental terrace wedge and by the growing wedge of imbricate thrust fault slices. The sediments that filled the foreland basin comprise the clastic outwash that was eroded from the emerging thrust slices, and was trapped in this moat (Price 1973; Beaumont 1981). Thus the history of development of the foreland thrust belt and the lithospheric plate movements that gave rise to the foreland thrust belt is recorded by the sediments deposited in the foreland basin.

Three general stages in the evolution of the foreland thrust belt can be identified on the basis of structural relations within it (Price 1981), and these coincide with three general phases in the evolution of the foreland basin (figure 15). The long history of growth of the continental terrace wedge ended in a mid-Jurassic (Bajocian?) 'collision' with a 'foreign' terrane comprising an oceanic volcanic arc assemblage that accumulated on a basement of Triassic and Upper Palaeozoic rocks, quite unlike those in the continental terrace wedge (Monger & Price 1979). During the late Jurassic and early Cretaceous Columbian Orogeny, while the continental terrace wedge was still situated outboard from the craton, it was compressed between the 'foreign' terrane and the North American craton, was tectonically thickened, and was thrust over both. The load imposed on the lithosphere by the tectonic thickening and overthrusting of the continental terrace wedge was responsible for the isostatic flexure and the subsidence of the foreland basin that is recorded by the late Jurassic – early Cretaceous non-marine to brackish deposits of the Kootenay – Nikannasin clastic wedge, and the early to mid-Cretaceous mainly non-marine deposits of the much more extensive Blairmore clastic wedge (figure 14). Mid-Cretaceous granitic plutons that cut the thrust faults along the west side and in the central part of the deformed continental terrace wedge define the upper limit for the time of deformation there. The granitic intrusions are part of a regional episode that coincided with the widespread mid-Cretaceous marine transgression over the foreland basin. They appear to be associated with a lull in the convergence between the accreted 'foreign' terrane and the North American craton. The final stage in the evolution of the foreland thrust belt, during the late Cretaceous and Palaeocene Laramide orogeny, involved renewed convergence during which: (1) the deformed continental terrace wedge was detached from its basement and was displaced over the margin of the craton, (2) the supracrustal rocks scraped off the continental craton were incorporated into the eastward prograding accretionary wedge of imbricate fault slices, and (3) the foreland basin subsided rapidly and expanded eastward as it was filled with the thick regressive sequence of clastic wedge deposits that mark the terminal stage in the evolution of the foreland basin.

Tejas sequence

The record of the Tejas sequence is fragmentary. It is limited to a few small erosional outliers on the craton and to local basin-fill deposits in the eastern Cordillera. It is a record of intermittent crustal extension and listric normal faulting during which thin sheets of gravel were spread across the Western Canada Sedimentary Basin.

Summary

The Palaeohelikian flat-lying unmetamorphosed basin remnants high on the craton show that the craton was established long before the commencement of the Phanerozoic cycle of basin development examined in this paper. The first significant event in this history was the development, over attenuated crust, of an Atlantic type down-to-ocean faulted margin in the Neohelikian, upon which was deposited a great wedge of clastic sediments across the structural grain of the Archaean platform, followed after a gap of 500 Ma by renewed faulting and subsidence and further loading in Hadrynian time. The subsidence of the continental margin as a result of this imposed load and thermal contraction extended far into the craton and set the stage for the first cycle of deposition represented by the Sauk sequence.

Sauk sequence. The Sauk sequence records progressive onlap with the Upper Cambrian strata extending high on the western side of the craton and the depositional strike running north-westerly, parallel to the probable old continental margin. There is a notable lack of arches or isolated basins. This initial sequence is dominated by clastic sediments from the shield.

Tippecanoe sequence. Arches interrupted the linear depositional pattern. Marked sedimentary attenuation over highs created a relative thickening in the area of the Williston Basin. This sequence shows thin seams of fine sand and silt suggesting episodic uplift of the burgeoning arches, but the dominance of carbonate rock indicates that they were for the most part covered. The lack of an onlapping relation, and widespread inundation of the craton indicate a very rapid transgression. Like the Cambrian seas, those of the Ordovician and Silurian transgressed from the western margin.

Kaskaskia sequence. The strongly developed pre-Devonian arches surrounding and segmenting the Devonian depositional area resulted in a transgression of over 3000 km from the northwest, rather than the west, over a tectonically disturbed and eroded surface, resulting in complex facies patterns.

The Mississippian seas covered almost as large an area as those of the Ordovician, onlapping completely the Western Alberta and Peace River arches. Arches in the southern and western part of the Williston Basin continued to grow into late Kaskaskia time, creating sills with salt deposits in the regressive part of the sequence that culminated with a clastic assemblage derived from the flanking arches.

Absaroka sequence. The Sweetgrass Arch isolated the Alberta Basin from the Williston Basin for the first time. These opened to the southwest and west respectively. The Williston Basin had a dominant red bed-evaporite assemblage, the Alberta Basin a marine clastic assemblage.

This interval is characterized by periods of non-deposition and erosion, with depositional limits of each system indicating much less encroachment on to the craton, which appears to have progressively increased its rate of dip to the west. The progressively increased rate of westward thickening from the Sauk to Kaskaskia to Absaroka in the undisturbed part of the basin also suggests an increasing westward tilt or subsidence of the basin margin.

Zuni sequence. The Zuni reflects the late Jurassic orogenic activity in the west. The convergence of allochthonous terranes with the continental margin resulted in the thrusting of the marginal deposits eastward, tectonically loading the lithosphere and providing the provenance for the continental clastic sediment that filled the foredeep which formed in response to the loading. The continental sediment spread across the basin with marine seaways to the north and south that gradually advanced to join together to form a seaway extending from the Arctic to the Gulf of Mexico in mid-Cretaceous time. A second plate convergence resulted in further tectonic thickening with thrusting extending further east, cannibalizing previously formed deposits and generating a second sequence of terrestrial sediments that covered the entire basin again in the late Cretaceous and Palaeocene.

Notably absent during the Zuni was the influence of arches that had previously segmented the basin. Also absent are carbonates and evaporites. Instead the basin fill was entirely of clastic sediments. The present structural configuration of the basin was established during this period as the foredeep encroached onto the craton.

Discussion

The evolution of the Western Canada Sedimentary Basin involves a long history of recurrent regional or inter-regional vertical movements that controlled the transgressive–regressive megacycles recorded by the regional stratigraphic sequences. These cyclic movements appear to be part of a larger cycle involving gradual overall transgression and the expansion of realms of marine deposition from late Proterozoic to late Devonian to early Mississippian time, and then gradual overall regression and the retraction of realms of marine deposition until mid-Jurassic time. The rate of eastward thinning in the undisturbed portion of the Alberta Basin, expressed as sedimentary condensation as well as erosional truncation, increases with each successive sequence (figures 5, 8 and 13), indicating a progressive increase in westward tilt. Superimposed upon these regional or inter-regional movements are the recurrent local or intra-regional variations in amount of subsidence and uplift that controlled the protracted history of growth of the basins and arches. There is an obvious synchronism, on a regional scale, in the uplift of some of the arches and subsidence of some of the basins. This may also hold on an inter-regional scale. It is a fact that must be accommodated in any comprehensive models for the development of the basins and arches.

The obvious correlations between the history of development of the Western Canada Sedimentary Basin and that of the Cordillera leads to convincing explanations for some aspects of the development of the Western Canada Sedimentary Basin, but the explanations for some other fundamental aspects remain elusive. The subsidence of the basin during the final stage of its development can be attributed to isostatic flexure of the lithosphere in response to the tectonic loading imposed on it by the development of the foreland thrust belt, and to the ensuing loads imposed by the sediments that were trapped by the subsidence; on the other hand, the subsidence of the basin during the initial stage in its evolution, when the Sauk sequence was deposited as a simple pericratonic wedge that onlapped the crystalline basement, can be attributed to thermal contraction following late Proterozoic rifting and lithospheric stretching, and to isostatic flexure of the lithosphere in response to the load imposed on it by the accumulation of a thick wedge of clastic sediment in the continental terrace wedge. However, the development of cratonic arches and basins, which occurred during the intervening stages, remains enigmatic, because it does not appear to fit any of the popular contemporary models for the origin of sedimentary basins (Bally 1980).

A satisfactory model for the origin and development of the epeirogenic arches and basins on the craton will have to accommodate several important facts about their nature and evolution.

1. The Williston Basin and most of the cratonic arches originated after the Sauk sequence had been deposited as a relatively uniform northeasterly tapering wedge that covered the flank of the craton. They were superimposed on this simple pericratonic wedge, and it was deformed as they developed; but this deformation has not provided any evidence of significant crustal stretching when the Williston Basin began to form, nor is there any obvious indication of crustal heating or igneous intrusions preceding the initial subsidence of the basin.

2. The cratonic basins occupy the spaces within a network of intersecting linear arches (figure 1), and the development of the basins is as much a result of uplift and erosion of the cratonic arches between the transgressive–regressive depositional cycles as it is a result of differential subsidence of the basins during the cycles. The individual arches can be assigned to one of several groups that have similar histories of protracted intermittent uplift and erosion (figure 16).

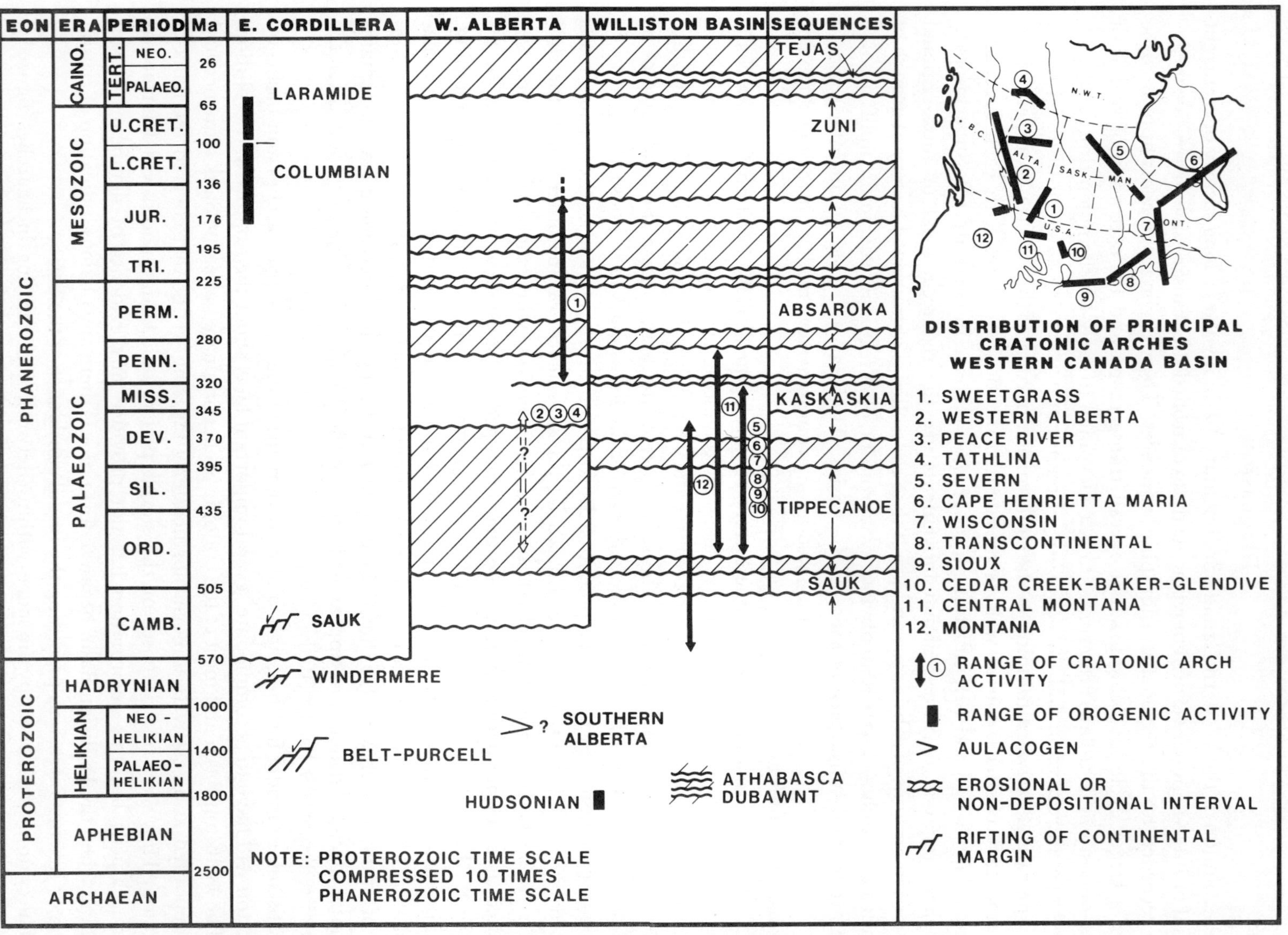

FIGURE 16. Tectonic history of the Western Canada Sedimentary Basin.

The Western Alberta, Peace River and Tathlina arches were uplifted and eroded mainly in early Devonian time, but they show some evidence of a previous history of Ordovician and Silurian uplift. These times of uplift coincide with the episodes of emergence of the continental terrace wedge. Uplift of these arches appears to have been related to deformation that was localized along zones of inherent structural weakness that follow the ancient rifted margin of the continent and the older northeasterly structural grain of the basement complex. Most of the other cratonic arches underwent synchronous intermittent uplift throughout the interval from Middle Ordovician to latest Mississippian time, and perhaps beyond; but uplift of the Sweetgrass River Arch did not begin until the Pennsylvanian, and it definitely persisted until the Middle Jurassic, isolating the Williston Basin from the continental terrace wedge. Mississippian and Pennsylvanian uplift of the Central Montana Arch was also important in isolating the Williston Basin. All of the arches were dormant during the compressive régime of the foreland basin (Zuni) deposition, although the Sweetgrass Arch appears to have been reactivated by Laramide compression in the Palaeogene.

3. The long history of intermittent differential subsidence of the Williston Basin spans more than 300 Ma from the Middle Ordovician to the Middle Jurassic. Although the geochronometric calibrations are quite imprecise, the average rates of accumulation in the basin during the various depositional episodes from the Middle Ordovician to the Mississippian appear to have been about the same, and to have been about an order of magnitude greater than during later episodes. We are not aware of any satisfactory explanation for the origin and development of these kinds of cratonic arches and basins; we consider this problem to be one of the major contemporary challenges in the study of the continents.

References (Porter *et al.*)

Aitken, J. D. 1966 *Bull. Can. Petrol. Geol.* **14**, 405–441.

Aitken, J. D. 1971 *Bull. Can. Petrol. Geol.* **19**, 557–569.

Bally, A. W. 1980 In *Dynamics of plate interiors* (ed. A. W. Bally *et al.*) (*Geodynamics Series*, vol. 1), pp. 5–20.

Beaumont, C. 1981 *Geophys. Jl R. astr. Soc.* **65**, 291–329.

Benvenuto, G. L. & Price, R. A. 1979 *Bull Can. Petrol. Geol.* **27**, 360–394.

Coney, P. L., Jones, D. L. & Monger, J. W. H. 1980 *Nature, Lond.* **288**, 329–333.

Davis, G. A., Monger, J. W. H. & Burchfiel, B. C. 1978 In *Mesozoic paleography of the western United States* (Society of Economic Paleontologists and Mineralogists, Pacific Coast Paleogeography Symposium) (ed. D. G. Howell & K. A. McDougall), pp. 481–502.

Douglas, R. J. W. 1969 *Geological map of Canada*. Geological Survey of Canada Map no. 1250A.

Fraser, J. A., Donaldson, J. A., Fahrig, W. F. & Tremblay, L. P. 1970 In *Symposium on Basins and Geosynclines of the Canadian Shield* (*Geol. Surv. Can. Pap.* no. 70–40) (ed. A. J. Baer), pp. 213–238.

Fuller, J. G. C. M. & Porter, J. W. 1962 *J. Alberta Soc. Petrol. Geol.* **10**, 455–481.

Gussow, W. C. (ed.) 1962 *Regional geological cross-sections of Western Canada Sedimentary cover*. Alberta Society of Petroleum Geologists and Geological Association of Canada.

Hoffman, P. F. 1973 *Phil. Trans. R. Soc. Lond.* A **273**, 547–581.

Kanasevich, E. R., Clowes, R. M. & McCloughan, C. H. 1969 *Tectonophysics* **8**, 513–527.

Lis, M. G. & Price, R. A. 1976 *Geol. Surv. Can. Pap.* no. 76–1A, pp. 135–136.

Mallory, W. W. 1972 In *Geological atlas of the Rocky Mountain Region* (ed. W. W. Mallory), pp. 111–127. Denver: Rocky Mountain Association of Geologists.

McMechan, M. E. 1979 *Geol. Soc. Am. Abstr. Programs* **17** (7), 477.

McMechan, M. E. 1980 Ph.D. thesis, Queen's University, Kingston, Canada.

McMechan, M. E. & Price, R. A. 1982 *Can. J. Earth Sci.* **19**. (In the press.)

Monger, J. W. H. & Price, R. A. 1979 *Can. J. Earth Sci.* **16**, 770–791.

Norris, D. K. & Price, R. A. 1966 *Bull. Can. Petrol. Geol.* **14**, 385–404.

Porter, J. W. & Fuller, J. G. C. M. 1959 *Bull. Am. Ass. Petrol. Geol.* **43**, 124–189.

Price, R. A. 1964 *Bull. Can. Petrol. Geol.* **12**, 399–426.

Price, R. A. 1973 In *Gravity and tectonics* (ed. K. A. De Jong & R. Scholten), pp. 491–502. New York: Wiley.

Price, R. A. 1981 In *Thrust and nappe tectonics* (ed. K. McClay & N. J. Price) (*Geol. Soc. Lond. spec. Pap.* no. 9), pp. 427–448.
Price, R. A. & Mountjoy, E. W. 1970 In *Structures of the southeastern Canadian Cordillera.* (ed. J. Wheeler) (*Geol. Ass. Can. spec. Pap.* no. 6), pp. 7–25.
Rascoe, B., Jr & Baars, D. L. 1972 In *Geological atlas of the Rocky Mountain Region* (ed. W. W. Mallory), pp. 143–165. Denver: Rocky Mountain Association of Geologists.
Sears, J. W. & Price, R. A. 1978 *Geology* **6**, 267–270.
Sloss, L. L. 1963 *Bull. geol. Soc. Am.* **74**, 93–113.
Stewart, J. H. 1976 *Geology* **4**, 11–15.
Wheeler, J. O. 1963 *Geol. Surv. Can. Pap.* no. 62–32.
Wheeler, J. O. 1965 *Geol. Surv. Can. Pap.* no. 64–32.

Bibliography

Apart from those specific references cited throughout the text, general information bearing on the tectonic elements and on the surface and subsurface distribution, thickness and lithology of systems comprising the sequences have been gleaned from the following sources.

Douglas, R. J. W. (ed.) 1968 *Geology and economic minerals of Canada.* Geological Survey of Canada.
King, P. B. 1969 *Tectonic map of North America* 1969, scale 1:5000000. U.S. Geological Survey.
Mallory, W. W. (ed.) 1972 *Geological history of the Rocky Mountain Region.* Rocky Mountain Association of Petroleum Geologists.
McCrossan, R. G. & Glaister, R. P. (eds) 1964 *Geological history of western Canada.* Alberta Society of Petroleum Geologists.
McCrossan, R. G. (ed.) 1973 *Future petroleum provinces of Canada.* Memoir no. 1, Canadian Society of Petroleum Geologists.
Muehlberger, W. B. 1967 *Basement map of North America*, scale 1:5000000. American Association of Petroleum Geologists.
Stockwell, C. H. 1969 *Tectonic map of Canada* 1969, scale 1:5000000. Geological Survey of Canada.

Phil. Trans. R. Soc. Lond. A **305**, 193–205 (1982) [193]
Printed in Great Britain

Evolution of sedimentary basins in the Canadian Arctic

BY J. W. KERR

J. William Kerr and Associates Ltd, Suite 407, 630 8*th Avenue S.W.,*
Calgary, Alberta, Canada T2P 1*G*6

Since middle Proterozoic time, two long-lasting phases affected the Canadian Arctic Archipelago, each forming a different sedimentary basin.

The Franklinian Basin, which was floored by continental or quasi-continental crust, received 10 km or more of clastic, carbonate and volcanic rocks from the mid-Proterozoic to Devonian. Internal parts of the basin were deformed, intruded and metamorphosed locally, and external parts were folded and thrust cratonward by compressional episodes of the Ellesmerian Orogeny, which culminated in the late Devonian. This marked the end of a phase, at which time the entire region may have been emergent. The nature of plate interactions that produced Ellesmerian deformation are unknown.

The second phase began in the early Carboniferous, when plate movements of the Boreal Rifting Episode created the proto-Canada Basin by left-hand transform motion of a plate along the modern continental margin and the location of the Kaltag Fault of northern Alaska. As a marginal side effect of that motion, the Sverdrup Basin developed as a peri-cratonic incipient rift. From the Carboniferous to late Cretaceous the basin received about 13 km of cratonic-derived clastic detritus.

From late Cretaceous to early Tertiary time, the Arctic Archipelago was disrupted by the interference of two plate movements originating in the Arctic and North Atlantic regions. Those events had three main effects: the craton was extended and a graben-filled depression formed in the southeastern part of the archipelago; the eastern and central parts of the Sverdrup Basin were compressed and uplifted (Eurekan Orogeny); and resultant clastics prograded northwestward toward the Canada Basin, to form the Arctic continental terrace wedge.

INTRODUCTION

Several of my recent papers have dealt with different aspects of the geology of the Canadian Arctic (Kerr 1980*a*, *b*, *c*, 1981*a*, *b*). Together they describe the evolution of the sedimentary basins there. That evolution is briefly summarized here by the use of figures published in the earlier papers.

Two long-lasting phases affected the Canadian Arctic. The constructional phase involved formation of a supercontinent, Pangaea. Nearly all of the present Arctic lands was part of Pangaea, including the present lands and marine areas. The subsequent fragmentation phase involved the break-up of that supercontinent by plate tectonic processes to form the present oceans, continents, islands and channels.

CONSTRUCTIONAL PHASE

The constructional phase included all of early Palaeozoic time and ended at the end of the Ellesmerian Orogeny in late Devonian or early Mississippian time. The configuration of the Canadian Arctic at the end of this phase is shown in figure 1 (from Kerr 1980*a*, fig. 3). The area of the present Canada Basin, Canadian Arctic Islands, and nearby seaways apparently was made

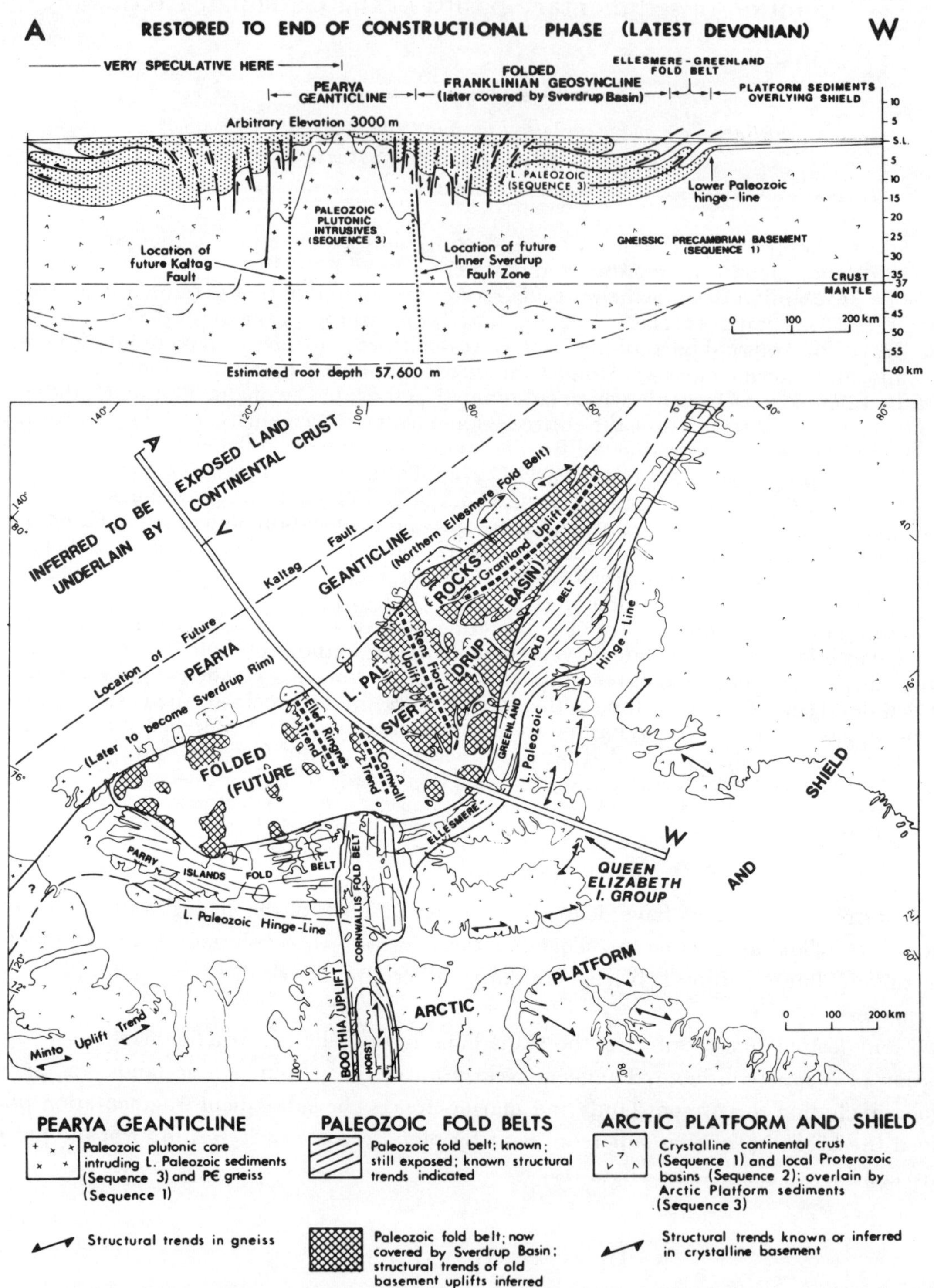

FIGURE 1. Structural configuration of the Canadian Arctic in latest Devonian time, at the end of the constructional phase. For location see figure 3.

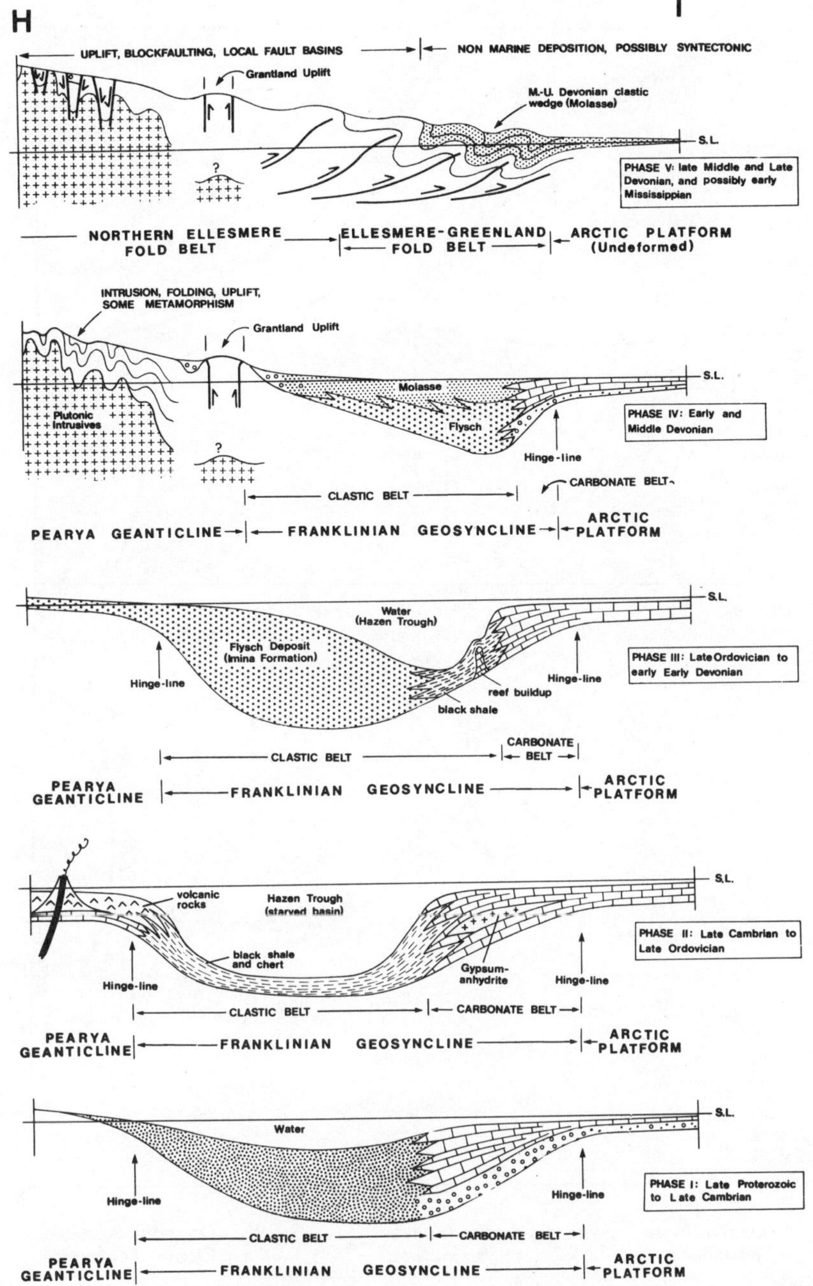

Figure 2. Evolution of the Franklinian Geosyncline (sequence 3, figure 1), showing the five phases of its development, with the youngest at the top.

up of a continental crust. A linear tectonic belt extending NE–SW through this region was the Pearya Geanticline. It had been active through much of previous Palaeozoic time. At the end of the constructional phase the geanticline was a topographically high region that had been deformed, intruded and metamorphosed by the Ellesmerian Orogeny. This thick rigid geanticline was to have a major control on structures that would develop in the later phase.

The Lower Palaeozoic Franklinian Geosyncline lay southeast of the Pearya Geanticline, and presumably formed on continental crust (figure 1). It developed simultaneously with the Pearya

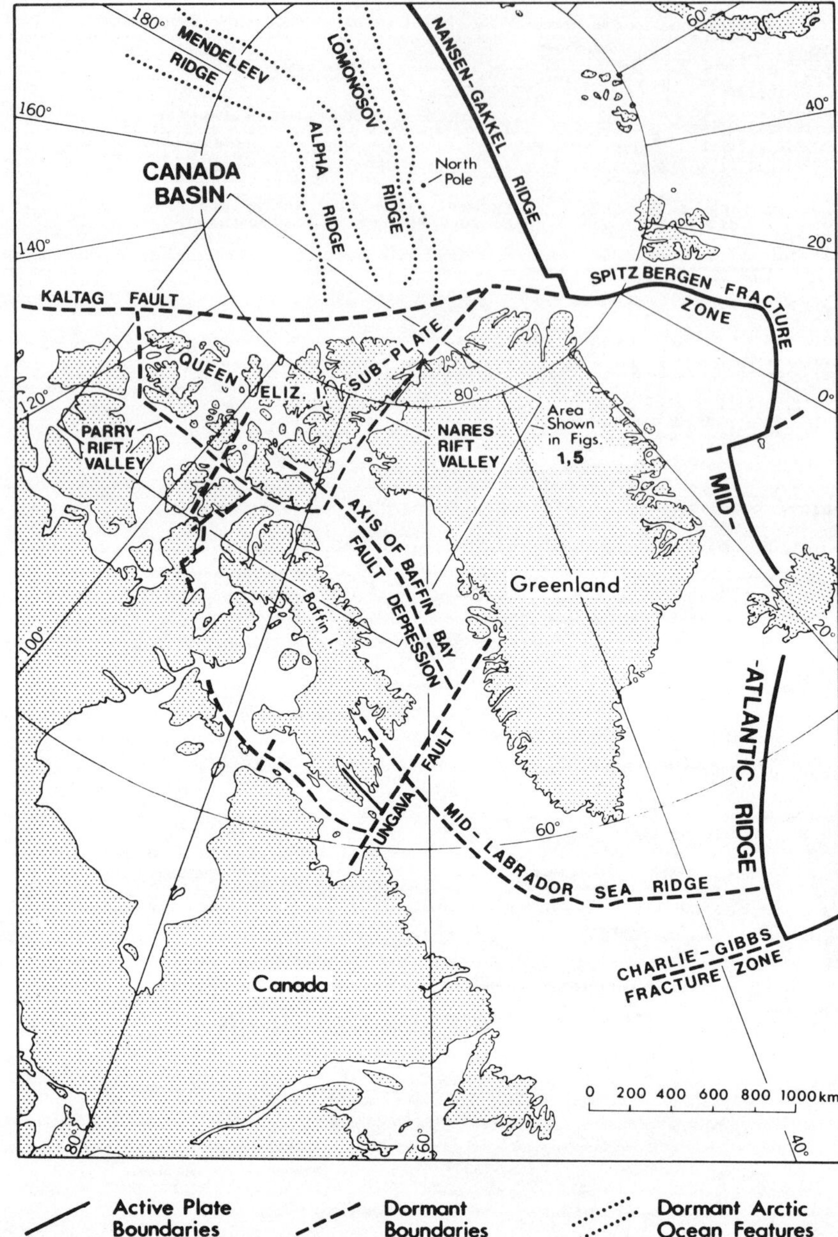

FIGURE 3. Present-day geography of the North American Arctic, showing the major plate-related structures.

Geanticline. The five phases in the development of the Franklinian Geosyncline are shown in figure 2 (from Kerr 1981 *a*, fig. 7).

FRAGMENTATION PHASE

The fragmentation phase of the Canadian Arctic involved plate tectonic processes that broke apart the supercontinent to form the present-day lands, seas and structures that are shown in figure 3 (after Kerr 1980 *a*, fig. 1). Fragmentation resulted from the interplay of two plate tectonic events that were largely restricted to opposite sides of the present continent. Each of these events

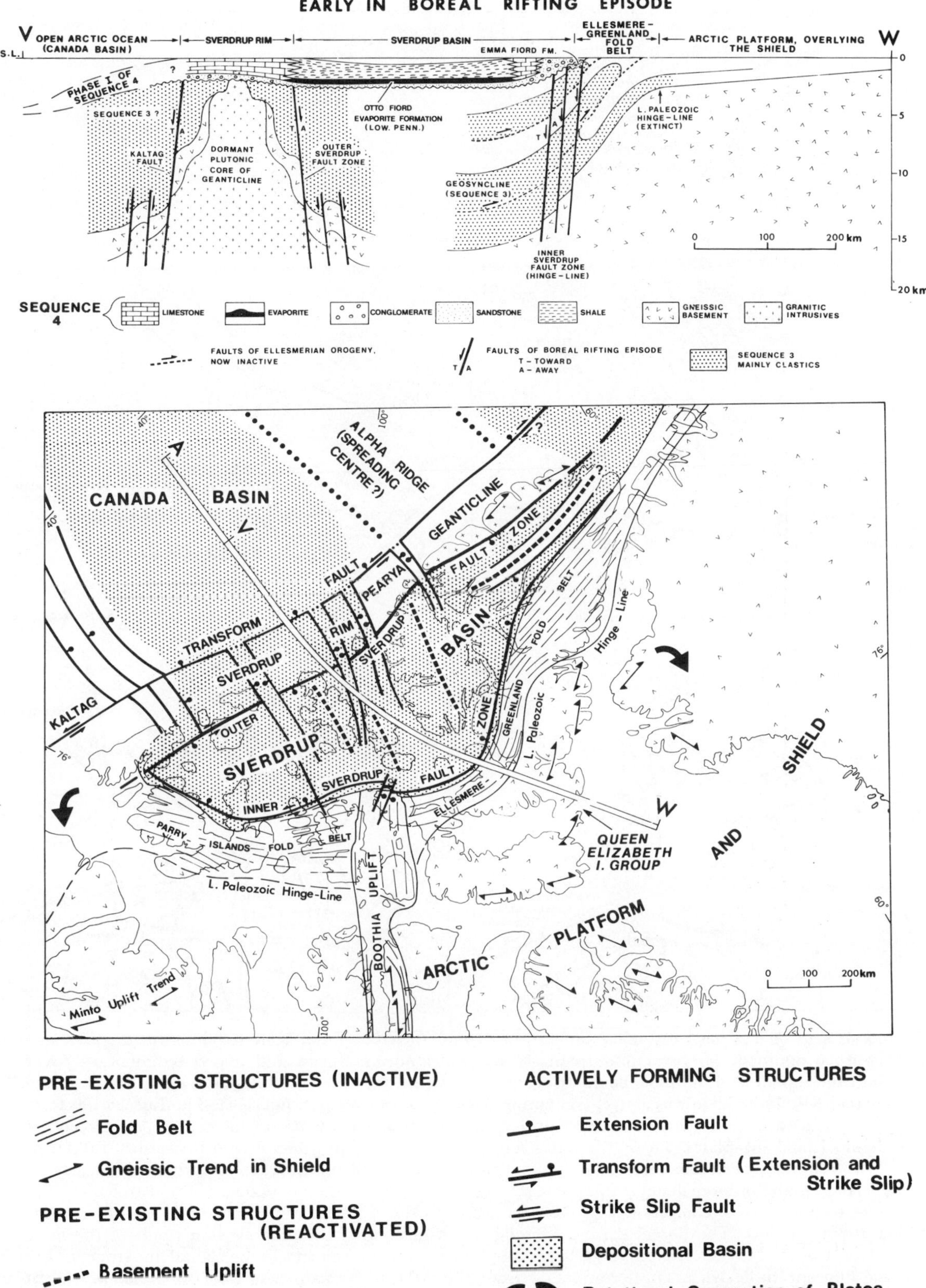

FIGURE 4. Initial fragmentation of the Canadian Arctic to form early stages of the Canada Basin and the Sverdrup Basin. This represents late Palaeozoic time (Carboniferous to Permian) in an early stage of the Boreal Rifting Episode. The islands had not yet formed (cf. figure 3). The structures in the map and cross section were controlled by the older structures shown in figure 1.

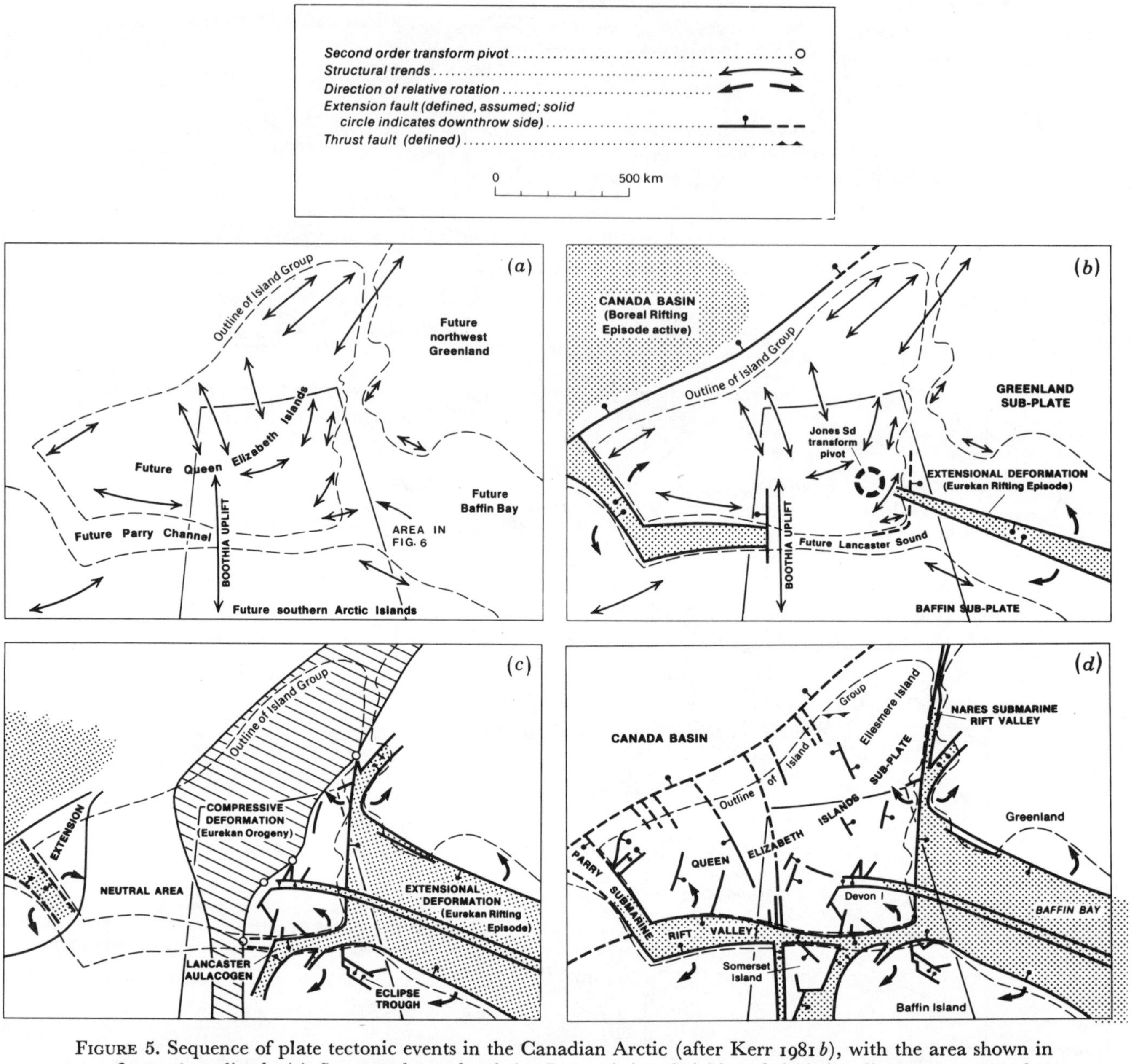

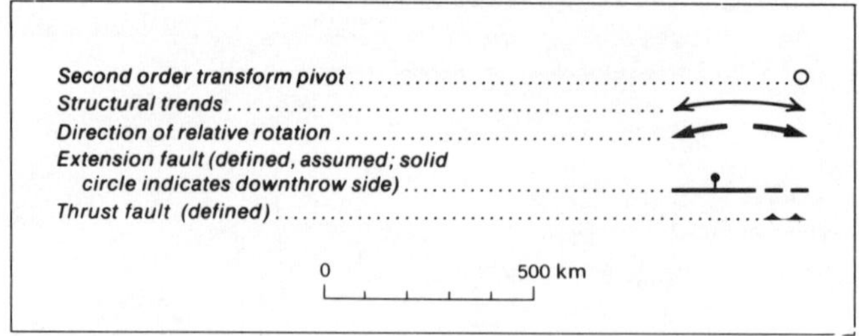

FIGURE 5. Sequence of plate tectonic events in the Canadian Arctic (after Kerr 1981 *b*), with the area shown in figure 6 outlined. (*a*) Structural trends of the Precambrian Shield and folded sedimentary cover that influenced plate and sub-plate boundaries and movements. (*b*) Latest Cretaceous to early Tertiary time. The Boreal Rifting Episode was active, emanating from the northwest (cf. figure 4). The Eurekan Deformation was in an early phase, and northwest propagating extension faults began to separate the Greenland and Baffin Island sub-plates. These sub-plates rotated apart about the Jones Sound first-order transform pivot in the future Queen Elizabeth Islands. (*c*) Climactic phase of the Eurekan Deformation in early to mid-Tertiary time (for more detail of this phase see figure 6). Extension in the southeast (Eurekan Rifting Episode) was transformed to compression in the northwest (Eurekan Orogeny) by means of several second-order transform pivots. The Boreal Rifting Episode was diminishing, presumably as a result of the increasing activity of the Eurekan Deformation. (*d*) Final phase of plate movements in the Canadian Arctic, in Miocene or Pliocene time, when faults emanating from the Atlantic Ocean were able to break northwestward through the continent to connect with older faults of the Arctic Ocean, obliterating the transform pivots. This structural configuration of sub-plates exists today, modified by erosion.

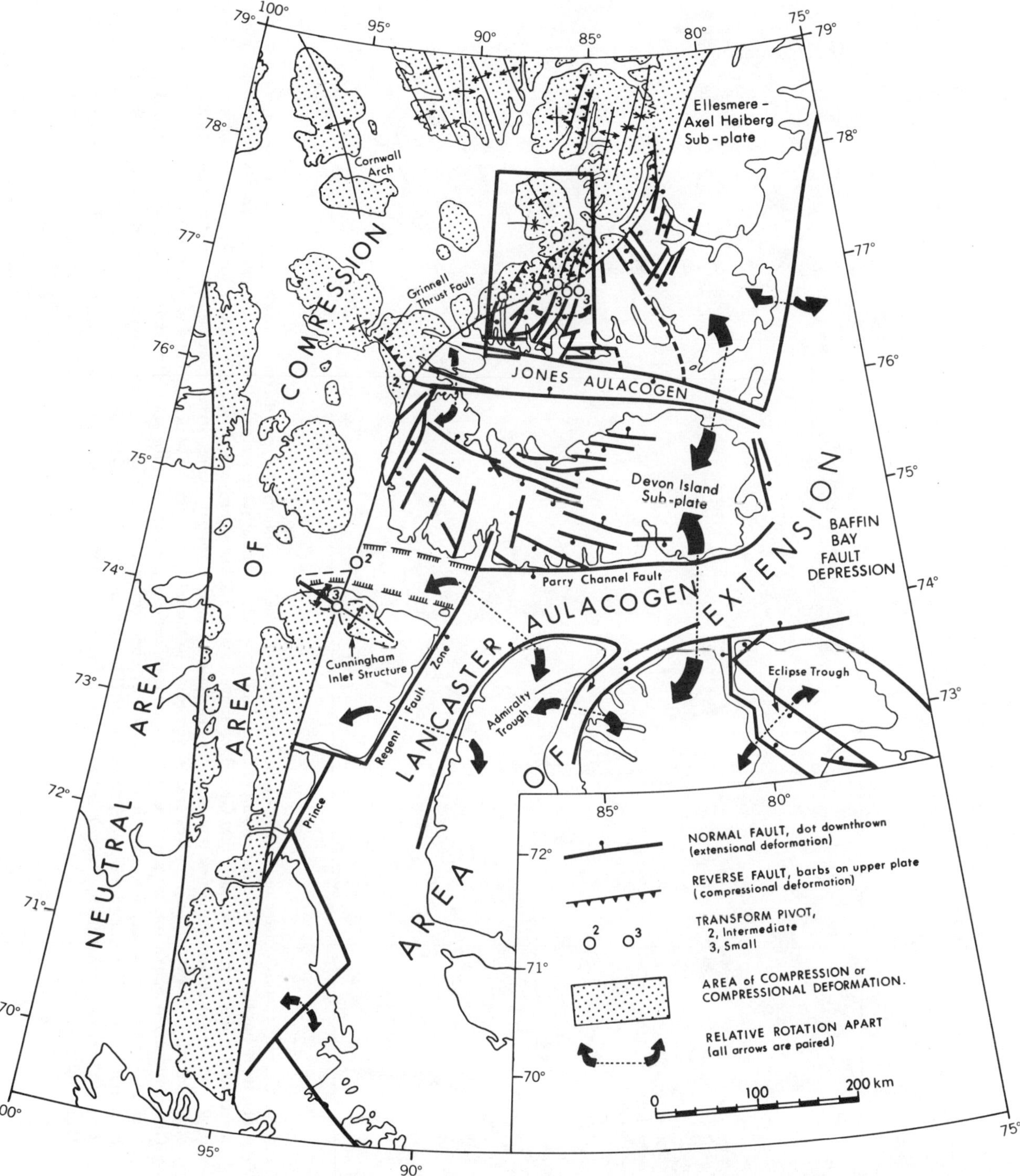

FIGURE 6. Movements that may have occurred at the terminus of the Canadian Arctic Rift System during the climactic phase of the Eurekan Deformation. Figure 5c shows the larger region during the same phase. For explanation of sub-plate movements see Kerr (1981 b).

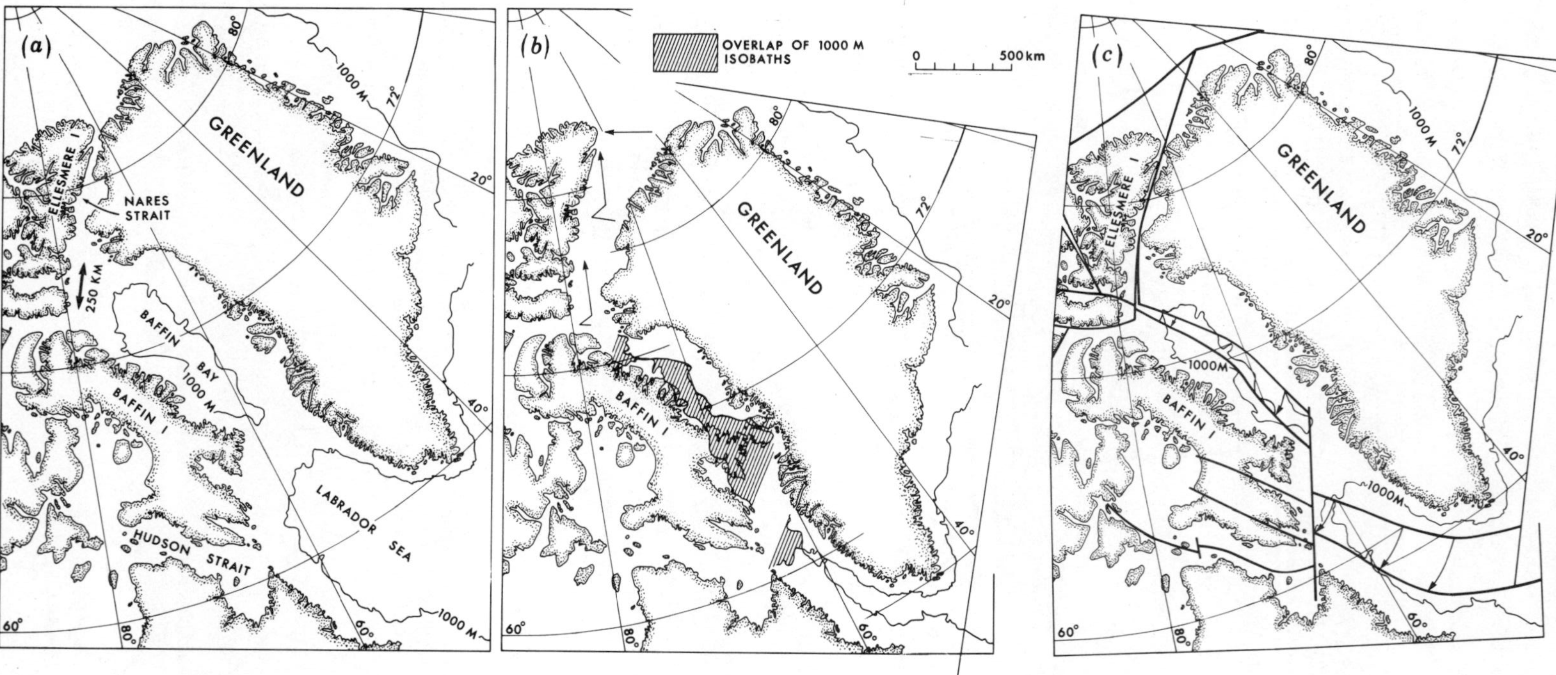

FIGURE 7. Current theories of tectonics that explain the relation of Greenland to the rest of North America (from Kerr 1981 *b*). (*a*) Fixist theories in which there has been no lateral movement of Greenland. (*b*) Conventional plate tectonic theories (moving continents or moving plates) in which most modern-day workers consider that Greenland moved about 250 km along Nares Strait. The conventional reconstruction shown (after Srivastava 1978) considers that Greenland first moved toward Ellesmere Island, closing up Nares Strait (horizontal arrows) and later had northward strike-slip displacement (vertical arrows). (*c*) Restoration according to the integrated theory of plate tectonics (Kerr 1967*a*, *b*, 1980*a*, 1981*a*, *b*, and herein). Greenland and Canada rotated apart, with the lateral movement increasing to the southeast from Nares Strait to Baffin Bay to the Labrador Sea. The amount of restoration made here is shown by curving arrows. Much of the great movement apart in Labrador Sea is accounted for by minor rotational opening in southern Nares Strait. The 1000 m isobaths in Baffin Bay and Labrador Sea cannot be brought back together. The gap between them that appears at first to be missing continental material can be accounted for by a foundered remnant of continental crust (Kerr 1967*b*).

was largely responsible for the formation of the oceanic area on one side of the present continent, as well as for certain deformations within the adjacent continental margin.

The Boreal Rifting Episode began in the present-day Arctic Ocean, apparently in about mid-Mississippian time, as shown in figure 4 (from Kerr 1980*a*, fig. 4). This resulted from movement of plates away from a spreading centre occupying the present Alpha Ridge. The Canada Basin and Makarov Basin formed on either side of the spreading centre. Extension faults of this spreading event apparently had a tendency to advance southeastward. Their advance was largely impeded by the thick plutonic core of the now-dormant Pearya Geanticline, and this caused the Kaltag Transform Fault to form along the northwest margin of that core. A small amount of extension affected the geanticline, which subsided slightly to form the Sverdrup Rim. The Sverdrup Basin formed southeast of the geanticline, located on an older folded geosyncline. The existence of that older basin may have made the crust more amenable to subsidence there, and was responsible therefore for the location of the Sverdrup Basin.

The sequence of events in the fragmentation history of the Canadian Arctic Islands is shown in figure 5 (after Kerr 1981*b*, fig. 7). The old structural grains that were present at the end of the constructional phase (figure 5*a*) had a major influence in controlling the pattern of fragmentation. The fragmentation phase began with the Boreal Rifting Episode, which apparently had a basically similar stress pattern from mid-Mississippian to late Cretaceous time (figure 5*b*). It appears that during this time the Canada Basin widened progressively, as the plate moving southwestward from the Alpha Ridge travelled progressively farther. The Sverdrup Basin formed on the adjacent continent as a side effect of events emanating from the Canada Basin. As the Canada Basin widened by the southwestward movement of a plate, a wider area of the adjacent continent became affected, and Parry Rift Valley began to form (figure 5*b*). The Eurekan Rifting Episode then began to develop, and in late Cretaceous time was in a very early stage (figure 5*b*). It began to rotate the Greenland and the Baffin Island sub-plates apart about a transform pivot. The rotation was accommodated by the northwestward advance of extension faults.

As the Eurekan Deformation progressed, it continued to separate and rotate apart the blocks on either side of Baffin Bay. The extension faults, however, could not readily advance northwestward because of the impediment caused by pre-existing structures. A bifurcation developed with branches deflected westward and northward. This episode reached its climactic phase (figure 5*c*) in mid-Tertiary time (between middle Eocene and early Miocene). Extension in the southeast (Eurekan Rifting Episode) caused compressional deformation farther northwest (Eurekan Orogeny).

As the Boreal Rifting Episode and the Eurekan Rifting Episode were being propagated toward each other, a plate tectonic contest developed (figure 5*c*, after Kerr 1980*a*, fig. 8). The Eurekan Rifting Episode was stronger and dominated. It produced a large area of compressional deformation in the central Canadian Arctic (figure 5*c*), which replaced the earlier extensional deformation that had emanated from the Canada Basin. Still farther west there was a large neutral area. In the extreme west, however, extension of the Boreal Rifting Episode continued uninterrupted, presumably because this was too far west to be affected by the Eurekan Deformation. The effects of the plate tectonic contest are shown in more detail in figure 6 (after Kerr 1981*b*, fig. 6). The Canadian Arctic was being fragmented into large sub-plates, each of which was itself fragmented into numerous smaller and smaller sub-plates. Extensional–compressional couples formed, in which areas of extensional deformation were transformed through pivots into areas of compressional deformation. The large area of continental lithosphere in the southeast was being stretched on a grand scale, by means of listric normal faults in upper levels of the crust and flow at greater depth. Lancaster Aulacogen developed in this stage as a westward projection into the continent, and has been documented in detail (Kerr 1980*b*).

The final severing of the North American Continent in the Arctic (figure 5*d*) resulted from the final pulse of the Eurekan Rifting Episode. This occurred in early Miocene or later time (figure 5*d*). For the first time the Queen Elizabeth Islands Sub-plate became completely surrounded by fault zones. This was achieved as faults of the Eurekan Rifting Episode broke northwestward through the continent to connect with faults that had formed earlier in the northwest by the Boreal Rifting Episode.

The Canadian Arctic Rift System is now dormant (figure 3). It became dormant when the Eurekan Deformation ceased after its final Miocene or younger pulse (figure 5*d*). Activity may have continued at sea in a minor way where the major zones of weakness of the rift system exist. However, the weak seismic activity along the system is not characteristic of plate margins, but rather is an expression of the readjustment of existing structures to the regional stress field.

Reconstructions

Reconstructions of the Canadian Arctic Rift System are shown in figure 7 (from Kerr 1981*a*, fig. 16). The conventional plate reconstruction (figure 7*b*; after Srivastava 1978) involves great displacement along Nares Strait. My own reconstruction (figure 7*c*; from Kerr 1981*b*, fig. 16) used my earlier suggestion (Kerr 1967*a*) that Nares Strait is a rift valley with minor rotational opening and minor strike-slip displacement. The history of the Nares Strait controversy was summarized recently (Kerr 1980*c*). Abundant new evidence (Christie *et al.* 1981) confirms that displacement was minor, less than 25 km.

The plate reconstruction of northeastern North America advocated here (figure 7*c*) is similar to a reconstruction made much earlier (Kerr 1967*b*). This reconstruction is part of an integrated theory of plate tectonics. It integrates two conflicting schools of thought, showing that the conventional plate tectonic theories and the oceanization theories of the fixists can be reconciled with one another. According to this integrated theory there was indeed some plate separation and it occurred by rotation; however, there also was major foundering of continental material to leave a continental remnant beneath Baffin Bay and Labrador Sea (Kerr 1967*b*, 1981*a*). The rotational separation and the related crustal foundering were accomplished by fragmenting, stretching and bending of the North American Plate. These three related processes took place on the Canadian Arctic Rift System and have recently been documented (Kerr 1981*b*). All three processes diminished and died out northwestward on that rift system. The bending occurred as the Greenland Sub-plate was rotated away from the rift system in a relative counterclockwise sense, while the main part of the North American Plate rotated clockwise relatively.

References (Kerr)

Christie, R. L. *et al.* 1981 *Nature, Lond.* **291**, 478–480.
Kerr, J. W. 1967*a* *Bull. Can. Petrol. Geol.* **15**, 483–520.
Kerr, J. W. 1967*b* *Earth planet. Sci. Lett.* **2**, 283–289.
Kerr, J. W. 1980*a* *Spec. Pap. Geol. Ass. Can.*, no. 20, pp. 457–486.
Kerr, J. W. 1980*b* *Bull. Geol. Surv. Canada*, no. 319.

Kerr, J. W. 1980*c* *Bull. Can. Soc. Petrol. Geol.* **28**, 279–289.
Kerr, J. W. 1981*a* In *The ocean basins and margins*, vol. 5 (*The Arctic*), pp. 105–199. New York: Plenum.
Kerr, J. W. 1981*b* *Can. Soc. Petrol. Geol. Mem.*, no. 7, pp. 245–278.
Srivastava, S. P. 1978 *Geophys. Jl R. astr. Soc.* **52**, 313–357.

Discussion

D. P. McKenzie, F.R.S. In the areas discussed by Price and by Kerr is there evidence of reactivation of faults during later stages of the deformation? Presumably in Price's area, if this occurred, thrusts would use the planes of pre-existing normal faults, whereas in the Canadian Arctic Islands the normal faults would follow earlier thrust or normal faults.

R. A. Price. A number of normal faults that were active in early Palaeozoic time within, rather than at the margin of, the Continental Terrace Wedge are now reverse faults, with some right-handed strike-slip motion, within the overthrust belt. Elsewhere in the overthrust belt some of the largest thrust faults juxtapose rocks of the same age, but with very different thicknesses and facies. It is probable that these thrust faults are reactivated normal faults that originally had their downthrown side towards the basin to the west.

J. W. Kerr. The present geometry of the sedimentary basins in the Canadian Arctic has been produced by the reactivation of older Precambrian faults that have moved many times. There are faults that have had as many as four phases of movement, with the sense of movement being different at different times.

J. A. Jackson. Where are the normal faults that Price has described now exposed in the field, and what do they look like in the overthrust belt? Are they to the west of the region where the seismic reflexion profiles suggest that the basement is not involved in the deformation?

R. A. Price. The normal faults are within what was the Continental Terrace Wedge. We know nothing about whether similar faults cut the Precambrian basement, except what we can discover from interpretations of gravity profiles. In the field, Upper Devonian rocks are superimposed on Middle Proterozoic in one place, yet a few kilometres away the same rocks will overlie a sequence of Lower Palaeozoic rocks up to 6 km thick. There is therefore 6 km of stratigraphic elief beneath the Sub-Devonian unconformity. This structural boundary in some places trends northeast, and can therefore be followed within individual thrust sheets. Elsewhere it trends northwest, parallel to the thrusts, and cannot be mapped.

A. W. Bally. I am inclined to agree with what Price has just said; however, it is not at all easy to observe the traces of pre-existing normal faults in the Canadian overthrust belt. They are never exposed as normal faults in an outcrop. Even if one saw a normal fault in an outcrop, it would be difficult to decide whether the fault had been formed within soft sediments and due to gravity gliding, like the growth faults of the Gulf Coast, or whether it was a tectonic feature associated with extension of the basement.

Kerr's interpretation of the Kaltag fault differs from the interpretation of other geologists now working in Alaska, who believe that the fault system that follows the Canadian Arctic slope may be a normal fault system associated with the opening of the Arctic Ocean, and not a transform fault as Kerr believes. I agree with Kerr. The structural geology of the region is complicated and there appear to be two fault systems, the Kaltag fault in Alaska and another one, which I will call the Boreal fault, which appears to form the northern boundary of the eastern Canadian Arctic. In my judgement, during the early Mesozoic the Boreal fault ran into the east side of the Richardson Mountains and did not join the Kaltag fault. The leading edge of the Innuitian fold belt subcrops between Prince Patrick and Banks Islands, and then continues into the Arctic Ocean, to be intersected by the Boreal fault. The same leading edge of the Innuitian fold belt reappears in the northern Richardson Mountains. Hence at some time in the early Mesozoic the Boreal fault continued into the east side of the Richardson Mountains and offset the leading edge of the Innuitian fold belt by more than 1000 km. The Richardson Mountains themselves, as P. Ziegler pointed out to me, were formed by a later inversion. The argument then suggests that the Richardson Mountain region had formed as a deep graben system in Cambrian-Precambrian time.

In early Mesozoic times, left-lateral strike-slip movements along the Boreal fault coincided with and reactivated the east Richardson Mountain fault system. Only later, during the Laramide orogeny, the fill of the earlier basin was inverted. Because of these late movements, the geometry is difficult to reconstruct and to argue which particular fault was reactivated.

A. B. Watts. Both Price and Kerr mentioned the existence of a hinge line or hinge zone in their areas. How far can this zone be followed along strike in the Canadian Cordillera?

R. A. Price. The west flank of this zone forms a large monocline in the Southern Canadian Cordillera. Across this monocline there is a stratigraphic displacement of 22 km, from Proterozoic on one side to Jurassic. I believe that this monocline was produced by draping the miogeocline over the ancient rifted margin. When one reconstructs the original geometry of the overthrust belt, the boundary between the platform rocks and the miogeocline rocks lies where this monocline now is. There is evidence of Lower Palaeozoic normal faulting within the continental terrace wedge (miogeocline), with the western sides being downthrown. Hence the boundary between the continental edge and slope was originally west of the present position of the monocline. The major boundaries trend northeast in Southern Canada, changing to north and then northwest as they extend northwards for about 300 km.

R. Stoneley. In view of the fact that Devonian to Jurassic, and even lowermost Cretaceous, sediments on the North Slope of Alaska were derived for the north, from the site of the Canada Basin, could Dr Kerr please give the evidence on which he postulated opening of the Canada Basin as early as the Mississippian?

J. W. Kerr. The main evidence that the Canada Basin began to form in Mississippian time and persisted through much of Late Palaeozoic and Mesozoic times is that the adjacent Sverdrup Basin of the Canadian Arctic Islands developed in this interval. The Sverdrup Basin formed on the outer side of a continent, being bordered by the stable continent on the southeast. It subsided because of extension faults beneath that were propagated there from a more deeply subsiding basin farther northwest. That more deeply subsiding basin farther northwest is ancestral to the Canada Basin.

The clearest facies evidence for the early existence of the Canada Basin is in the Triassic, where thick shallow-water clastic rocks grade northwest toward deeper water sediments in the direction of the Canada Basin. Faunal evidence is also supportive. Upper Palaeozoic faunas of the Sverdrup Basin are of Asiatic origin, and appear to have reached the basin via the Canada Basin.

D. G. Roberts. Recent work in the Nares Strait region to the southeast of the Sverdrup Basin shows that there has only been a limited amount of strike-slip motion between the two sides. This observation implies that Baffin Bay should be underlain by stretched continental crust. Is it possible that the pole of rotation between North America and Greenland lay close to the northern end of Baffin Bay at this time? This could explain why the deformation that Kerr has studied changes from extension in the southeast to compression in the northeast.

J. W. Kerr. The pole of relative motion certainly does lie close to where Roberts has suggested. The deformation is not, however, quite as simple as he proposes, and is complicated by the independent motion of a number of small blocks.

Phil. Trans. R. Soc. Lond. A **305**, 207–224 (1982) [207]
Printed in Great Britain

Carboniferous sedimentary basins of northern Europe and the nature of emergence around the margins of the Mesozoic rifted sedimentary basin of the North Sea

By A. M. Clarke

National Coal Board, Hobart House, Grosvenor Place, London SW1X 7AE, U.K.

The evidence of the infilling, shape, longevity and setting of the Carboniferous basins of northern Europe, long available in the public domain and now, with additions from the petroleum and coal-mining industries, is assembled. It suggests that their nature and origin are products of a sequentially and spatially nesting hierarchical mechanism of crustal change, within a framework that remains stationary within the bounds of the subcontinent. The observed dynamic morphology of subsidence in the coal-mining basins suggests an alternating sequence of downwardly narrowing linearly elongated, generally rectilineal zones, across which vertical movement takes place in the crust. The narrowing sequence of vertical movement is elliptically basinal or domal in the near surface, with subordinately horizontal tensional and compressional zones within it; rotational and counter-rotational respectively, below the wider, overlying, tensional and compressional zones; and finally, a vertical plane lying beneath the transition point between the two rotational zones. It is concluded that, to produce the observed pattern, the earliest cycles in the mechanism have cumulatively underprinted their successively more primary lines of disturbance, to produce the upward diminishing hierarchy of sizes of rectilineal block motions, responsible for the Carboniferous, preceding and succeeding basins.

Introduction

Long before it attended the birth of geology in Europe, successful mining had always had to embrace internally generated cumulative views of the architecture conditioning the unfolding morphology in many of the very varied European mineral bodies mined. These views have always been expressed, not in scientific terms, but in the stepwise chronological record of the morphology of previous workings, together with each new plan for their advance. The scientific view of the relations between successive mountain chains and the rising swells or subsiding sedimentary basins in what is now NW Europe has also been cumulatively advanced in successive steps from those originally presented by the founding fathers such as Lyell and E. Suess. The rapidly branching chain of supporting references stretching back for more than a century can be entered from the exhaustive lists accompanying the two most recent major compilative works here cited, i.e. Ziegler (1980) and Bless *et al.* (1977). Much of the substance of the review by Bless *et al.* is devoted to an examination of the limits of error in the stratigraphical controls that determine true comparability of the geological sections selected in each basin.

Interest in first the gasfields and then the oilfields of the North Sea, and renewed interest in the potential of the surrounding land areas, generated very large quantities of new evidence in NW Europe on the areal scale required to study the evolution of sedimentary basins in general. These syntheses gave a fresh geological, as opposed to solely geophysical, view of the up to multi-megayear and near gigayear mechanism responsible for the palimpsests of a general process of change in the upper crust: a process that we may observe, often apparently 'frozen' – because it is so slow, at different steps in the process, in different places within the upper crust. Only by the release of such syntheses and supporting data is there any increase in the traditional geological 'depth of focus' brought to bear within the tens of kilometres thick field of the general crustal mechanism responsible for generating a succession of basins and swells, horst and graben, etc. Traditionally, it was necessary to make the best of the products of such deep-seated processes as were exposed within a slice of only 1 km or so thick.

Examination of the implications of a succession of uniquely complete large-scale stratigraphic palaeogeographic reconstruction has had to await petroleum geologists such as Kent (1975, 1977, 1980) and Ziegler (1975, 1978, 1980) over the mid to late 1970s. Ziegler (1980) states that 'the Permo-Carboniferous fault system of NW Europe was time and time again reactivated, particularly during the Mesozoic, and played a pre-eminent role in the development of post-Variscan sedimentary basins'.

Field geologists' constructions of 'deep' cross sections of the crust below rift valleys, etc., may often 'look right' for two reasons: firstly, because, for lack of evidence to the contrary, they imply that the upper crust may act mechanically as a single entity; secondly, within that entity, the mechanics of the system conditioning its movement and stasis has been made mechanically analogous to one of the systems so familiar on all kinds of scales, in the open situations observed everywhere at the surface. Such systems underlie the stability of Roman (and natural) arches, mantelpieces cantilevered over fireplaces, shelves between their supporting brackets and 'concertina-like' horizontal stacks of books when being passed from one pair of hands to another. As will be argued below, however, even when in precisely analogous boundary conditions, the 'structural blocks' in an upper crustal cross section do not necessarily operate in the familiar form in which they can be intuitively 'seen to work' when observed at the surface. Some of the 'space problems' and the difficulties arising from implied tensional or compressional conditions are only brought about by the assumption that the mechanical systems so familiar at the surface must also condition the deeper movement and rest position of structural blocks.

Comparative aspects of the mechanics of basin and graben formation obtainable from parallel observations of their mining origin in the coal basins of northern Europe

The state of geomechanics in upper crustal studies and soft-rock mechanics in coal-mining

For the upper crust it is possible to make suggestive-looking overall abstractions from the general patterns within the lineaments – cumulatively modified by successive episodes of translational, rotational and differential strain movement, both of and within 'blocks'. Although certain limiting overall proportionalities in the statics and dynamics from the physics of the system must contribute, there is as yet no quantitative, all-embracing 'mechanics of upper crustal movements' to explain the observed patterns on all scales of occurrence.

Thus a possibly fourfold, upwardly propagated succession of changes in the upper crust may be observed. The four primary constituents of such a succession may be as follows. (1) The (perhaps bottom-most) zones of either old–cold–thick or young–hot–thin now crystalline crust whose essential attribute is that they are much longer than they are wide, and hence form a lineament. (2) The corresponding, perhaps middle depth, orogenic 'zones of disturbance' of the once simply stratified, and then more extended, layers of the upper crust. (3) The adjacent, perhaps shallower systems of macro-fault lines. These may be in the form of either (horizontally) 'translational fault-and-end-splay' or major (vertically) rotational fault or paired vertical fault systems (the major horst and graben). They may possibly overlie the one locally plan parallel member of the set of earlier deeper (once bottom-most but subsequently eroded) lineaments imposed in each earlier cycle of change. If this is so, then movement on the parallel much earlier basal lineament might well be expected, when contemporaneous examples of (1) and (2) are being originated nearby. (4) The corresponding very slightly more elongate and wider overlying basins and swells. These are of a flexural nature and across them, more or less at the contemporary surface, sedimentation or erosion takes place. These basins and swells in many recent 'greatest depth of focus' studies appear to be the final members of such a vertical sequence of upwardly propagating original and reactivated movements in the upper crust (that is, wherever the longer axes of the wider zones of flexural movement both vertically succeed, and lie parallel to identifiable, narrower, underlying forms of lineament).

The analogous vertical and lateral sequence of patterns within, seen in the products of both upper crustal and soft rock mechanics, may be due to a rheologically similar mechanism. In both domains, the mechanism produces the two observed linked phenomena. Firstly, subsidence basins and 'upsidence' swells over a flexural plane parallel to the Earth's surface. Secondly, highly localized sub-vertical planar discontinuities, across which movement of the strata takes place. Both phenomena have been quantitatively observed, and the ratios between their forms, in different cross section setting and dynamic relations, quantitatively characterized, as a result of innumerable observations of them in the mining domain.

A similar overall set of ratios in the domains of upper crustal and soft-rock mechanics may condition the internal patterns geologically observed, although on a completely different scale.

It is necessary for the earth scientist, interested in the data on the relations between trans- and intralineament and corresponding flexural movement within the mining domain, to appreciate the origin and nature of certain essential configurations that obtain in deep coal mining. These data are primarily obtained by empirical observation of the movements themselves (Peng 1978). Although perpetually in use within the mining industry for precise prediction of surface subsidence and form of strata movement, the mass of empirical data involved is almost entirely unpublished.

A general terminology for descriptions within a hierarchy of scales, of disturbance, horst and graben, basins and swells in upper crust

Many special terminologies are employed to identify morphological features of the general tectonic and depositional framework seen from a different standpoint. The special terminologies originate when the features are seen in the detail obtainable in field stratigraphic, petrographic, etc., studies. These special terminologies may imply that the morphological features occur on a particular scale, or they may carry some connotations of origin, particularly when considered in isolation from its predecessors and successors. Examples of this may be found in palaeogeographical terms such as 'shoal', 'shore line', 'St George's Land' and 'Northern Continent', in palaeoecological descriptions such as those implying a broaching of old 'land barriers', to explain the palaeogeographical disposition of a marine fauna containing new elements, and in sedimentological terms particularly those contrasting certain kinds of facies. Such sedimentary facies are those that imply a change in the rate of tilt of the depositional surface lying between two foci or loci of contrasting maxima in rate of vertical movement of the Earth's surface, e.g. slow or rapid, transgressional or regressional, shallow water or deepwater facies.

A more generalized and strictly non-scalar morphological terminology for the analysis of patterns within a general system of lineaments and flexures appears desirable. It would enable any repetitions to be observed, within a nesting hierarchical set of scales – parts of which are observed in different places throughout the upper crust – and which may also operate within the rheological framework of coal-mining strata control and subsidence. It is also desirable if the evidence obtainable from (*a*) analysis of essentially descriptive field geological observations of local patterns in the internal structure of lineament and flexure *on all scales* is to be added to (*b*) the strictly geophysical measurements of the modifications to the Earth's energy and stress fields, and the deeper-seated changes in the physical properties to the crust across basins and grabens, etc.

The deep coal-mining engineer tends to place all geological phenomena locally disrupting or impeding the (revenue earning) advance of his highly capitalized longwall panels in the one category of 'geological disturbances' to the continuity of mining. It seems appropriate to borrow from this terminology and categorize any linear feature occurring on any scale that the rheological continuity of the zone in the enclosing crustal layer, when viewed on the same scale as 'a disturbance'. It may be of linear or areal flexural form.

In order to 'disturb' the enclosing zone, firstly, the zone must be of a width and thickness that is determined by the minimum overall scale on which 'the disturbance' is observed and, secondly, the disturbance must (usually) post-date what is the (otherwise) *undisturbed* enclosing zone. Thus, the frame of reference within which a zone, thus defined, may be referred to as 'undisturbed' must always be within a particular range of areal scales, and hence occur at a particular level in the hierarchy of 'lines of disturbance'.

At one end of the scale, the areal framework of the enclosing layer may be of 'continental' size and of 'whole crustal plate thickness'. This is appropriate when viewing, say, the belt formed by the Caledonian orogeny as a 'line of disturbance' within its surrounding and, at this level in the hierarchy, 'undisturbed' zone. At the other end of the scale, it may be as small as is appropriate to the vertically and laterally dying line of disturbance caused by a small fault that dies out both upwards and downwards within the boundaries of the crustal layer containing only a few neighbouring and, at this level in the hierarchy and otherwise undisturbed, Westphalian C. On the smallest scale in the nesting processes to be considered, the line of disturbance may be a strata control engineer's induced 'fault' disturbing the overlying strata for only a few tens of metres over the gateroad of his longwall mining panel, before the extension he desires passes into a block rotational form, and finally into the tensional zone within a surface subsidence curve.

From a bird's-eye view, at some considerable distance from the thus shrunken frame of reference of the 'undisturbed' zone containing it, a major 'rift valley' can thus become a single lineament of 'disturbance'. This is only true at one of the higher levels in the hierarchy of such lines of disturbance to the continuity of the hierarchy of 'crustal layers' of diminishing thickness or extent, or both, within any major zone on the continental surface. It can be seen intuitively that there may be a nesting hierarchy in the range of areal sizes and thicknesses of the 'undisturbed' zones, implied at each level containing a member of the hierarchy of correspondingly longer and thicker 'lines of disturbance'.

The term 'subsidence basin' can have a purely morphological connotation. The term appears to be applicable on all scales of occurrence of any possible general nesting hierarchical mechanism which, *inter alia*, produces it within the surface of 'undisturbed zones'. Although the term 'upsidence' has been used in the mining domain as a corresponding term for flexurally continuous surface but of opposite polarity, a 'swell' may be more appropriate: that is, always provided that the scale of the frame of reference within which the term 'swell' is employed is precisely the same as that on which neighbouring basins are being identified. A 'swell within a swell' (or within a basin) is of a lower scalar order in frame of reference, and it distinguishes a form of local interruption to the continuity of a much larger scale basin containing it. In this case, the term 'hump' for such a swell within a basin, e.g. above deep-seated 'horst' block, might be appropriate.

Finally, when distinguishing an intervening zone at any level in the hierarchy which neither appears to be involved in any movement, nor to contain any flexural or linear 'disturbances' within it, it might be appropriate to use the term neutral 'block' or, when viewed only superficially, neutral 'zone'.

Foundation morphology of the Carboniferous basin floor in northern Europe

Palimpsests of pre-Caledonian tectonic elements within the Erian–Hibernian floor of the Upper Palaeozoic

Many European authors since Cogné (1976) describing the structural architecture of Laurasia (NW Europe before the opening of the Atlantic) have drawn attention to the underlying influence of a foundational 'core shape', in the form of an inverted T or inverted, flattened Y. In most Palaeozoic 'plate' reconstructions the 'triple point' of the inferred suture lines at the T junction between the belts is usually placed somewhere in what is now the Atlantic just off SW Ireland.

However, it is possible to describe them solely in the morphological terminology expounded above. The belts then shrink to 'lines disturbance' crossing what then become 'undisturbed' areas. As expected, part of the southeastern 'thick-pencil' edge of the indisputable Caledonian 'line of disturbance' curves across Britain between Pembroke and the top of East Anglia, and part of the northern edge of the probably pre-Caledonian line sweeps across Britain between the Bristol Channel and Dungeness.

With one 'triangle of exception', both of these trends are nicely reflected in the bimodal fault patterns now impressed within each of the coalfields of England and South Wales. The belt of exceptional coalfields lie within a tall right-angled triangle with its apex in Morecambe Bay, its perpendicular dropping down to the top of the Severn Estuary and its hypotenuse dropping down to North Foreland (NW Kent). The line of the hypotenuse parallels the east-northeastern edge of the London–Brabant Massif in Ziegler's (1978, fig. 2, p. 592) reconstruction of the late Caledonian framework of NW Europe.

When taken in conjunction with inferences, discussed below, from the location and morphology of the North Sea Cainozoic basin, it is tempting to infer the existence of two other and similarly fundamental 'lines of disturbance'. By using this terminology, there is no need to seek evidence of whether they are or are not 'suture lines' of some earlier plate accretions. Their supposed existence is implied purely from the profundity and enduring nature of their effects on all the later observed features.

The earlier and deeper of these two 'other' successively imposed inferred 'foundation lines' of disturbance would run north–south, through the centre of the northern North Sea. The position of the shallower of the two inferred 'foundation lines' would run northnorthwesterly; it could be located in a number of alternative places, so its position is unknown. Part of it *might* be off the northnorthwestern edge of the London–Brabant Massif. Certainly the continuation southwards of this inferred north–south 'foundation line' of disturbance to a point in the North Sea halfway between Lincolnshire and Schleswig-Holstein would appear to terminate at a 'junction' and to swing southsoutheast, parallel to this edge of the Massif. Such an inferred southsoutheastern foundation line would then continue under present day Amsterdam, to form another 'triple-point' with the well known Variscan line where the latter changes direction, in the vicinity of the junction between the Ruhr Graben and the Rhine–Leme Graben.

The late Caledonian framework of northwestern Europe has been reconstructed by Ziegler (1978, fig. 2, p. 592). The succession of primary mobile belts, underlying all stratigraphically later crustal eras, can be thus shrunk into 'foundation lines': lines disturbing the otherwise undisturbed European crust. The existence of two earlier lines of disturbance originating in lower (earlier) crustal eras is inferred. The slightly more definitive of the two inferred lines of disturbance, running northnorthwest–southsoutheast, taken together with the east–west probably pre-Caledonian line and the northeast–southwest Caledonian lines, are responsible for the existence of the triangular London–Brabant Massif in this pre-Variscan continental framework. The largely triangular London–Brabant Massif is shown by Ziegler with its three apices located in Pembroke, in the North Sea just north of East Anglia, and in Belgium not far to the northwest of Brussels.

The central hypothesis underlying this commentary on evidence available from mining and geoscientific consideration of Carboniferous basins in northwestern Europe

It is suggested that, apart from the 'foundation lines', the crust contains sets of intersecting macro-lines of disturbance, together forming an irregular grid of intersecting lines. Each successively imposed set would be parallel to each of these successive actual or inferred primary 'foundation lines of disturbance' located as an orogenic belt. Some of these primary lines were associated with plate collisions. The subordinate, largely parallel, members of the grid of associated macro-lines would be successively superimposed on the 'undisturbed' palaeocontinental zones. These zones would contain both (*a*) the successive earlier foundations lines (where not completely eroded, either at the surface or by absorption into the underlying mantle), and (*b*) all their associated parallel 'macro-lines' of disturbance.

Thus, on these hypotheses, on each successive occasion, both a new set of macro-lines of disturbance was imprinted and movement on the earlier sets would be reactivated in a manner analogous to the soft-rock strata control engineer's fracture-faulted and, much wider, induced cleavage zones. In the 'upper crustal case', however, reactivation takes place within the 'undisturbed' or 'less viscoplastic' phases below the upper crust.

In each cycle of change, it would then be the successive layers of foundation blocks that controlled the position of all overlying subsequent graben, asymmetrical graben and trap door 'hinge-line and fault systems' and the shape and dynamics of sedimentation in all overlying surface subsidence basins of sedimentation. In each cycle, these features all lie within those wide flat-bottomed basins of deposition recognized as 'the' Palaeozoic and Mesozoic and Tertiary 'sediments and metamorphosed sediments' of NW Europe as a whole.

The other constituent of the floor of the coalfield basins of northwestern Europe: Devonian and Lower Carboniferous subsidence and depositional basins and eroded swells

Kent (1980), in his earlier work on the structural framework conditioning the subsidence of the North Sea basin as a whole, emphasizes that the subsidence began early in the Upper Palaeozoic and has been virtually continuous ever since. He states that '... initial subsidence of the basins of the NW European Shelf... began shortly after the end of the Caledonian Orogeny...' (Kent 1980, p. 282). He draws attention to the migration, outwards from the centre, of certain segments of the hinge lines between the vertically negative areas that were subsiding and the vertically positive uplifting areas, on the other side of these segments. He describes '... overlap in England of Lower Devonian across Cambrian and Silurian in the southern Midlands, by the overlap in turn of the Upper Devonian onto the older rocks of the Welsh Highland and East Anglian Massif. ... Thus except for Cornubia (which is essentially Hercynian†) each of the Palaeozoic blocks was a positive structure during the Upper Palaeozoic, as indeed through the Mesozoic'. He states (p. 282) that this observation on the positive nature of earlier Palaeozoic blocks also applies to the 'Fyn–Rinkøbing High in Denmark and the other Massifs which bound the basin of NW Germany and of the Central Massif in France'.

Ziegler makes the same point in his wider palaeogeographic reconstruction of the late Devonian framework of the combination of NW Europe (Ziegler 1978, fig. 2, p. 594). He also gives details of the shape, maximum thickness and nature of the Devonian sedimentation in the basins of Ireland, Scotland, W Norway and Svalbard, the Midland Valley Graben, the Orcadian and the Northumberland basins. He describes the accompanying post-orogenic emplacements of granites in the (inferred) simultaneous uplifting of the Scottish Highland and S Upland Massifs.

Contemporaneously in the south, in the predominantly marine part of the Devonian basin of sedimentation, Ziegler (1978, fig. 4, p. 596) illustrates the further emergence of two dissected broadly east–west lines of 'highs', flanking their associated Cornwall–Rhenish and Central Amorican–Saxonian 'troughs'. He points to the large reef-fringed carbonate platforms established in the mid-Devonian (during Givetian times) 'in the shelf areas and on local highs' (p. 597), and to the general vertically continuous and in this sense 'stationary' tectonic setting of sedimentation.

Ziegler (1978, p. 598) draws attention to the 'extension tectonics' to the north of the lines formed, firstly, by the Normanian geanticline and the mid-German High and, secondly, by the Averno–Lugian geanticline and the Bohemian Massif, which 'persisted throughout late Devonian times into early Carboniferous times'. (Within coal-mining subsidence basins, such 'extension and compression tectonics', recognized by upward movement of such features as kerb-stones to relieve horizontally compressive stress, are a corollary of outward propagation of a subsidence basin. They are also preceded by a zone of tensional strain, such as would allow graben and basin formation in the upper crust. However, in this case the northwardly propagating 'subsidence waves' would have to originate deep within in the mantle.) He describes the onset of the Variscan orogeny in the Upper Visean and the start of underthrusting along the margins of the Normanian mid-German high: '... With the general advance of the deformation front during Namurian and Westphalian, the basin axis of the Variscan foredeep migrated

† It is here suggested that because its lineaments are parallel to east–west margins of earlier foundation blocks, Cornubia also has a pre-Caledonian (? Cadomian), or earlier, foundation.

northwards and the newly folded-up sediments were subject to erosion.' He shows, on his palaeogeographic reconstruction of the Lower Carboniferous (Dinantian) of NW Europe and surrounds (Ziegler 1978, fig. 5, p. 597), both the areas of Flysch sedimentation, abutting the deep basin 'black shale' areas, and the classical Carboniferous Limestone deposited around the (then relatively neutral) Caledonian 'highs' in the remainder of the basin, to the north.

In Britain, local areas of lateral change, overlying certain earlier foundation blocks, are clearly seen – within the subsiding basin of deposition of more widespread Lower Carboniferous facies. They are reflected in modern stratigraphic recorrelation of the original (largely litho-stratigraphic) division of the Carboniferous, into 'Carboniferous Limestone', 'Millstone Grit' and 'Coal Measures'. In one immediately illuminating stratigraphic–lithostratigraphic cross section of Britain in the Carboniferous, Ramsbottom *et al.* (1978, fig. 1, p. 6) show clearly (*a*) the early onset of 'Millstone Grit' facies, in the uppermost Dinantian of the southwest around Bristol, and (*b*) the persistence of 'Carboniferous Limestone' facies, into the Namurian of the 'Derbyshire dome' (then one of the basins within the Pennine basin), followed northwards by (*c*) the greater and greater persistence of 'Carboniferous Limestone' facies higher and higher into the Namurian until the Midland Valley of Scotland is reached, and the northward record terminates. Without oil industry data, it is the whereabouts of present-day exposure that selects two-dimensional vertical ribbons of exposure, illustrating a northward advance of the fluviatile 'Millstone Grit' element over the Namurian chronological interval.

Reading (1969) reviews certain Carboniferous sedimentation sequences in selected (largely Namurian and Westphalian) depositional basins of western Europe, north of the Variscan front. He draws attention to the sedimentary facies succession, infilling the basins, initially black shale; then greywacké turbidite (down the front of the fluviatile platform, which is usually spreading out, and thus filling and covering the basin); and finally the fluviatile facies (retained on the top of the platform only to the extent that the bottom of the basin continues to subside). He then states (p. 1410) that 'basins of considerable depth were filled with great rapidity'. He concludes that, by late Namurian times – when fluviatile conditions predominate – that, with one exception, successions over NW Europe do not substantially differ. The exception is in the Cornubian trough, where turbidites persist into the Westphalian A; this is also seen in Ramsbottom *et al.* (1978, fig. 1).

Depositional conditions on the floor of the coal-bearing sedimentary basins of northern Europe

By the time the lithostratigraphic interval formerly known by such terms as the Productive (coal) Measures had been reached, the depositional floor of most of the overall basin of north-western Europe (north of the Variscan front), the sedimentation *surface* of the local basins of deposition, was generally at or above 'sea level'. The local basins were 'full up' and sediments were 'washed across the top of them' by the widespread and repeatedly avulsing system of fluviatile sedimentation long known as the 'Coal Measures'. Subsidence across the shelf no longer exceeded sedimentation anywhere.

In general, a vast sheet, over which fluviatile sedimentation persisted, occupied the immensely long east–west shelf between them covering most of northern Europe. It was bounded to the north by the eroded, but presumably rejuvenated, grassless, 'bad lands' of 'the uplands of the Laurasian shield'. To the south, it was bounded by the Variscan trough. Sedimentation exceeded subsidence. The preserved column was thinner over the tops of certain relatively much more slowly subsiding intervening blocks across which the fluviatile sediments travelled to reach the various 'sinks', in which a greater rate of downward movement of the base persisted over the Westphalian time interval.

Thus, by the beginning of the Coal Measures, the pattern of basins, swells and neutral blocks were well established to the north of the Variscan belt. The period of both (*a*) 'still-stand' (giving carbonate sedimentation) on the swells above the more neutral blocks, and (*b*) 'rapid subsidence' (giving 'flysch' in the foredeep and 'greywacké turbidite' sedimentation) in many of the minor basins had largely ceased. Because an avulsing fluviatile system now covered the vast majority of northern Europe, depositional conditions at the floor of the coal-bearing sedimentary basins of Europe did not differ substantially.

Below this floor, the rate of preservation of the fluviatile components of the sediments was still controlled by the hierarchical movement of the blocks making up the tectonic framework underlying them. Thus, where they exist – because the edges of the basin have not been sliced off and eroded by the next (Mesozoic and Tertiary) cycles – the subaerial components of the sediments (the coal seams) are generally all seen to converge on one another towards the edges of the basins, as the intervening fluviatile components thin towards swells.

The macro-'basin' of Upper Carboniferous deposition in northwestern Europe

The palaeogeography of the macro-'basin'

Ziegler (1978, fig. 9) shows the extent of the main 'basin' of deposition in the Upper Carboniferous, being greater than 2000 km long and about 500 km wide but only, on average, about 1–2 km deep. It is shown stretching from the westnorthwest to eastsoutheast across northern Europe, from beyond the west of Ireland to eastern Russia and to be cut off to the north and south respectively by two very different types of land area. To the north, the 'basin' is margined by the ancient Upper Palaeozoic land area over the Scottish–Norwegian Caledonides and the Fenno-Scandian shield. To the south the main 'basin' is shown terminating against the Variscan line of disturbance, with its newly emerging mountain chains, within which are the very subordinate Plessis, Saar, Salle, Upper Silesian and Carpathian basins of deposition. On this map, the sediment fill is shown to be shale, sandstone and coal (flysch sedimentation having largely ceased at the end of the Lower Carboniferous) (Ziegler 1978, fig. 8).

A number of small land areas, low humps, largely over (here inferred) Caledonoid blocks, within the main 'basin' are shown (figure 1). It will be seen that the larger triangular area of the Welsh Massif just touches S Ireland. The inherited near 'neutral block', here exposed as the London–Brabant Massif, is also shown in its Carboniferous setting.

The location and aspects of the geometry of the subsiding 'basins proper' and less subsiding 'swells' within the Upper Carboniferous macro-'basin'

Ziegler (1978, fig. 9) shows 'spot' thicknesses of 'intra-Westphalian' sediment at the centre of the larger of the smaller depositional 'basins proper' within the main 'basin'. The subsiding 'basins proper' are the ones with a distinct centre-point or line and their margins (shown by isogonals over the largest interval between a lowermost and an uppermost stratigraphic horizon present right across the basin) everywhere dipping into it in all directions. The deepest are the NW German (3.5 km) and the Central North Sea (3.2 km) basins. The other deep

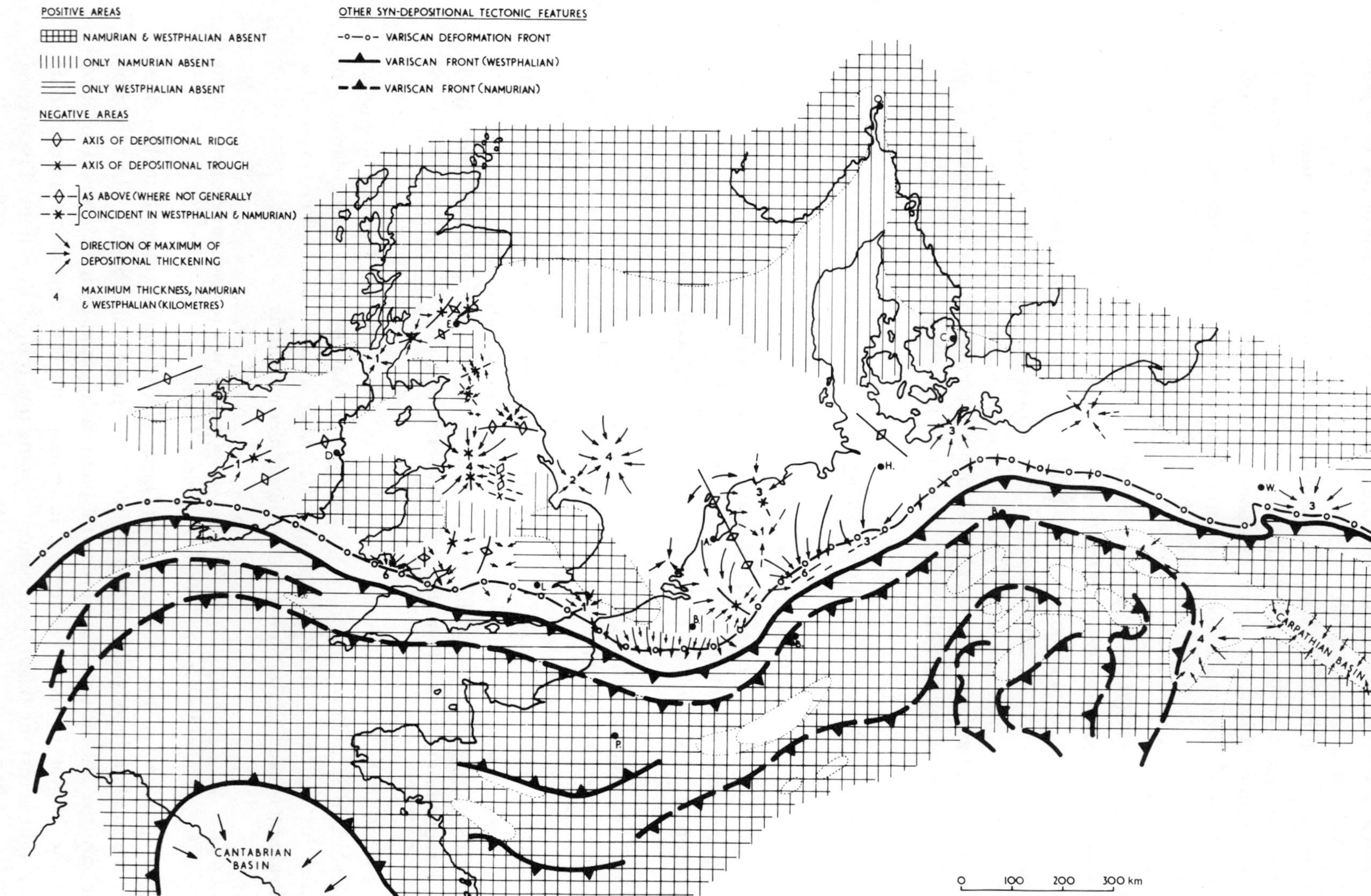

FIGURE 1. The Carboniferous basins of Europe, based on Ziegler's (1978) figs 5 and 6. D = Dublin, E = Edinburgh, L = London, P = Paris, B = Brussels, etc. Sets of isogonal dip lines have been added around the major basins, and the positions of the more minor depositional basins: those between four short Caledonoid trending 'humps' in Central Ireland, added from Ramsbottom *et al.* (1978); the five minor, but comparatively steep-sided, elongate basins of the Midland Valley of Scotland, added from Francis, in Craig (1965); the Campine–Brabant and the remaining six, shallower, mid to eastern European basins identified, along with most of the others, added from Bless *et al.* (1977).

depositional basins are the Pennine basin (2.7 km), the Kent – northern France – southern Belgium 'ex-foredeep' basin (2.5 km), the Saar (2.1 km) (intra-montane) basin, the South Wales (1.8 km) and the Lublin (1.2 km) basins. Although Ziegler shows both those 'humps' that (probably) remained above seam level in the Upper Carboniferous, and also the dispositions and maximum depths of the main basins, he does not show the dispositions of the suite of humps between the basins that did not pierce the surface of sedimentation, but only attenuated it. These may be inferred from the pattern shown by the isogonals in the basins.

In the Namurian, the flattish-centred Pennine basin with a north–south-trending axis centred north of Manchester, almost rectilinear relatively steep sides, and with east–west and north–south-trending margins is flanked, beyond its east side 'subsiding monocline', by a series of three equally steep-sided westnorthwesterly pitching elongate minor 'semi-basins' possibly underlain by 'trap-door' graben. By the Westphalian A, B and C times, this 'depositional monocline' had ceased to move. It then becomes a low semi-conical shaped basin (as shown by Calver 1969, fig. 1*b*, p. 3). Its eastern quadrant covers the underlying minor pitching basins. Were the record complete, it must surely have terminated against a line overlying that of the major change of facies in the underlying Namurian and the present-day Craven fault belt. This, in turn, overlies a deep-seated east–west lineament in the Upper Caledonian foundation, running through S Denmark and the southern Irish Sea.

Bless *et al.* (1980) give selected cross sections of the total Carboniferous over the whole of northwestern Europe. They show both (*a*) the draping off and the thickening towards the centre of the isochronous sedimentation interval that takes place between the margins and the centre of underlying block-controlled basins, and (*b*) the corresponding thinning of the isochronously identical sedimentation sequences over the 'swells'. This style of illustration does not bring out periods of slight overlap or offlap, or both, over the transition points between neighbouring underlying blocks.

As observed when mining in the coal basins of northern Europe, the morphology of 'underlying-graben' (formed by coal extraction in successive panels) and overlying induced subsidence basins appears to be directly analogous with the morphology of the products of this process observed in nature, in the upper crust. In mining, the morphology of the mechanism (controlling 'more rigid layer' graben-induced subsidence flexure, in the overlying more rheid layer) is innately associated with a horizontal tensional zone over the convex upward part of the shoulders of the corresponding subsidence flexure. In nature, whether overlap or offlap predominates in the sedimentation at the shoulders of the basin, this may be critical in determining the relative rate of deceleration of the underlying centre and shoulder blocks, in the underlying more rigid layer. Because, in nature, the blocks in plan view appear to exhibit differences in their structure, investigation by means of a succession of coincident palaeogeographic reconstructions of basin formation, built up successively above a post-orogenic tectonic cycle foundation map, appears necessary for such critical features.

The sedimentary facies infilling the basins and draping the swells

During the deposition of the avulsing fluviatile Coal Measures facies, the centres of the bottoms of the basins were obviously subsiding at up to ten times the overall rate of subsidence of the centre of the tops of the intervening 'swells'. It was often more the rate of preservation of the parts of the sediment column, than the rate of sedimentation *per se*, that increased in the middle of the basins, and decreased over the swells and the more neutral zones.

A detailed lithostratigraphic cross section of the preserved parts of the thick succession in the 'Sulzerpacher Schichten' and 'Geisheck Schichten' of the Saar basin (Kneuper 1971, fig. 2, p. 155; fig. 3, p. 159) shows the typical pattern of 'Coal Measures' cross sections. Cross sections of wide sheet sandstone bodies, and occasionally successively stacked lenticular cross sections of ribbon-like sandstones of all sizes, are surrounded by envelopes of finer sediments, the whole being threaded by a complex, anastomosing, succession of relatively thin coal seams with their underlying seatearths. Collectively, the coal seams form not more than a small percentage of the total thickness of the section. (Where exceptionally thick and apparently vertically continuous, the sections of sandstone grade sediment are now easily broken down into the fluvial cycles within them (see, for example, Clarke 1963, fig. 11, p. 682).) Even though they portray only sediment grade and not sedimentary structure as well, it is salutary to compare such lithostratigraphic cross sections with those generated by Bridge & Leeder (1979) from their mathematical simulation model of alluvial sedimentation.

Bridge & Leeder (1979) use a range of flood plain geometries, of compactional relief and of aggradation rates and optional tectonic fault controlled marginal movement to show the small effect of change in these parameters on the character of sedimentation in the resulting detailed lithostratigraphic cross sections. The overall character of the sedimentation produced is identical to that in the Coal Measures, both over the flat depositional upper surface overlying large 'less negative' blocks, and in central parts of flattish-bottomed basins overlying the flat depositional upper surface of large underlying, more negative blocks.

The only discernible difference in sedimentation sequence over the surface of the less negative (more positive) zones or, in the northern part of the NW German basin (Hedeman & Teichmuller 1971, p. 137), apart from the expected diminution in the overall thickness, is the resulting winnowing away of the silty and muddy constituents of the many sandstones in the succession, leaving them white near orthoquartzite, instead of grey, nearer in greywackés mineral composition. This can only be due to the lower overall rate of aggradation of sediment for the same total amount of aqueous flow across the two kinds of submerged block.

The difference in lithostratigraphic cross section between Coal Measures sedimentation in (*a*) a 'flat-bottomed' or a 'flat-topped' (subsidence) situation, and that in (*b*) the smaller, narrower, steep-sided but comparably deep, elongate elliptical or circular, whole- or 'hinge line terminated semi'-basins has been demonstrated by both Duff & Walton (1964), in an example from the Pennine depositional basin, and Read & Dean (1976), in examples from the Midland Valley of Scotland depositional basin. They show that when sediment thickness between the subaerial deposits diminishes and the coal seams converge, the number of 'cycles' between them decreases. The number of 'cycles' decreases (in proportion to the diminishing overall thickness of the column) as the margining 'humps' and 'swells' of the basin are approached. Where this convergence ultimately gives rise to an extensive 'plateau' of very thick coal, it could be said that the successive subaerial deposits in each cycle overlap the successive fluviatile (or marine) deposits at the edge of the basin, e.g. against 'St George's Land' (a contemporary expression of the Wales–Brabant Massif) on the south side of the Pennine basin.

Duff & Walton (1962) have also made an extensive statistical analysis of cycles. In doing so, they found that the classical 'cyclothem' (Wanless & Weller 1932) is generally the exception rather than the rule in the (virtually Carbonate-free) Westphalian 'Coal Measures'.

The rare very thin, but very persistent, marine 'black shale' bands (with the 'Tonsteins') are the only laterally extensive deposit in the 'Coal Measures', other than the subaerial coal seams. With regard to the marine bands, Clarke (1963, p. 690) has suggested that 'the whole northward mountain chain may have suffered depression, causing the rivers to aggrade in their valleys (and thus sedimentation on the coastal plain to diminish). If the coastal plain macro-'basin' as a whole continued to subside during these rare intervals in which the land area (macro-'swell') to the north, then the rare very thin layers of marine deposits would spread sometimes across the whole of the macro-'basin' shelf, e.g., the Aegiranum Marine Band.' The only other explanation appears to be the more commonly invoked worldwide eustatic rise (and sudden subsequent fall) of sea level.

The termination of sedimentation across the shelf was heralded by the macro-'basin' wide advance of the east–west aligned 'front' of the subaerial 'red-bed' facies. It occurred in successively higher horizons in all basins southward, has often been analysed in European palaeogeographic reviews of the close of the Carboniferous (e.g. by Hedeman & Teichmuller (1971)). Very late (?)Stephanian, very small, coal-bearing, intra-montane limnic basins were formed at a number of scattered localities over a large land area to the south of the Variscan front. They have been preserved, for example in France, wherever the Alpine orogeny has not sliced them up (or otherwise removed them from contemporary means of geological detection).

Tectonic phenomena in the European Coal Measures basins

On evidence of the diachronous spread of 'red beds' (see above), apart from very minor localized heralding movements, the overall depositional phase of the Carboniferous in northern Europe terminated with the slow rise, along a (presumably) east–west, mega-axis lying in the land area to the north of the Upper Carboniferous macro-'basin'. This would have been the horizontal hinge (or perhaps rotation) line along which the macro-'basin', as a whole, tilted upwards – everywhere, at least, where the record is preserved, i.e. to the north of the Variscan 'line of disturbance'. The now reddened alluvial sedimentation with occasional thin carbonate layers (Spirobis Limestones) continued to flow across both the local (?graben underlain) basins, of relatively greater – and across the intervening swells and neutral zones of relatively greater and lesser – subsidence, respectively. No large 'upper-Stephanian' sequence of deposition is preserved in this now tectonically positive, and hence erosive, phase in northern Europe.

The smaller-scale constituent 'basement' blocks within those larger blocks underlying the basins, began to move positively, probably no more continuously in the time scale than is evidenced by the changes they had earlier brought about in subsidiary basin and hump topography both within, and between, the Namurian and Westphalian. Being smaller-scale constituent blocks, e.g. within the Pennine or within the Scottish basin as a whole in space scale, the effects they produced were smaller on the areal scale, as well as on the time scale. Ziegler (1980, fig. 1, p. 251) shows only a few tiny scattered Autunian sedimentary basins in NW Europe, with most of the area shown as land.

Ziegler (1980, fig. 1, p. 251) shows the pattern of mega-faults, including graben and 'hinge line and fault' systems (producing 'trap door' depositional semi-basins above, when reactivated) with the dyke swarms and major volcanics that together sliced up the newly deposited Carboniferous during this late-Stephanian–Autunian time of emergence and erosion.

Evidence of the amount of locally elliptical uplift and extension, from the mine plans within the worked coal-bearing basins, suggests that the pattern of uplift followed the same pattern as the subsidence. During subsidence, the relief of the broad elongate elliptical basins, swells and near neutral zones (both of them containing their subordinate basins and humps) flooring the depositional basin of the Upper Carboniferous (as a whole) has been suggested wherever cross sections can be made, but not fully mapped. A similar, but subaerial, pattern may have emerged, had the palaeogeographer been present to map the surface of the crust in this period of hierarchically scaled differential N European *uplift* – as opposed to the same overall pattern of N European *subsidence*.

Additional evidence on the nature of block control of uplift around the North Sea basin available within detailed coalfield mining records

The worked coalfields of northern Europe contain multiple overlapping detailed mine plans on which are to be seen both (*a*) very nearly all the 'lines of disturbance' and, because the samples are so complete, (*b*) the statistics of very nearly all of the undisturbed areas between these lines. Within these 'undisturbed' areas, many of these plans show, at least sample, and often detailed, levelling by the mine surveyors on each of these overlapping plans of the worked areas of the coal seams.

Thus there is a unique, nearly complete, sample of both the amount and the inferred shape of crustal extension produced on the major 'swells'. The two directions in which the aggregate fault 'want' per unit horizontal length of coalfield are at a maximum and a minimum, together with the total amount of this horizontal tensional strain in these two directions, are both obtainable from the statistical distributions of fault orientation, strike length, hade, throw and intensity. These are obtainable from analysis of the normal faulting shown on the mine plans in each coalfield.

In the NE Coalfield of Britain, the increase in surface area, shown by the normal faulting, is about $\frac{1}{2}$ %, and 'This is near the order that might be expected of an elliptical regional uplift or doming of the crust in N. England to remove (or not deposit) 4/5,000 ft. [*ca.* 1.2–1.5 km] of Mesozoic and 5,000 ft. of Paleozoic rocks' (Clarke 1962, p. 212).

From the statistical distributions of strike lengths and throws given in the relevant part of the NE Coalfield, the elliptical uplift would have been locally more nearly circular on the constituent Alston block, with its longer axis east–west. Over the uplifted area occupying the southeastern part of the Northumberland Trough, it would have been less 'nearly circular', but still with a direction of principal horizontal tensional strain east–west. In other words, the same 'blocks' as gave slight differences in vector amount of subsidence also acted to give slight differences in vector amount of uplift.

Over the now emergent sub-block on the northeast quadrant of what was earlier the Pennine basin, from the normal fault pattern shown in the Yorkshire Coalfield uplift would have been more strongly elliptical, with its long axis northeast–southwest. As the (exceptional) Upper Palaeozoic 'foundation triangle' is approached, semi-conical uplift gives an internally radially fractured 'arrowhead' shape, with its north–south centre line on the mega-line of disturbance now seen as the Pennine fault – Namurian depositional monocline/Malvern Hills lineament (see Wills 1956, fig. 3, p. 96).

A triangular 'trap door' uplift is inferred to have operated in the Bristol coalfield to produce the areas of only just subvertical coal seams, once 'stope' mined, in this area. The hypoteneuse of this particular triangle is part of the fan of lineaments between the inferred mega-line of disturbance, at the southeastern edge of the Welsh Massif, curving down and southwestwards, to join the westward margins of the ((?)shallower) Variscan 'mega-line', the two continuing together as the well known Variscan–(?)Cadomian line across southern Ireland.

With regard to differential movement of the smallest constituent foundation blocks (at the highest in level in the crust, but lowest in the hierarchy of size) beneath coalfields, casual studies have been made of the fault statistics recorded in the mine plans in the areas (*a*) over, and (*b*) surrounding a few of these 'micro'-blocks. The elongate, but only a few kilometres long, Harton (tectonic) Dome in the Durham Coalfield and the Ashby anticline in the S Derbyshire and Leicestershire Coalfield, both clearly show smaller crustal extension over these small 'humps' than in the annular ring-like area around them. This is interpreted as implying regional uplift followed by a slight additional tensional strain due to subsequent, very local, slight drawing down of (1) a ring-like area around the dome, and (2) along the two synclinal lines flanking the anticline. In other words, no 'plie-de-couverture' or 'compressional folding', either after or before (or both) a period of 'tensional faulting', is involved. All that need be involved is slight differential vertical movement in a hierarchy of small upper crustal blocks.

'Mine plan' studies of the east–west trending *en échelon* systems of the faulted monoclines that partition the Alston and Askrigg blocks in the northern Pennine hills have shown that the here abnormal want due to the here abnormal (45°) hade of the fault member of this type of disturbance is exactly compensated for by the crustal shortening produced by the associated folding (Clarke 1962, p. 213). Morphologically, but here at a gigantic increase in scale, these mirror the 'rotational slips' observed, and quantitatively analysed, by the civil engineering soil-mechanist.

Similar but more detailed studies have been made of the tectonics of the Ruhr coalfields. Mine plan studies of the southern Belgian – northern French coalfield, being so greatly 'thrust and napped', are likely to defy all but the most persistent analyst.

The northern European coalfield mine plans show the integrated effects of both the Variscan, Laramide and the Alpine periods of emergence or less widespread subsidence. However, the intensity of drilling and shaft sinking within the coalfields is usually sufficient to identify the differential or incremental amount of crustal extension that is attributable to the major re-activation of swells and their associated tensional fault patterns during succeeding late-Mesozoic–Tertiary tectonic hiatus. Thus to enable this later movement to be subtracted from the integrated post-Variscan position sampled in all NW European mine plans, e.g. as shown in the NE Coalfield of Britain by Clarke (1963, pl. 2, p. 308). The correlations between pre- and post-Permian, etc., movements established in the coalfields may be carried between coalfields when the areas sampled by deep reflexion seismic surveys by the oil industry become similarly available.

It is by these means that the amount of uplift around the margins of the Mesozoic depositional and tectonic basin in the North Sea may now be given some additional quantification.

Conclusions

1. With the recent input of data into the public domain from the oil and coal industries and on the basis of a study of the origin, nature and setting of the Carboniferous basins of the northern European subcontinent, it is hypothesized that they may be the products of a nesting hierarchical subcontinentally fixed mechanism of crustal change. The evidence that has been assembled is primarily that of an observed succession of morphologies, and does not rest on the morphologies giving rise to measured or measurable properties generated through processes that lie essentially within the physics or applied mathematics of change in the upper crust.

2. The *geological* evidence available on the middle-sized depositional basins within the Carboniferous 'basin' of northern Europe is most easily observed north of the Variscides. The data available on the basins and their intervening less rapidly subsiding swells and neutral zones show that (*a*) they were generally of elliptical or semi-elliptical shape and lie in a nesting hierarchy of sizes whose location is conditioned by the underlying tectonic framework; (*b*) the bottom of the middle sizes of basin subsided at a diminishing rate and the bottom of most of the intervening swells and planar zones subsided at a more steady rate; (*c*) except in the neutral blocks, this pattern of movement of the same nesting hierarchy of elliptical, etc., basins and swells was inverted.

3. The *mining* evidence in the coal-bearing basins of northern Europe shows that (*a*) on propagation, the graben formations induced at depth by mining, each give rise to a *wider* overlying zone of block tilting (involving initial rotation and counter-rotation on a horizontal axis, within a wavelike set of translational movements in the vertical plane, as the axial length of the propagating graben grows. (*b*) Concurrently, this overlying zone, in turn, gave rise to a still *wider* overlying subsiding basin in the less self-confined strata at the surface level. Each side of the generally symmetrical basin has, at the near surface, a concentric pair of zones. The outer zone is one of intrinsically horizontally extensional and the inner zone of intrinsically horizontally compressional movement.

The dynamically changing *morphology* of the both horizontally and vertically propagating system retains almost fixed width: depth ratios. These width:depth ratios, except where the subsiding layer is very thin compared with depth, are almost independent of depth. The mining industry quantifications of the evidence are entirely from observation and quantification of the morphology of graben and predicted subsidence basin formation and *not* from the physics of the system.

4. On the evidence of relative rates of initial infilling during the Upper Carboniferous and provenance of the sediments (more than basin morphology) it is necessary to distinguish between the Carboniferous basins lying along the line of the Variscan Front (Flysch) in the supposed area of active plate collision, and those elsewhere. It would appear, from the pattern within the morphology of the 'elsewhere' type of Carboniferous basins, that a mechanism for propagation of vertical movement existed in the crust of continental Europe that is comparable to that seen in coal-mining in the basins. On the hypothesis given in this paper, an original line of active plate collision left both a suture line and a set of parallel lines of lesser disturbance in the crust. Later collisions left both suture lines at different angles to the original line, and parallel lines of lesser disturbance higher in the crust. On such a mechanism, with further collisions, the more primary lines would either be preserved (but stiffened beyond movement in the 'old shield' areas) or be absorbed in the mantle during 'plate tilting'. Thus in the Carboniferous, the northern European crust, collision and orogenesis at the Variscan front reactivated movements on one of the sets of parallel-oriented, more deeply imposed, earlier lines of disturbance. This initiated rectilineal and triangular block movements propagating upwards towards the surface (but downwards in hierarchical size sense). Propagation was upward through a successively more dense array of earlier imposed, successively more criss-crossing hierarchy of blocks, whose relations to one another are fixed within the subcontinental crust. These control the fixed configuration of the overlying contemporary basin and swell movements. The results of these movements are seen on the reconstructed surface of the Carboniferous in areas away from the Variscan orogeny. The final movement was a 'whole

plate tilt'. Subsequent block movements were often in the opposite sense to basin or swell forming movement.

5. Because of the stationary scalar nesting hierarchical nature of the single mechanism that, in this hypothesis, is involved, for analytical purposes it has been necessary to devise a terminology independent of horizontal and vertical scale, i.e. to employ the term 'line of disturbance' within (at a comparable level of disturbance, and hence (because of convergence) comparable level in the crust) 'undisturbed' surrounding surfaces. In this terminology, to make the frame of the undisturbed surface the same at each level in the crust, the area covered by the frame containing any given number of lines of disturbance becomes smaller and smaller at higher and higher levels within the lithosphere.

References (Clarke)

Bless, M. J. M., Bouckaert, J., Calver, M. A., Graunlich, J. M. & Paproth, E. 1977 *Meded. Rijks geol. Dienst.* **28**, 101–147.

Bless, M. J. M., Bouckaert, J. & Paproth, E. 1980 *Meded. Rijks geol. Dienst.* **32**, 1–14, enc. 1.

Bridge, J. S. & Leeder, M. R. 1979 *Sedimentology* **26**, 617–644.

Calver, M. A. 1969 In *C.r. 6e Cong. Int. Strat. Geol. Carb., Sheffield* 1967, vol. 1, pp. 233–254.

Clarke, A. M. 1962 *Trans. Instn Min. Engrs* **122**, 209–231.

Clarke, A. M. 1963 *Trans. Instn Min. Engrs* **122**, 667–706.

Craig, G. Y. 1965 In *Geology of Scotland*, pp. 311–354. Edinburgh: Oliver & Boyd.

Cogné, J. 1976 *Coll. Int. C.N.R.S.*, no. 268 (*Géologie de l'Himalaya*), pp. 111–129.

Duff, P. McL. D. & Walton, E. K. 1962 *Sedimentology* **1**, 235–255.

Duff, P. McL. D. & Walton, E. K. 1964 *Devl Sedimentol.* **1**, 114–122.

Hedemann, H. A. & Teichmuller, R. 1971 *Fortschr. Geol. Rheinld Westf.* **19**, 132–145.

Kent, P. E. 1975 In *Petroleum and the continental shelf of north west Europe* (ed. A. W. Woodland), pp. 3–28. London: Applied Science Publishers.

Kent, P. E. 1977 *J. Geol. Soc. Lond.* **134**, 1–18.

Kent, P. E. 1980 *Phil. Trans. R. Soc. Lond.* A **294**, 125–135.

Kneuper, G. 1971 *Fortschr. Geol. Rheinld Westf.* **19**, 152–162.

Peng, S. Y. 1978 In *Coal mine ground control*, 208–232 and 281–341. New York: Wiley.

Ramsbottom, W. H. C., Calver, M. A., Eager, R. M. C., Hodson, F., Holliday, D. W., Stubblefield, C. J. & Wilson, R. B. 1978 *J. geol. Soc. Lond.* Report no. 10.

Read, W. A. & Dean, J. M. 1976 *Sedimentology* **23**, 107–120.

Reading, H. G. 1969 In *C.r. 6e Cong. Int. Strat. Geol. Carb., Sheffield* 1967, vol. 4, pp. 1401–1411.

Wanless, H. R. & Weller, J. M. 1932 *Bull. geol. Soc. Am.* **3**, 1003–1016.

Wills, L. J. 1956 *Concealed coalfields.* Glasgow: Blackie & Son.

Ziegler, P. A. 1975 *Bull. Am. Ass. Petrol. Geol.* **59**, 1073–1079.

Ziegler, P. A. 1978 *Geol. Mijnb.* **57**, 589–626.

Ziegler, P. A. 1980 *Mem. B.R.G.M.* **108**, 249–280.

Discussion

A. W. Bally. What is the resolution of the seismic lines that Mr Clarke has shot across the coalfields? Do they show details of the sedimentary structures within the coal measures?

A. M. Clarke. We use interface waves trapped on the boundaries of the coal seams themselves if we need a vertical resolution of 1 m or less. When we are working on the surface we are only interested in the structure in the first $\frac{1}{2}$ s of recording time and use a geophone spacing of a few tens of metres. This spacing, combined with accurate static corrections, gives us a vertical resolution of between 5 and 10 m. The smallest fault displacements we map as diffractions.

We have done some work on modelling facies variations statistically. The seismic reflexions from these variations show good lateral continuity and we can follow features like shoestring sands, and distinguish them from sheet sands. Various papers describing such records have been published (Buchanan *et al.* 1981; Gebala & Nowak 1980; Jackson 1981; Ziolkowski 1981).

References

Buchanan, D. J. *et al.* 1981 *Geophysics* **46**, 994–1002.
Gebala, W. & Nowak, J. 1980 *Przegl. górn.* **36**, 531–536.
Jackson, L. 1981 *Geophysical examination of coal deposits.* (116 pages.) London: I.E.A. Coal Research.
Ziolkowski, A. M. 1981 *Min. Engr* **140**, 605–615.

Phil. Trans. R. Soc. Lond. A **305**, 225–247 (1982) [225]
Printed in Great Britain

Evolution and subsidence of early Precambrian sedimentary basins

By M. J. Bickle† and K. A. Eriksson‡

† *Department of Geology, University of Western Australia, Nedlands, W.A.* 6009, *Australia*
‡ *Department of Geological Sciences, Virginia Polytechnic Institute, and State University, Blacksburg, Virginia* 24060, *U.S.A.*

Many of the models for modern sedimentary basins postulate two-stage subsidence; a rapid initial subsidence due to thinning or loading of the crust, followed by a more protracted thermal stage as the lithosphere, which is thinned during the initial stage, relaxes to equilibrium thickness. The geology of a number of Archaean greenstone belts and early Proterozoic cratonic basins in South Africa may be explained by such a model. Rapidly erupted shallow marine or subaerial volcanic rocks predominate in the lower parts of sedimentary–volcanic sequences. These are thought to relate to initial subsidence as (1) accommodation of relatively thick volcanic sequences requires substantial and rapid subsidence, and (2) marginal uplift following these early volcanic intervals is consistent with viscous relaxation following the initial elastic response of the lithosphere to localized loading. Sedimentary sequences overlying initial volcanic dominated intervals may have been deposited during the ensuing phase of more widespread subsidence related to thermal relaxation of the thinned lithosphere. If so, sediment-filled subsidence of *ca.* 5.5 km in greenstone terrains at 3.5 and 2.6 Ga and of 7–10 km in cratonic shelf basins between 2.7 and 2.1 Ga require increases of lithosphere thickness between *ca.* 60 and 90 km. These minimum estimates of early Precambrian lithosphere thickness, although crude, are similar to estimates of present lithosphere thickness. In some early Precambrian basins, the cause of subsidence may have been crustal extension with development of faulted grabens that evolved into continental margins, but in cratonic shelf basins faulting did not occur during or after the initial subsidence, and some less obvious causal mechanism must be sought.

Introduction

Mechanisms for sedimentary basin formation and preservation are still both controversial and puzzling. A variety of mechanisms have been proposed (see, for example, Sleep *et al.* 1980), but the similarity of rates of subsidence to those expected from thermal relaxation of a thinned lithosphere suggests that the main tectonic control is thermal and on the scale of the lithosphere (see, for example, Sleep (1971), Watts & Ryan (1976), Steckler & Watts (1978), Keen (1979), Haxby *et al.* (1976), Sclater & Christie (1980) and Sleep *et al.* (1980)). This mechanism is the same as the subsidence of the ocean floor where, with increasing age, the lithosphere cools, becomes denser and sinks to maintain isostatic equilibrium.

McKenzie *et al.* (1980) suggested that the amounts of thermally controlled subsidence in Precambrian sedimentary basins might allow estimates of lithospheric plate thickness, estimates not easily derived from other evidence but of critical significance to interpretation of Archaean tectonic processes (Bickle 1978; Burke & Kidd 1978). In this paper, we re-examine models for sedimentary basin formation dependent on subsidence of a cooling lithosphere to identify those aspects relevant to the early Precambrian. This is followed by a discussion of a few of the better studied early Precambrian basins, to analyse the genetic significance of different facies associations in terms of the different models for basin formation, and to evaluate the significance of sedimentary thicknesses in terms of lithosphere thickness.

Sedimentary basin formation: models

A permanent, isostatically compensated, sedimentary basin will only be preserved on continental crust if the crust is thinned, or if the density of the crust or underlying lithosphere is changed. Permanent density changes without associated compositional changes within the lithosphere require special pleading, especially as critical garnet-producing reactions are probably near isobaric (O'Hara *et al.* 1971; Newton 1977).

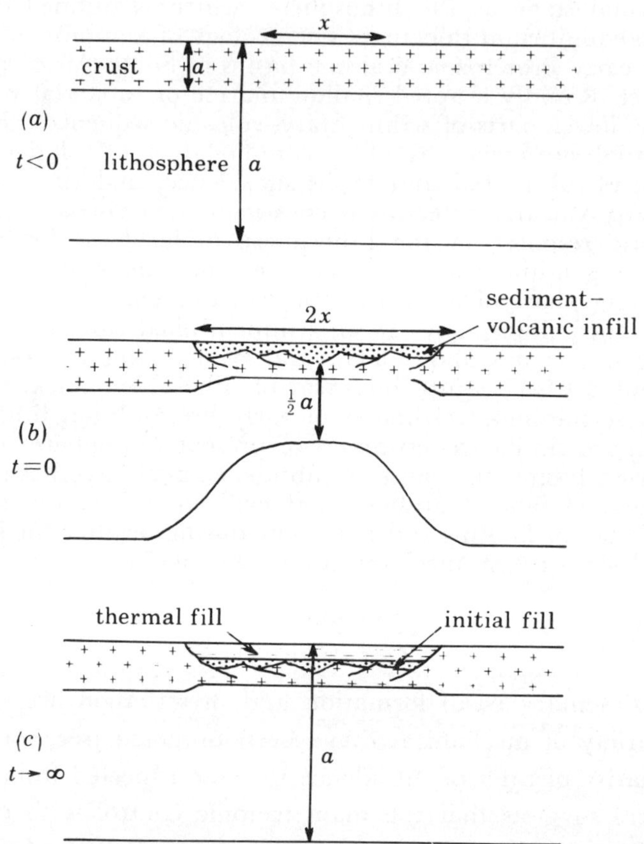

Figure 1. Schematic evolution of a sedimentary basin following crustal extension, after McKenzie (1978). (*a*) Before stretching. (*b*) Initial subsidence. Sediment or volcanic fill, or both, deposited during faulting of basement. (*c*) Thermal subsidence related to relaxation of lithosphere to original thickness. Any mechanism that rapidly thins the lithosphere and thins or loads the crust will give a similar pattern of subsidence.

McKenzie (1978) has examined the tectonic implications of stretching continental crust and underlying lithosphere, and the evolution of several modern sedimentary basins can be successfully explained by this model (see, for example, Sclater & Christie 1980) even though available data do not necessarily exclude alternative models. McKenzie's model is essentially simple (figure 1). Continental crust and lithosphere in thermal equilibrium are stretched during a rifting phase (the lithosphere is defined as a thermal boundary layer). Subsidence will occur in two stages. The first, the 'initial stage', which may also result in uplift, results from the immediate isostatic compensation of thinned crust and lithosphere, and the second, the 'thermal stage', involves subsidence at an exponentially decreasing rate as the lithosphere cools and returns to equilibrium thickness.

McKenzie *et al.* (1980) show how initial subsidence and total subsidence depend on the crustal thickness, amount of stretching, and lithosphere thickness. Total subsidence depends only on the amount of crustal thinning and the density of the sedimentary fill. The thermal subsidence depends only on the increase in lithosphere thickness after the initial stretching.

Alternative models for basin development that postulate relatively rapid crustal thinning, or effective crustal thinning with associated thinning of the lithosphere, can result in rather similar subsidence histories. For example, rapid dyke injection and lithosphere thinning, a phase change triggered by a hot mantle diapir, or even rapid erosion with an associated thermal event, will all result in rapid isostatic adjustment to the new crustal thickness or density followed by the slower thermal adjustment of the lithosphere. In these cases, a direct relation between crustal and lithospheric thinning will not be be attained, as it is in the rifting model. Even with rifting, if there is associated volcanicity, the heat transported from depth might thin the lithosphere destroying the simple relation.

For a large class of basin formation models, the subsidence history can therefore be divided into an initial phase directly relating to the mechanism causing crustal subsidence and a thermal phase relating to thermal relaxation of the thinned lithosphere. The discussion of early Precambrian basins below will be concerned with two main aspects: (1) separation of the initial and thermal subsidence, and (2) estimates of subsidence during both the initial and the thermal phases. Separation of initial and thermal subsidence can only be attempted on the basis of the interpretation of sedimentary–tectonic environments, and these in turn should place some constraints on the fundamental cause of basin formation.

Early Precambrian sedimentary basins

Early Precambrian tectonic processes, and consequently the setting of the sedimentary sequences, are much debated with little or no general consensus. We propose to discuss the evolution of early Precambrian sedimentary basins in terms of three main sedimentary–tectonic associations recognized in well preserved and well studied areas. Additional associations undoubtedly exist, and the affinities of many well known areas are not yet clear. This simple subdivision provides a convenient way to present a discussion of basin evolution in terms of phenomena that are at least adequately documented and were apparently relatively widespread: if basin development can be related to some simple physical model, the recognition of re-occurring features must initiate elucidation of the model.

The three associations are: (1) Archaean greenstone associated sediments, which probably encompasses more than one environment; (2) cratonic shelf basins, such as the Witwatersrand and Transvaal supergroups in South Africa; and (3) cratonic rift basins, of which an example is the Ventersdorp in South Africa.

Archaean greenstone association

All of the well preserved sedimentary sequences older than 3000 Ma, and many of the sequences formed between 2500 and 3000 Ma, occur in the Archaean greenstone association.

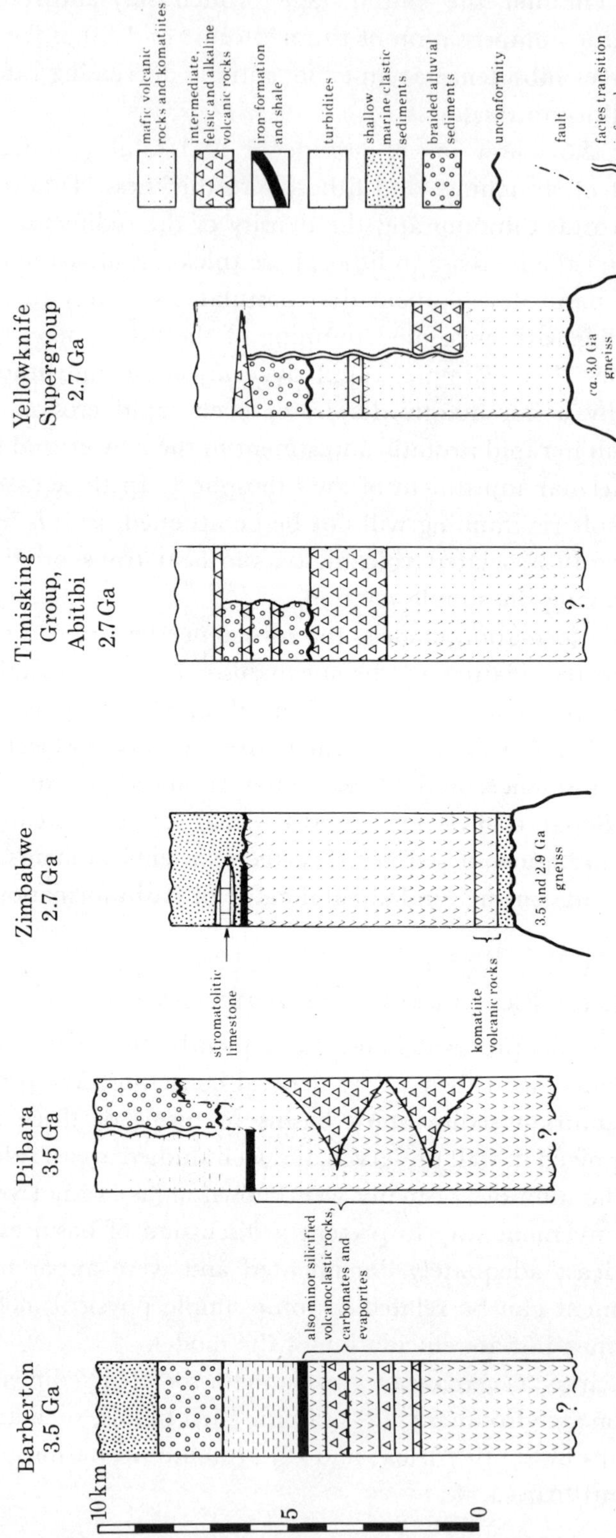

FIGURE 2. Schematic greenstone belt stratigraphic sections (based on references given in table 1).

Many greenstone belts are characterized by thick mafic volcanic sequences with subsidiary felsic and ultramafic volcanics, which are overlain by major sedimentary units. The volcanics and sediments are preserved between domal or elongate–domal granite–gneiss batholiths, which dominate the structural setting.

(*a*) *Tectonic setting of Archaean greenstone belts*

A variety of sedimentary–tectonic environments is recognised in greenstone terrains. Interpretation of virtually all greenstone areas is frustrated by uncertainty as to the nature of the basement to the volcanic sequences, especially whether the basement was a mafic–oceanic or granitic–continental type crust (Anhaeusser 1973; Hunter 1974), together with stratigraphic and structural complexities. In a few areas, geochronology and mapped unconformities have proven continental basement, but in most areas the nature of the basement is unknown, being obscured by deformation, metamorphism and granitic plutonism. Despite these problems study of sedimentary facies relations in a number of Archaean areas is sufficiently advanced to provide insight into Archaean sedimentary–tectonic processes. These studies have led to the recognition of continental margin sequences in both the *ca.* 3500 Ma Barberton Mountain Land (South Africa) and Pilbara (Western Australia) greenstone belts, continental shelf sequences in the *ca.* 2600 Ma Zimbabwe greenstone belts and volcanic arc related depositional environments in the Canadian Superior and Yellowknife provinces.

(*b*) *Basin evolution*

Among the best preserved and exposed greenstone sequences are the relatively old 3500 Ma Barberton Mountain Land and Pilbara Block. In both areas, a thick volcanic sequence is overlain by the main sedimentary sequence (figure 2, table 1). The intensity of deformation and metamorphism is laterally variable. Parts of the stratigraphy are preserved in simple synclinoria or even as flat-lying undeformed units. Elsewhere, particularly in the lower parts of the volcanic sequence, deformation and metamorphism are more intense.

The volcanic units (Onverwacht Group in Barberton, Warawoona Group in the Pilbara) range from komatiitic to felsic volcanics, although pillowed komatiitic and tholeiitic basalts predominate. Thin and often cherty metasediments are interbedded with the volcanic rock. These represent silicified carbonates, volcanoclastic sediments and evaporative sulphates (Barley *et al.* 1979; Lowe & Knauth 1977; Dunlop & Buick 1981). This sedimentary evidence indicates that the volcanic rocks accumulated in very shallow marine environments. It is impossible to give precise estimates of the thicknesses of the volcanic units in the Pilbara and at Barberton, as the bases of the sequences are not preserved. In the Pilbara, relatively undeformed sections 5–8 km thick may be measured (Barley *et al.* 1979), and gravity anomalies preclude much greater thicknesses. In Barberton, estimates of 14 km by Viljoen & Viljoen (1969) may include unrepresentative sections, and do not take into account potential structural complexities (Williams & Furnell 1979).

The volcanic sequences are overlain by thick clastic sedimentary units in both areas, and Eriksson (1981 *a*) has constructed rather similar palaeogeographic models in both Barberton and the Pilbara (figure 3). Laterally persistent iron-formation–shale units overlie areas of the volcanics, suggesting early rapid subsidence of the shallow water volcanic sequence. Thick and aerially extensive turbidites pass either laterally or vertically into braided alluvial sequences. In the Pilbara, this transition is entirely lateral and abrupt, with no intervening

TABLE 1

basin	basement	age and duration	initial phase: rock types	initial phase: thickness	initial phase: density assumed	thermal phase: rocks types	thermal phase: thickness	thermal phase: density assumed
Barberton, S. Africa	not known; sialic crust inferred from sediments	3540[1] ± 30 Ma; duration not known	komatiitic, tholeiitic and felsic volcanic rocks, carbonate, minor volcanoclastics, chert, evaporites: shallow marine[15,16]	not known	2.9	iron-formation, shale, turbidites overlain by braided alluvial and shallow marine[17]	5.5 km	2.55
Pilbara, W. Australia	not known; sialic crust inferred from sediments	3556[2] ± 32 Ma; duration not known	komatiitic, tholeiitic and felsic volcanics. Minor volcanoclastics, silicified carbonate, evaporites: shallow marine[18]	> 5 to 8 km	2.9	iron-formation, shale overlain by turbidites or braided alluvial[19]	max. *ca.* 5 km; 3 km of turbidites with water depth ≳ 500 m	2.55
Pongola, S. Africa	3160[3] Ma granites	3090[4] ± 90 Ma on volcanics; intruded by 2819[5] Ma mafic dyke	basal arkosic, arenites, mainly tholeiitic basaltic lavas, subordinate stromatolitic dolomites and shale: shallow marine[20]	5.5 km	2.7	shale, siltstones, conglomerates: intertidal + shallow marine[20,21]	3 km	2.55
Younger greenstones, Zimbabwe	3500[6] and 2900[7] Ma granite, gneiss and greenstones	2640[8] ± 140 Ma; intruded by 2570[33] ± 15 Ma granites	basal shallow marine and intertidal sediments; mainly komatiites and tholeiitic basalts: shallow marine[22,23]	5–7 km	2.9	conglomerates, siltstones, minor iron-formation stromatolitic limestones, quartzite: shallow marine[22,24]	2 km	2.55
Abitibi Supergroup & Timisking Group, Abitibi Greenstone Belt, Canada	not known	*ca.* 2700[9] Ma	mafic and intermediate lavas: mainly submarine[25]	not known	2.9	turbidites overlie and are laterally equivalent to braided alluvial; minor volcanic horizons[25]	3.5 km turbidites at top in water depths ≳ 500 m	2.55

TABLE 1 (*cont.*)

			initial phase			thermal phase		
basin	basement	age and duration	rock types	thickness	density assumed	rock types	thickness	density assumed
Yellowknife Supergroup, Slave Province, Canada	*ca.* 3000 Ma granitic gneiss[10]	2665[11] ± 15 Ma on volcanic rocks; intruded by 2500 Ma granites[11]	mainly tholeiitic submarine basalts, minor felsic volcanics[10]	< 10 km	2.9	turbidites laterally equivalent to and overlie braided alluvial sediments; minor felsic volcanics[26]	5.5 km turbidites at top of section. Water depths > 500 m	2.55
Witwatersrand Supergroup, S. Africa	*ca.* 2800[34] Ma granitic rocks	*ca.* 2700 Ma; overlain by 2643[12] ± 80 Ma lavas	basal braided alluvial subaerial basaltic volcanic rocks[27]	2 km	2.83	intertidal quartz arenites and shales overlain by braided alluvial arkoses and conglomerates[27]	7 km	2.55
Ventersdorp Supergroup, S. Africa	2700[3] Ma granitic crust and Witwatersrand sequence	2643[12] ± 80 Ma; overlain by 2240 Ma Transvaal Supergroup	subaerial basaltic volcanic rocks, alluvial fan and lacustrine sediments related to graben development[28,29]	3 km	2.75	not well defined; braided alluvial overlain by basic volcanic rocks[28,29]	500 m	2.55
Transvaal Supergroup, S. Africa	2480[13] Ma granites	2200[5,35] Ma on volcanic interval; cut by Bushveldt, 2100[13,14] Ma	braided alluvial, basaltic volcanic rocks, intertidal argillite, arenite[30]	2 km	2.66	shallow marine and intertidal dolomites overlain uncomformably by shallow marine arenites and shales[31,32]	10 km	2.66

References: (1) Hamilton *et al.* (1979), (2) Hamilton *et al.* (1980), (3) Allsopp *et al.* (1962), (4) Burger & Coertze (1973), (5) Davies *et al.* (1970), (6) Moorbath *et al.* (1977), (7) Hawkesworth *et al.* (1979), (8) Hamilton *et al.* (1977), (9) Goodwin (1977), (10) Padgham (1980), (11) Henderson (1981), (12) Van Niekerk & Burger (1978), (13) Coertze *et al.* (1978), (14) Hamilton (1977), (15) Viljoen & Viljoen (1969), (16) Lowe & Knauth (1977), (17) Eriksson (1980), (18) Barley *et al.* (1979), (19) Eriksson (1981*b*), (20) Mathews (1967), (21) von Brunn & Hobday (1976), (22) Bickle *et al.* (1975), (23) Nisbet *et al.* (1977), (24) Martin *et al.* (1980), (25) Hyde (1980), (26) Henderson (1972, 1975), (27) Pretorius (1976), (28) Winter (1976), (29) Buck (1980), (30) Button (1973*a*), (31) Eriksson *et al.* (1976), (32) Button & Vos (1977), (33) Hickman (1978), (34) Allsopp (1964), (35) Button (1976).

shallow marine shelf facies yet recognized. In Barberton, turbidites of the Fig Tree Group pass vertically into fluvial sediments of the Moodies Group. Eriksson (1981 *a*) interprets this relation in terms of a steep rift-type continental margin, with alluvial sediments prograding directly onto fan-slope deposits (figure 3). Shallow marine facies only become well developed in the Moodies Group higher in the stratigraphy at Barberton, after the establishment of a continental shelf. The scale of the turbidite deposits in the Pilbara (tens of thousands of square kilometres) and evidence for a substantial tidal range at Barberton have been taken to suggest that these basins were marginal to major oceans. The sediments in both areas contain a high proportion of potash feldspar. Eriksson (1981 *a*) cites this as evidence that a

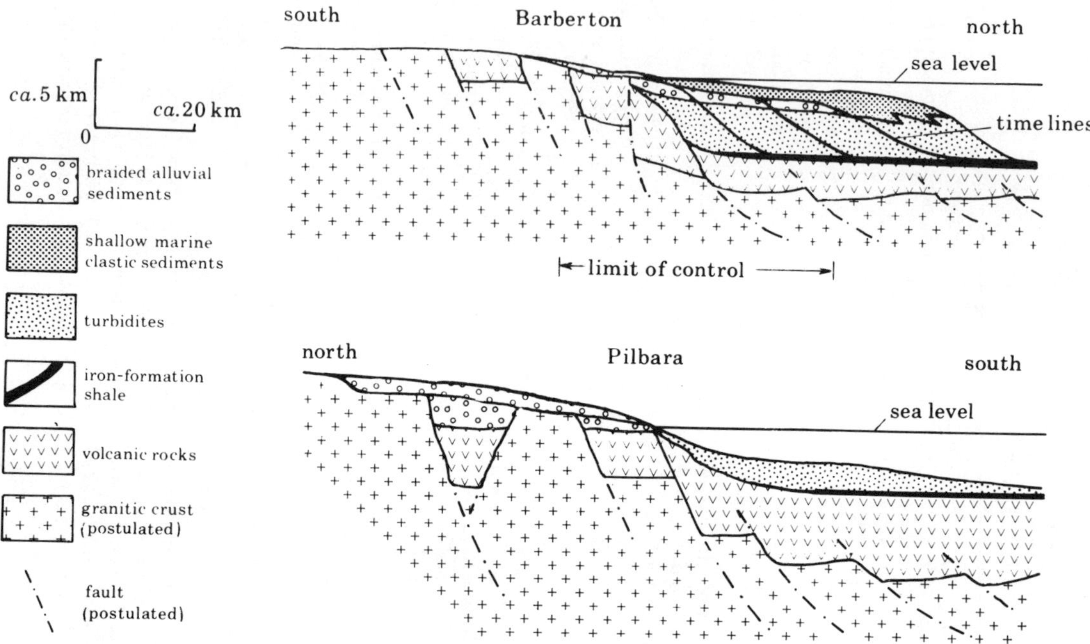

FIGURE 3. Schematic palaeogeographic reconstructions in Barberton and Pilbara greenstone belts, after Eriksson (1980, 1981 *b*).

granite–gneiss terrain formed a significant part of the source area for the sediments, and was part of a continental type basement to the greenstone sequences.

Younger *ca.* 2600 Ma greenstone sequences exhibit some similarities to the 3550 Ma Barberton and Pilbara belts, but perhaps not surprisingly exhibit more diversity. It is probable that a range of tectonic–sedimentary environments is preserved and caution should be exercised when trying to formulate models for their tectonic evolution.

Zimbabwe contains younger greenstone belts (2640 ± 140 Ma) (Hamilton *et al.* 1977), which developed over an area of *ca.* 700 km × 200 km on basement containing older sialic and greenstone material. It is possible to correlate the greenstones lithostratigraphically across much of the area (Wilson *et al.* 1978), and in two areas well preserved unconformities are preserved between basal sedimentary units and the older granite–gneiss basement (Bickle *et al.* 1975; Orpen & Wilson 1981). Distinctive features of the stratigraphy are a thin basal clastic sedimentary formation, a very uniform 5–7 km thick sequence of komatiitic and tholeiitic basalts, and an overlying clastic sedimentary sequence (figure 2). Above these widely distributed lower units, the younger greenstone stratigraphy shows regional variation with the development of a major calc-alkaline belt along the *ca.* 700 km northwest margin of the craton. The sedimentology of the Rhodesian greenstone sequences has not been studied in adequate detail, but the preservation of basal unconformities does allow the early subsidence history to be studied, providing information not available in other Archaean areas.

In the Belingwe Greenstone Belt (Bickle *et al.* 1975), the basal sedimentary formation is in unconformable contact with both 3500 Ma gneisses and older greenstone sequences. The unconformity shows only minor relief. The basal sediments pass up from intertidal facies into deeper water, sub-wave base facies, and are capped by a laterally persistent iron formation; subsidence was initiated before the ensuing volcanicity. The overlying volcanic sequence comprises a monotonous 5–7 km sequence of pillowed and massive basaltic flows, with a major komatiite horizon at its base. It contains only minor cross-bedded volcanoclastic sediments and rare cherts. The volcanic rocks are overlain by a second major sedimentary interval of conglomerates derived from the underlying volcanic rocks as well as shallow marine clastic facies, iron-formation and stromatolitic limestones (Martin *et al.* 1980). The volcanic rocks are overlain directly by iron formation in the northwest, and by locally derived conglomerates in the southeast. Both subsidence and localized uplift must have occurred on termination of the volcanicity.

In Belingwe, the sediments are *ca.* 2 km thick, and the top is not preserved. Elsewhere, they pass into a further volcanic association. Facies variation within the Belingwe Greenstone Belt, coupled with uniform north-northwest trending current directions based on cross-bedding measurements, suggests that the currently preserved sequence was not the site of a localized basin, but rather part of a more widespread sedimentary and volcanic blanket.

Canadian greenstone terrains are of similar age to the Zimbabwe younger greenstones (*ca.* 2600 Ma). The greenstones of the Superior Province form a major area of mafic volcanics, felsic volcanics and sediments similar in general features to other greenstone terrains. Like most Archaean areas, evidence for the basement to the greenstone sequences is largely lacking. The stratigraphy is complex, and unconformities occur within the greenstones, with several cycles of volcanicity and sedimentation. A number of unconformities are reviewed by Baragar & McGlynn (1976) but the regional significance of these is still uncertain.

Hyde (1980) has summarized the sedimentology of the Timiskaming Group (figure 2), and points to similarities with sedimentation in other parts of the Superior Province (see, for example, Walker 1978). Sedimentary sequences overlie submarine mafic to felsic volcanic sequences. Two main clastic sedimentary associations are identified: non-marine braided alluvial and deeper water turbidite associations. In contrast to both Barberton and the Pilbara, the marine turbidite association characteristically *overlies* the non-marine association, or is laterally equivalent to it. Hyde suggests non-marine sedimentation on transitory volcanic islands, which ultimately subsided back into the surrounding deeper water turbidite basins. Palaeocurrent directions indicate variably orientated palaeoslopes, analogous to a variety of modern ocean volcanic island environments. The sedimentary provenance is largely volcanic with some granitic detritus derived from subvolcanic plutons. There is no evidence of a granitic provenance as in Barberton or the Pilbara.

The total thickness of sediments with minor intercalated volcanics is between 2 and 3 km in the Timisking area and is similar in several other areas in the Superior Province (Hyde 1980). In all these areas, tops to sequences are not preserved, or terminate with more volcanism. Turbidite sediments predominate near the tops of sequences, implying substantial water depths (at least *ca.* 500 m and possibly were more than 1 km). Henderson (1972, 1975) records *ca.* 5 km of braided alluvial and turbidite deposits in a similar sedimentary association, in the Yellowknife greenstone province.

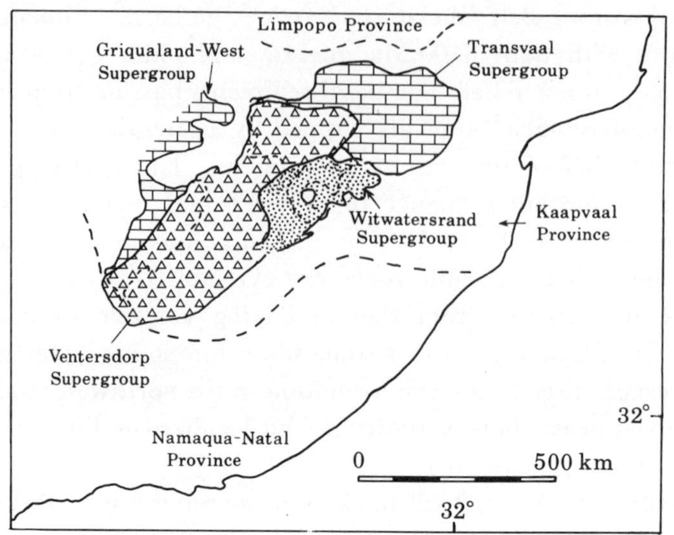

FIGURE 4. Distribution of cratonic basins (3.0–2.0 Ga) in South Africa, stripped to show oldest basin.

Cratonic basins

Major sedimentary basins developed on stabilized Archaean granite–gneiss and greenstone 'cratons' in all the major continents, and are preserved with little deformation. In South Africa, the Kaapvaal province (figure 4) had stabilized by *ca.* 3000 Ma, and four major basins developed in the period 3000–2000 Ma, the Pongola, Witwatersrand, Ventersdorp and Transvaal–Griqualand West supergroups; their distribution on the province is shown in figure 4, and their thicknesses, stratigraphy and available age constraints in figure 5 and table 1.

These four early Precambrian stratigraphic units are separated by province-wide unconformities similar to those documented by Sloss & Speed (1974) for the Phanerozoic history of the United States. The sedimentary basins evolved during *ca.* 200 Ma periods of cratonic downwarping with or without submergence, whereas the unconformities represent periods of cratonic emergence.

(*a*) *Tectonic setting of the early Precambrian cover sequences*

The four early Precambrian cover sequences developed in progressively larger basins and in response to the northward migration of depositional axes; the Pongola Supergroup accumulated in the southeastern corner of the Kaapvaal Province, whereas the Transvaal–Griqualand West Supergroup developed on the northern and western margin of the province. The migration of depositional axes has been related by Hunter (1974) to a complementary migration of loci of granite emplacement north of the corresponding sedimentary basin.

The Pongola, Witwatersrand and Transvaal–Griqualand West supergroups are terrestrial and shallow water cratonic shelf sequences that developed in stable foreland basins fronting uplifted source terrains to the north. In particular, the Pongola Supergroup has been interpreted as the post-orogenic response to closure of a northward-opening passive margin along which the Fig Tree and Moodies groups in the Barberton Mountain Land were deposited (Eriksson 1981*a*). The Ventersdorp Supergroup accumulated in intracratonic grabens and thus, in contrast to the sheet-like geometry of the aforementioned stratigraphic sequences, displays abundant wedging of lithologies. The middle Ventersdorp sediments in particular thicken dramatically towards fault scarps (fig. 7 in Buck 1980) and, as a result of reactivation of faults, numerous regional surfaces of unconformity occur within the supergroup.

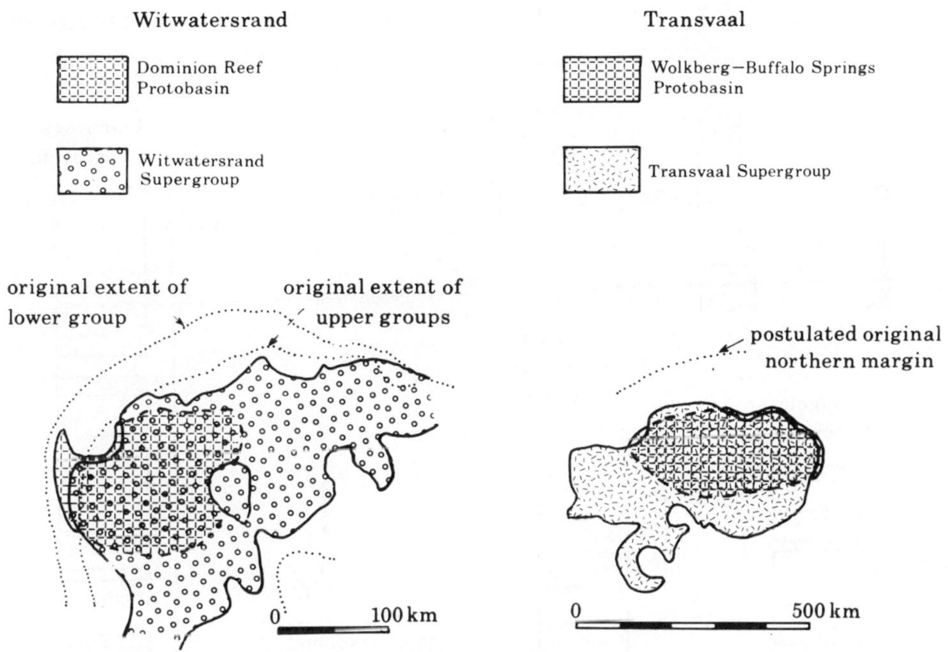

Figure 5. Extent of Witwatersrand and Transvaal proto-basinal and main basin phases, after Pretorius (1976), Button (1973*b*) and Tyler (1979).

(*b*) *Basin evolution*

(i) *Protobasinal stage.* Of the four early Precambrian stratigraphic sequences occurring on the Kaapvaal Province, the Witwatersrand and Transvaal–Griqualand West Supergroups contain recognizable protobasinal phases. These are represented respectively by the *ca.* 2000 m thick and 150 km wide Dominion Reef Group (figures 5 and 6) and the *ca.* 1600 m thick and 200 km wide Wolkberg–Buffalo Springs Group (figures 5 and 6) (Button 1973*a*; Tyler 1979). The lithologies of these two protobasinal units are dominated by volcanics and coarse-grained arkosic sediments. The volcanic rocks are mainly continental basalts having some oceanic affinities, although the Dominion Reef and Buffalo Springs volcanics do display bimodal characteristics. Tyler (1979) has documented potassium-rich rhyolites from the Buffalo Springs Group. The Dominion Reef volcanic rocks and the Wolkberg–Buffalo Springs volcanic rocks are predominantly subaerial. The associated arkosic sediments are of braided alluvial origin with palaeocurrent patterns indicating transport both normal and parallel to the axes of the protobasins (Button 1973*a*; Tyler 1979). The basaltic and bimodal composition of the volcanic rocks in the Dominion Reef and Wolkberg–Buffalo Springs groups favour extensional conditions during the initial stages of basin development. There is, however, no field evidence of graben development associated with either of the protobasins. This may be a function of preservation with the original fault-controlled margins of the protobasins beyond the present outcrop limits of the stratigraphic units.

The Pongola Supergroup does not contain a recognizable protobasin, but begins with arkosic rocks overlying the basal unconformity and passes up into a relatively thick sequence of calc-alkaline volcanic rocks (N. V. Armstrong, personal communication 1979; Kroner *et al.* 1980). The Ventersdorp stratigraphy is dominated by volcanic rocks. The lower Ventersdorp volcanic rocks are conformable on sediments of the Witwatersrand Supergroup, are laterally continuous and are considered to be the the terminal phase of the Witwatersrand Supergroup.

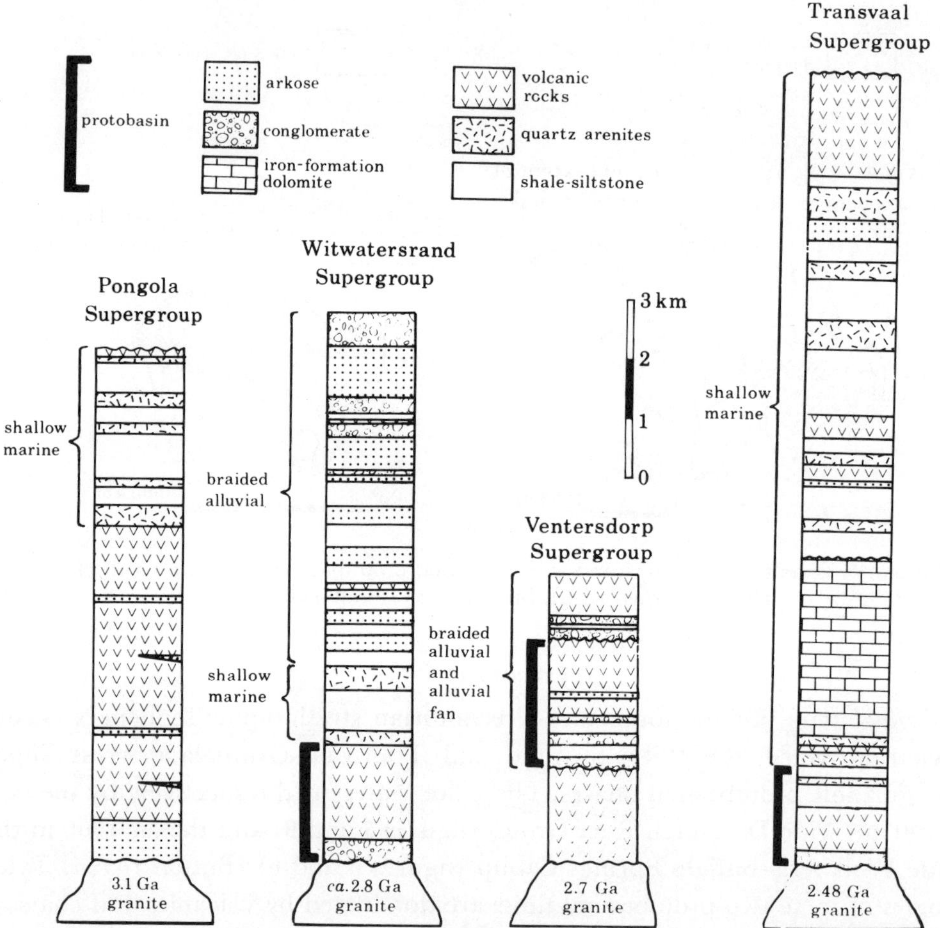

FIGURE 6. Stratigraphy of Pongola, Witwatersrand, Ventersdorp and Transvaal basins (references in table 1).

Graben formation occurred above these volcanics (figure 7), and may represent the initial subsidence phase of the Ventersdorp Supergroup. Alluvial fan braided alluvial and lacustrine sediments interbedded with volcanic horizons developed in the grabens.

(ii) *Subsidence stage.* The protobasinal to subsidence stage transition is well displayed in the Witwatersrand and Transvaal Supergroups. Although unconformable along the margins of the protobasins, the transition becomes conformable and gradational towards the centre of the basins implying limited erosion during the transition (figure 8). Rather, subsidence occurred over a much greater area to produce depositional basins with minimum widths of 200 km for the Witwatersrand Supergroup and 300 km for the Transvaal Supergroup (figure 5). On the basis of palaeocurrent and facies patterns, Button (1973*b*) and Eriksson *et al.* (1976)

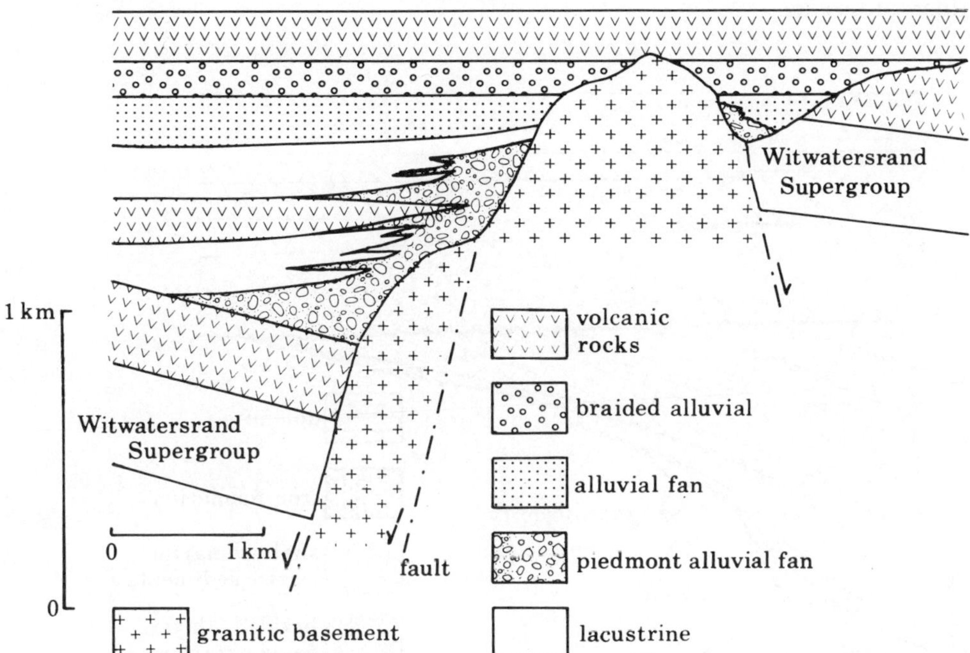

FIGURE 7. Schematic section through Ventersdorp Supergroup, after Buck (1980).

proposed that the post-Wolkberg Transvaal depository may have been as much as 600 km in width.

In both the Witwatersrand and Transvaal supergroups, the subsidence stage began with a marine transgression, resulting in the drowning of the Kaapvaal Province. The earliest sediments in both these basins are considered to have accumulated under conditions of tidal reworking (Button 1973*b*; Eriksson *et al.* 1981). In the Witwatersrand Supergroup, the lower tide-dominated sequence of quartz arenites and shales passes upwards into conglomerates and arkoses of braided alluvial origin (figure 6). Shallow marine conditions prevailed almost throughout the history of the Transvaal–Griqualand West Supergroup with the 10 km thick sequence of chemical and terrigenous clastic sediments (figure 6) accumulating in various tidal environments (Eriksson *et al.* 1976; Button & Vos 1977). The shallow marine sediments in both supergroups are characteristically arranged in progradational, upward-coarsening sequences, indicating an excess of sediment supply over basin subsidence. The sequences are terminated along surfaces of transgression, which developed upon cessation of sediment supply.

Volcanic horizons are present within both the Witwatersrand and Transvaal supergroups, and compose *ca.* 7% of the latter (figure 6). Tholeiitic basalts are again predominant but the uppermost volcanic horizon in the Transvaal Supergroup does contain acidic phases (Button 1973*b*). Superimposed on the overall subsidence history of the Witwatersrand and Transvaal–Griqualand West supergroups were periods of emergence during which internal surfaces of unconformity developed. Particularly striking is the regional unconformity at the top of the chemical sedimentary unit in the Transvaal Supergroup. Volcanicity continued throughout the Ventersdorp Supergroup, and it is not possible to separate thermal subsidence from initial subsidence unambiguously. The relatively thin, but laterally more continuous, upper fluvial sediments (figure 7) may represent thermal subsidence, but these are terminated by further continental basalts.

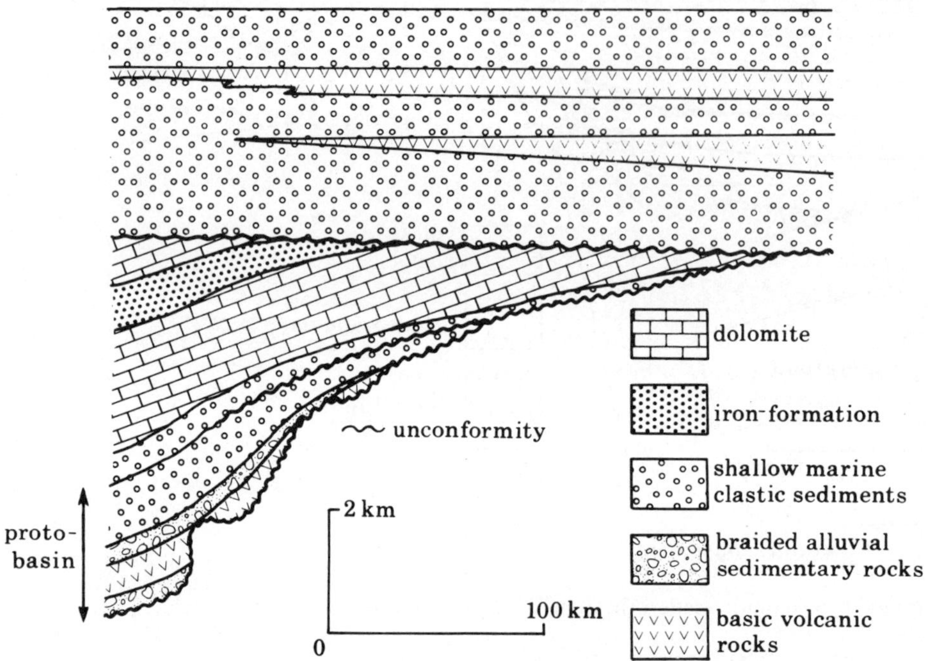

FIGURE 8. Schematic section across Transvaal Supergroup, after Button (1976).

TECTONIC IMPLICATIONS

Distinction of initial and thermal subsidence phases

Recognition and separation of the initial from the thermally controlled subsidence is the basis for our interpretation of early Precambrian sedimentary basins. In all the basins discussed above, the volcanic rocks compose a major interval low in the sequence, and in two examples (Witwatersrand and Transvaal supergroups) these volcanics are confined to a spacially restricted protobasin. There are a number of reasons to associate the volcanicity with the initial phase of subsidence, and we shall argue that the termination of volcanicity in the majority of the basins discussed above marks the cessation of the initial subsidence phase.

Modern volcanic rocks are invariably erupted in tensional environments, and there are physical constraints on the transport of magmas that require this (Weertman 1971). Volcanic intervals may therefore provide *a priori* evidence for extension (although it should be noted that the volcanic and sedimentary facies do not always indicate graben development indicative of crustal extensions). Where volcanic horizons occur higher in the sedimentary sequence, these may indicate renewed tectonic activity, and models based on one initial instantaneous tectonic event may not be appropriate. Transport of magmas from mantle depths may also provide sufficient heat to thin the lithosphere and thus subsidence associated with the volcanic event would be followed by subsidence related to thermal relaxation of the lithosphere.

The most compelling reason to associate the volcanic-dominated intervals with initial subsidence is that, to accommodate up to 7 km of volcanic rocks, substantial and rapid subsidence must have occurred. Subsidence in greenstone belts cannot be due to loading an excess thickness of volcanic rocks, as the volcanic rocks were either erupted in shallow water as evidenced by intercalated cherty sediments (Barberton and Pilbara) or are immediately underlain and overlain by shallow marine sediments (Zimbabwe). In the Witwatersrand, Transvaal–Griqualand West and in the Ventersdorp supergroups, volcanic rocks are mainly subaerial, although in the Witwatersrand and the Transvaal shallow marine sediments overlie the volcanics requiring net subsidence during the initial phase. This is also inconsistent with subsidence being only due to loading by the sedimentary–volcanic pile. The small proportion of interbedded sediment within the volcanic intervals, relatively minor horizons of reworked volcanoclastic material and lack of weathered or soil horizons within the subaerial volcanic intervals, all attest to the rapidity of the volcanism and thus of the initial subsidence phase. The shallow marine setting of greenstone volcanic intervals allows accurate estimates of total subsidence during the initial phase. The subaerial to shallow marine transition across the initial subsidence phase in the Witwatersrand and Transvaal may have required more subsidence than implied from sedimentary–volcanic thicknesses alone, although deposition of major alluvial deposits more than a few hundred metres above sea level is uncommon today.

Marginal unconformities passing into continuous depositional sequences in basin centres may mark the termination of the initial subsidence phase. In all of the greenstone-associated basins, the initial volcanic phase is terminated by the uplift and erosion of parts of the volcanic sequence. In the protobasin underlying the Transvaal, shallow marine sediments *ca.* 400 m thick overlie the volcanics, but along the margins of the basin, both these and the volcanics were eroded. This uplift is consistent with volcanicity being associated with initial subsidence, as Beaumont (1978) shows that marginal recovery over *ca.* 10 Ma would reduce the area of depression caused by the elastic response to a load on the lithosphere. The only constraint is that the viscous recovery must be more rapid than the ensuing thermal subsidence, which resulted in deposition over a larger area than that occupied by the protobasin.

Additional evidence to support this assertion of volcanic-dominated initial subsidence might be expected from two other sources. First, the tectonic setting of the volcanics would be expected to provide evidence of the tectonic mechanism for initially thinning or loading the crust and, secondly, subsidence following the initial subsidence should take place at an exponentially decreasing rate. In the early Precambrian and lacking precise geochronology, estimates of sedimentation rates are virtually impossible. Only in the 3550 Ma Barberton and Pilbara greenstone areas is there evidence of rapid thermal subsidence in the form of deeper water iron-formation, shale and then major turbidite sequences deposited on shallow-water volcanic rocks. In both these areas, there is evidence of prograding sedimentary sequences, best displayed in Barberton by the transition from turbidite to braided alluvial and shallow marine deposits. In the other basins, shallow marine and alluvial sediments fill the entire sequence; presumably, sediment supply was sufficient to keep the basin filled, and excess sediment was transported across the basin.

In some basins, there is evidence of active faulting and graben formation during the initial subsidence phase consistent with an extensional origin for the basins. In both Barberton and the Pilbara, Eriksson (1981*a*) proposes steep-rift type continental margins dominated by major fault scarps to account for the abrupt turbidite to braided alluvial transition (figure 3). However, there is less evidence for active faulting during the volcanicity; volcanoclastic sediments are of minor importance except where directly related to felsic volcanic centres, and the lack of granitic detritus within the volcanic sequences implies that basement was not exposed during this phase (Lowe 1980). Similarly, in Zimbabwe, both the thin basal sediments and the overlying volcanic stratigraphy are laterally continuous.

In the Witwatersrand and Transvaal protobasins (the Dominion Reef and Wolkberg–Buffalo Springs groups), there is no evidence for fault-controlled volcanicity or sedimentation. Much of the Dominion Reef Group is hidden under the deeper undrilled parts of the Witwatersrand basin, but much younger (late Karoo) uplift has exposed a cross-section through the central parts of the Transvaal Basin. Button (1973*a*) records considerable topographic control on early deposition of the Wolkberg Group, but no syn-depositional faulting (figure 8). It is difficult to reconcile this with an extensional origin for the basin. The rapid lateral facies variation and the thickness changes within the Ventersdorp (Buck 1980) (figure 7) demonstrates the ease with which fault control on sedimentation may be recognized.

Implications for lithosphere thickness

The discussion above shows that the sedimentary history of a number of early Precambrian basins can be related to a relatively rapid initial subsidence phase followed by more widespread, mainly volcanic-free, sedimentation. Sedimentation in the initial phase in all the basins studied is either shallow marine or fluvial, allowing reasonable estimates of the initial subsidence to be made. In several of the basins, the second subsidence phase also terminates with shallow marine or alluvial sediments. On the basis of modern analogues, we shall assume that the second phase can be related to thermal relaxation of the lithosphere and that the initial phase must reflect the prime tectonic control on basin subsidence, either crustal extension or some other crustal loading process. The total basin subsidence, initial plus thermal, depends only on the ultimate cause, either the amount of crustal thinning or the magnitude of crustal loading. The thermal subsidence relates directly to the increase in lithosphere thickness. We have estimated sedimentary thickness, water depths, and densities for the basins discussed above (table 1), and have calculated the equivalent crustal thinning for probable crustal thickness from the formula

$$1/\beta = 1 - \frac{S(\rho_s - \rho_o)}{t_c(\rho_c - \rho_o)},$$

where β is the amount of crustal extension, ρ_o is mantle relative density (3.33), ρ_c is the relative density of the continental crust (2.8), ρ_s is the relative average density of sediment–volcanic infill and S is the total subsidence.

We have then calculated lithosphere thickness from the thermal subsidence on the assumption that lithosphere thinning is inversely proportional to crustal extension. Lithosphere thickness, a, is given by

$$a = \frac{2S_t(\rho_o - \rho_s)}{\alpha T_1(1 - 1/\beta)\rho_o},$$

where S_t is the thermal subsidence, T_l the temperature of the base of the lithosphere (1333 °C) and α the coefficient of thermal expansion (3.28×10^{-5} K^{-1}). This calculation assumes a linear thermal gradient through the lithosphere, crust and sediments, and ignores heat production, but given uncertainty on α (*ca.* $\pm 1 \times 10^{-5}$) and T_l ($\pm$ *ca.* 100 °C) a more sophisticated treatment seems unjustified.

TABLE 2. STRETCHING MODEL PARAMETERS FOR SELECTED EARLY PRECAMBRIAN BASINS

(Note that increase in lithosphere thickness gives minimum lithosphere thickness if relation between initial equivalent crustal thinning and initial lithosphere thinning is not valid.)

	thickness of continental crust† (if initially at sea level)/km	extension† (β)	equilibrium† lithosphere thickness km	increase in lithosphere thickness for thermal subsidence km
Greenstone belts, 3.5 and 2.6 Ga	13.7 (minimum)	∞	59	59
total subsidence 12.5 km, $\bar{l}$ = 2.75	20	3.2	86	
thermal subsidence 5.5 km, $\bar{l}$ = 2.55	30	1.8	133	
	40	1.5	177	
Pongola	10.9 (minimum)	∞	32	32
total subsidence 8.5 km, $\bar{l}$ = 2.65	20	2.2	59	
thermal subsidence 3 km, $\bar{l}$ = 2.55	30	1.6	86	
	40	1.4	112	
Witwatersrand, 2.7 Ga	12.2 (minimum)	∞	75	75
total subsidence 9 km, $\bar{l}$ = 2.61	20	2.6	121	
thermal subsidence 7 km, $\bar{l}$ = 2.55	30‡	1.7	182	
	40‡	1.4	262	
Ventersdorp, 2.64 Ga	4.0 (minimum)	∞	5.4	5.4
total subsidence 3.5 km, $\bar{l}$ = 2.72	20	1.25	27	
thermal subsidence 0.5 km, $\bar{l}$ = 2.55	30‡	1.15	41	
	40‡	1.11	54	
Transvaal, 2.2 Ga	15.2 (minimum)	∞	92	92
total subsidence 12.0 km, $\bar{l}$ = 2.66	20	4.1	121	
thermal subsidence 10.0 km $\bar{l}$= 2.66	30‡	2.0	184	
	40‡	1.6	245	

† Assumes initial subsidence is a result of thinning crust and lithosphere by stretching only.
‡ Likely estimates for initial crustal thickness.

The lithosphere thickness estimated depends directly on the assumption that the initial thinning was proportional to the effective thinning of the crust. As discussed above, even if an extensional model is appropriate, transport of heat by magmas may over-thin the lithosphere. The only reliable figure is the minimum lithosphere thickness appropriate to infinite extension in table 2, which gives the amount by which lithosphere thickness must increase to cause the observed thermal subsidence.

The results of these calculations are given in table 2. The Ventersdorp Supergroup, although possessing a well developed volcanic–sedimentary initial phase, apparently developed only a very limited phase of thermal subsidence. Calculated minimum lithosphere thicknesses range from *ca.* 60 km in the greenstone areas to *ca.* 90 km for the Transvaal. In Zimbabwe,

Witwatersrand and the Transvaal, there is evidence for a significant thickness of continental crust preserved beneath the basins, and for the crust (*ca.* 30 km in South Africa) to be preserved, lithosphere thickness must not have been reduced to less than *ca.* 40 km (including the crust). Realistic minimum equilibrium lithosphere thicknesses must have ranged from *ca.* 75 km in greenstone terrains to 110–130 km under the cratonic basins in South Africa at 2.6–2.2 Ga. For the two South African basins, the present crustal thickness of *ca.* 36 km (Gane *et al.* 1956) enables constraints to be placed on the amount of extension, if this was the fundamental cause. Factors of between 1.4 and 1.7 for Witwatersrand and 1.6–2.0 for the Transvaal would require *ca.* 200 km thick lithospheres. The total time required for thermal relaxation of such thick lithosphere (*ca.* 500 Ma) is far greater than the *ca.* 200 Ma available and suggests that simple extension is incapable of explaining the thermal subsidence alone. If rifting was the cause, some mechanism must have caused over-thinning of the lithosphere, as suggested by Steckler & Watts (1980).

The validity of these conclusions depends on the correct identification of the initiation of thermal subsidence and, as noted by McKenzie *et al.* (1980), precise estimates of the duration of subsidence will provide the only possible independent verification of these figures. Uncertainty in densities and the coefficient of thermal expansion also cause uncertainties on the estimates of the order of $\pm 30\,\%$.

Conclusions

Early Precambrian sedimentary basins occur in a number of different tectonic settings, including greenstone belts, cratonic shelf and cratonic rift environments. All the basins contain major volcanic horizons low in the sedimentary sequences. There are a number of reasons why these volcanics are thought to be associated with initial subsidence of the basins: (1) volcanicity was restricted to protobasins in some examples; (2) volcanicity was followed by marginal uplift and erosion, but subsidence continued in basin centres; (3) thick volcanic sequences erupted entirely in shallow marine conditions imply significant and rapid subsidence; (4) there are theoretical reasons why volcanicity might be associated with basin formation: eruption of volcanics requires a tensional environment and transport of heat by volcanics provides a mechanism to thin the lithosphere, allowing the subsequent protracted thermal subsidence. We postulate that the initial volcanic-dominated subsidence was a direct result of the tectonic mechanism, which thinned or loaded crust, permitting permanent preservation of the basin. Sedimentary facies relations in some greenstone terrains and in the Ventersdorp cratonic rift sequence are consistent with an extensional tectonic cause for basin formation in these areas. Cratonic shelf basins such as the Witwatersrand and the Transvaal, although containing volcanic-dominated initial phases preserved in protobasins, did not apparently develop above a rifted crust, and some other mechanism must be sought

Distinction of the initial subsidence phase from the subsequent 'thermal' subsidence phase related to thermal relaxation of a thinned lithosphere enables constraints to be put on minimum lithosphere thicknesses. Such estimates imply equilibrium lithosphere thicknesses greater than *ca.* 75 km in greenstone terrains at 3.5 Ga, *ca.* 75 km in greenstone terrains at *ca.* 2.6 Ga, greater than *ca.* 110 km in cratonic shelf basins at 2.6 Ga, and greater than *ca.* 130 km in cratonic shelf basins at 2.2 Ga. If our distinction of initial subsidence from thermal subsidence is correct, then the lithosphere in the early Precambrian was comparable in thickness with that of the present day.

K.A.E. acknowledges financial assistance by Earth Sciences Section, National Science Foundation Grant no. EAR 7842307. M.J.B. acknowledges financial assistance under A.R.G.C. grant no. E78-15665.

References (Bickle & Eriksson)

Allsopp, H. L. 1964 *Nature, Lond.* **204**, 361–363.

Allsopp, H. H., Roberts, H. R., Schreiner, G. D. L. & Hunter, D. R. 1962 *J. geophys. Res.* **67**, 5307–5313.

Anhaeusser, C. R. 1973 *Phil. Trans. R. Soc. Lond.* A **273**, 359–388.

Baragar, W. R. A. & McGlynn, J. C. 1976 *Geol. Surv. Can. Pap.* no. 76–14 (21 pages.)

Barley, M. E., Dunlop, J. S. R., Glover, J. E. & Groves, D. I. 1979 *Earth planet. Sci. Lett.* **43**, 74–84.

Beaumont, C. 1978 *Geophys. Jl R. astr. Soc.* **55**, 471–497.

Bickle, M. J. 1978 *Earth planet. Sci. Lett.* **40**, 301–315.

Bickle, M. J., Martin, A. & Nisbet, E. G. 1975 *Earth planet. Sci. Lett.* **27**, 155–162.

von Brunn, V. & Hobday, D. 1976 *J. sedim. Petr.* **46**, 670–679.

Burger, A. J. & Coertze, F. J. 1973 *Bull. geol. Surv. Dep. Min. S. Afr.* **58**, 1.

Burke, K. & Kidd, W. S. F. 1978 *Nature, Lond.* **272**, 240–241.

Buck, S. G. 1980 *Precamb. Res.* **12**, 311–330.

Button, A. 1973*a* *Trans. geol. Soc. S. Afr.* **76**, 15–25.

Button, A. 1973*b* Ph.D. thesis, University of Witwatersrand, Johannesburg. (352 pages.)

Button, A. 1976 *Minerals Sci. Engng* **8**, 262–293.

Button, A. & Vos, R. G. 1977 *Sedim. Geol.* **18**, 175–200.

Coertze, F. J., Burger, A. V., Walraven, F., Marlow, A. G. & MacCaskie, D. R. 1978 *Trans. geol. Soc. S. Afr.* **81**, 1–12.

Davies, R. D., Allsopp, H. L., Erlank, A. J. & Manton, W. I. 1970 *Spec. Publ. geol. Soc. S. Afr.* no. 1, pp. 576–593.

Dunlop, J. S. R. & Buick, R. 1981 *Spec. Publ. geol. Soc. Aust.* No. 7. (In the press.)

Eriksson, K. A. 1980 *Precamb. Res.* **12**, 141–160.

Eriksson, K. A. 1981*a* *Tectonophysics*. (In the press.)

Eriksson, K. A. 1981*b* *Spec. Publ. geol. Soc. Aust.* no. 7. (In the press.)

Eriksson, K. A., Truswell, J. F. & Button, A. 1976 In *Stromatolites* (*Developments in sedimentology*, vol. 20) (ed. M. R. Walter), pp. 635–643. Amsterdam: Elsevier.

Eriksson, K. A., Turner, B. R. & Vos, R. G. 1981 *Sedim. Geol.* **29**, 309–325.

Gane, P. G., Atkins, A. R., Sellschop, J. P. F. & Seligman, P. 1956 *Bull. seism. Soc. Am.* **46**, 293–316.

Goodwin, A. M. 1977 *Can. J. Earth Sci.* **14**, 2737–2759.

Hamilton, J. 1977 *J. Petr.* **18**, 24–52.

Hamilton, P. J., Evensen, N. M., O'Nions, R. K., Smith, H. S. & Erlank, A. J. 1979 *Nature, Lond.* **279**, 298–300.

Hamilton, P. J., Evenson, N. M., O'Nions, R. K., Glikson, A. Y. & Hickman, A. H. 1980 In *Extended Abstracts Second Int. Archaean Symp. Geol. Soc. Aust., Perth*, pp. 11–12.

Hamilton, P. J., O'Nions, R. K. & Evensen, N. M. 1977 *Earth planet. Sci. Lett.* **36**, 263–268.

Hawkesworth, C. J., Bickle, M. J., Gledhill, A. R., Wilson, J. F. & Orpen, J. L. 1979 *Earth planet. Sci. Lett.* **43**, 285–297.

Haxby, W. F., Turcotte, D. L. & Bird, J. M. 1976 *Tectonophysics* **36**, 57–75.

Henderson, J. B. 1972 *Can. J. Earth Sci.* **9**, 882–902.

Henderson, J. B. 1975 *Bull. geol. Surv. Can.* no. 246.

Henderson, J. B. 1981 In *Plate tectonics in the Precambrian* (ed. A. Kröner), pp. 213–233. Amsterdam: Elsevier.

Hickman, M. H. 1978 *Geology* **6**, 214–216.

Hunter, D. R. 1974 *Precamb. Res.* **1**, 259–294.

Hyde, R. S. 1980 *Precamb. Res.* **12**, 161–195.

Keen, C. E. 1979 *Can. J. Earth Sci.* **16**, 505–522.

Kröner, A., Tegtmeyer, A., Hegner, E. & von Brunn, V. 1980 In *Extended Abstracts Second Int. Archaean Symp. Geol. Soc. Aust., Perth*, p. 96.

Lowe, D. R. 1980 *A. Rev. Earth planet. Sci.* **8**, 145–167.

Lowe, D. R. & Knauth, L. P. 1977 *J. Geol.* **85**, 699–723.

McKenzie, D. 1978 *Earth planet. Sci. Lett.* **40**, 25–32.

McKenzie, D., Nisbet, E. & Sclater, J. G. 1980 *Earth planet. Sci. Lett.* **48**, 35–41.

Mathews, P. E. 1967 *Trans. geol. Soc. S. Afr.* **70**, 39–63.

Martin, A., Nisbet, E. G. & Bickle, M. J. 1980 *Precamb. Res.* **13**, 337–362.

Moorbath, S. W., Wilson, J. F., Goodwin, R. & Hamm, M. 1977 *Precamb. Res.* **5**, 229–239.

Newton, R. C. 1977 In *Thermodynamics in geology* (ed. D. G. Fraser), pp. 29–46. Dordrecht: D. Reidel.

Nisbet, E. G., Bickle, M. J. & Martin, A. 1977 *J. Petr.* **18**, 521–566.

O'Hara, M. J., Richardson, S. W. & Wilson, G. 1971 *Contr. Miner. Petr.* **32**, 48–68.

Orpen, J. L. & Wilson, J. F. 1981 *Nature, Lond.* (In the press.)
Padgam, W. A. 1980 *Spec. Publ. geol. Soc. Aust.* (In the press.)
Pretorius, D. A. 1976 In *Handbook of strata-bound and stratiform ore deposits* (ed. K. H. Wolf), vol. 2 (*Regional and specific studies*), pp. 29–88. Amsterdam: Elsevier.
Sclater, J. G. & Christie, P. A. F. 1980 *J. geophys. Res.* **85**, 3711–3739.
Sleep, N. H. 1971 *Geophys. Jl R. astr. Soc.* **24**, 325–350.
Sleep, N. H., Nunn, J. A. & Chon, L. 1980 *A. Rev. Earth planet. Sci.* **8**, 17–34.
Sloss, L. L. & Speed, R. C. 1974 *Tectonics and Sedimentation* (*Soc. Econ. Paleontologists and Mineralogists Spec. Publ.* no. 22).
Steckler, M. S. & Watts, A. B. 1978 *Earth planet. Sci. Lett.* **41**, 1–13.
Steckler, M. S. & Watts, A. B. 1980 *Nature, Lond.* **287**, 425–429.
Tyler, N. 1979 *Trans. geol. Soc. S. Afr.* **82**, 215–226.
Van Niekerk, C. B. & Burger, A. J. 1978 *Trans. geol. Soc. S. Afr.* **81**, 155–163.
Viljoen, M. J. & Viljoen, R. P. 1969 *Spec. Publ. geol. S. Afr.* **2**, 7–28.
Walker, R. G. 1978 *Can. J. Earth Sci.* **15**, 1213–1218.
Watts, A. B. & Ryan, W. B. F. 1976 *Tectonophysics* **36**, 25–44.
Weertman, J. 1971 *J. geophys. Res.* **76**, 1171–1183.
Williams, D. A. C. & Furnell, R. G. 1979 *Precamb. Res.* **9**, 325–347.
Wilson, J. F., Bickle, M. J., Hawkesworth, C. J., Martin, A., Nisbet, E. G. & Orpen, J. L. 1978 *Nature, Lond.* **271**, 23–27.
Winter, H. de la R. 1976 *Trans. geol. Soc. S. Afr.* **79**, 31–48.

Discussion

D. G. Roberts. How do the authors obtain the lithospheric thickness in the Archaean? The basin subsidence was presumably produced partly by the load of sediments infilling an existing basin and perhaps partly by thermal contraction. How can one distinguish between these two mechanisms?

M. J. Bickle. We assumed a sediment density and corrected for the sediment load in the usual way to obtain the subsidence history the basin would have had if it had been filled by water only. The principal error is the uncertainty in the sediment density at the time of loading.

D. Hastings. I tend to agree with the authors' hypothesis that at least some early Precambrian sedimentary basins (greenstone belts) formed as a result of rifting. Many scientists, however, argue that the tightly folded rocks of many such features require a strongly compressive environment. Are the authors' arguments for a tensile origin of the belts compatible with theirs?

Incidentally, do the authors see their tensile environment in terms of parallels to Cainozoic (postulated) early stages of ridge-like extension which failed to develop very far, to postulated back-arc environments (do they see any examples at all of early Precambrian subduction?), or another possible model?

M. J. Bickle. The intense folding that commonly deforms the rocks of greenstone belts has been produced by later events. In the Pilbara the sedimentary volcanic sequence formed at 3.55 ± 0.02 Ga. The next event was the intrusion of granodiorite bodies at about 3.35. Ga. The original sediments were still essentially undeformed when these rocks were intruded. The sediments were then strongly folded, in places even recumbantly folded. We did not use thicknesses obtained in such places to estimate the subsidence. We therefore do not believe basin formation and subsequent tight folding are directly related.

B. Chadwick. I wish to describe some results of recent field investigations in collaboration with Dr M. Ramakrishnan and Mr M. N. Viswanatha, Geological Survey of India, in the Chitradurga region (76° 20′ E; 14° 10′ N) in the Karnataka craton of southern India which appear to support certain aspects of the models for the evolution of late Archaean basins proposed by Dr Bickle & Dr Eriksson.

The late Archaean Dharwar Supergroup (*ca.* 2600 Ma) in southern Karnataka occurs in a series of belts and irregular basins resting unconformably or with tectonic contact on a basement of tonalitic–granitic Peninsular Gneiss (*ca.* 3000 Ma) which contains enclaves and locally extensive tracts of an older supracrustal association called the Sargur Group or Sargur high-grade schists. Detailed descriptions are provided by Swami Nath & Ramakrishnan (1981).

The evolution of the western part of the late Archaean basin in the Chitradurga region is shown schematically in figure 1. The succession in the Dharwar Supergroup begins with basal

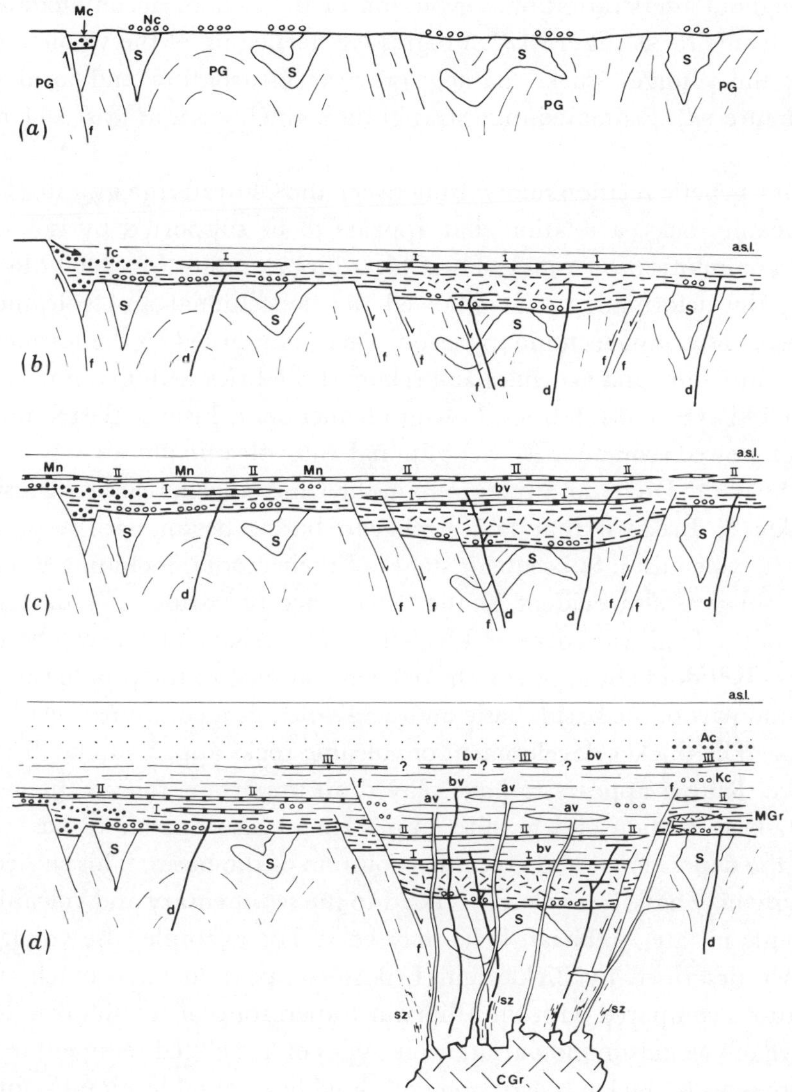

FIGURE 1. Schematic representation of the stratigraphic evolution of the Dharwar Supergroup in the Chitradurga region, Karnataka. S, Sargur Group; PG, Peninsular Gneiss; Mc, Mayakonda polymict conglomerate; Nc, Neralekatte quartz conglomerate; Tc, Talya polymict conglomerate; I–III, banded iron formations; bv, basic volcanic rocks; av, acid volcanic rocks; Kc, Kurmeredikere conglomerate; Ac, Aimangala polymict conglomerate. Thick dashed lines, basic volcanic rocks; thin lines, siliceous phyllites and cherts; dotted ornament, greywacké-turbidites, a.s.l. approximate sea level; sz, f, schematic representation of shear zones and faults active during the depositional and volcanic phases. Estimated width of section 25–30 km.

quartz pebble conglomerates and quartzites (figure 1*a*) which appear to have been deposited after peneplanation and mature weathering of the basement. Basic volcanism and sporadic introduction of cross-bedded, clean-washed sands characterize the initial platformal environment that we believe to have covered much of the Karnataka craton at this time. Thicker basic volcanic rocks in a more internal part of the basin (figure 1*b*) indicate greater subsidence of the basement, which we relate to rifting and graben formation. Polymict conglomerates (Tc, figure 1*b*) containing clasts of basement and Dharwar cover sediment and volcanic rocks formed near the margin of the basin. Continued subsidence accommodated siliceous phyllites, cherts and banded iron formations (figure 1*c*). Certain iron formations are Mn-bearing in the marginal parts of the basin. Greater rates of subsidence in the more internal parts appear to have been maintained throughout the stratigraphic evolution of the belt to accommodate sequences of basic and acid volcanic rocks (figure 1*d*). Progressive deepening of the whole basin took place to accommodate the youngest rocks, namely greywacké-turbidites and local polymict conglomerates (Ac, figure 1*d*). Estimated total stratigraphic thickness is at least 10 km in the deeper parts of the basin.

We suggest that a genetic relation may exist between the Chitradurga granite (CGr, figure 1*d*) and the acid volcanic rocks, a relation that appears to be supported by isotope data (P. N. Taylor, personal communication 1981). The Chitradurga granite (*ca.* 2600 Ma) rose through the cover during the deformation that followed the depositional and volcanic phases. The granite now appears as a post-tectonic intrusion. Deformation led to the formation of a series of mainly upright anticlines and synclines and related LS fabrics with extreme variations in the plunge of coaxial fold axes and L fabrics. Low-grade metamorphism outlasted the deformation. The deformation followed soon after the depositional and volcanic phases.

The stratigraphy of the Dharwar Supergroup in the Chitradurga region is consistent with the proposal by Bickle & Eriksson that late Archaean basins began with a protobasin phase dominated by basic volcanism. Subsequent stages of their model, including erosion and uplift of part of the basin, are also evident in the occurrence of various polymict conglomerates containing clasts of the Dharwar cover and its basement gneisses. However, the volcanic phase in the Chitradurga belt does not appear to have been confined to the protobasin phase because in the more internal part of the basin, basic and acid volcanism continued relatively late in the depositional phase. The thicker development of volcanic rocks appears to be directly related to greater subsidence. Rifting appears to have played an important role in the evolution of the Chitradurga belt. Deformation and granite emplacement also appear to have followed soon after the close of the depositional and volcanic evolution of the basin. This subsequent tectonic and igneous activity may have been closely related to the sedimentary and volcanic events.

Patterns of events in late Archaean belts may vary. For example, the younger greenstone belts of Zimbabwe described by Bickle and Eriksson appear to have much thicker basalt–komatiite associations compared with the Dharwar Supergroup of Karnataka. The differences in type and degree of volcanism and variations in types of associated sedimentary rocks suggest that while the proposals by Bickle & Eriksson will provide a valuable stimulus for research into the evolution of Archaean sedimentary basins, much detailed investigation of Archaean supracrustal associations remains to be done to establish the general validity of their models.

Reference

Swami Nath, J. & Ramakrishnan, M. 1981 *Mem. geol. Surv. India*, no. 112. (350 pages.)

A. M. Clarke. Have the authors seen any evidence in the Precambrian for a clear separation between sediments laid down during the rifting phase and those deposited during the thermal and flexural phases of basin formation? Have they found any examples of the superposition of a later on a younger basin, as Kerr has described during the development of the Sverdrup basin?

K. A. Eriksson. The three basins in South Africa cover a time interval of 600 Ma, and in places are superimposed on each other to produce a sediment thickness of 15 km. Each basin shows an area of limited extent, the protobasin, of initial subsidence followed by subsidence over a larger region. These protobasins are, however, aerially separated.

P. A. Ziegler. The stratigraphic record of the basins discussed by the authors indicates that volcanic activity persisted, albeit with intensity variations, throughout their evolution. As volcanic activity cannot be reconciled with the lithospheric cooling and contraction model of inactive rifts and continental margins it is questioned whether these basins had ever entered a clear post rifting stage.

Could Dr Bickle provide further information about the late volcanism in these basins such as chemistry, mode of emplacement of volcanics and possible tectonic causes of volcanism?

M. J. Bickle. Only about 7% of the sequence consists of volcanic material. If the thermal structure of the lithosphere is to be strongly affected by volcanism, volumes of magma comparable with that of the continental crust must be intruded. Unless there was further crustal thinning even such large amounts of volcanism would only delay the onset of the thermal subsidence phase.

K. A. Eriksson. The volcanics are separated by well developed soil horizons, which perhaps could have resulted from pauses in the thermal subsidence produced by the heat from the magmas.

M. F. Osmaston. Could the Precambrian basins have been produced by doming, followed by erosion of the dome, as has been argued happened in the North Sea during the Mesozoic?

M. J. Bickle. The existence of superimposed basins suggests that uplift and erosion did not play an important part in basin formation.

Phil. Trans. R. Soc. Lond. A **305**, 249–281 (1982) [249]
Printed in Great Britain

Lithospheric flexure and the evolution of sedimentary basins†

By A. B. Watts, G. D. Karner and M. S. Steckler
Lamont-Doherty Geological Observatory and Department of Geological Sciences of Columbia University, Palisades, New York 10964, *U.S.A.*

The sediments that have accumulated in sedimentary basins during geological time represent a load on the lithosphere that should respond by flexure. Simple elastic and viscoelastic (Maxwell) plate models have been used to examine quantitatively the contribution of flexure to basin formation. The models have been used to predict the stratigraphy and gravity anomalies associated with basins for different thermal and loading histories. The predictions of the models have been compared with observed stratigraphy and free-air gravity anomalies from interior and cratonic basins. The best overall fit to the observations is for an elastic plate model in which the flexural strength of the lithosphere increases with age. A similar model has recently been used successfully to explain observations from continental margin basins, oceanic islands and seamounts, and deep-sea trench–outer rise systems. This model explains the increase in the overall width of basins during their evolution as well as the stratigraphy of the basin edges. The apparent decrease in the widths of some basins through time can be explained by the model if sediment deposition is followed by erosion of the basin and its edges. The models results suggest that the flexural properties of continental and oceanic lithosphere are generally similar and that flexure is an important factor to consider in backstripping studies in which the tectonic subsidence of a basin is isolated and stratigraphic studies in which relative changes of sea level are estimated.

1. Introduction

Sedimentary basins are dominated during their evolution by epeirogenic or vertical movements of the Earth's crust. Although an individual basin may change its tectonic setting during its evolution, most basins can be classified as occurring in either a rifted or an orogenic setting (Sloss & Speed, 1974; Dickinson & Yarborough 1976; Bally & Snelson 1980). Rifted basins (Bally & Snelson 1980) are associated with divergent plate boundaries where extension is dominant, for example, the U.S. Atlantic margin basins (Baltimore Canyon Trough, South Carolina Trough), which are located on transitional crust between ocean and continent, and possibly the North Sea Basin, which occurs on pre-Mesozoic continental crust. Orogenic basins (Bally & Snelson 1980), on the other hand, are associated with convergent plate boundaries where compression is dominant, for example fore-arc basins (Cook Inlet, Alaska) and foreland basins (Appalachian, Alberta and Ganges).

Most early studies of sedimentary basins were concerned with the role of sediment loading in basin formation (Barrell, 1914; Bowie 1922; Lawson 1938; Walcott 1972). The maximum thickness of sediment that can accumulate in a basin is unlikely on isostatic grounds to exceed about 2.5 times the available water depth. Thus the large thickness of shallow-water sediments observed in some basins cannot be attributed to sediment loading alone, and Bowie (1922, p. 282), for example, recognized that forces other than sediment loading were required.

Recent studies have attempted to isolate the effects of these forces by correcting observed sediment accumulations for the consequences of sediment loading (Watts & Ryan 1976; Steckler

† Lamont-Doherty Geological Observatory Contribution no. 3265.

& Watts 1978; Keen 1979; Watts & Steckler 1981). That part of the subsidence of a basin not caused by sediment loading can be isolated by 'backstripping' sediment and water loads progressively through time. Backstripping consists of reconstructing the stratigraphy of a basin for different intervals of time and then progressively unloading the sediments through time. The resulting 'tectonic subsidence' of a basin can then be directly compared with subsidence curves calculated from different geophysical models.

There is now general agreement that the main cause of the tectonic subsidence of basins is thermal contraction after heating of the lithosphere at the time of basin initiation (Sleep 1971; Falvey 1974; Sleep & Snell 1976; Haxby *et al.* 1976; McKenzie 1978; Middleton 1980; Sleep *et al.* 1980). Sleep (1971) and Sleep & Snell (1976) showed that the tectonic subsidence in deep wells in the Michigan and Illinois basins and the U.S. Atlantic continental margin decreased exponentially with time in a manner similar to a mid-ocean ridge. Sleep (1971) proposed that the tectonic subsidence of continental margin basins was caused by crustal thinning after uplift and erosion at the time of continental rifting. There is, however, generally little evidence of extensive erosion at the time of basin initiation, either at continental margins (Kent 1976) or at interior and cratonic basins (Sleep *et al.* 1980); other processes must therefore have caused the crustal thinning in these basins. McKenzie (1978) proposed a model that addressed this problem in which the lithosphere undergoes extension at the time of rifting. The extension causes thinning of the crust and heating of the lithosphere, which subsequently subsides with time. Sclater & Christie (1980) and Sclater *et al.* (1981*a*) have shown that the tectonic subsidence of the North Sea and Pannonian basins can be explained by this model for values of the lithospheric stretching factor, β, in the range 1.5–3.0 (50–200 %).

The determination of the tectonic subsidence of a basin by using backstripping techniques requires, however, detailed information on compaction, palaeobathymetry, sea level, and the lithospheric response to sediment loads for different times during basin evolution. The effects of compaction and palaeobathymetry can usually be estimated from down-hole geological and geophysical well logs (see, for example, Steckler & Watts 1978; Sclater & Christie 1980) and simple models now exist for long-term sea level changes through time (Pitman 1978; Bond 1978; Watts & Steckler 1979). The response of the lithosphere to sediment loads is a more difficult problem. Most backstripping studies have been based on the Airy model of isostasy, in which sediments are locally supported, rather than on the flexure model, in which sediments are supported by the lateral strength of the crust.

Flexure studies in continental and oceanic regions show that in a number of cases the response of the lithosphere to surface loads such as ice sheets, river deltas, and seamounts and oceanic islands can be modelled by a thin elastic or viscoelastic (Maxwell) plate overlying a weak fluid. Those studies that have considered the role of flexure in basin formation (Walcott 1972; Watts & Ryan 1976; Sweeney 1976; Sleep & Snell 1976; Steckler & Watts 1978; Beaumont 1978, 1981; Karner & Watts 1982) are usually based, however, on a preferred model for the lithospheric response to loading. Only Beaumont & Sweeney (1978) have considered the relative importance in basin formation of both elastic and viscoelastic (Maxwell) models.

The purpose of this paper is to examine critically the role of lithospheric flexure in sedimentary basin formation. We shall apply similar sedimentary loads to different mechanical models for the response of the lithosphere. Both elastic and viscoelastic (Maxwell) models will be considered, since these models have been used widely to interpret geological and geophysical observations associated with other types of loads on the Earth's surface. The stratigraphy and gravity anomaly for basins will be calculated by using different thermal and loading histories, and comparisons will be made with observations to estimate the best-fitting flexural model. We shall then examine the implications of the best-fitting model to studies of the rheology of continental lithosphere, the tectonic subsidence of basins, and variations of sea-level through time.

2. Lithospheric flexure

The principal evidence for a strong lithosphere has come from studies of its responses to surface loads such as ice sheets (Walcott 1970*a*), seamount chains (Walcott 1970*c*; Watts & Cochran 1974; Watts 1978), isolated seamounts and oceanic islands (Watts *et al.* 1975; McNutt & Menard 1978; Cazenave *et al.* 1980), sediments (Walcott 1972; Cochran 1973; Sleep & Snell 1976; Beaumont 1981), an oceanic or continental plate as it approaches a deep-sea trench (Walcott 1970*c*; Hanks 1971; Watts & Talwani 1974; Caldwell *et al.* 1976; Caldwell 1979) and the relation of topography and gravity anomalies in the interiors of continental (McGinnis 1970; Walcott 1970*b*; Banks *et al.* 1977; McNutt & Parker 1978) and oceanic plates (McKenzie & Bowin 1976; Detrick & Watts 1979; McNutt 1979). These studies suggest that the response to long-term geological loads can be satisfactorily modelled as either an elastic or a viscoelastic (Maxwell) plate overlying a weak substratum.

The parameters that characterize the flexure of thin elastic or viscoelastic (Maxwell) plates are the flexural rigidity, D, and the Maxwell relaxation time, τ. The elastic thickness of the plate, T_e, is related to D by

$$D = ET_e^2/12(1-\sigma^2), \tag{1}$$

where E is the Young modulus and σ the Poisson ratio. In the elastic model the elastic thickness T_e determines the amplitude and wavelength of flexure and does not depend on the age of the load. The viscoelastic (Maxwell) model, in contrast, is described by T_0, the initial elastic thickness, and τ, the Maxwell relaxation time. The amplitude and wavelength of flexure in the viscoelastic model varies as a function of the age of the load and is given by T_0 and τ. The elastic model can therefore be considered as a viscoelastic model with $\tau \rightarrow \infty$.

The most applicable model for studies of the mechanical properties of the lithosphere is the subject of much debate at present. In general, most evidence for the elastic model has come from studies in the oceans, while most evidence for the viscoelastic (Maxwell) model has come from studies in the continents. The effects of viscoelasticity should be more apparent in the continents since the oceans are characterized by younger loads (less than about 55 Ma old), whereas the continents are characterized by loads as old as several hundreds of megayears. The elastic model, however, has been most successful in explaining a wide range of geological and geophysical observations in the oceans. Since sedimentary basins occur on both continental and oceanic lithosphere, we shall consider both models in this study.

(*a*) *Elastic model*

The results of most recent flexure studies in the oceans are summarized in table 1 and figure 1. Figure 1 shows the estimates of elastic thickness, T_e, plotted as a function of the age of the plate at the time it was loaded. This figure is similar to an earlier plot by Watts *et al.* (1980) but includes one additional estimate. The estimates of T_e in figure 1 are based on different loads on oceanic lithosphere, which include the topographic relief of oceanic layer 2 (solid squares), river deltas (open circles), seamounts and oceanic islands (solid circles), and deep-sea trench–outer rise systems (solid triangles). The age of the plate at the time of loading was estimated for each load by subtracting the age of the load from the age of the underlying sea floor.

The main result shown in figure 1 is that T_e *increases* with the age of the oceanic lithosphere at the time of loading. Thus loads emplaced on young lithosphere are associated with small values of T_e while loads formed on old lithosphere are associated with large values. Figure 1 shows that there is good general agreement between T_e and the 300–600 °C oceanic isotherms based on the cooling plate model. This result was interpreted by Watts (1978) as indicating that as the oceanic lithosphere increases in age it becomes more rigid in its response to surface loads.

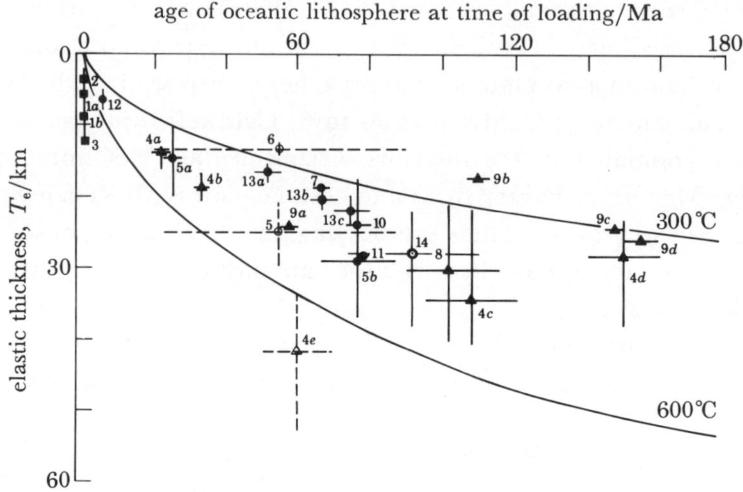

FIGURE 1. Plot of elastic thickness, T_e, against age of the oceanic lithosphere at the time of loading. The sources of the data are summarized in table 1. The broken lines indicate estimates that either represent broad regions rather than an individual geological feature (5, 6) or are considered unreliable (4*e*). The solid lines are the 300 and 600 °C oceanic isotherms based on a cooling plate model.

Figure 1 shows that the estimates of T_e in oceans based on long-term (1–55 Ma) geological loads are one-half to one-third of the seismic thickness based on short-term (*ca.* 150 s) seismic wave periods. Thus on loading there must be rapid relaxation of the oceanic lithosphere from its short-term seismic thickness to its long-term elastic or mechanical thickness. Subsequent relaxation is not observed, but may occur on very long timescales. T_e is therefore acquired soon after loading (less that 1 Ma) and does not appear to change significantly with time.

The results of flexure studies in the oceans are generally compatible with inferences of the rheology of oceanic lithosphere based on data from experimental rock mechanics (Goetze & Evans 1979; Bodine *et al.* 1981). Goetze & Evans (1979) developed a yield stress envelope based on the rheology of olivine in which the yield stress increases, then decreases, with depth. They showed that the oceanic lithosphere could be characterized by an upper part in which deformation occurs by cataclastic flow, and a lower part in which deformation occurs by ductile flow. Between these two regions is a central core that bends elastically but does not develop any permanent deformation. The yield stress model successfully predicts the flexural response for different age oceanic lithosphere while predicting reasonable values for the stresses developed during loading (Bodine *et al.* 1981). The deflexions produced by the yield stress model are in close agreement with those produced by the simple elastic model.

Flexure studies in the oceans are therefore in general support of an elastic plate model for the lithosphere. It is important to point out, however, that the model is referred to as an 'elastic'

TABLE 1. SUMMARY OF OCEANIC FLEXURE STUDIES

geological feature	age of load t_l / Ma B.P.	age of underlying sea-floor t_{sf} / Ma B.P.	age of sea-floor at time of loading $(t_{sf}-t_l)$ / Ma	effective flexural rigidity, D / 10^{21} Nm	effective elastic thickness, T_e/km ($E = 10^{11}$ Pa, $\sigma = 0.25$)	symbol in figure 1	reference
East Pacific Rise Crest	0	0	0	0.07–20	2–6	1*a*	Cochran (1979)
Juan de Fuca Ridge Crest	0	0	0	4.9	3.8	2	McNutt (1980)
Mid-Atlantic Ridge Crest	0	0	0	3–19	7–13	1*b*	Cochran (1979)
Mid Atlantic Ridge Crest	0	0	0	16	12.2	3	McKenzie & Bowin (1976)
Western Walvis Ridge	*ca.* 5	0	*ca.* 5	1.1–4.5	5–8	12	Detrick & Watts (1979)
Nankai Trough	0	19–25	19–25	19–40	13–16.5	4*a*	Caldwell (1979)
Emperor Seamounts north of 40° N	52–58.5	80†	22.5–28	10–71	10.5–20.0	5*a*	Watts (1978)
Middle America Trench	0	30–35	30–35	40–80	16.5–20.8	4*b*	Caldwell (1979)
Pacific Ocean atolls	0‡	20–90	20–90	17–25	12.4–14.1	6	McNutt & Menard (1978)
Marquesas Island	1–4	*ca.* 44–46	*ca.* 40–45	40	16.5	13*a*	Cazenave *et al.* (1980)
Aleutian Trench	0	55–60	55–60	120	24.3	9*a*	Caldwell *et al.* (1976)
Aleutian Trench	0	50–70	50–70	370–1400	24.6–54.6	4*e*	Caldwell (1979)
Great Meteor Seamount	11–16§	80	64–69	60	18.9	7	Watts *et al.* (1975)
Society Islands	0.5–4	*ca.* 70	*ca.* 66–69	80	20.8	13*b*	Cazenave *et al.* (1980)
Hawaiian Ridge south of 40° N	3–48.5	80–115	66.5–87	39–450	17.5–37.0	5*b*	Watts (1978)
Hawaii Island	0–3	80	77–80	200	28.2	11	Walcott (1970*c*)
Hawaiian Ridge	0–6	80	74–80	90–200	20–28.2	10	Suyenaga (1977)
Crozet Island	5	*ca.* 80	*ca.* 75	100	22.5	13*c*	Cazenave *et al.* (1980)
Amazon Cone	*ca.* 20–25‖	100–130	75–110	100–500	22.5–38.4	14	Cochran (1973)
Kuril Trench	0	90–115	90–115	91–590	21.7–40.5	8	McAdoo *et al.* (1978)
Bonin Trench	0	145–150	145–150	120	24.3	9*c*	Caldwell *et al.* (1976)
Mariana Trench	0	150–160	150–160	140	25.1	4*d*	Caldwell *et al.* (1976)
Mariana Trench	0	140–160	140–160	110–140	23.4–38.1	9*d*	Caldwell *et al.* (1976)
Kuril Trench	0	95–120	95–120	91–600	21.7–40.7	4*c*	Caldwell (1979)
Kuril Trench	0	95–120	95–120	50	17.5	9*b*	Caldwell *et al.* (1976)

† Age uncertain. Based on a model in which the Emperor Trough is a fracture zone offset.
‡ Ages may range to about 15 Ma (Jarrard & Turner 1979).
§ Age based on Wendt *et al.* (1976).
‖ Age based on Kumar (1978).

plate model even though T_e changes with time. For a single loading event the lithosphere relaxes rapidly from its short-term seismic thickness to an asymptotic value that does not change with time. Subsequent loading events, however, are emplaced on a lithosphere with a greater value of T_e due to its increase in flexural strength with age.

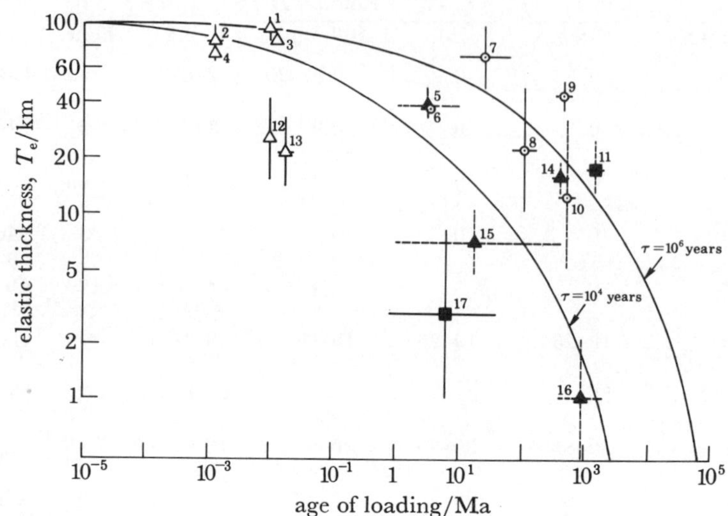

FIGURE 2. Plot of elastic thickness, T_e, of the continental lithosphere against age of load. The sources of data are summarized in table 2. The broken lines indicate estimates in which the age of load is uncertain due either to the broad extent of the region (14–16) or unreliable age data (10–11, 5–6). The solid lines show the calculated T_e based on a viscoelastic (Maxwell) plate with values of the relaxation time $\tau = 10^4$ and 10^6 years.

(*b*) *Viscoelastic* (*Maxwell*) *model*

The results of most recent flexure studies in the continents are summarized in table 2 and figure 2. Figure 2 shows the elastic thickness of the continental lithosphere, T_e, plotted as a function of the age of the load. This figure is therefore similar to an earlier plot by Walcott (1970*c*), but differs in including more recent estimates of T_e. The estimates of T_e in figure 2 are based on different loads on the continental lithosphere, including ice sheets (open triangles), sediments (open circles), mountain ranges (solid triangles) and continental rifts (solid squares). The age of the load was estimated from the available geological data. For the East Africa rift system we assumed an age of 0–30 Ma for the load because of the relatively recent uplift of the region. Furthermore, we assumed an age for the large continental regions (such as U.S. and Australia) corresponding to that of the most recent major orogeny (McNutt & Parker 1978), since orogenic regions dominate the continental topography spectra. The age range of loads in figure 2 therefore varies from nearly 0 Ma for East Africa to 1400 Ma for Australia.

The main result shown in figure 2 is that, in contrast to figure 1, T_e appears to *decrease* with age. Thus loads that have been emplaced on the lithosphere for relatively short periods of time are associated with large values of T_e, whereas loads that remain on the lithosphere for longer periods are associated with low values. This suggests that there is a time-dependent stress relaxation that results in a softening of the lithosphere with time. For short times (less than 10^5 years), the elastic thickness of the lithosphere supporting a load is 80–100 km, while for long times (more than 1 Ma), the thickness is much smaller. In the limit ($t \to \infty$), the response approaches that of the Airy model of isostasy.

The data in figure 2 were interpreted by Walcott (1970*c*) as indicating that the lithosphere response to loads as a viscoelastic rather than an elastic plate. The solid lines in figure 2 show the variation of T_e with age that would be expected if the lithosphere responds to loads as a viscoelastic (Maxwell) plate. Figure 2 shows that T_0 in the range 100–110 km and τ in the range 10^4–10^6 years generally constrains the estimates of T_e from the continents. The actual value for τ

TABLE 2. SUMMARY OF CONTINENTAL FLEXURE STUDIES

geological feature	age of load Ma B.P.	basement estimated age Ga B.P.	effective flexural rigidity 10^{21} N m	effective elastic thickness ($E = 10^{11}$ Pa, σ = 0.25)	symbol in figure 2 and figure 16‡	reference
		cratonic setting				
Lake Agassiz	0.007–0.014	1.6–1.8	590–13200	87–114	1	Walcott (1970*a*)
Lake Algonquin	0.00115–0.00125	1.2–1.5	2900–11000	69–104	2	Walcott (1970*a*)
Fennoscandia	0.012	2.0–2.8	5900	87	3	McConnell (1968)
Lake Superior	0–0.0012	1.2–1.5	3600	74	4	S. Dutch (unpublished)
Caribou Mountains	1–10?	1.6–1.8?	360–940	34–47	5	Walcott (1970*a*)
Interior Plains	1–10?	1.6–2.5?	598	44	6	Walcott (1970*a*)
Ganges Basin	10–50	1.0–2.0	1000–10000	48–104	7	Karner (1981)
Odaho/Wyoming Thrust	50–150	0.225–0.325	10–1000	10–48	8	T. E. Jordan (unpublished)
Michigan Basin	370–460	1.2–1.5?	480–1200	38–52	9	Haxby *et al.* (1976)
Boothia Uplift	460–550?	1.6–1.8?	1.2–300	5–32	10	Walcott (1970*b*)
mid-continent gravity high	1100?	1.2–1.5	20–150	13–26	11	McGinniss (1970), Cohen & Meyer (1976)
		orogenic setting				
Lake Hamilton	0.009–0.019	< 0.2	35–600	16–40	12	Fulton & Walcott (1975)
Lake Bonneville	0.015–0.02	< 0.2	24–290	14–32	13	Crittenden (1967, 1970), Walcott (1970*a*)
Northern Great Dividing Range	400–500?	0.4–0.5?	19–60	13–19	14	Wellman (1979)
		uncertain setting				
Continental United States	< 500?	< 0.5?	1.0–10	5–10	15	Banks *et al.* (1977)
Australia	400–1400?	0.4–1.4?	< 0.071	< 2	16	McNutt & Parker (1967)
east Africa	0–30†	0.5–1.4	0.01–50	1–8	17	Banks *et al.* (1977)

† Age of uplift.
‡ Note that only values for which both the age of load *and* basement are considered reliable are plotted in figure 16.

preferred by Walcott (1970*c*) was 10^5 years, which corresponds to an effective viscosity of the lithosphere of 10^{24} P†. Sleep & Snell (1976) and Beaumont (1978, 1981) subsequently used this estimate in their modelling studies of the stratigraphy of the Michigan, North Sea, Canadian foreland and U.S. Atlantic coastal plain basins.

† 1P = 10^{-1} Pa s.

There has been no attempt similar to that of Bodine *et al.* (1981) for the oceans, to establish whether the observations of flexure in the continents are compatible with results from experimental rock mechanics. The main problem is that the continental rheology is too poorly known at present. Thus there is little observational evidence available at present in support of a viscoelastic model for the lithosphere.

3. Simple models of flexure

The stratigraphy of a sedimentary basin is the result of the interaction of a number of different geological processes during time. Backstripping studies have shown that the principal processes affecting the stratigraphy of a basin are thermal contraction and sedimentary loading. Sedimentary factors such as compaction, palaeobathymetry and changes in sea level modify the stratigraphy, but their effects are generally small compared with those of thermal contraction and sedimentary loading. The main problem, therefore in modelling the stratigraphy of basins is determining the form of the thermal contraction and the nature of the basement's response to sediment loads. Thermal contraction controls the overlap shape of the basin that is available for sediments to infill, while the basement response is the main control on the geometry of a basin and the stratigraphy of the basin edges.

We shall first consider the simplest form for the tectonic subsidence of a basin in which the depth available for sedimentation is given by the square root of time since basin initiation. A similar model has been used to describe the subsidence of sea floor less than 70 Ma in age in the oceans (Parsons & Sclater 1977). The shape of the basin is given by

$$Y(x,t) = \begin{cases} 0 & |x| > a, \\ (350/b)\,t^{\frac{1}{2}}(|x|-a) & a \leqslant |x| \leqslant b, \\ 350t^{\frac{1}{2}} & |x| < b, \end{cases} \tag{2}$$

where t is the time since basin initiation and a and b define the basin width and slope respectively. We assume that sediments infill a broad basin (figure 3) during each $t^{\frac{1}{2}}$ time interval so that the load is constant for any one interval. The sediment load at time t_i is therefore

$$P(x) = (\rho_s - \rho_w)\, g Y_i(x, t_i), \tag{3}$$

where ρ_s is the density of the infilling sediment, ρ_w is the density of the displaced water and g is average gravity.

The form of the tectonic subsidence in figure 3 implies the following sedimentary loading history for a basin:

(1) the basin depth available for sediments decreases with time;

(2) the basin shape is constant during each major depositional cycle (100 Ma);

(3) during each depositional cycle, sediments are transported rapidly from the basin flanks to the basin centre;

(4) peripheral bulges that develop during flexure are eroded to sea level.

The differential equation relating the sedimentary load and basement response for an elastic plate model is given by

$$D\partial^4 y/\partial x^4 + (\rho_m - \rho_{\text{infill}})\, gy = P(x), \tag{4}$$

and for a viscoelastic plate model by

$$D\partial^4 \dot{y}/\partial x^4 + (\rho_m - \rho_{\text{infill}})\, g[\dot{y} + y/\tau] = P(x)/\tau, \tag{5}$$

where y is the deflexion due to the load $P(x)$, $\dot{y}$ is the first derivative of y with respect to time, ρ_{infill} is the density of the material infilling the deflexion, and ρ_m is the density of the mantle.

The solution to equations (4) and (5) for a load represented in the frequency domain can be expressed as

$$y(k) = \Phi P(k)/(\rho_{\mathrm{m}} - \rho_{\mathrm{infill}})\, g, \tag{6}$$

where k = wavenumber and Φ = basement response function. Thus y can be calculated by inverse Fourier transforming (F^{-1}) the product of the response function and the load spectrum.

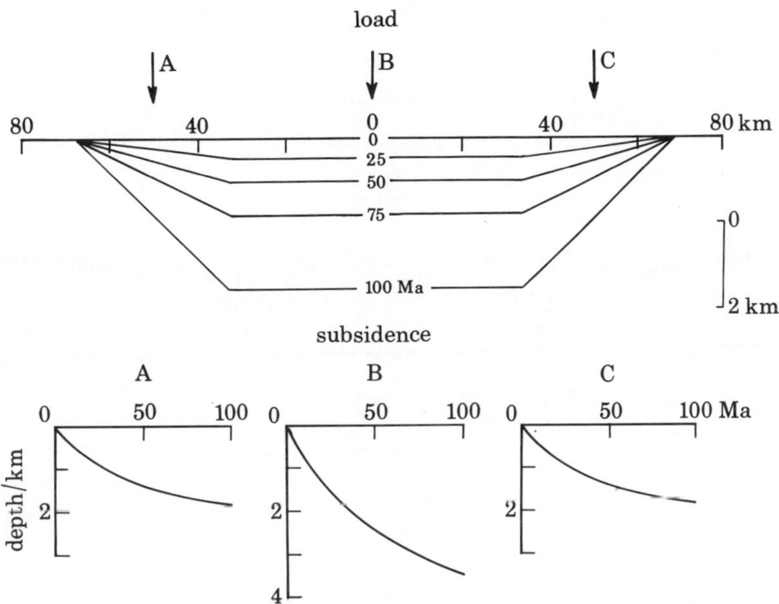

FIGURE 3. Simple model for the tectonic subsidence of a sedimentary basin as a function of age. The subsidence is assumed to decrease exponentially with time, in a manner similar to a mid-ocean ridge. The average width of a basin available for sediments to infill is assumed to remain constant during basin evolution. Vertical exaggeration × 10.

The response functions for both the elastic, $\Phi_{\mathrm{e}}(k)$, and viscoelastic, $\Phi_{\mathrm{v}}(k,t)$, plate models can be obtained (see, for example, Walcott 1976):

$$\Phi_{\mathrm{e}}(k) = 1/(1+C); \tag{7}$$

$$\Phi_{\mathrm{v}}(k,t) = 1 - C\Phi_{\mathrm{e}}(k)\, \mathrm{e}^{-t/\tau(1+C)}, \tag{8}$$

where $C = Dk^4/(\rho_{\mathrm{m}} - \rho_{\mathrm{infill}})$. For sediments infilling the basin to sea level we then have

$$S_i(x,t_i) = Y_i(x,t_i) + y_i(x,t_i),$$

where $S_i(x,t_i)$ is the sediment accumulation, $y_i(x,t_i)$ is the deflexion, and $Y_i(x,t_i)$ is the tectonic subsidence associated with the ith load and the ith time interval since basin initiation. For an elastic plate we have, for the total sediment thickness,

$$S(x,t_i) = \sum_{i=1}^{N} \left\{ Y_i(x,t_i) + F^{-1}[\Phi_{\mathrm{e}}(k)\, Y_i(k,t_i)] \frac{(\rho_{\mathrm{s}} - \rho_{\mathrm{w}})}{(\rho_{\mathrm{m}} - \rho_{\mathrm{infill}})} \right\}, \tag{9}$$

where $Y_i(k,t_i)$ is the load spectrum, $\Phi_{\mathrm{e}}(k)$ the elastic plate response function for each load increment, and N is the total number of loads. For a viscoelastic plate we have

$$S(x,t_i) = \sum_{i=1}^{N} \left(Y_i(x,t) + F^{-1}\left[\Phi_{\mathrm{v}}(k,t_0)\, Y_i(k,t_i) + \sum_{j=0}^{N-i} \{\Phi_{\mathrm{v}}(k,t_{i+1}) - \Phi_{\mathrm{v}}(k,t_j\} Y_{i+j}(k,t_{i+j}) \right] \frac{(\rho_{\mathrm{s}} - \rho_{\mathrm{w}})}{(\rho_{\mathrm{m}} - \rho_{\mathrm{infill}})} \right), \tag{10}$$

where t_j and t_{j+1} is the age range of the ith load. Thus $\Phi_e(k)$ and $\Phi_v(k, t)$ must be evaluated for each load applied. For the elastic model, $\Phi_e(k)$ is only evaluated once for each load applied, since once $\Phi_e(k)$ is acquired it does not change subsequently with time. For the viscoelastic model, in contrast, $\Phi_v(k, t)$ has to be evaluated for each load as well as for each underlying load.

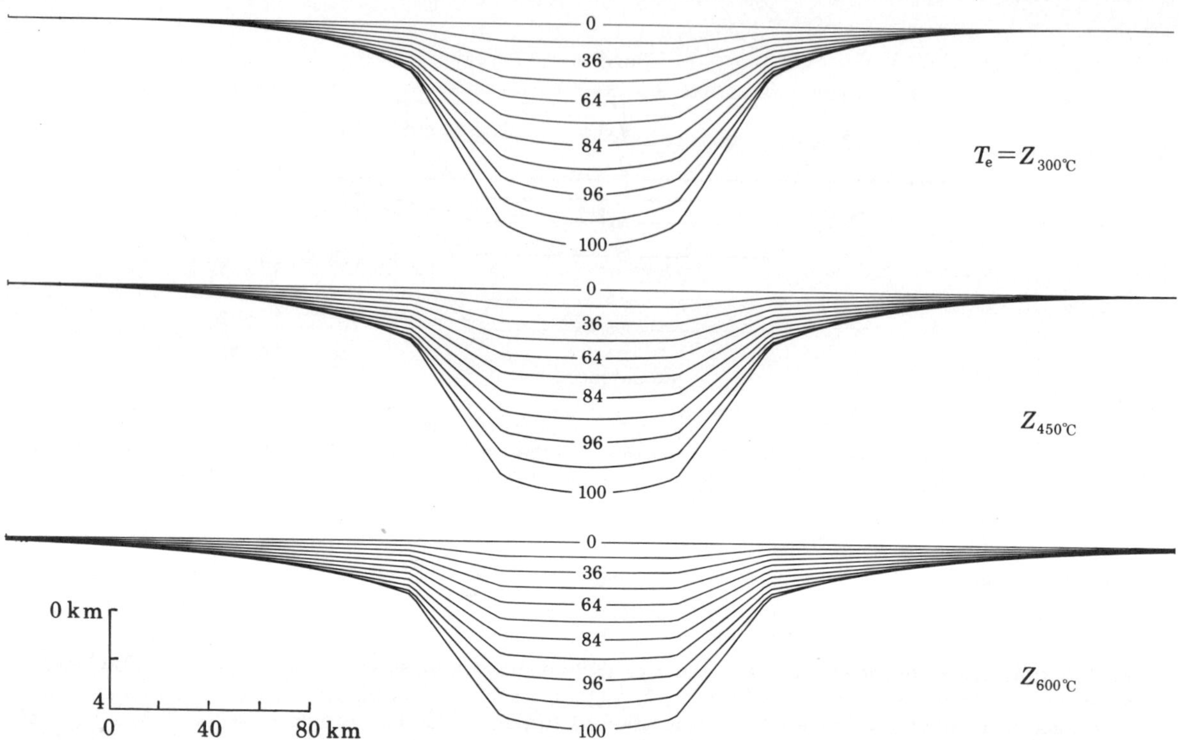

FIGURE 4. Calculated stratigraphy of a sedimentary basin, based on the tectonic subsidence in figure 3 and a simple elastic plate model. The calculations are based on different values of T_e, depending on whether T_e is given by the depth to the 300, 450 or 600 °C oceanic isotherm (figure 1). T_e = 300 °C corresponds to the weakest plate and gives the smallest basin width. T_e = 600 °C corresponds to the strongest plate and gives the widest basin. All models show a progressive overstepping of younger strata at the margins of the basin, due to the increase of T_e with age. Vertical exaggeration × 10.

The results of the calculations for the stratigraphy of a basin formed on an elastic and viscoelastic plate assuming the load shape in figure 3 are shown in figures 4 and 5. For the elastic model, values of T_e given by the depth to the 300, 450 and 600 °C oceanic isotherms have been used, because these values best describe the flexure observations in figure 1. For the viscoelastic model (figure 5), values of T_0 = 104 km ($D_0 = 10^{25}$ N m), and τ of 10^4, 10^5 and 10^6 years have been chosen, because these values apparently best describe the observations in figure 2.

The overall shapes of the basins in figures 4 and 5 are similar for each model. For the elastic models the widths of the basins increase while the amplitudes in the centre decrease as the flexural strength of the basement increases. Similarly for the viscoelastic models, the widths increase while the amplitudes decrease for increasing τ. Both models show a similar overall shape for the most 'rigid' cases with a broad basin overlying a narrow basin.

The elastic and viscoelastic models differ, however, in the stratigraphy predicted at the basin edges. The elastic models (figure 4) show progressive overstepping of younger strata onto basement due to the increase in T_e with age since basin initiation. Sediments that infill the basin soon after formation load a relatively weak plate, whereas sediments that infill the basin later in its evolution load a relatively strong plate. The viscoelastic models (figure 5), in contrast, show older strata outcropping at the edges of the basin with younger sediments restricted to the basin centre owing to the increasing stress relaxation with age of the load. Thus the youngest sediments in the basin centre show more subsidence and associated peripheral uplift and erosion in flanking regions than the oldest sediments.

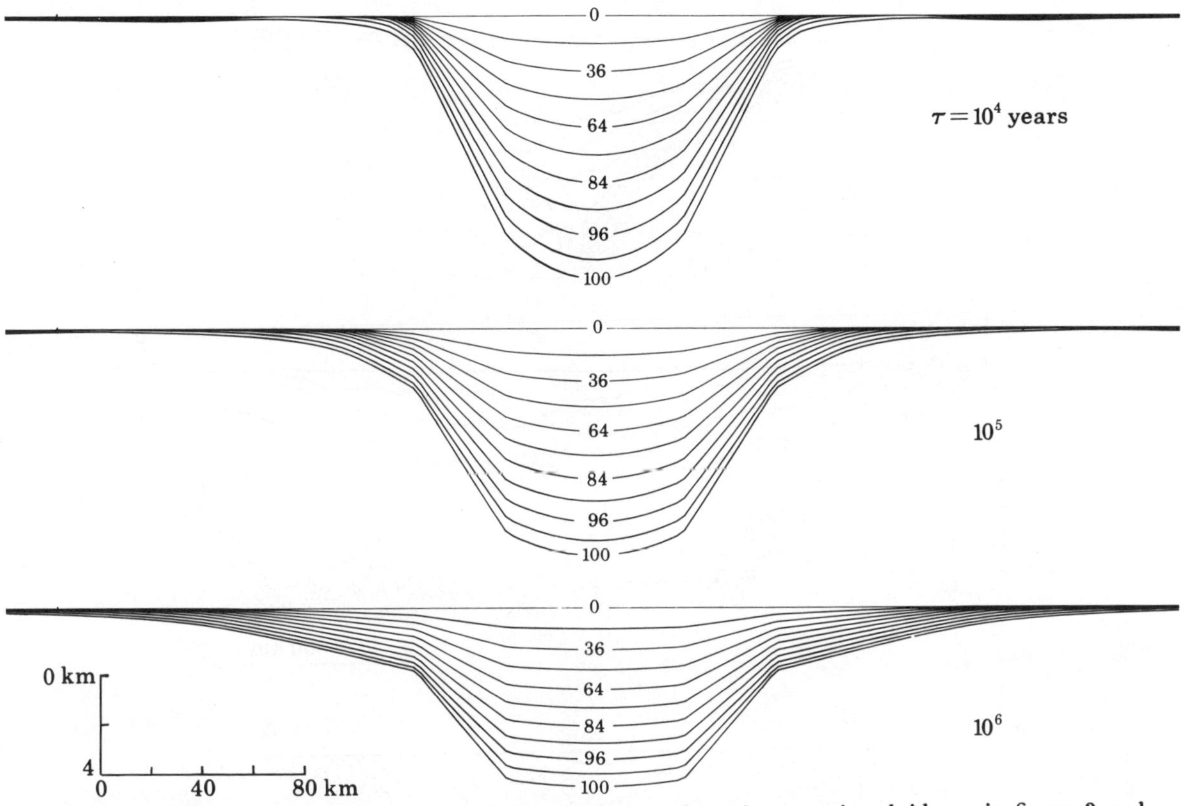

FIGURE 5. Calculated stratigraphy of a sedimentary basin based on the tectonic subsidence in figure 3 and a simple viscoelastic plate model. The calculations are based on different values of τ, in the range 10^4–10^6 years: 10^4 years is associated with the most rapid relaxation and smallest basin width; 10^6 years is associated with the least relaxation and the widest basin. All models show the youngest sediments restricted to the basin centre. Vertical exaggeration $\times 10$.

The direct comparison of the model results to observations is, of course, complicated by the varying ages and geometries of basins that occur in the geological record and the differing effects of sedimentary processes such as compaction, eustasy, transport, and sediment supply in basin formation. The general features of the basin edge stratigraphy can, however, be compared with observations. Figure 6 shows, for example, a stratigraphic cross section of the Gippsland Basin in southeast Australia (Hocking 1976). This basin consists of a late Cretaceous to Eocene formation of terrigeneous sandstones and siltstones overlain either conformably or unconformably by Oligocene to Pliocene marine limestones and calcareous siltstones. The edges of the basin show a progressive onlap of younger sediments onto the basement, similar to the predictions of the elastic model (figure 4). There are a number of other examples in the geological record of basins that show onlap at their edges; the western margin of the West Siberia Basin during the Jurassic and Cretaceous (Zhabrev *et al.* 1975) (figure 6), the coastal plain of northwest Africa (Vail *et al.*

1977), and the North Sea Basin during the Permian (Ziegler 1978). Each example shows a progressive widening of the basin with time after its initiation. Figure 6 shows, in contrast, a stratigraphic cross section of the Michigan Basin in the central U.S. This basin consists of a lower formation of Cambrian–Ordovician sandstones and shales, a middle formation of Silurian–Devonian limestones, dolomites and evaporites, and an upper formation of Carboniferous shales

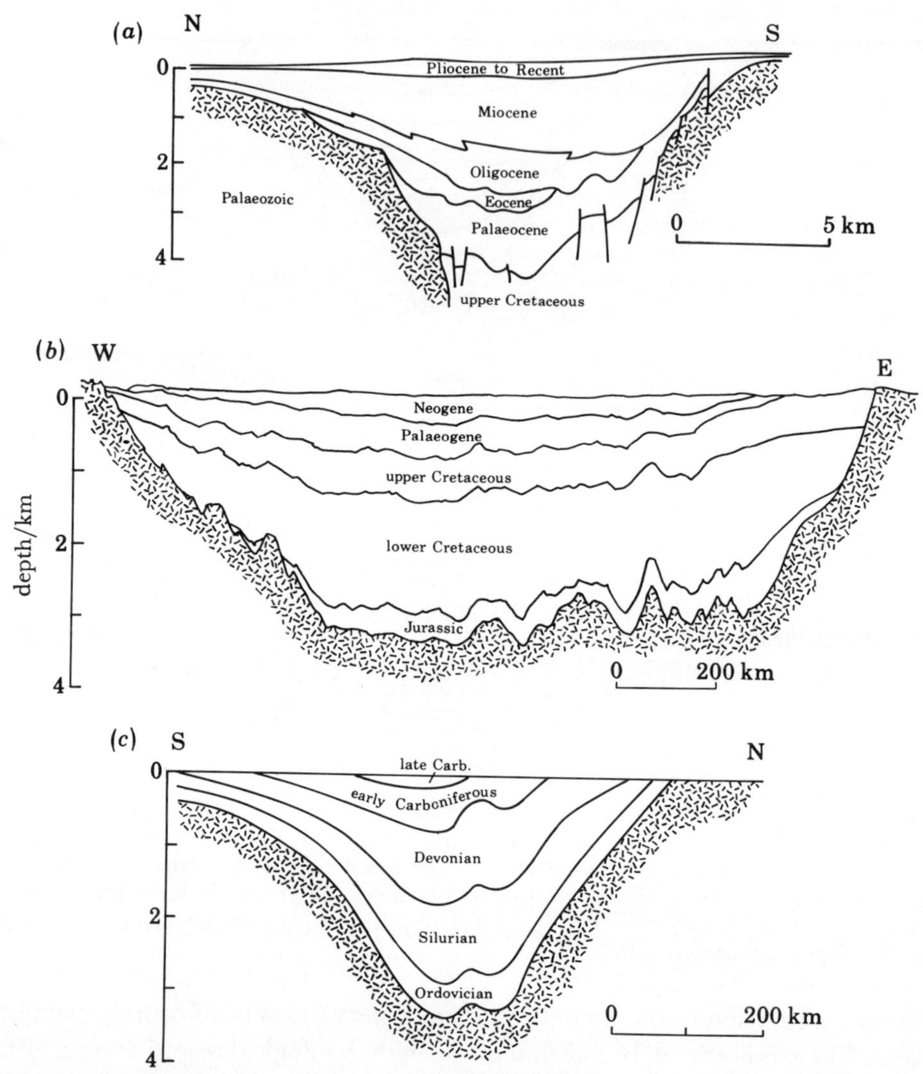

FIGURE 6. Stratigraphic cross section of (*a*) the Gippsland Basin (Hocking, 1976), (*b*) the West Siberia Basin (Zhabrev *et al.* 1975) and (*c*) the Michigan Basin (Sleep & Snell 1976). The Gippsland Basin is an example of a basin that shows progressive onlap of younger sediments on the basement. The Michigan Basin, on the other hand, is an example of a basin in which the younger sediments occur in the centre. The West Siberia Basin shows both types of stratigraphy at its edges. Vertical exaggerations: (*a*) × 2; (*b*) × 150; (*c*) × 100.

and sandstones. The edges of the basin (which are covered in part by a Pleistocene drift) are characterized by younger sediments toward the basin centre, similar to the predictions of the viscoelastic model (figure 5). There are a number of other examples of basins that illustrate this feature; the eastern margin of the West Siberia Basin during the Upper Cretaceous (Zhabrev *et al.* 1975) (figure 6), the North Sea during the late Tertiary (Ziegler & Louwerens 1977), the

Aquitaine Basin during the Neocomian (Winnock 1971), the Paris Basin (B.R.G.M. 1980), and the U.S. Gulf coastal plain (Weaver 1951). Thus both types of model can be recognized in the geological record.

The geometry of stratigraphic horizons in sedimentary basins is determined by the history of subsidence and uplift associated with an individual basin during its evolution. Sloss & Scherer (1975), from a consideration of the geometry of individual stratigraphic horizons, the distance to

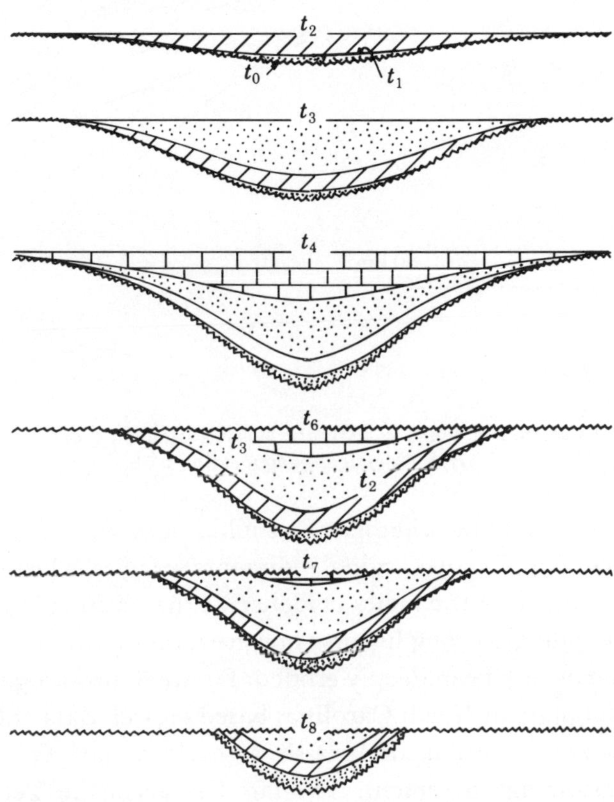

FIGURE 7. Stages in the subsidence, filling and erosion of a sedimentary basin (after Sloss & Scherer 1975).

basin hinge 'lines' (the position of maximum rate of change of sediment thickness), and the thickness of strata, concluded that the Michigan, Williston, and Moscow basins each increased in overall width during their evolution. They proposed a general model for basin evolution in which the basin history changes from an erosional to a depositional cycle and then back to an erosional cycle (figure 7). At time t_0, the depositional cycle is initiated on an essentially planar surface after the completion of a preceding erosional cycle. Deposition between time t_0 and t_1 is confined mainly to the basin centre, but spreads to a greater area between time t_1 and t_2. Differential subsidence continues until time t_4, with subsidence rates decreasing between the basin centre and margins. At time t_4, then, a broad basin exists in which younger sediments progressively onlap basement. After time t_4, the depositional cycle ceases and at some time between t_4 and t_6 the erosional cycle is initiated. By time t_8, a narrow basin exists in which the youngest sediments are restricted to the basin centre.

Figure 7 suggests that an elastic model in which the flexural strength of the lithosphere increases with time could explain configuration of, for example, the Michigan Basin, if the main deposition cycle in this basin was followed by erosion. Haxby *et al.* (1976), in fact, suggested from detailed stratigraphic studies that T_e increased from the Ordovician to the Devonian in the Michigan Basin. Thus although the present outcrop patterns of the Michigan basin suggest a viscoelastic model, such a model is not required to explain the evolution of this basin.

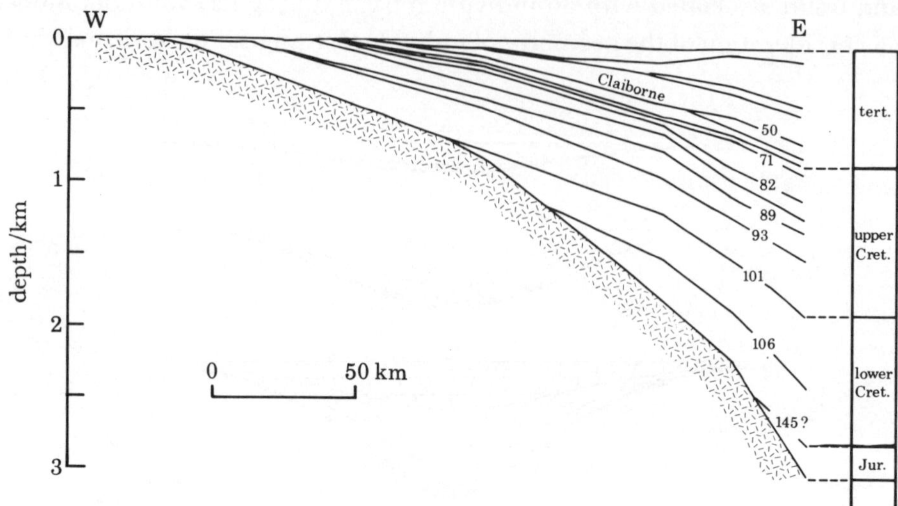

FIGURE 8. Stratigraphic cross section of the U.S. Atlantic coastal plain in North Carolina (after Sleep & Snell 1976). Vertical exaggeration × 50.

A more precise way of estimating whether a basin has been widening during a major depositional cycle is by detailed mapping of marginal sedimentary facies. The problem with basins such as the Michigan, is that much of the facies evidence at the basin edges has been subsequently removed by erosion. Despite their long history, however, the coastal plain of Atlantic-type continental margin basins have not been deeply eroded. Figure 8 shows a geological cross section of the U.S. Atlantic coastal plain in North Carolina, based on well data (Sleep & Snell 1976). The cross section shows progressive onlap of Jurassic to early or late Cretaceous sediments onto Precambrian or Palaeozoic age basement. The late Cretaceous to Tertiary sediments do not onlap basement, and younger sediments progressively subcrop in a seaward direction beneath a Pleistocene cover. Sleep & Snell (1976) interpreted the overall geometry of the coastal plain in terms of a viscoelastic model. Studies of sedimentary facies within the Claiborne (38–49 Ma B.P.), however, show a transgressive sequence of coarse sandstones and silstones of the Conagree formation overlain by fossiliferous highly calcareous limestones of the Santee formation (figure 8) (Colquhoun & Johnson 1968) suggesting the present outcrop pattern of the Claiborne is due to subsequent erosion of the coastal plain (due probably to seaward tilting of the coastal plain). Thus an elastic model in which T_e increases with age followed by erosion could explain the stratigraphy of the U.S. coastal plain.

We have also computed the gravity anomaly predicted by the elastic and viscoelastic models to determine whether observed free-air gravity anomalies can be used to distinguish between these models. The gravity anomaly depends on both the loading history and the basement rheology and so may be a useful model constraint. Figure 9 compares the gravity anomaly predicted by the elastic and viscoelastic models. The calculations assume that the tectonic subsidence of the basin is locally compensated (Airy model), since most thermal models consider basin initiation and rifting to be in local compensation, and that the sediments infilling the basin load the crust flexurally. Figure 9 shows there is a large spatial variation of the calculated gravity anomalies, which depend on the flexural parameters assumed in the models. The largest amplitude positive anomalies occur over the centre of basins that form on relatively 'rigid' plates ($T_e = Z_{600°C}$; $\tau = 10^6$ years), whereas the largest amplitude negative anomalies occur over the centre of basins that form on relatively weak plates ($\tau = 10^4$ years).

There is a large variation in both amplitude and wavelength of free-air gravity anomalies observed over interior or cratonic basins. For example, the Michigan Basin is associated with generally positive gravity anomalies and the Gippsland Basin is associated with generally

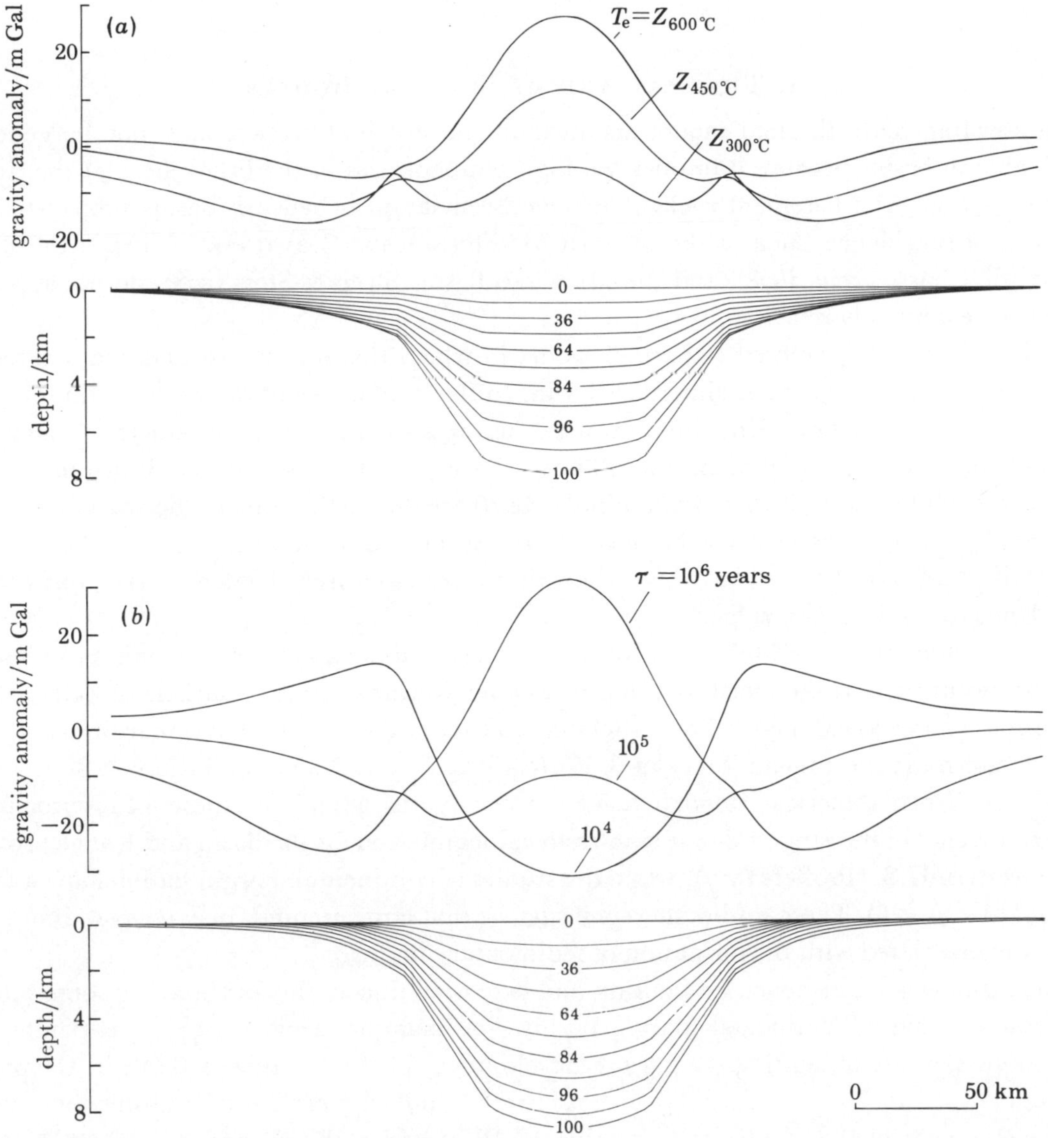

FIGURE 9. Calculated gravity anomalies associated with the sedimentary basins in figures 4 and 5. The upper curves are (*a*) based on the elastic plate model with T_e given by the depth to 300, 450 and 600 °C isotherm, and the lower curves (*b*) are based on the viscoelastic plate model with $T_0 = 104$ km and $\tau = 10^4$, 10^5 and 10^6 years. The tectonic subsidence of the basin is assumed to be due to crustal thinning at the time of basin formation. The crustal thinning has been calculated from the tectonic subsidence assuming Airy isostasy and an initial crustal thickness of 30 km. Vertical exaggeration × 10.

negative anomalies (up to -60 mGal). Furthermore, large-amplitude negative gravity anomalies (up to -60 mGal) characterize basins in the continental shelf north of Scotland (Bott & Watts 1970) and off Nova Scotia (Loncarevic & Ewing 1967), while low-amplitude positive anomalies and nearly zero anomalies characterize the North Sea Basin (Collette 1960), the eastern U.S. continental margin basins (Grow *et al* 1979; Karner & Watts 1982) and major river deltas (Walcott 1972). The model studies suggest that the large variation of observed gravity anomalies over these basins may be due to differences in their tectonic setting on either 'weak' lithosphere (negative anomalies) or 'rigid' lithosphere (positive anomalies), although it is not possible to use the observed gravity anomalies to distinguish between the different flexure models.

4. Thermal and mechanical models

The cooling plate thermal model assumed in the previous section may not be generally applicable to basins because it implies too high temperatures in the lithosphere at the time of their initiation. The basement rocks that have been sampled beneath basins subject to large amounts of subsidence, such as the western Mediterranean (Cravatte *et al.* 1974), the Bay of Biscay (de Charpal *et al.* 1978), and the Michigan Basin (Sleep & Sloss 1978), do not appear to have been extensively reheated.

McKenzie (1978) proposed a thermal model in which the lithosphere undergoes extension at the time of formation of a sedimentary basin. In this model, extension produces thinning and heating of the lithosphere. The amount of extension is defined by a parameter β such that $(1-1/\beta)$ defines the amount of thinning. The model assumes local isostatic equilibrium throughout basin evolution so that there is an initial subsidence due to thinning of the crust followed by a thermal subsidence as heat is conducted to the surface and the lithosphere cools. The case of infinite extension ($\beta \rightarrow \infty$), in which the lithosphere is entirely heated, corresponds to the subsidence of a mid-ocean ridge.

The stretching model of McKenzie (1978) has been used to explain the tectonic subsidence of interior basins such as the North Sea Basin (Sclater & Christie 1980; Christie & Sclater 1980; Wood 1981) and the Pannonian Basin (Sclater *et al.* 1981*a*), and continental margin basins such as the western Mediterranean (Steckler & Watts 1980), Bay of Biscay (Le Pichon & Sibuet 1981) and eastern North America (Royden & Keen 1980; Watts 1981). The studies of interior basins suggest β values in the range 1.3–2.0, similar to values inferred for the Basin and Range province of the western U.S. (Proffett 1977), while the studies of continental margin basins show a larger range β $(2 < \beta < 6)$. These studies therefore suggest that large amounts of extension (more than 30%) are associated with the formation of sedimentary basins.

There is now good evidence for faulting and extension during the formation of some interior basins and rifting of continental margin basins. For example, Ziegler (1977) has pointed out two major periods of faulting during the development of the North Sea Basin in Devonian–Carboniferous and Jurassic–early Cretaceous times. Similarly, periods of faulting characterize the development of the continental margins off West Africa, Brazil and northwest Australia (Kent 1976) Jurassic–early Cretaceous in time. Seismic reflexion profiling in basins in the western U.S. (Effimoff & Pinezich 1981) and the Bay of Biscay (de Charpal *et al.* 1978), have shown that the faults are in part listric so that fault movements result in rotation of individual crustal blocks.

We have constructed a model in which $Y(x, t)$ in the centre of the basin is given by the stretching model with $\beta = 2.0$ (figure 10). The amount of initial subsidence given by his model, using the parameters in table 3, is 1.73 km and the thermal or tectonic subsidence is given by McKenzie (1978) as

$$Y(x,t) = \begin{cases} 0 & |x| > a \\ (2.095/b)\,(1-\mathrm{e}^{-t/62.8})\,(|x|-a) & a \leqslant |x| \leqslant b \\ 2.095(1-\mathrm{e}^{-t/62.8}) & |x| < b \end{cases} \tag{11}$$

The depth of the basin is calculated for each $t^{\frac{1}{2}}$ time interval. The shape of the water-filled basin assumed is shown in figure 10 for different times after formation. We used similar assumptions for the loading history of the basin as used in the models in figures 4 and 5.

The results of the calculations for the stratigraphy of a basin based on the stretching model are shown in figure 11 for an elastic and viscoelastic plate. For the elastic model (upper basin) values of T_e are given by the depth to the 450 °C isotherm. For the viscoelastic model, the value of

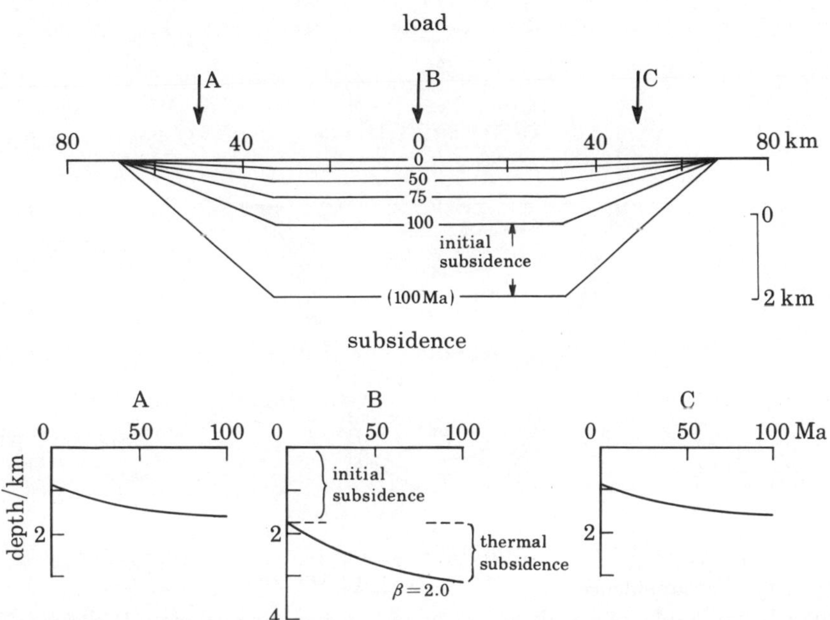

FIGURE 10. Simple model for the tectonic subsidence of a sedimentary basin, based on the stretching model of McKenzie (1978) with $\beta = 2.0$. The subsidence in this model is characterized by an initial subsidence followed by a thermal subsidence that exponentially decreases with time. Vertical exaggeration × 10.

TABLE 3. SUMMARY OF PARAMETERS USED IN MODEL CALCULATIONS

$\rho_w = 1.03\ \mathrm{g\ cm^{-3}}$

$\rho_s = 2.5\ \mathrm{g\ cm^{-3}}$

$\rho_{infill} = 2.5\ \mathrm{g\ cm^{-3}}$

$\rho_c = 2.8\ \mathrm{g\ cm^{-3}}$

$\rho_m = 3.33\ \mathrm{g\ cm^{-3}}$

$E = 10^{11}\ \mathrm{Pa}$

$\sigma = 0.25$

$L = 125\ \mathrm{km}$

$T_c = 31.2\ \mathrm{km}$

$\alpha = 3.4 \times 10^{-5}\ \mathrm{K^{-1}}$

$\kappa = 8.0 \times 10^{-3}\ \mathrm{cm^2\ s^{-1}}$

$T_l = 1333\ \mathrm{°C}$

$g = 981\ \mathrm{cm\ s^{-2}}$

T_0 is given by the depth to the 1109 °C (T_0 = 104 km at $t = \infty$) isotherm and the value of τ is given as 10^5 years. The flexural parameters in both models are obtained by 'tracking' the depths of individual isotherms based on the stretching model. The amplitudes of the basins are similar for each model but the widths are strikingly different. For the elastic model, the sediments that form during the thermal subsidence overstep by about 125 km the sediments that infill the basin formed by the initial subsidence due to the relatively high initial value of T_e and its subsequent increase with age. For the viscoelastic model, in contrast, the magnitude of the overstep is significantly reduced to about 5–10 km owing to the lower initial value of T_0 associated with the heating event and the subsequent increase in stress relaxation with time.

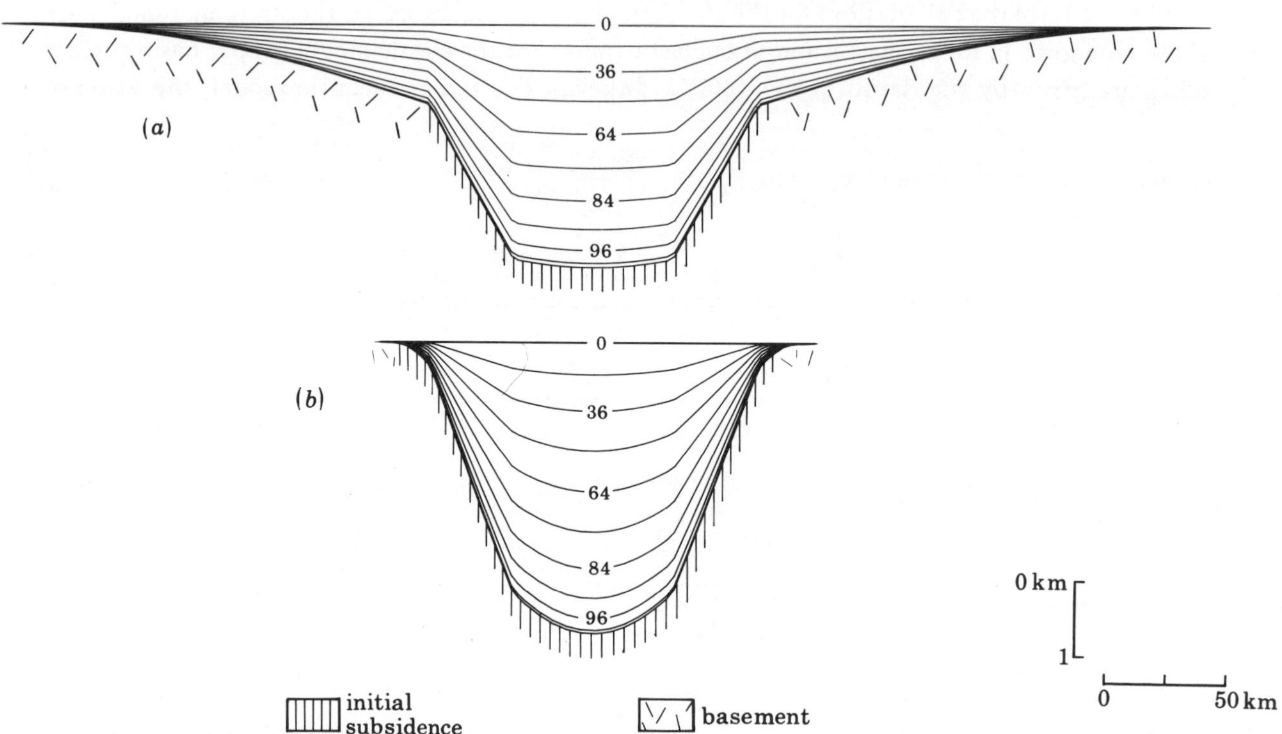

FIGURE 11. Calculated stratigraphy of a sedimentary basin based on the tectonic subsidence in figure 10 and an elastic (*a*) and viscoelastic (*b*) plate model. The elastic model is based on $T_e = Z_{450\,°C}$ and the viscoelastic model is based on $\tau = 10^5$ years and $T_0 = Z_{1109\,°C}$. The sediments that infill the initial subsidence (figure 10) are assumed to load an Airy-type crust, whereas the sediments that infill the thermal subsidence are assumed to load either an elastic plate with T_e increasing with age (*a*) or viscoelastic plate with a constant viscous relaxation time during basin evolution and an initial elastic thickness that increases with age (*b*). Vertical exaggeration × 33.

We have compared the stratigraphy predicted by the elastic and viscoelastic models to interior and cratonic basins that were characterized during their evolution by both fault controlled and thermal or flexural controlled subsidence. Figure 12 shows a stratigraphic cross section of the central North Sea Basin (Ziegler 1977; Ziegler & Louwerens 1977). Rifting was initiated in the basin during the Permian and Trias and continued through the Jurassic and Lower Cretaceous. The greatest subsidence occurred over a relatively narrow region (about 100 km) in the Central graben between the mid-North Sea high and the Stavanger platform. Minor faulting during the Upper Cretaceous was associated with infilling of the earlier graben system, but the Tertiary basin developed as a simple syncline over a broad region. The maximum subsidence occurred over the region of the fault-controlled central graben. Subsidence extended to flanking regions so that younger Tertiary sediments overstep the older sediment-filled graben by up to 200 km. The overall Late Mesozoic to Tertiary development of the central North Sea Basin therefore generally follows the prediction of the elastic model (figure 11). The main difference with the model is that at the eastern margin of the basin, sediments decrease in age progressively towards the basin centre. The eastern margin appears, however, to have been truncated by late Pleistocene uplift and erosion after ice loading of the Fennoscandian shield (Holtedahl & Bjerkji 1975) so that subsequent sedimentary processes have modified the stratigraphy of one of the edges of this basin.

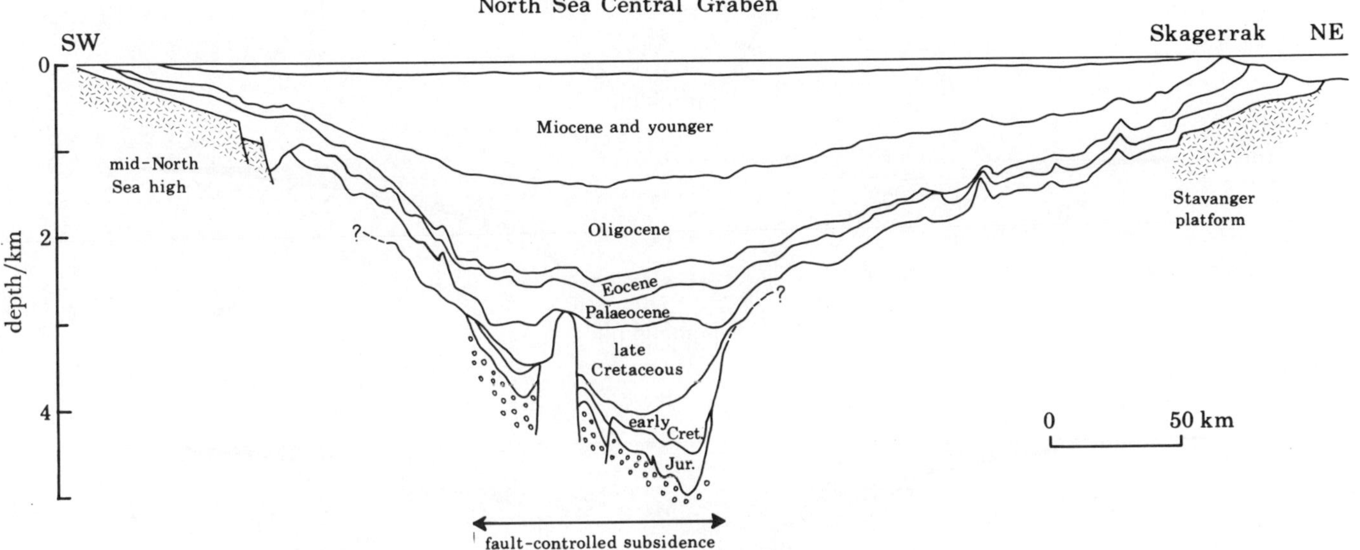

FIGURE 12. Stratigraphic cross section of the North Sea Basin in the region of the Forties and Maureen oil fields, based on Ziegler & Louwerens (1977) and Ziegler (1977). Vertical exaggeration × 33.

There are other examples in the geological record of basins that show a period of fault-controlled subsidence followed by a flexural-controlled subsidence. These include Atlantic-type continental margin basins such as those off eastern North America, Brazil and northwest Australia, and interior or cratonic basins such as the Anadarko, Val Verde, and West Texas in North America (Dewey & Pitman, this symposium) and the Great Artesian Basin in Australia (Tanner 1966). Thus an elastic model in which the basement rigidity increases with time according to the temperature structure defined by the stretching model appears able to explain the overall shape of those basins that were characterized by both oscillatory (Sloss & Speed 1974), or rifting, and submergent, or flexural, depositional cycles. The viscoelastic model, in contrast, does not appear to be able to explain the shape of these basins.

We have not considered in the simple elastic model of figure 11 the effects of lateral variations in the elastic thickness T_e across a basin, or the lateral flow of heat from the stretched region in the centre of the basin to the unstretched basin edges. These effects are difficult to assess because they require a complete knowledge of the thermal structure across a basin. The one-dimensional stretching model can be modified, however, to permit an examination of these effects (see appendix).

The simple elastic model in figure 11 assumes a single constant value for T_e based on the thermal history of the basin centre. In fact, if T_e, corresponds to the depth of the 450 °C isotherm as suggested by figure 1, then T_e will vary *across* the basin. The centre of the basin will be associated with a value for T_e similar to that assumed in the simple model, but at the edge of the basin, T_e will correspond to that of cold, unstretched lithosphere. In order, therefore, to model correctly the flexure of the lithosphere due to sediment loads, a T_e that varies as a function of time as well as position should be used.

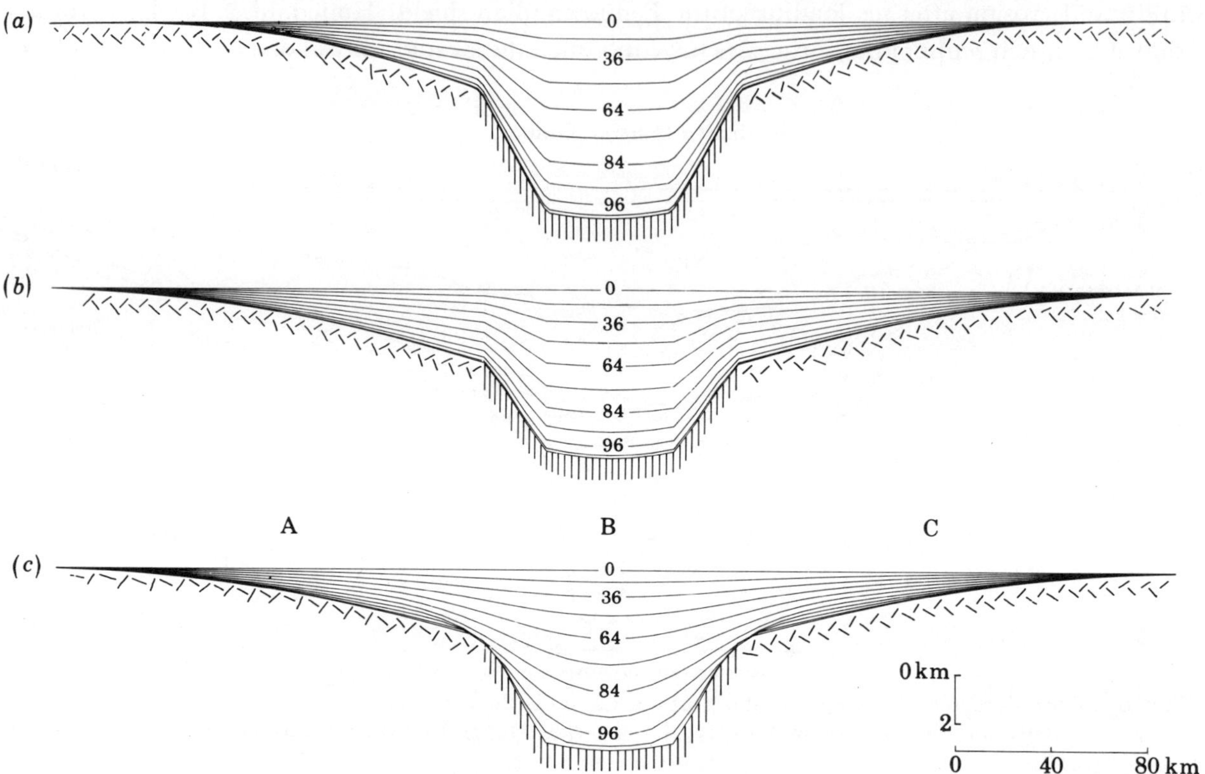

FIGURE 13. Calculated stratigraphy of a sedimentary basin, based on a stretching model with $\beta = 2.0$ (figure 10) and different elastic plate models. (*a*) Flexure is a function of time and T_e is given by the depth to the 450 °C isotherm in the centre of the basin (this model is identical to that in the upper model of figure 11 and is shown for comparison). (*b*) Flexure is a function of time and position and T_e is given by the depth to the 450 °C isotherm across the basin. (*c*) Flexure is a function of time and position, T_e is given by the depth to the 450 °C isotherm across the basin, and the effects of lateral heat conduction are included. Vertical exaggeration × 10.

The appropriate form of the differential equation relating the sediment load and basement response for a variable flexural rigidity is given by

$$\frac{\partial^2}{\partial x^2}\left\{D(x)\frac{\partial^2 y}{\partial x^2}\right\} + (\rho_m - \rho_{infill})\, gy = P(x). \tag{12}$$

This equation can be solved numerically by using a finite difference technique (Bodine 1981). T_e is given by the depth to the 450 °C isotherm at selected points across the basin. By using this technique, the restoring force can also be varied across the basin so that the correct value for $\rho_m - \rho_{air}$ can be used in the region of the flexural bulge.

The stratigraphy of a basin based on a T_e that varies as a function of both time and position across a basin is shown in figure 13*b*. This figure shows that, compared with the simple model of flexure (figure 13*a*), the basin is wider and the progressive onlap of sediments onto the basement is not as well developed. These features are due to the large T_e associated with the flanks of the basin, which also results in a larger sediment thickness in the flanks relative to the centre of the basin. A greater proportion of the sediments in the centre of the basin occurs early in basin history, when the contrast in T_e across the basin is largest.

The estimate of T_e used to calculate the stratigraphy of the basin in figure 13*b* was based on a model that only considered the vertical cooling of the lithosphere during basin formation. In fact, the heat within the stretched lithosphere will flow laterally as well as vertically, cooling the centre of the basin but heating the unstretched lithosphere of the basin flanks.

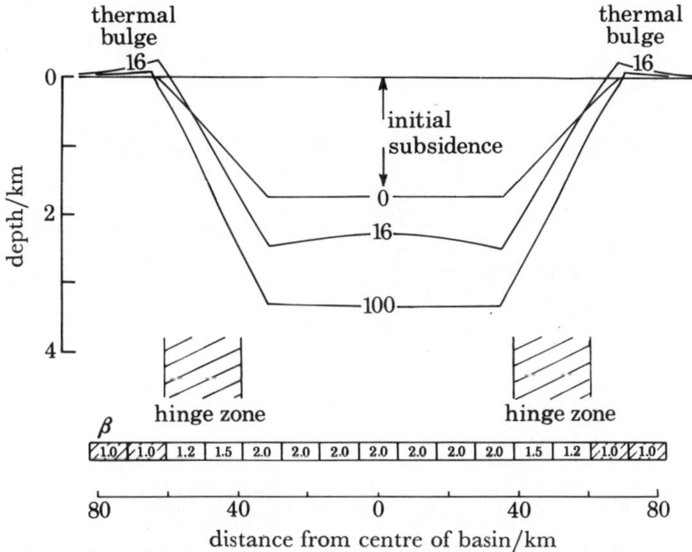

FIGURE 14 Simple model for the variation of the stretching factor β (McKenzie 1978) across a sedimentary basin. The curves show the tectonic subsidence of the basin for 0, 16 and 100 Ma after stretching. The initial subsidence ($t = 0$), which is assumed to occur instantaneously, is similar to the simple model in figure 10. For greater times the lateral flow of heat from the basin produces thermal bulges at the edges of the basin. A corresponding region of increased subsidence occurs in adjoining regions of the basin. Vertical exaggeration × 20.

Figure 14 shows the effects of lateral flow of heat on the overall basin shape. Unlike the simple model in figure 10, the shape of the basin available for sediments does not remain constant during basin evolution. Furthermore, while the initial subsidence is identical, at later times heat flows laterally out of the basin, producing a thermal bulge in flanking regions. The maximum uplift exceeds 250 m and occurs early in the history of the basin. As time progresses, the uplift decreases but also broadens as the heat diffuses outwards. The uplift in the flanks of the basin is accompanied by rapid subsidence on the basin side of the hinge zone. This region loses heat and therefore subsides more rapidly than the centre of the basin. As heat diffuses laterally, the basin flattens out but the position of the basement remains a few hundred metres deeper than the simple model, even after 100 Ma.

We show the effect of including the lateral flow of heat on the stratigraphy of the basin in figure 13*c*. Since heat flows laterally as well as vertically, the lithosphere in the flanking regions is weakened. Thus T_e in flanking regions will decrease so that, compared with figure 13*b*, which only includes flexure varying as a function of time and position, the width of the basin decreases early in its evolution. Near the edge of the unstretched lithosphere, the thermal bulge and flexure compete, resulting in erosion during the first 16 Ma of basin evolution. In the centre of the basin, the lateral flow of heat results in rapid subsidence and smoothes individual strata. The maximum sediment thickness in figure 13*c* is less than the model in figure 13*a* but a greater proportion of the subsidence occurs early in basin evolution.

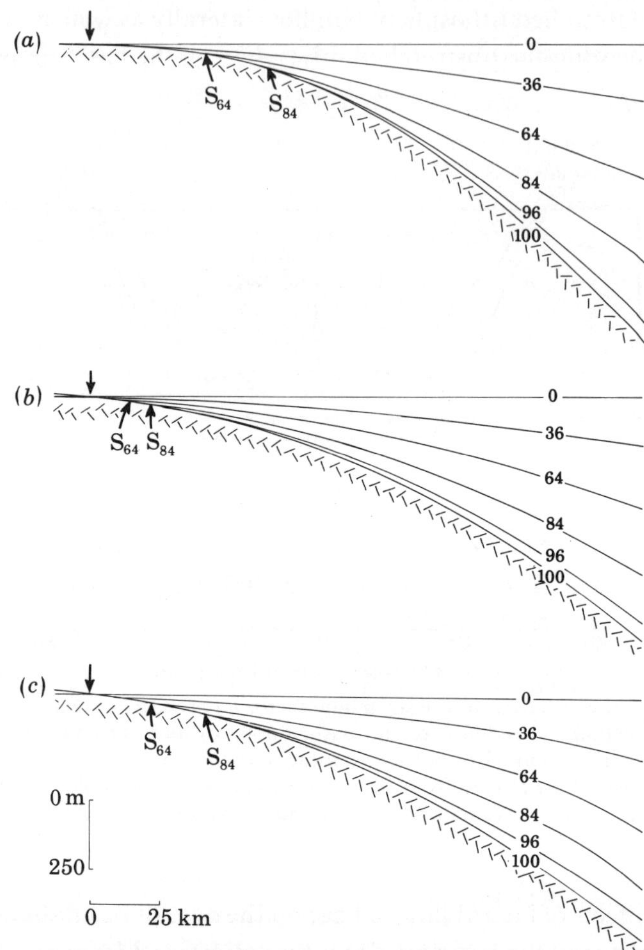

FIGURE 15. Calculated stratigraphy of the edge of the sedimentary basins in figure 14. The heavy sloping arrows indicate the subcrop of the 84 and 64 Ma horizons due to overstepping of younger strata onto the basement. Vertical exaggeration × 100.

Figure 15 shows the details of the stratigraphy at the basin edges for each of the three elastic models. The position of the subcrop of the 64 and 84 Ma horizons and the final stratigraphy at the edge of the basin are shown in the figure. The effect of varying flexure as a function of both time and position (figure 15*b*) causes an increase in the width of the basin and a decrease in the prominence of onlap at the basin edges. The effect of lateral flow of heat (figure 15*c*), however, competes with flexure, causing a decrease in the width and an increase in the prominence of onlap. Thus the stratigraphy of basins that include both the lateral effects of flexure and heat flow (figure 15*c*) closely resembles that of the simple elastic model (figure 15*a*) based only on the one-dimensional heat flow equation.

5. Discussion

We have compared in this study the predictions of simple elastic and viscoelastic (Maxwell) models of loading with observations of the geometry and stratigraphy of sedimentary basins. The best overall fit to these observations is for an elastic model in which the flexural strength of the lithosphere increases with age. We examine in this section the implications of this model to studies of the rheology of continental lithosphere, the tectonic subsidence of basins, and sea level changes through geological time.

(a) *Rheology of the continental lithosphere*

The principal evidence for the rheological behaviour of the Earth has come from studies of the manner in which it responds to applied loads. For deformations of the duration of seismic wave periods (a few seconds to a few minutes) the Earth generally behaves as an elastic material. The Earth is not completely elastic at these periods, however, as indicated by finite 'Q' values associated with free oscillation and surface wave data. For deformations of much longer duration (of the order of the age of the Earth) large spatial wavelengths are fully relaxed so that the Earth deforms essentially as a fluid. The duration of loads associated with ice sheets and geological processes (a few thousand to several hundred megayears) is intermediate between those of seismology and the age of the Earth.

A rheology that has been widely used in glacial loading studies is that of a linear viscoelastic (Maxwell) solid. This approximation appears satisfactory because for load durations shorter than the Maxwell relaxation time the response is essentially elastic, while for longer times it becomes viscous. This model can therefore be used to explain both the seismic phenomena at short durations and viscous phenomena at longer durations.

Early studies (for example McConnell 1968) suggested, however, that in order to explain the spectral characteristics of post-glacial rebound some rheological stratification for the Earth was necessary. McConnell (1968) and Peltier (1980) considered models in which there was an upper layer that behaves elastically for the duration of loading overlying a layer of relatively low viscosity. The contribution from an upper elastic layer depends on the spatial wavelength of the applied load (see, for example, Peltier 1980). For relatively long wavelengths the relaxation time is dominated by the viscous properties of the relatively low viscosity layers. For relatively short wavelengths, however, the elastic layer suppresses the viscous response so that the load is completely supported by the upper layer.

The study of loads of small spatial wavelengths provide the best evidence for the long-term mechanical properties of the upper layers of the Earth. Because of the dependence on k^4 in the flexural models, there is a sharp cutoff between short-wavelength loads that are supported by the upper layer and longer-wavelength loads that are comparatively unaffected by the layer.

Walcott (1970*c*) studied the response of the lithosphere to geological loads longer in duration than the Pleistocene ice sheets and concluded that the lithosphere was viscoelastic rather than elastic on long time-scales (1–1000 Ma). He inferred that the lithosphere had a relaxation time in the range 10^4–10^6 years (figure 2). This rheology is consistent with the glacial loading models because for times less than the relaxation time the lithosphere would behave elastically.

We have shown in this study, however, that the overall shape and stratigraphy of sedimentary basins, which represent geological loads of 1–100 Ma on the continental lithosphere, can be satisfactorily explained by an elastic rather than a viscoelastic model. This conclusion is therefore in contradiction to the results of Walcott (1970*c*). Walcott (1970*c*) noted, however, that a viscoelastic model could not explain the anomalous flexural rigidity associated with Lake Bonneville. This load, which lasted about 10^4 years, is more characteristic of flexural rigidities associated with loads as old as 1 Ma (figure 2). Walcott (1970*c*) suggested that the 'anomalous' Lake Bonneville value could be due to its location within the Basin and Range province of the western U.S., which is characterized by high heat flow, active volcanism and shallow zones with low shear wave velocities.

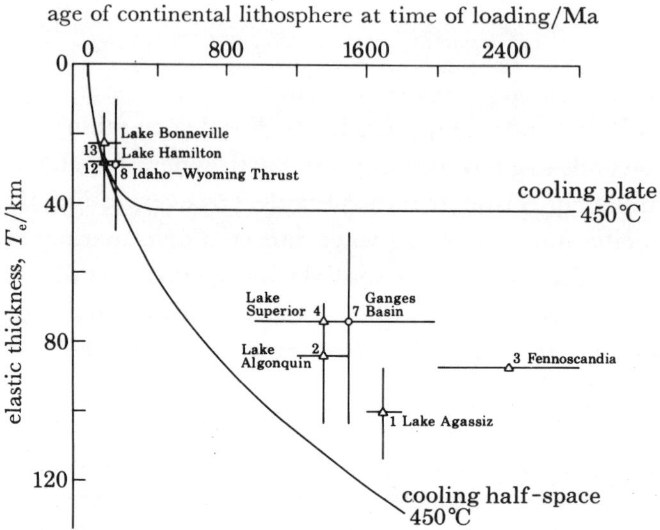

FIGURE 16. Plot of elastic thickness, T_e against age of the continental lithosphere at the time of loading. The sources of the data are summarized in table 2. Estimates 5–6, 10–11 and 14–16 have not been plotted because the age of the load is uncertain, and estimate 9 has not been plotted because the effective age of the basement is uncertain. The solid lines show for comparison the depth to the 450 °C oceanic isotherm based on a cooling plate (upper curve) and cooling half-space model (lower curve).

The low Lake Bonneville value suggests that the flexural rigidity from continents may be influenced by local variations in the thermal structure of continental lithosphere. We therefore need to consider the rigidity of the basement *at* the time of loading. Usually this is obtained by subtracting the age of the load from the age of the basement. However, subsequent thermal events, such as orogeny, basin evolution and heating associated with hot spots, may reset the basement rigidity of continents.

We have examined this possibility by plotting estimates of T_e from the continents against the estimated age of the continental lithosphere at the time of loading (figure 16). We have plotted in figure 16 only those estimates for which both the age of the load and the underlying basement are known (table 2). For example, in the Idaho–Wyoming thrust the age of the load is latest Jurassic to early Eocene (50–150 Ma) and the reset age of the underlying basement is Mississippian to Permian (225–325 Ma) (Jordan 1982). Thus the age of the continental lithosphere at the time of loading for this feature is 75–275 Ma. Furthermore, since the Pleistocene ice sheets are of relatively short duration, the age of the continental lithosphere at the time of loading for these loads is given approximately by the age of the underlying basement.

The main result shown in figure 16 is that there is an apparent increase in T_e from the relatively small values associated with thermally reset continental regions to higher values associated with the more stable interiors of the craton. The solid lines in figure 16 represent the 450 °C isotherms, based on the cooling plate and half space models, and are shown for comparison with figure 1. Comparison of figures 1 and 16 therefore suggests that T_e for the continental lithosphere increases with age in a manner similar to T_e for the oceanic lithosphere. The small values of T_e in figure 16 can be attributed to the relatively weak, hot lithosphere associated with a thermal event, while the high values of T_e can be attributed to the relatively strong, cold lithosphere associated with the stable craton.

These inferences for the thermal structure of continental lithosphere are in general agreement with the study by Sclater *et al.* (1981*b*) based on reduced heat flow estimates in the continents. They concluded that in order to explain reduced heat flow values in the continents the thermal perturbations associated, for example, with orogeny should decay at a rate similar to that of the oceans. They concluded that with small modifications the cooling plate model can, in fact, be used to predict the thermal structure of the continents.

The plot in figure 16 therefore suggests that, as well as its response to sediment loading, the mechanical behaviour of continental lithosphere may be generally similar to that of oceanic lithosphere. In addition, an elastic plate model in which T_e increases with age after a thermal event helps to explain the existance of large-amplitude long-wavelength gravity anomalies in stable Precambrian and Palaeozoic terrains, as well as the contrasts in structural styles along some orogenic belts (see, for example, Molnar & Tapponnier 1980). Further work, however, is required to determine whether these inferences of the rheology of continental lithosphere are compatible with data based on experimental rock mechanics.

(*b*) *Tectonic subsidence of sedimentary basins*

We have shown in this study that flexure is an important factor to consider in the development of a sedimentary basin. The role of flexure during basin evolution is controlled by the thermal structure of the basin and varies both as a function of time and position. For example, the cooling plate model predicts that T_e increases from near zero soon after basin formation to 30 km at 100 Ma, whereas the stretching model with $\beta = 2.0$ predicts that T_e increases from about 23 km soon after basin formation to 37 km at 100 Ma. The differences between the models arise because, in contrast to the cooling plate model, the upper lithosphere is not as extensively heated at the time of basin formation in the stretching model.

Flexure should therefore be taken into account in backstripping studies that attempt to isolate the tectonic subsidence of a sedimentary basin. For example, figure 17 shows the total sediment accumulation in the centre of the model basin (figure 13*c*) in which the lithosphere increases its flexural strength with time. The broken line shows the tectonic subsidence that would have been calculated if these sediments were unloaded by using an Airy rather than a flexure model. The solid line for $\beta = 2.0$ represents the tectonic subsidence actually assumed in the calculations. Because the Airy model neglects the effect of the lateral strength of the lithosphere, the difference between the assumed and calculated tectonic subsidence is large (figure 17). In fact, if an Airy model had been used, a best fitting $\beta = 1.6$ would have been estimated, which is significantly smaller than the $\beta = 2.0$ assumed in the calculations. The corresponding amount of extension inferred in the centre of the basin would therefore have been underestimated by about 40 %.

Sclater & Christie (1980) have estimated the amount of stretching in the North Sea Basin by using the tectonic subsidence calculated at selected wells. Although they reconstructed the stratigraphy of the basin at the wells taking into account the effects of compaction and palaeo-bathymetry, they backstripped the sediments by using an Airy rather than a flexure model. We suggest that flexure is important, even for the relatively wide North Sea Basin, and that the β values obtained by Sclater & Christie (1980) may have been significantly underestimated.

Clearly, additional data, such as fault geometry and seismic reflexion and refraction data, are required to better constrain stretching estimates of sedimentary basins that are based only on well data.

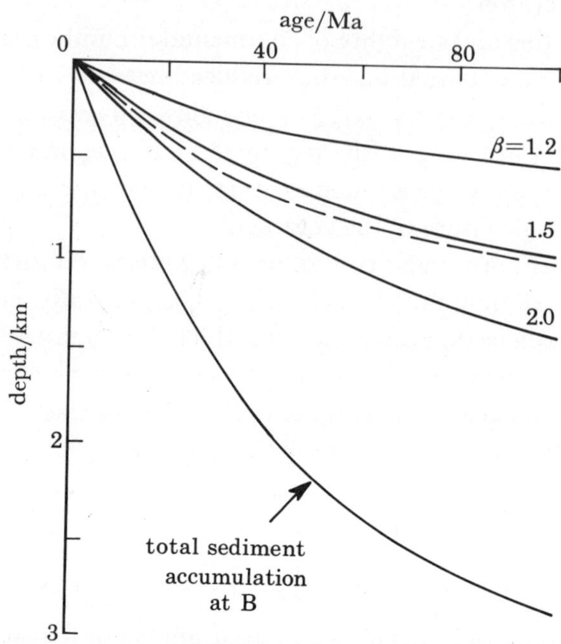

FIGURE 17. Sediment accumulation through time at B in the centre of the sedimentary basin in figure 13*c*. The solid lines indicate the subsidence that would have occurred at B if sediments had not infilled the basin. The broken line is the subsidence at B that would have been inferred from the sediments if they had been 'backstripped' by using an Airy rather than a flexure model. The best fit to the 'backstripped' curve is for $\beta = 1.6$ which is about a 40% smaller extension factor than actually assumed at B ($\beta = 2.0$).

(*c*) *Sea level changes through geological time*

The main problem in determining sea level changes through geological time is in separating tectonic from eustatic effects (see, for example, Fairbridge 1961). Relatively long-term changes of sea level have been estimated, based on changes in the volume of mid-ocean ridges through time (Hays & Pitman 1973; Pitman 1978), percentage estimates of continental flooding (Wise 1974), and tectonic processes in continental interiors (Bond 1978) and margins (Watts & Steckler 1979). These studies are in general agreement that there was a rise of sea level from the late Jurassic to the early Cretaceous and a relative fall since the late Cretaceous, although they differ in size by as much as a factor of 2 to 3. Vail *et al.* (1977) have suggested, based on the recognition of sedimentary sequences on seismic reflexion profiles, that superimposed on these long-term changes are a number of short-term fluctuations. They recognized slow rises in sea level followed by abrupt falls. The major periods of sea level rise were grouped by Vail *et al.* (1977) into supercycles, varying in age from about 5 Ma in the Tertiary to about 100 Ma in the Palaeozoic.

There is considerable debate at present about whether the supercycles identified by Vail *et al.* (1977) represent eustatic changes in sea level or widespread tectonic events. As Donovan & Jones (1979) have pointed out, the supercycles inferred by Vail *et al.* (1977) are based on seismic reflexion and well data that are largely unpublished. It is therefore difficult to determine whether

Vail *et al.* (1977) have taken into account those other factors that affect the stratigraphic record, such as compaction, palaeobathymetry and tectonic subsidence (Watts & Steckler 1979). Bally (1980) has argued, in fact, that many of the unconformities separating the cycles of Vail *et al.* (1977) may be due to major plate tectonic reorganizations rather than eustatic sea level changes.

The principal method by which Vail *et al.* 1977) and Vail & Todd (1982) estimate sea level rise in the stratigraphic record is by recognition of coastal onlap. They use seismic reflexion profiles at the edges of sedimentary basins, in conjunction with lithological and palaeontological data from nearby wells and seismic facies analysis, to estimate the vertical component of coastal onlap (coastal aggradation) for different geological time intervals. They then use a weighted average of coastal onlap estimates from various basins to infer eustatic sea level rise.

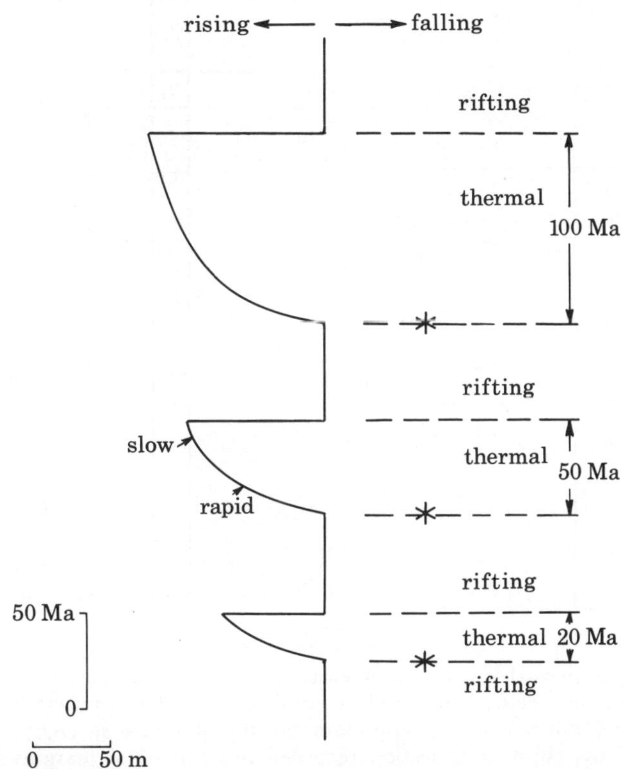

FIGURE 18. 'Apparent' sea-level changes inferred from the stratigraphy at the edge of the basin in figure 13*c*, by assuming that the progressive onlap of sediments at the edge of the basin represents a sea level rise. The changes have been calculated for assumed 20, 50 and 100 Ma durations of thermal subsidence in the basin.

We have shown in this paper that coastal onlap of the type used by Vail *et al.* (1977) and Vail & Todd (1982) to estimate sea level rise can be produced by sedimentary loading of a lithosphere that increases its flexural strength with time. For example, figure 18 shows the 'apparent' sea level rise that would be predicted from the stratigraphy at the edge of the basin in figure 13*c*, for different times after its initiation. The sea level rise is initially rapid and then slows, due to the increase of T_e with age. There is a striking resemblance between the overall shape of the apparent sea level rise and the shape of the sea level rise inferred by Vail *et al.* (1977) from the stratigraphic record.

This study suggests that some of the cycles identified by Vail *et al.* (1977) as due to a sea level rise may in fact have a tectonic rather than a eustatic control. Figure 19 compares the relative changes of sea level inferred by Vail *et al.* (1977) since the Triassic with major tectonic events associated with the break-up of the Pangaea supercontinent. There is a good correlation between the unconformities separating the supercycles with the age of the rift–drift transition recorded in the stratigraphy of continental margins formed as a result of break-up. The rift–drift transition

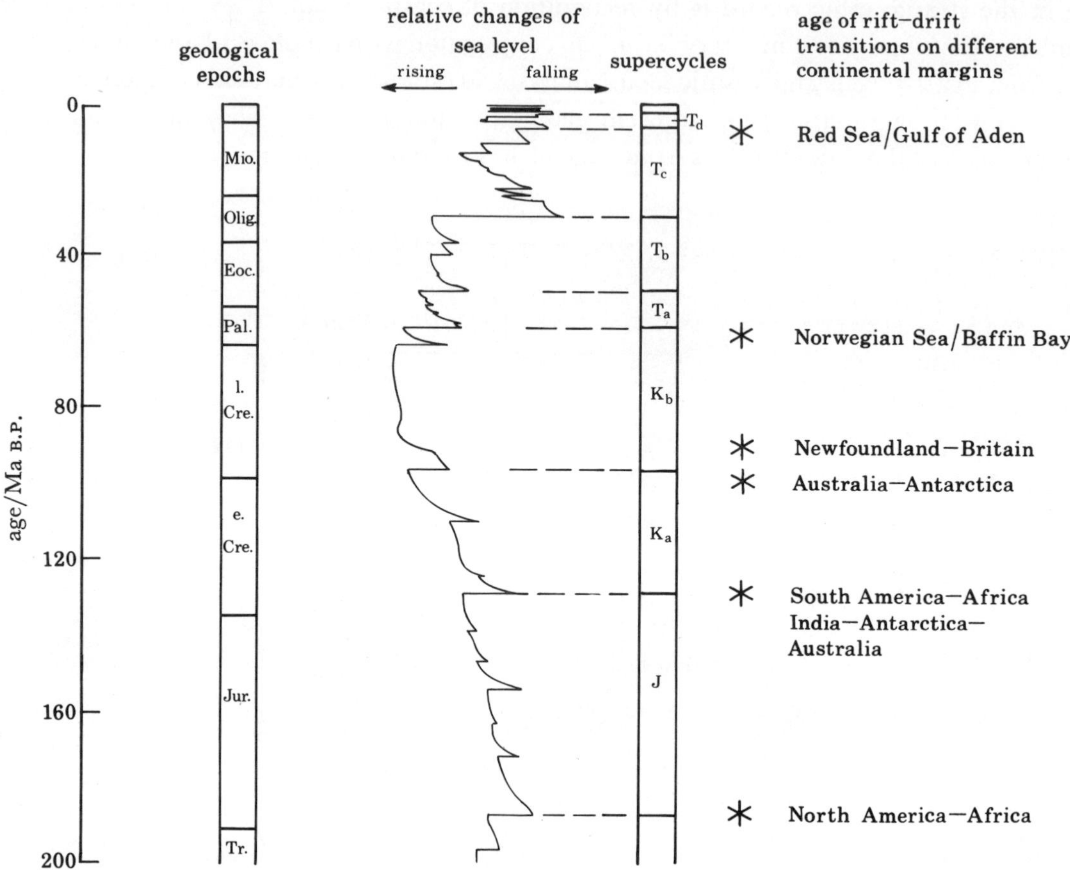

FIGURE 19. 'Relative changes in sea-level' based on Vail *et al.* (1977). A similar curve has recently been referred to by Vail & Todd (1982) as 'relative changes in coastal onlap'. Their preferred new Jurassic and early Cretaceous sea level curve, however, closely approximates that of Vail *et al.* (1977). The asterisks represent the approximate time of the rift-drift transition recorded in continental margins formed by the break-up of Pangaea. There is a good correlation between the beginning of some of the major supercycles of Vail *et al.* (1977) and major tectonic events in the ocean basins. The main exception is the beginning of supercycle T_c which, as pointed out by Vail *et al.* (1977), cannot in fact be correlated worldwide.

corresponds at most margins to the age separating fault-controlled from flexural-controlled subsidence. Thus, each margin after rifting would be expected to show coastal onlap as the lithosphere cools and increases its flexural strength with age. The supercycles identified by Vail *et al.* (1977) may therefore be widespread because the age of the rift–drift transition is similar for a number of widely separated margins (figure 19). We suggest, however, that the supercycles are not necessarily worldwide.

These considerations suggest that in order to estimate short-term variations in sea level it is necessary to consider sedimentary sequences that are not dominated by flexure, such as sediments associated with the rifting phase of continental margin basins or the oscillatory phase of interior and cratonic basins. The separation of eustatic and tectonic effects, however, will require detailed studies of the subsidence and uplift history of widely separated sedimentary basins in continental margin, interior, and cratonic basins. The data required for these studies include multichannel seismic reflexion profiles and down-hole geological and geophysical well logs. Unfortunately, many of these data are not widely available at present.

6. Conclusions

We conclude the following from this study of lithospheric flexure and the evolution of sedimentary basins.

1. The dominant mechanisms affecting basin subsidence are thermal contraction following heating and thinning of the lithosphere at the time of their formation, and sedimentary loading. Thermal contraction controls the overall shape of basin that is available for sedimentation, whereas sedimentary loading is the main control on the stratigraphy of a basin.

2. The best fit to the overall basin geometry and stratigraphy is for an elastic plate model in which the flexural strength increases with time after basin initiation as the lithosphere cools.

3. The elastic plate model predicts a number of tectonic–stratigraphic features of basins including the increase in their overall width during evolution (West Siberia, Gippsland, North Sea during the Permian, coastal plains of Atlantic-type continental margins) and the progressive onlap of younger sediments onto basement at the basin edges.

4. The elastic plate model can also explain basins in which the youngest sediments are restricted to the basin centre (Michigan, Paris, Gulf coastal plain) if the sedimentary, or flexural controlled, cycle is followed by erosion of the basin or its edges, or both.

5. The elastic plate model is consistent with observations of flexure in the region of other geological loads on the lithosphere such as seamounts and oceanic islands, ice sheets, and an oceanic or continental plate as it approaches a deep-sea trench. In the oceans the flexural flexural strength of the lithosphere increases with age away from a mid-ocean ridge crest, while in the continents, the flexural strength appears to increase with age after a thermal event.

6. The role of flexure varies as a function of both time and position during the evolution of a basin and is an important factor to consider in 'backstripping' sedimentary loads through geological time. The use of an Airy model rather than a flexure model may, for example, lead to a significant underestimate of the amount of extension across a basin.

7. The elastic plate model predicts coastal onlap at the edges of basins of a type similar to that used by Vail *et al.* (1977) to infer relative rise in sea level through geological time. Thus many of the supercycles identified by Vail *et al.* (1977) may have a flexural (or tectonic) rather than a eustatic control.

8. The boundaries of most of the supercycles of Vail *et al.* (1977) appear to correlate with major tectonic events associated with the break-up of Pangaea. Thus the supercycles identified by Vail *et al.* (1977) may be widespread because a number of widely separated continental margins are characterized by similar ages for the transition from fault-controlled to flexural-controlled subsidence, but they are unlikely to be world wide.

We thank Jim Cochran, Dan McKenzie, F.R.S., and an anonymous reviewer for their helpful comments on the manuscript. Major support for this work was provided by National Science Foundation grants numbers OCE 79-18917 and OCE 79-26308. Additional support was provided by an Australian Public Service Post-Graduate Scholarship to G.D.K. and a Phillips Petroleum Fellowship to M.S.S.

APPENDIX. TWO-DIMENSIONAL STRETCHING MODEL

In order to model accurately the subsidence and sediment accumulation in sedimentary basins, it is necessary to include horizontal as well as vertical conduction of heat in model calculations. Variations in extension across basins can produce large horizontal thermal gradients and substantial lateral heat flow, altering the temperature distribution and therefore the elastic thickness T_e.

The appropriate equation to solve is the two-dimensional heat-flow equation,

$$\frac{\partial^2 T}{\partial x^2}+\frac{\partial^2 T}{\partial z^2}=\frac{1}{\kappa}\frac{\partial T}{\partial t}, \tag{A 1}$$

where $T = T(x,z,t)$ is the temperature distribution within the lithosphere, x is distance, z is depth (positive downwards), t is time, and k is the thermal diffusivity of the lithosphere. In general, the initial vertical temperature distribution will vary as a function of x as well as z and therefore an analytic solution cannot be guaranteed. If, however, the lithosphere is divided into a series of vertical blocks with a uniform initial vertical temperature distribution within each block and varying independently from block to block, then $T(x,z,t)$ can be expressed as a product of solutions of simpler differential equations than (A 1).

Following Carslaw & Jaeger (1959, p. 33), the temperature distribution in the lithosphere due to the thermal perturbation in one block is

$$T(x,z,t) = X(x,t)\,Z(z,t) \tag{A 2}$$

where

$$\frac{\partial^2 X}{\partial x^2}=\frac{1}{\kappa}\frac{\partial X}{\partial t} \quad \text{and} \quad \frac{\partial^2 Z}{\partial z^2}=\frac{1}{\kappa}\frac{\partial Z}{\partial t}. \tag{A 3}$$

Taking the z coordinate first, we have the one-dimensional heat-flow equation, the solution of which is the simple stretching model (McKenzie 1978)

$$Z(z,t) = T_0\frac{z}{a}+\sum_{n=1}^{\infty}\frac{2}{n\pi}\left\{\frac{\beta}{n\pi}\sin\left(\frac{n\pi}{\beta}\right)\right\}\sin\left(\frac{n\pi z}{a}\right)\mathrm{e}^{-n^2\pi^2\kappa t/a^2}. \tag{A 4}$$

Equation (A 4) consists of two parts, a steady-state temperature gradient and a thermal transient due to the instantaneous stretching, $\zeta(z,t)$, which has the form

$$\zeta(z,t) = \sum_{n=1}^{\infty}\frac{2}{n\pi}\left\{\frac{\beta}{n\pi}\sin\left(\frac{n\pi}{\beta}\right)\right\}\sin\left(\frac{n\pi z}{a}\right)\mathrm{e}^{-n^2\pi^2\kappa t/a^2}. \tag{A 5}$$

In the x direction, the boundary conditions on each block are such that the lithosphere is continuous and the solution becomes (Carslaw & Jaeger 1959)

$$X(x,t) = \tfrac{1}{2}(\pi\kappa t)^{-\frac{1}{2}}\int_{-\infty}^{\infty} X(x',0)\,\mathrm{e}^{-(x-x')^2/4\kappa t}\,\mathrm{d}x'. \tag{A 6}$$

The initial conditions, $X(x',0)$, corresponding to a single block, is $X = 1$ between a and b, the limits of the block, and $X = 0$ otherwise. This gives the result

$$X(x,t) = \frac{1}{2}\left[\operatorname{erfc}\left\{\frac{x-a}{(4\kappa t)^{\frac{1}{2}}}\right\}-\operatorname{erfc}\left\{\frac{x-b}{(4\kappa t)^{\frac{1}{2}}}\right\}\right]. \tag{A 7}$$

The full solution for the temperature distribution in the lithosphere is therefore

$$T(x,z,t) = T_0\frac{z}{a}+\frac{1}{2}\left[\operatorname{erfc}\left\{\frac{x-a}{(4\kappa t)^{\frac{1}{2}}}\right\}-\operatorname{erfc}\left\{\frac{x-b}{(4\kappa t)^{\frac{1}{2}}}\right\}\right]\zeta(z,t), \tag{A 8}$$

where it is assumed that the steady-state temperature gradient is the same for the entire lithosphere. The temperature structure due to the stretching of a single block is essentially the same as (A 4), but now the transient solution spreads laterally according to the error function distribution.

Since the heat-flow equation is linear, the law of superposition can be applied to give the temporal temperature structure of lithosphere containing a number of stretched blocks. The temperature in any block is then a weighted sum of the one-dimensional solution of the temperature field of that block and all the surrounding blocks where the weighting is given by the error function

$$T(x,z,t) = T_0\frac{z}{a}+\sum_{i=1}^{n_{\text{box}}}\frac{1}{2}\left[\operatorname{erfc}\left(\frac{x-x_i}{(4\kappa t)^{\frac{1}{2}}}\right)-\operatorname{erfc}\left(\frac{x-x_{i+1}}{(4\kappa t)^{\frac{1}{2}}}\right)\right]\zeta_i(z,t), \tag{A 9}$$

where x_i and x_{i+1} are the position of edges of block i, $\zeta_i(z,t)$ is the transient temperature solution for that block, and n_{box} is the number of stretched blocks.

The thermal expansion can be calculated from the vertical integral of (A 9). Thus the thermal expansion can be computed in a similar way to the temperature structure as the solution for the one-dimensional equation diffusing laterally according to an error function.

Bibliography (Watts *et al.*)

Artemjev, M. & Artyushkov, E. V. 1971 *J. geophys Res.* **76**, 1197–1971.

Bally, A. W. 1980 In *Dynamics of plate interiors* (*Geodynamics series*, vol. 1) (ed. A. W. Bally, P. L. Bender, T. R. McGretchin & R. I. Walcott), pp. 5–10.

Bally, A. W. & Snelson, S. 1980 In *Facts and principles of world petroleum occurrence* (*Can. Soc. Petrol. Geol. Mem. no.* 6) (ed. A. Mull), pp. 9–94.

Banks, R. J., Parker, R. L. & Huestis, S. P. 1977 *Geophys. Jl R. astr. Soc.* **51**, 431–542.

Barrell, J. 1914 *J. Geol.* **22**, 425–443.

Beaumont, C. 1978 *Geophys. Jl R. astr. Soc.* **55**, 471–498.

Beaumont, C. 1981 *Geophys. Jl R. astr. Soc.* **65**, 291–329.

Beaumont, C. & Sweeney, J. F. 1978 *Tectonophysics* **50**, T19–23.

Bodine, J. H. 1981 *Lamont-Doherty Geol. Obs., tech. Rep. no.* 1, CU-1-80 (36 pages).

Bodine, J. H., Steckler, M. S. & Watts, A. B. 1981 *J. geophys Res.* **86**, 3695–3707.

Bond, G. 1978 *Geology* **6**, 247–250.

Bott, M. H. P. 1973 In *Implication of continental drift to the earth sciences* (ed. D. H. Tarling & S. K. Runcorn), vol. 2, pp. 675–683. London and New York: Academic Press.

Bott, M. H. P. & Watts, A. B. 1970 *Nature, Lond.* **225**, 265–268.

Bowie, W. 1922 *Bull. geol. Soc. Am.* **33**, 273–286.

Bureau de Recherches Géologique et Minières (B.R.G.M.) 1980 *Synthèse géologique du Bassin de Paris*, vol 2. (Atlas). *Mem. B.R.G.M.* no. 102.

Caldwell, J. G. 1979 Ph.D. thesis, Cornell University. (151 pages.)

Caldwell, J. G., Haxby, W. F., Karig, D. E. & Turcotte, D. L. 1976 *Earth planet Sci. Lett.* **31**, 239–246.

Carslaw, H. S. & Jaeger, J. C. 1959 *Conduction of heat in solids.* (510 pages.) Oxford University Press.

Cazenave, A., Lago, B., Dominti, K. & Lambeck, K. 1980 *Geophys. Jl R. astr. Soc.* **63**, 233–254.

Christie, P. A. F. & Sclater, J. G. 1980 *Nature, Lond.* **283**, 729–732.

Cochran, J. R. 1973 *Bull. geol. Soc. Am.* **84**, 3244–3268.

Cochran, J. R. 1979 *J. geophys. Res.* **84**, 4713–4729.

Cohen, T. J. & Meyer, R. P. 1976 *Geophys. Monogr.* no. 19, pp. 431–438. Washington, DC.: American Geophysical Union.

Colquhoun, D. J. & Johnson, M. S. 1968 *Palaeogeog. Palaeoclimat. Palaeoecol.* **5**, 105–126.

Cravatte, J., Dufaure, P., Prim, M. & Rouaix, S. 1974 *Compagnie fr. Pétrol. Notes Mém.* **11**, 209–274.

Crittenden, M. D., Jr. 1967 *Geophys. Jl R. astr. Soc.* **14**, 261–279.

Crittenden, M. D., Jr 1970 *Can. J. Earth Sci.* **7**, 727–729.

Collette, B. J. 1960 *The gravity field of the North Sea, Gravity Expeditions, 1948–1958*, vol. 5, pt 2. Delft: Netherlands Geodetic Commission.
de Charpal, O., Guennoc, P., Montadert, L. & Roberts, D. G. 1978 *Nature, Lond.* **275**, 706–711.
Detrick, R. S. & Watts, A. B. 1979 *J. geophys. Res.* **84**, 3637–3653.
Dickinson, W. R. & Yarborough, N. 1976 In *Plate tectonics and hydrocarbon accumulation* (*Am. Ass. Petrol. Geol., Short Course, New Orleans*), pp. 1–56.
Donovan, D. & Jones, E. J. W. 1979 *J. geol. Soc. Lond.* **136**, 187–192.
Effimof, I. & Pinezich, A. R. 1981 *Phil. Trans. R. Soc. Lond.* A **300**, 435–442.
Fairbridge, R. W. 1961 In *Physics and chemistry of the Earth* (ed. L. N. Ahrens, F. Press, K. Rankama & S. K. Runcorn), pp. 99–115. London: Pergamon Press.
Falvey, D. A. 1974 *J. Aust. Petrol. Explor. Ass.* **14**, 95–106.
Fulton, R. J. & Walcott, R. I. 1975 *Mem. geol. Soc. Am.* no. 142, 163–173.
Goetze, C. & Evans, B. 1979 *Geophys. Jl R. astr. Soc.* **59**, 463–478.
Grow, J. A., Mattrick, E. R. & Schlee, J. S. 1979 In *Geological investigations of continental margins* (*A.A.P.G. Mem.* no. 29) (ed. J. S. Watkins, L. Montadert & P. W. Dickerson), pp. 65–83.
Hanks, T. 1971 *Geophys. Jl R. astr. Soc.* **23**, 173–189.
Haxby, W. F., Turcotte, D. L. & Bird, J. M. 1976 *Tectonophysics* **36**, 57–75.
Hays, J. & Pitman, W. C. 1973 *Nature, Lond.* **246**, 18–22.
Heiskanen, W. A. & Vening Meinesz, F. A. 1958 *The Earth and its gravity field.* (470 pages.) New York: McGraw-Hill.
Hocking, J. B., 1976 In *Geology of Victoria* (ed. J. G. Douglas & J. A. Ferguson) (*Geol. Soc. Aust. spec. Publ.* no. 5).
Holtedahl, H. & Bjerkji, K. 1975 *Norg. geol. Unders.* **316**, 241–252.
Jarrard, R. D. & Turner, D. L. 1979 *J. geophys. Res.* **84**, 5691–5694.
Karner, G. D. 1981 *Eos, Wash.* **62**, 390.
Karner, G. D. & Watts, A. B. 1982 *J. geophys. Res.* (In the press.)
Keen, C. E. 1979 *Can. J. Earth Sci.* **16**, 505–522.
Kent, P. 1976 *Tectonophysics*, **36**, 87–92.
Klemme, H. D. 1975 In *Petroleum and global rectonics* (ed. A. G. Fisher & S. Judson), pp. 251–305. Princeton University Press.
Kumar, N. 1978 *Bull. Am. Ass. Petrol. Geol.* **62**, 273–294.
Lawson, A. C. 1938 *Bull. geol. Soc. Am.* **49**, 401–416.
Le Pichon, X. & Sibuet, J.-C. 1981 *J. Geophys. Res.* **86**, 3708–3720.
Loncarevic, B. C. & Ewing, G. N. 1967 *Geophysical study of the Orpheus Gravity anomaly* (*Proc. 7th World Petrol. Cong.*). (828 pages.)
McAdoo, D. C., Caldwell, J. G. & Turcotte, D. L. 1978 *Geophys. Jl R. astr. Soc.* **54**, 11–26.
McConnell, R. K., Jr 1968 *J. geophys. Res.* **73**, 7089–7105.
McGinnis, 1970 *J. geophys. Res.* **75**, 317–331.
McKenzie, D. P. 1978 *Earth planet. Sci. Lett.* **40**, 25–32.
McKenzie, D. P. & Bowin, C. 1976 *J. geophys Res.* **81**, 1903–1915.
McNutt, M. K. 1979 *J. geophys. Res.* **84**, 7589–7598.
McNutt, M. K. & Menard, M. W. 1978 *J. geophys. Res.* **83**, 1206–1212.
McNutt, M. K. & Parker, R. L. 1978 *Science, Wash.* **199**, 773–775.
Molnar, P. & Tapponnier, P. 1980 *Earth planet. Sci. Lett.* **52**, 107–114.
Middleton, M. F. 1980 *Geophys. Jl R. astr. Soc.* **62**, 1–14.
Parsons, B. & Sclater, J. G. 1977 *J. geophys. Res.* **83**, 803–827.
Peltier, W. R. 1980 In *Dynamics of plate interiors* (*Geodynamics Series*, vol. 1) (ed. A. W. Bally, P. L. Bender, R. McGetchin & R. I. Walcott), pp. 111–129.
Pitman, W. C., III 1978 *Bull. geol. Soc. Am.* **89**, 1389–1406.
Profett, J. M., Jr. 1977 *Bull. geol. Soc. Am.* **88**, 247–266.
Royden, L. & Keen, C. E. 1980 *Earth planet. Sci. Lett.* **51**, 343–361.
Royden, L., Sclater, J. G. & von Herzen, R. P. 1980 *Bull. Am. Ass. Petrol. Geol.* **64**, 173–187.
Sclater, J. G. & Christie, P. A. 1980 *J. geophys. Res.* **85**, 3711–3739.
Sclater, J. G., Royden, L., Horváth, F., Burchfiel, B. C., Semken, S. & Stegena, L. 1981*a* *Earth planet. Sci. Lett.* **51**, 139–162.
Sclater, J. G., Parsons, B. & Jaupart, C. 1981*b* (In the press.)
Sleep, N. H. 1971 *Geophys. Jl R. astr. Soc.* **24**, 325–350.
Sleep, N. H., Nunn, J. A. & Chou, L. 1980 *A. Rev. Earth planet. Sci.* **8**, 624–000.
Sleep, N. H. & Snell, N. S. 1976 *Geophys. Jl R. astr. Soc.* **45**, 125–154.
Sleep, N. H. & Sloss, L. L. 1978 *J. geophys. Res.* **83** (B12), 5815–5819.
Sloss, L. L. & Scherer, W. 1975 *Mem. geol. Soc. Am.* no. 142, 71–88.
Sloss, L. L. & Speed, N. H. 1974 In *Tectonics and sedimentation* (ed. W. R. Dickinson), pp. 98–119.
Steckler, M. S. & Watts, A. B. 1978 *Earth planet. Sci. Lett.* **41**, 1–13.
Steckler, M. S. & Watts, A. B. 1980 *Nature, Lond.* **287**, 425–529.

Suyenaga, W. 1977 Ph.D. thesis, University of Hawaii. (147 pages.)
Sweeney, J. 1976 *Tectonophysics* **36**, 181–196.
Tanner, J. J. 1966 *J. Aust. Petrol. Explor. Ass.* **6**, 116–120.
Turcotte, D. L., Ahern, J. L. & Bird, J. M. 1977 *Tectonophysics* **42**, 1–28.
Vail, P. R., Mitchum, R. M. & Thompson, S. 1977 *Mem. Am. Ass. Petrol. Geol.* no. 26, pp. 83–97
Vail, P. R. & Todd, R. G. 1982 In *Proc. Conference on Petroleum Geology of the Continental Shelf of Northwest Europe.* (In the press.)
Walcott, R. I. 1970*a* *Can. J. Earth. Sci.* **7**, 716–727.
Walcott, R. I. 1970*b* *Can. J. Earth Sci.* **7**, 931–937.
Walcott, R. I. 1970*c* *J. geophys. Res.* **75**, 3941–3954.
Walcott, R. I. 1972 *Bull. geol. Soc. Am.* **83**, 1845–1848.
Walcott, R. I. 1976 In *International Woollard Symposium* (*A.G.U. Monograph* no. 19), pp. 431–438. Washington, D.C.: American Geophysical Union.
Watts, A. B. 1978 *J. geophys. Res.* **83**, 5989–6004.
Watts, A. B. 1981 Am. Ass. Petrol. Geol., Short Course, Atlantic City, New Jersey.
Watts, A. B. Bodine, H. J. & Ribe, N. M. 1980 *Nature, Lond.* **283**, 532–537.
Watts, A. B. & Cochran, J. R. 1974 *Geophys. Jl R. astr. Soc.* **38**, 119–141.
Watts, A. B., Cochran, J. R. & Selzer, G. 1975 *J. geophys. Res.* **80**, 1391–1398.
Watts, A. B. & Ryan, W. B. F. 1976 *Tectonophysics* **36**, 25–44.
Watts, A. B. & Steckler, M. S. 1979 In *Maurice Ewing Symposium Series*, vol. 3, pp. 218–234. Washington, D.C.: American Geophysical Union.
Watts, A. B. & Steckler, M. S. 1981 *Oceanological Acta* (*Proc. of* 26*th Int. Cong., Paris*), **4**, 143–153.
Watts, A. B. & Talwani, M. 1974 *Geophys. Jl R. astr. Soc.* **36**, 57–90.
Weaver, P. 1951 *Bull. Am. Ass. Petrol. Geol.* **35**, 393–398.
Wellman, P. 1979 *B.M.R. Jl Aust. Geol. Geophys.* **4**, 373–382.
Wendt, I., Kreuzer, H., Muller, P., Von Rad, U. & Raschka, 1976 *Deep Sea Res.* **23**, 849–862.
Winnock, E., 1971 In *Histoire structurale du Golfe de Gascogne*, vol. 1 (*Collection collogues et seminaires*), p. 22. Institut Français du Pétrole, 1-1-1-30.
Wise, D. U. 1974 In *The geology of continental margins* (ed. C. A. Burke & C. L. Drake), pp. 45–58.
Wood, R. J. 1981 *Earth planet. Sci. Lett.* **54**, 306–312.
Zhabrev, I. P., Zubov, I. P., Krylov, N. A. & Semenovich, V. V. 1975 In *Proceedings of the Ninth World Petroleum Congress*, vol. 2 (Geology), pp. 83–91.
Ziegler, P. A. 1977 *Geol. J.* **1**(1), 7–32.
Ziegler, P. A. 1978 *Geologie Mijnb.* **57**, 589–626.
Ziegler, P. A. & Louwerens, C. J. 1977 *Acta Univ. Ups. Symp.*, vol. 2, pp. 7–22.

Phil. Trans. R. Soc. Lond. A **305**, 283–294 (1982) [283]
Printed in Great Britain

Thermal mechanisms of basin formation

By D. L. Turcotte and C. L. Angevine
Department of Geological Sciences, Cornell University, Ithaca, New York 14853, *U.S.A.*

Thermal subsidence of the sea floor explains the observed bathymetry of ocean ridges. A similarity solution for a one-dimensional cooling model successfully predicts bathymetry, heat flow and geoid anomalies under a wide range of conditions. This similarity solution can be modified to predict the thermal subsidence of sedimentary basins. For older sedimentary basins it is necessary to consider an input of heat to the base of the lithosphere that places a limit on subsidence. The similarity solution for thermal subsidence is in quite good agreement with the observed subsidence history of a variety of sedimentary basins. Some basins subside freely and in others the flexural rigidity of the elastic lithosphere inhibits subsidence.

An empirical model is proposed for the conversion of kerogen to oil and the subsequent conversion of oil to gas. This model is then used in conjunction with the thermal evolution predicted by the similarity solution in order to determine the oil window and relative volume of oil as a function of the age of the basin.

Introduction

Sedimentary basins can be formed in a variety of ways. The subsidence history of many sedimentary basins appears to be consistent with a thermal time constant (Sleep 1971; Sleep & Snell 1976). As the lithosphere cools and thickens, the lithospheric rocks become more dense owing to thermal contraction. Isostatic subsidence follows and a sedimentary basin is formed. The time constant for thermal subsidence is of the order of 50 Ma. For smaller basins the flexural rigidity of the elastic lithosphere restricts the subsidence.

If the lithosphere initially has zero thickness, a similarity solution for the thermal structure of the lithosphere and sedimentary basin as well as the subsidence of the basin can be obtained (Turcotte & Ahern 1977). The primary advantage of a similarity solution is that an analytic solution is obtained. This similarity solution has been used to predict the thermal subsidence of the Los Angeles Basin (Turcotte & McAdoo 1979) and the Baltimore Canyon Trough (Angevine & Turcotte 1981) and is in quite good agreement with observations.

Lithospheric thinning must precede thermal subsidence. One mechanism for lithospheric thinning is the diapiric replacement of cold lithospheric rock by hot mantle rock. A variation on this hypothesis is the convective heating of the lithosphere due to the upward migration of magma (Lachenbruch 1978). An alternative hypothesis is lithospheric delamination (Bird 1979). In this hypothesis the lower part of the dense lithosphere breaks away and founders into the asthenosphere. The cold rocks of the delaminated lithosphere are replaced by hot asthenospheric rocks.

Lithospheric thinning would be expected to lead to crustal uplift through isostasy. Continental domes such as the Ethiopian Swell and the East African Swell on the East African Rift system, the Tibesti uplift, and the uplift associated with the Rhine Graben are attributed to this mechanism.

In many cases lithospheric thinning is associated with continental rifting. The horizontal strain associated with rifting can lead to crustal thinning (Bott 1976; McKenzie 1978; Jarvis & McKenzie 1980). Crustal thinning results in isostatic subsidence and the formation of a sedimentary basin. Subsequently, thermal subsidence can occur. Partial rifting can lead to a sedimentary basin within a continent such as the North Sea Basin. If rifting leads to the formation of an ocean basin, sedimentary basins can form on the passive continental margins. There are a number of examples along the margins of the Atlantic Ocean (Keen 1979).

In some cratonic basins there is no evidence for tectonic thinning or thermal uplift before the initiation of subsidence, even though the subsidence exhibits a thermal time constant. An example is the Michigan Basin. The subsidence of this basin requires an increase in density of the rocks within the elastic lithosphere. One hypothesis that explains this increase in density is that a phase change is occurring in the deep crustal rocks; an example of such a phase change is the transformation of basalt to eclogite (Joyner 1967; Haxby *et al.* 1976).

One purpose for studying the subsidence and thermal structure of sedimentary basins is to determine the location of oil in a sedimentary basin. If the thermal history is known, the conversion of kerogen to oil and the subsequent conversion of oil to gas can be studied. We first consider this problem and then introduce several similarity solutions.

Petroleum generation

In the upper few metres of subsiding sedimentary basins, organic molecules such as lignin, carbohydrates and proteins are broken down as a result of microbial action. The remaining portion of organic sediment undergoes a chemical condensation that produces molecules of high molecular mass. With increasing depth these macromolecules become even more insoluble; the end result is the petroleum precursor kerogen. As temperature increases progressively with depth the molecular structure of kerogen is altered, giving off medium and high mass hydrocarbons or oil in the process (Tissot & Welte 1978).

Tissot & Espitalie (1975) proposed a quantitative model for the rate of thermal alteration of kerogen to petroleum (petroleum includes both oil and gas):

$$\frac{\mathrm{d}n_{\mathrm{k}i}}{\mathrm{d}t} = -n_{\mathrm{k}i} A_i \exp\left(-\frac{E_i}{RT}\right) \quad i = 1, 2, \ldots, 6, \tag{1}$$

where $n_{\mathrm{k}i}$ is the mass fraction of the ith kerogen, E_i is its activation energy and A_i is a pre-exponential factor, R is the universal gas constant and T is thermodynamic temperature. The mass fraction of petroleum created at a particular horizon is

$$n_{\mathrm{p}} = n_{\mathrm{p0}} + \sum_{i=1}^{6} (n_{\mathrm{k}i0} - n_{\mathrm{k}i}) \tag{2}$$

where n_{p0} represents the fraction of petroleum due to microbial action (such as the synthesis of lipids) and $n_{\mathrm{k}i0}$ is the mass fraction of the ith kerogen that was present initially. Values of $n_{\mathrm{k}i0}$, A_i and E_i are given in table 1 for the six type I kerogens of Tissot & Espitalie 1975); they also give $n_{\mathrm{p0}} = 0.05$. This model has been used with good results to estimate the thermal maturity of the Los Angeles Basin (Turcotte & McAdoo 1978) and the Baltimore Canyon Trough (Angevine & Turcotte 1981).

If the altered kerogens and their liquid petroleum by-products are buried to even greater temperatures, carbon–carbon bonds will crack to produce gas. Oil is destroyed during the cracking process. We have constructed a simple model based on a rate equation of the form (1) to account for the generation of oil and its subsequent destruction. Figure 1 illustrates the conversion

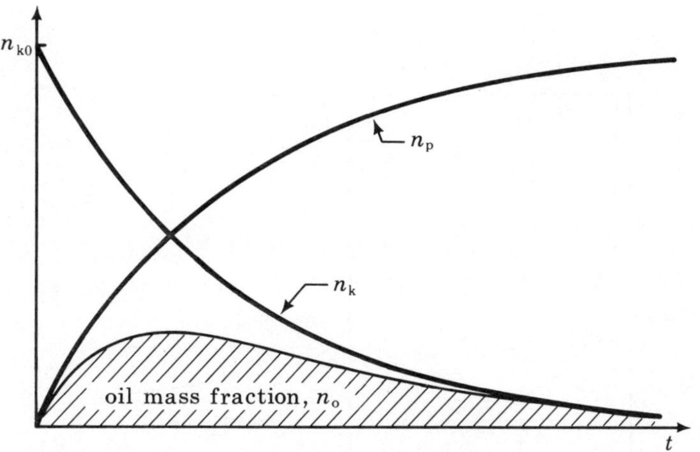

FIGURE 1. Mass fractions of kerogen, petroleum and gas as a function of time.

TABLE 1. PROPERTIES OF TYPE I KEROGEN

(From Tissot & Espitalié (1975).)

i	$\frac{E}{\text{kJ mol}^{-1}}$	$\frac{A_i}{\text{Ma}^{-1}}$	n_{ki0}
1	42	4.75×10^{4}	0.024
2	125	3.04×10^{16}	0.064
3	209	2.28×10^{25}	0.136
4	251	3.98×10^{30}	0.152
5	293	4.47×10^{31}	0.347
6	335	1.10×10^{34}	0.172

of kerogen to oil and gas at a single sedimentary horizon. We assume that the mass fraction of kerogen decreases at the rate

$$\frac{dn_k}{dt} = -n_k A \exp\left(-\frac{E}{RT}\right), \tag{3}$$

while the petroleum mass fraction increases at the rate

$$\frac{dn_p}{dt} = -\frac{dn_k}{dt}, \tag{4}$$

where n_k and n_p are related by

$$n_k + n_p = n_{k0}. \tag{5}$$

We further assume that the mass fraction of oil n_o present at any time is given by

$$n_o = n_k n_p / n_{k0}. \tag{6}$$

This is an empirical relation that is consistent with the thermally activated conversion of kerogen to oil and the subsequent cracking of oil to produce gas. Substituting (5) into (6) yields

$$n_o = n_k(n_{k0} - n_k)/_{k0}. \tag{7}$$

The rate of petroleum formation may be obtained by differentiating (7) with respect to time and substituting for dn_k/dt from (3)

$$\frac{dn_o}{dt} = \left(2\frac{n_k}{n_{k0}} - 1\right) n_k A \exp\left(-\frac{E}{RT}\right). \tag{8}$$

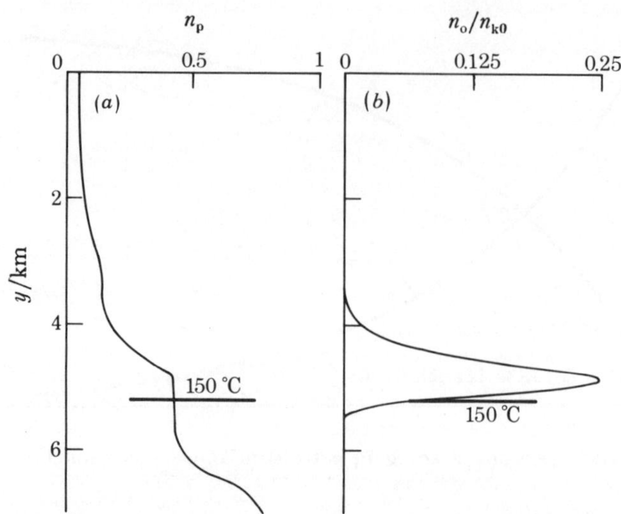

FIGURE 2. (a) Mass fraction of petroleum as a function of depth after 50 Ma with $T_0 = 20$ °C and a constant thermal gradient $\mathrm{d}T/\mathrm{d}y = 25$ K km^{-1} from (1) and (2) and values given in table 1. (b) Corresponding mass fraction of oil from (8) with $A = 1.3 \times 10^7$ Ma^{-1} and $E = 230$ kJ mol^{-1}.

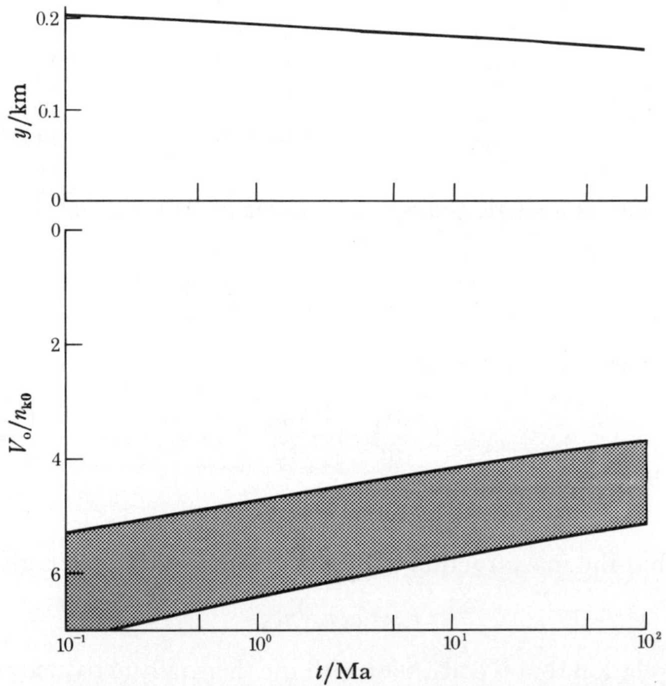

FIGURE 3. The oil window and volume of oil as a function of age in a sedimentary basin with $T_0 = 20$ °C and a constant thermal gradient $\mathrm{d}T/\mathrm{d}y = 25$ K km^{-1}.

If the constants A and E of the rate equation are known, then (7) may be evaluated over the depth of the sedimentary basin to determine the position of the oil window.

To calibrate our petroleum model with that of Tissot & Espitalie we consider a 10 km thick sequence of sediment that has been exposed to a constant thermal gradient of 25 K km^{-1} and a surface temperature of 20 °C for 50 Ma. Figure 2a shows the mass fraction of kerogen converted to petroleum as a function of depth from (1) and (2). The 150 °C isotherm is thought to be the oil limit, that is the temperature above which oil cannot survive for long periods. For this thermal history the zone of maximum petroleum production lies between 3.8 and 5.2 km in depth. In comparison, figure 2*b* shows the oil window predicted by our model. Here we take the rate constants to be $A = 1.3 \times 10^{27}\ \mathrm{Ma}^{-1}$ and $E = 230\ \mathrm{kJ\ mol}^{-1}$; these values will be used in all subsequent calculations. The bulk of production occurs between the depths of 4.0 and 5.2 km. It is possible to calculate the area under this curve and to determine the volume of oil in the sediment. The volume obtained is only important in a relative sense because the volume of organic material in the sediment is generally not known. Also, we do not consider the upward or lateral migration of oil. A significant fraction of the oil produced may be lost to the surface.

The position of the oil window and the volume of oil in the producing zone depends upon the thermal history of the sediments. Before proceeding to more complicated thermal histories it is worthwhile to understand how our oil model behaves in a simple situation. Once again we consider a 10 km thickness of sediment. The thermal gradient is held at $25\ \mathrm{K\ km}^{-1}$ while the surface temperature is fixed at 20 °C. Figure 3 shows the oil volume and location of the oil window as a function of time. The oil window is defined by the condition $n_o/n_{ko} > 0.01$ and the volume of oil V_o is given by

$$V_o = \int_0^\infty (n_o/n_{ko})\,\mathrm{d}y. \tag{9}$$

Time is plotted on a logarithmic scale between 0.1 and 100 Ma.

Depth to the oil window (the stippled region) decreases rapidly in the first 10 Ma, and more gradually afterwards. As shown in the upper part of figure 3, the petroleum volume also increases at first rapidly and then more slowly. Below the 150 °C isotherm, at depths greater than 5.2 km, petroleum is rapidly created and then destroyed. Above this limit more time is needed to create petroleum but its subsequent destruction occurs much more slowly.

Since it is unlikely that 6 km of sediments will be deposited in 100 000 years the production of petroleum in young sedimentary basins is likely to be dominated by the time-dependent history of sedimentation and the thermal history of the basin. In the next section we shall develop a similarity solution for this problem.

Similarity solution

A similarity solution for the subsidence of sedimentary basins can be obtained if several assumptions and approximations are made. These include the following.

(i) It is assumed that initially at $t = 0$ the temperature of the basement rock is a constant T_m and that a constant melt fraction χ_m is present. This is equivalent to assuming that a uniform asthenosphere reaches the surface.

(ii) It is assumed that sediments fill the basin caused by subsidence. At $t = 0$ basement rock fills the region $y > 0$. The position of basement as a function of time is $y_b(t)$. Sediments of density ρ_s fill the region $0 < y < y_b$. The model is illustrated in figure 4. In order to simplify the analysis we further assume that the physical properties of the sediments and basement rock are constant, and we neglect compaction. These assumptions are not necessary in order to obtain a similarity solution.

As the lithosphere cools and thickens, the original surface subsides owing to isostatic subsidence.

The surface is covered by a thickening layer of sediments. The temperature within the sediments satisfies the equation (Carslaw & Jaeger 1959, p. 148)

$$\frac{\partial T_s}{\partial t}+v\frac{\partial T_s}{\partial y}=\kappa\frac{\partial^2 T_s}{\partial y^2}, \tag{10}$$

where v is the subsidence velocity and κ is the thermal diffusivity. The temperature within the basement satisfies the equation

$$\frac{\partial T_b}{\partial t}+v\frac{\partial T_b}{\partial y}=\kappa\frac{\partial^2 T_b}{\partial y^2}. \tag{11}$$

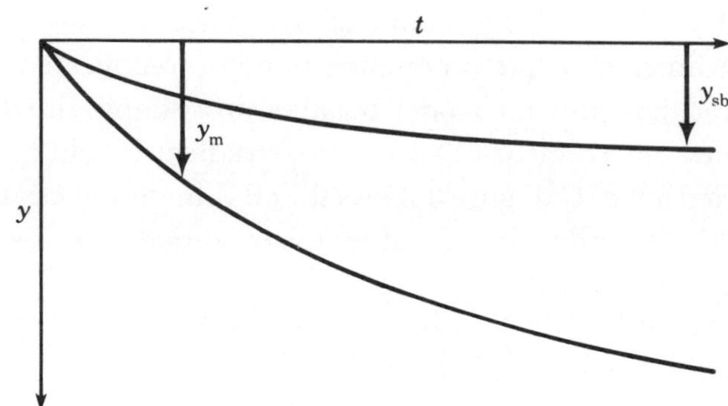

FIGURE 4. Illustration of the similarity model for thermal subsidence.

The boundary condition on the temperature at the surface is $T = T_0$. The temperature and heat flow must be continuous at the sediment–basement interface. Thus we have

$$T_s = T_b \quad \text{and} \quad \frac{\partial T_s}{\partial y}=\frac{\partial T_b}{\partial y} \quad \text{at} \quad y = y_{sb}. \tag{12}$$

The base of the lithosphere is a solidication front migrating through the asthenosphere (Oldenburg 1975). The heat flux at the lithosphere–asthenosphere boundary must match the heat produced by the solidification. Thus we have

$$T_b = T_m, \; \chi_m \rho_m L\frac{dy_m}{dt}=k\frac{\partial T_b}{\partial y} \quad \text{at} \quad y=y_m, \tag{13}$$

where χ_m is the melt fraction present in the asthenosphere.

The governing equations are reduced to total differential equations by introducing the similarity variable

$$\eta = y/(\kappa t)^{\frac{1}{2}}. \tag{14}$$

Substitution of (14) into (10) and (11) gives

$$(\lambda_s - \tfrac{1}{2}\eta)\frac{dT_s}{d\eta}=\frac{d^2T_s}{d\eta^2} \tag{15}$$

and

$$(\lambda_s - \tfrac{1}{2}\eta)\frac{dT_b}{d\eta}=\frac{d^2T_b}{d\eta^2}. \tag{16}$$

In order to obtain a similarity solution we further require that

$$y_{sb} = 2\lambda_s(\kappa t)^{\frac{1}{2}}, \tag{17}$$

where λ_s is a constant that must be determined. The velocity of subsidence is

$$v = dy_{sb}/dt = \lambda_s(\kappa/t)^{\frac{1}{2}}. \tag{18}$$

We further require that

$$y_m = 2\lambda_m(\kappa t)^{\frac{1}{2}}, \tag{19}$$

where λ_m is a constant that must be determined.

The solution of (15) that satisfies the boundary condition $T_s = T_0$ at $\eta = 0$ is

$$T_s - T_0 = C\{\operatorname{erf}\lambda_s - \operatorname{erf}(\lambda_s - \tfrac{1}{2}\eta)\}. \tag{20}$$

The solution of (16) that satisfies the boundary condition $T_b = T_m$ at $\eta = 2\lambda_m(y = y_m)$ is

$$T_m - T_b = B\{1 - \operatorname{erf}(\tfrac{1}{2}\eta - \lambda_s)/\operatorname{erf}(\lambda_m - \lambda_s)\}. \tag{21}$$

In order to satisfy the matching condition on temperature at $\eta = 2\lambda_s(y = y_s)$, we require

$$T_0 + C\operatorname{erf}\lambda_s = T_m - B. \tag{22}$$

In order to satisfy the matching condition on the heat flux at $\eta = 2\lambda_s$ $(y = y_s)$ we require

$$C = B/\operatorname{erf}(\lambda_m - \lambda_s). \tag{23}$$

Substitution of (22) and (23) into (20) and (21) gives

$$\frac{T_s - T_0}{T_m - T_0} = \frac{\operatorname{erf}\lambda_s - \operatorname{erf}(\lambda_s - \frac{1}{2}\eta)}{\operatorname{erf}(\lambda_m - \lambda_s) + \operatorname{erf}\lambda_s} \tag{24}$$

and

$$\frac{T_m - T_b}{T_m - T_0} = \frac{\operatorname{erf}(\lambda_s - \lambda_s) - \operatorname{erf}(\frac{1}{2}\eta - \lambda_s)}{\operatorname{erf}(\lambda_m - \lambda_s) + \operatorname{erf}\lambda_s}. \tag{25}$$

We next satisfy the condition on the heat flux at $\eta = 2\lambda_m(y = y_m)$ given in (13) and from (14), (19) and (25) we obtain

$$\frac{\chi_m \pi^{\frac{1}{2}}}{c_p(T_m - T_0)} = \frac{e^{-(\lambda_m - \lambda_s)^2}}{\lambda_m[\operatorname{erf}(\lambda_m - \lambda_s) + \operatorname{erf}\lambda_s]}. \tag{26}$$

In order to complete the formulation of the problem we must consider the isostatic subsidence of the sedimentary basin. Assuming that $\rho = \rho_m$ at $t = 0$, the condition of isostasy requires that

$$\int_0^\infty (\rho - \rho_m)\, dy = 0. \tag{27}$$

For our problem this can be written

$$\int_0^{y_b} (\rho_m - \rho_s)\, dy = \int_{y_b}^{y_m} (\rho_b - \rho_m)\, dy. \tag{28}$$

The densities of the sediments, basement rocks and mantle are given by

$$\rho_s = \rho_{s0}[1 - \alpha(T_s - T_0)], \tag{29}$$

$$\rho_b = \rho_{m0}[1 + \alpha(T_m - T_b)] \tag{30}$$

and

$$\rho_m = \rho_{m0}(1 - \chi_m) + \rho_1 \chi_m, \tag{31}$$

where ρ_{s0} is the surface density of the sediments, ρ_{m0} is the density of basement rocks at $T = T_m$, and ρ_l is the density of the magma fraction in the asthenosphere. Substitution of (29) to (31) into (28) gives

$$(\rho_{m0}-\rho_{s0})\, y_b + \alpha\rho_{s0} \int_0^{y_{sb}} (T_s - T_0)\, dy = (\rho_m - \rho_l)\, \chi_m y_m + \alpha\rho_{m0} \int_{y_b}^{y_m} (T_m - T_b)\, dy; \qquad (32)$$

substitution of (24) and (25) yields

$$(\rho_{m0}-\rho_{s0})\, \lambda_s - (\rho_m - \rho_l)\, \chi_m \lambda_m = \frac{\alpha(T_m - T_0)}{\pi^{\frac{1}{2}}} \frac{\{\rho_{m0}(1-e^{-(\lambda_m-\lambda_s)^2}) - \rho_{s0}(1-e^{-\lambda_s^2})\}}{\{\operatorname{erf}(\lambda_m - \lambda_s) + \operatorname{erf}\lambda_s\}}. \qquad (33)$$

Values of λ_s and λ_m are obtained by solving the coupled transcendental equations (26) and (33).

If the solidification of the asthenosphere is neglected, it is appropriate to assume that $\lambda_m \to \infty$. This approximation is usually made when considering the thermal structure of the lithosphere. In addition, if the depth of the sedimentary basin is small compared with the thickness of the thermal lithosphere, it is appropriate to assume that $\lambda_s \ll 1$. In these two limits the above results simplify considerably. The depth of the sedimentary basin is

$$y_{sb} = \frac{2\rho_{m0}\,\alpha(T_m - T_0)}{(\rho_{m0}-\rho_{s0})} \left(\frac{\kappa t}{\pi}\right)^{\frac{1}{2}}. \qquad (34)$$

Neglecting compaction, the depth of a sedimentary layer deposited at time t_s after the initiation of subsidence can be obtained as a function of time. At time t_s the depth of basement is obtained from (34) by setting $t = t_s$. The sediment deposited at time t_s will always be this distance above basement. However, the depth of the basement at time t is given directly by (34). Thus the depth to the sediment deposited at time t_s at a later time t, denoted by y_s, is given by the difference between the depth to basement at t and t_s,

$$y_s = \frac{2\rho_{m0}\,\alpha(T_m - T_0)}{(\rho_{m0}-\rho_{s0})} \left(\frac{\kappa}{\pi}\right)^{\frac{1}{2}} (t^{\frac{1}{2}} - t_s^{\frac{1}{2}}). \qquad (35)$$

Turcotte & McAdoo (1979) have shown that this subsidence relation is in good agreement with the subsidence record in the southwestern block of the Los Angeles Basin.

As long as the thickness of sediments is small, the presence of the sediments has a small influence on the temperature distribution in the lithosphere. Also, the temperature distribution in the sediments can be taken to be linear. Thus the temperature distribution in the sediments is

$$T_s = T_0 + \frac{(T_m - T_0)}{(\pi\kappa t)^{\frac{1}{2}}}\, y. \qquad (36)$$

The temperature of a sedimentary layer deposited at time t_s at a subsequent time t is obtained by substituting (34) in (35) with the result

$$T_s = T_0 + \frac{2}{\pi} \frac{\rho_{m0}\,\alpha(T_m - T_0)^2}{(\rho_{m0}-\rho_{s0})} \left\{1 - \left(\frac{t_s}{t}\right)^{\frac{1}{2}}\right\}. \qquad (37)$$

This result along with (8), can be used to determine the location of petroleum in a sedimentary basin for which the similarity solution is valid.

We consider the development of oil in a sedimentary basin for which the similarity solution is valid. Taking $T_0 = 20\,°C$, $T_m = 1200\,°C$, $\alpha = 3\times10^{-5}\,K^{-1}$, $\rho_{m0} = 3300\,kg\,m^{-3}$, $\rho_s = 2500\,kg\,m^{-3}$, $A = 1.3\times10^{27}\,Ma^{-1}$, and $E = 230\,kJ\,mol^{-1}$, the oil window and the volume of oil as a function of age are given in figure 5. The volume of oil increases with age and the oil window deepens. It should be emphasized that kerogens must be present for the production of oil and that oil will often migrate upwards in the section.

The results obtained above predict that the depths of sedimentary basins increase with the square root of time. For many sedimentary basins this is a good approximation for about the first 80 Ma of their evolution. The depth of older sedimentary basins appears to remain approxi-

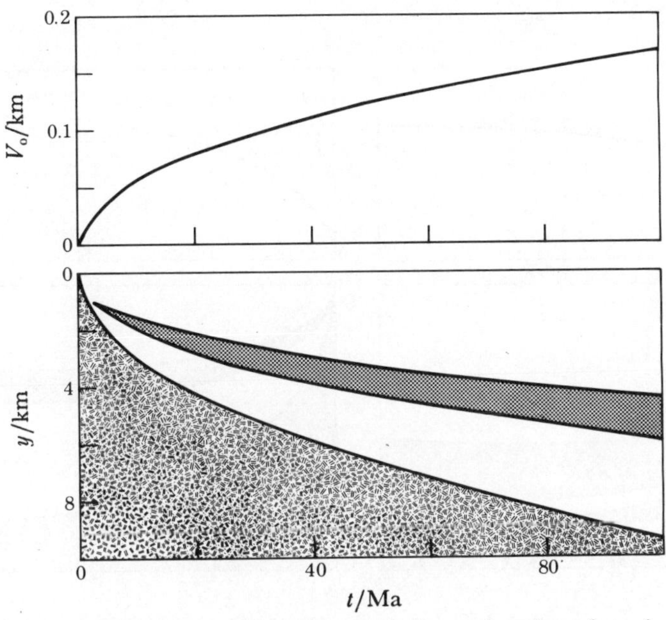

FIGURE 5. The oil window and the volume of oil as a function of age in a sedimentary basin for which the similarity solution (37) is valid.

mately constant. This deviation from the cooling model given above is probably due to the input of heat to the base of the thermal lithosphere. Possible causes of this heating are mantle convection beneath the rigid lithosphere or frictional heating. An empirical equation that predicts the flattening of the subsidence curve with time has been proposed by Turcotte (1980) and has the form

$$y_{sb} = y_{sbo}(1 - e^{-(t/\tau)^2})^{\frac{1}{4}}, \tag{38}$$

where y_{sbo} is the depth of the sedimentary basin at large times and τ is a characteristic subsidence time. For small times $t \ll \tau$ this reduces to (34) if

$$\tau = \frac{\pi}{\kappa}\left\{\frac{y_{sbo}(\rho_{mo} - \rho_{so})}{2\rho_{mo}\alpha(T_m - T_0)}\right\}^2. \tag{39}$$

The depth to sediments deposited at time t_0 is given by

$$y_s = y_{sbo}\{(1 - e^{-(t/\tau)^2})^{\frac{1}{4}} - (1 - e^{-(t_s/\tau)^2})^{\frac{1}{4}}\}. \tag{40}$$

Turcotte (1980) has shown that this semi-empirical relation predicts with reasonable success the subsidence record of the Michigan, Appalachian, Atlantic Coast and Gulf Coast sedimentary basins.

The near-surface thermal gradient β in the basin can be modelled in a similar way by assuming that

$$\beta = \beta_0(1 - e^{-(t/\tau)^2})^{-\frac{1}{4}} \tag{41}$$

where β_0 is the thermal gradient at large times. The temperature distribution in the sediments is

$$T_s = T_0 + \beta_0 y(1 - e^{-(t/\tau)^2})^{-\frac{1}{4}}. \tag{42}$$

The temperature of a sedimentary layer deposited at t_s at a subsequent time t is obtained by substituting (39) into (41) with the result

$$T_s = T_0 + \beta_0 y_{sbo} \left\{1 - \frac{(1 - e^{-(t_s/\tau)^2})^{\frac{1}{4}}}{(1 - e^{-(t/\tau)^2})^{\frac{1}{4}}}\right\}. \tag{43}$$

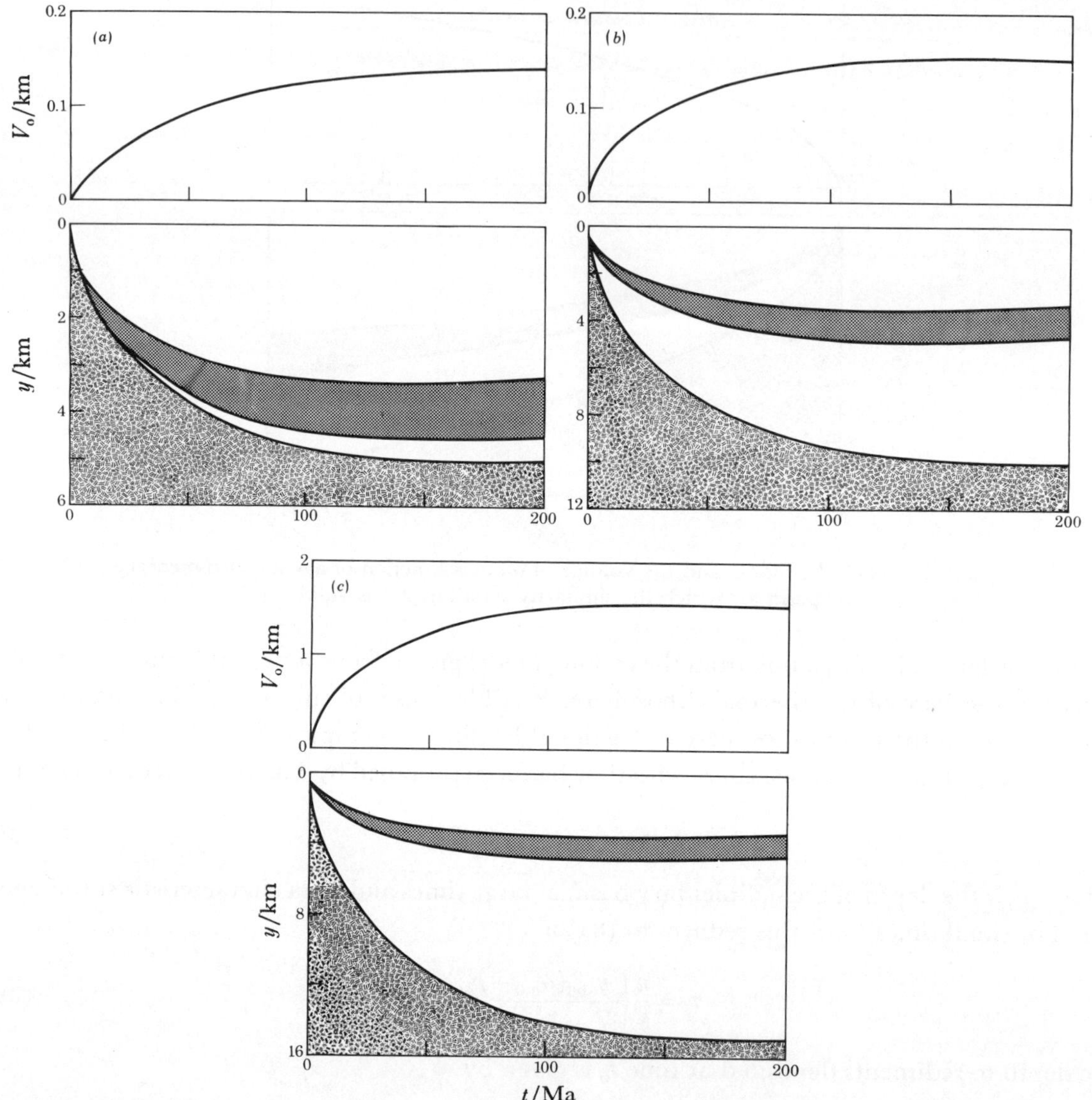

FIGURE 6. The oil window and the volume of oil as a function of age in a sedimentary basin for which the empirical thermal subsidence model (42) is applicable. (*a*) y_{sbo} = 5 km, (*b*) y_{sbo} = 10 km, (*c*) y_{sbo} = 15 km.

This result can be used along with (8) to determine the location of petroleum in a sedimentary basin for which the empirical thermal subsidence model is applicable.

The oil window and volume of oil are given as a function of age in figure 6 for $\beta_0 = 25\,\text{K}\,\text{km}^{-1}$, $\tau = 80\,\text{Ma}$, $A = 1.3 \times 10^{27}\,\text{Ma}^{-1}$, $E = 230\,\text{kJ}\,\text{mol}^{-1}$ and $y_{sbo} = 5$, 10 and 15 km. For the deep basins half the oil is produced in about the first 20 Ma whereas in the shallow basin (5 km) half of the oil is produced after about 30 Ma. The total amount of oil produced in each of the basins is about the same and the depth to the oil window as a function of age is about the same.

Conclusions

The direct applicability of the similarity solution for thermal subsidence requires a number of assumptions. However, experience with similarity solutions in other applications indicates that the results are relatively insensitive to some of the requirements. For example, the condition of constant temperature at $t = 0$ may be a rather poor approximation and the similarity solutions will still be a good approximation at large times.

For many sedimentary basins a significant fraction of basin subsidence may not be of thermal origin. For example, oceanic crust is normally formed at a depth of about 2.5 km. If this crust is formed near a continental margin with high sedimentation rates, 5 km or more of sediments may be deposited in a very short period. As the oceanic lithosphere subsequently cools, the similarity solution would be expected to be a satisfactory approximation for later sedimentation. If sedimentation does not keep pace with subsidence, deep water can develop and the sedimentary record is irregular. Also, changes in sea level can have a significant effect on the rate at which sediments accumulate.

This research was supported by the Earth Sciences Section, National Science Foundation, under grant no. EAR-7919414.

References (Turcotte & Angevine)

Angevine, C. L. & Turcotte, D. L. 1981 *Bull. Am. Ass. Petrol. Geol.* **65**, 219–225.
Bird, P. 1979 *J. geophys. Res.* **84**, 7561–7571.
Bott, M. H. P. 1976 *Tectonophysics* **36**, 77–86.
Carslaw, H. S. & Jaeger, J. C. 1959 *Conduction of heat in solids*, 2nd edn. Oxford University Press.
Haxby, W. F., Turcotte, D. L. & Bird, J. M. 1976 *Tectonophysics* **36**, 57–75.
Jarvis, G. T. & McKenzie, D. P. 1980 *Earth planet. Sci. Lett.* **48**, 42–52.
Joyner, W. B. 1967 *J. geophys. Res.* **72**, 4977–4998.
Keen, C. E. 1979 *Can. J. Earth Sci.* **16**, 505–522.
Lachenbruch, A. H. 1978 *Pure appl. Geophys.* **117**, 34–50.
McKenzie, D. 1978 *Earth planet. Sci. Lett.* **40**, 25–32.
Oldenburg, D. W. 1975 *Geophys. Jl R. astr. Soc.* **43**, 425–451.
Sleep, N. H. 1971 *Geophys. Jl R. astr. Soc.* **24**, 325–350.
Sleep, N. H. & Snell, N. S. 1976 *Geophys. Jl R. astr. Soc.* **45**, 125–154.
Tissot, B. & Espitalié, J. 1975 *Inst. fr. Pétrole Revue* **30**, 743–777.
Tissot, B. & Welte, D. 1978 *Petroleum formation and occurrence.* New York: Springer-Verlag.
Turcotte, D. L. 1980 In *Dynamics of plate interiors* (*Geodynamics Series*, vol. 1), pp. 21–26. Washington, D.C.: American Geophysical Union.
Turcotte, D. L. & Ahern, J. L. 1977 *J. geophys. Res.* **82**, 3762–3766.
Turcotte, D. L. & McAdoo, D. C. 1979 *J. geophys. Res.* **84**, 3460–3464.

Discussion

G. Karner. In the oceans, the mass that is loading the oceanic plate appears to be more easily recognized, for example seamounts and their respective moat infill material. However, within the continents, it is not obvious what constitutes the load acting on the continental plate. In relation to the Appalachian Basin, is it possible that the load giving rise to the foreland basin is within either the basin or crust?

D. L. TURCOTTE. As Beaumont argued earlier, the best method of detecting loads is to use the gravity observations. Over the Appalachians there is a gravity low and high, the low corresponding to the flexure and the high occurring towards the east. But the load involved is small and can only account for perhaps 30 % of the observed subsidence that produced the basin. In the Michigan Basin there is also a gravity high along what is believed to have been a Keweenawan Rift, flanked by lows on either side. But here also neither the high nor the flanking lows are large enough to account for the subsidence. In both these areas there must be some process other than loading involved, such as metamorphic reactions in the crust.

G. DUNGWORTH. Dewey and Watts and the authors have all used Lopatin's expression to relate the thermal history of a source bed to its maturation. This relation requires the reaction rate to double for every 10 K increase in temperature. In some of the figures in this paper this relation leads to a rather restricted depth range in which oil generation can occur. The authors also mentioned that they used Tissot's expression, with a pre-exponential factor and an activation energy. Of Tissot's three kerogen types only type 2 is relevant, and those reactions, which have activation energies of 42 or 125 kJ/mol^{-1}, will occur at low temperatures. Only the oil generation, which has activation energies of 210 and 250 kJ/mol^{-1} in Tissot's model, is of importance here. The authors used 230 $kJ\ mol^{-1}$, the average of these. Tissot also argued that the pre-exponential factor is linearly related to the activation energy, an argument for which there is no theoretical justification. Presumably this is why the authors used a pre-exponential factor of 1.3×10^{27}. Because both the activation energy and the pre-exponential factors are large the oil generation occurs in the authors' model over a rather limited depth range. Have they examined the width of the oil generation window when they make different assumptions?

D. L. TURCOTTE. What we attempted to do was to match the observed maturation by using the simplest model.

Phil. Trans. R. Soc. Lond. A **305**, 295–317 (1982) [295]
Printed in Great Britain

A comparison of foreland and rift margin sedimentary basins

By C. Beaumont†, Charlotte E. Keen‡ and R. Boutilier†

† *Department of Oceanography, Dalhousie University, Halifax, Nova Scotia, Canada B3H 4J1*

‡ *Atlantic Geoscience Centre, Geological Survey of Canada, Bedford Institute of Oceanography, Dartmouth, Nova Scotia, Canada B2Y 4A2*

Foreland and rift margin basins are compared on the basis of (1) their tectonic setting, (2) reasons for subsidence, and (3) their large-scale geophysical and geological characteristics. The thermal and mechanical properties of the underlying lithosphere are shown to be fundamental to the form of tectonic subsidence. The lithosphere beneath foreland basins is flexurally downwarped by the loading of the adjacent fold–thrust belt, whereas tectonic subsidence at rifted margins is caused by mass replacement at depth, during lithospheric extension on rifting, and subsequent thermal contraction as the lithosphere cools.

The effects of rheology, thermal maturity, and lateral changes in properties of the lithosphere are outlined for foreland basins, as is the topographic effect of possible phase changes beneath the fold–thrust belt. The thermal and rheological consequences of lithospheric extension at rift margins makes flexural subsidence relatively less important than in foreland basins. Flexure may, however, be partly responsible for uplift landward of the hinge line that is associated with rifting. Other mechanisms that could cause such uplift include depth-dependent extension and thermal expansion due to the lateral diffusion of heat.

The models describing the evolution of these basins are shown to predict characteristics that are in accord with observations. The superposition of foreland and rift margin basins as a result of ocean closure can lead to an overall basin stratigraphy that is complex. Such phases of basin subsidence must be separated according to the tectonic environment in which they formed in any analysis of the cause and consequences of basin evolution.

Introduction

The intent of this paper is to provide a descriptive geological interpretation of some recent theoretical models of the origin and evolution of foreland and rifted margin sedimentary basins, a discussion of their largest-scale characteristics, and a description of the geological and geophysical evidence that is needed to test these theoretical models. The models, based on simple concepts of continuum mechanics and conductive heat transport, appeal to the fundamental properties of the lithosphere under compression and tension to explain the formation of these two basin types. The models are also shown to fit within the plate tectonic framework and to be a direct consequence of horizontal motions and interactions of lithospheric plates that is central to the plate tectonic model. The only additional assumption is that the plate boundaries must be deformable.

The theme of the paper is to propose two archetypal models for the tectonic subsidence of these basins and to develop the models with regard to lithospheric rheology, lateral changes in lithospheric properties, and the response to sediment and water loading. 'Tectonic subsidence' implies subsidence that would occur without loading of the lithosphere by sediment or water infill of the basin. Predicted characteristics of each basin type and underlying lithosphere are then compared, some simple examples listed, and the consequences of superposition outlined. The content of this paper is partly based on earlier papers by Beaumont (1978, 1979, 1981), Beaumont *et al.* (1982) and Keen *et al.* (1981 *a*, *b*), to which the reader is referred for details of the methodology. It is intended as an introduction to, and an amplification of, concepts and results presented in those papers.

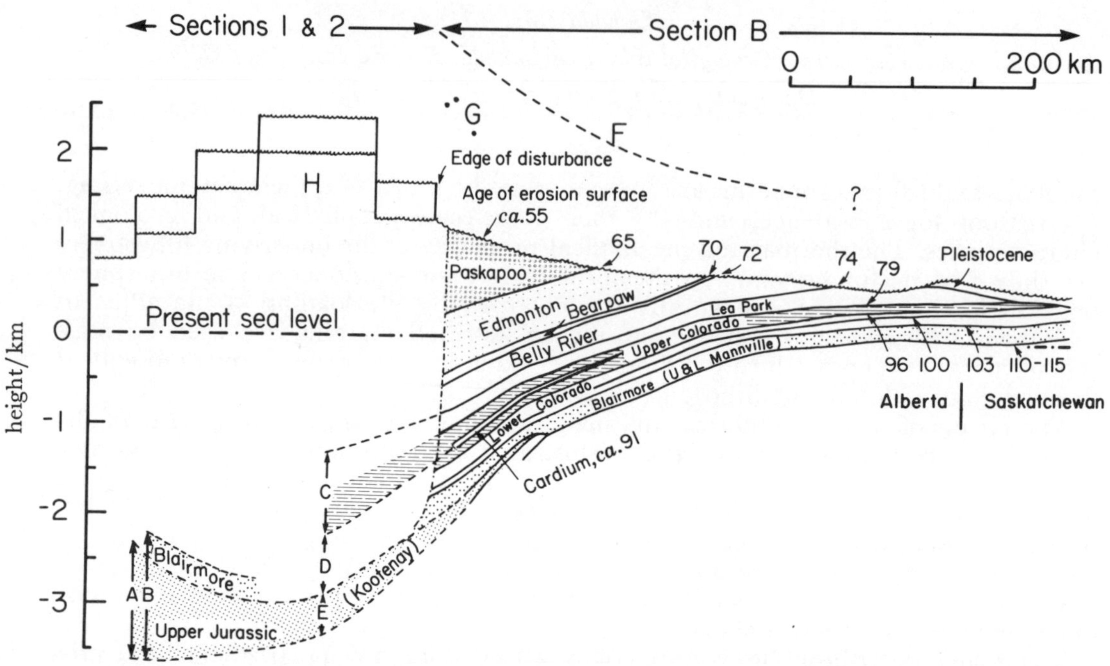

FIGURE 1. Simplified cross section of the post-mid-Jurassic foreland basin sequence of the Alberta Basin and adjoining Canadian Cordillera. The sections 1 and 2 and B extend from approximately 52° N, 118° W to 55° N, 105° W, as shown in Figure 6 of Beaumont (1981). A–E are palinspastically restored stratigraphic units; F and G are estimates of the position of the depositional baseline before erosion from coal moisture content of near-surface coals and shale compaction. The distance from F to the present surface is therefore an estimate of the amount of post-Laramide erosion that must be taken into account in any quantitative model of basin evolution. H are upper and lower bounds on topography of the adjacent 300 km of the Cordillera averaged on 50 km wide strips. The numbers are approximate ages in megayears of the important stratigraphic horizons. Note that much of the basin is above present sea level. The thickness of the crust beneath the sediments is not shown but is believed to be relatively uniform, particularly when compared with that at rifted continental margins.

Figures 1 and 2 illustrate cross sections of the Alberta Basin and Nova Scotian continental margin, which may be regarded as mature examples of foreland and rifted margin basins. Both basin types are asymmetric, a property inherited from the plate interactions responsible for their formation.

Foreland basins, which reach a maximum depth of *ca.* 6 km, are always bordered by an overthrust belt, a region of allochthonous sedimentary or crystalline crustal rocks, or both, that has been emplaced on the neighbouring lithosphere during overthrusting. This belt forms a topographically high mountain range of the Cordilleran or Himalayan type. The basin is wedge-shaped in cross section, with maximum depth adjacent to the mountains. The stratigraphy usually comprises similar laterally thinning units that may be truncated by erosion to expose sediments of increasing age with increasing distance from the mountains. In other examples successive sedimentary units appear either to thin to a common point, or to overstep each other, or are disrupted by the underlying sedimentary or basement movements that are believed to be contemporaneous with the emplacement of the fold–thrust belt. Along strike the basin depth and height of mountains vary, but the relation between basin and fold–thrust belt is maintained to give many examples of a sinuous, fundamentally two-dimensional character. Another universal feature is the lateral advance of the fold–thrust belt driving the foreland

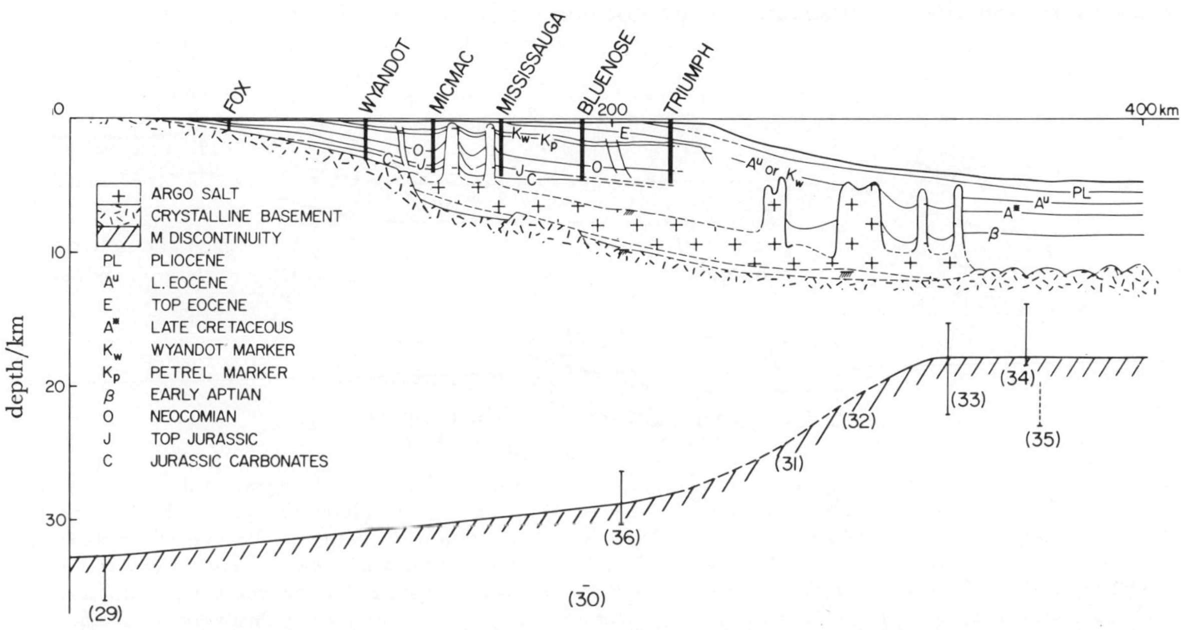

FIGURE 2. Simplified cross section of the Nova Scotian margin extending from approximately 45.5° N, 60.5° W to 42° N, 58.5° W, as shown in Beaumont *et al.* (1982). The depth to the Moho at the ends of seismic refraction lines adjacent but normal to the profile are shown projected onto the profile as vertical bars, which therefore represent the variability of crustal structure along strike. Regions where the positions of interfaces remain somewhat uncertain are shown as broken lines. The position of the basement–sediment interface beneath the basin is specified on the basis of wells that intersect it (Fox and Wyandot), seismic reflexion data, and seismic refraction data (shown as baseline with hatching). This basement is thought to be that of syn- and post-rift sediments; older sediments may extend to greater depth. Stratigraphy of the basin is based on evidence from deep exploratory wells (shown as vertical bars) and seismic reflexion data. Lithologies are regional averages representing the dominant component. Where unshaded, the sediments are thought to be mainly clastics. Stratigraphic and seismic marker horizons are labelled according to the key.

subsidence in front of it and incorporating successive basin units as the advance continues. These units are detached and transported basinward as thrust sheets. Detachment may be confined to the sediment column or extended into the underlying crystalline crust. Unroofing of the fold–thrust belt and uplift of the basin are also common features once the thrusting has ceased.

Rifted margin basins are usually deeper than foreland basins, sometimes reaching depths of *ca.* 16 km. They occupy tensional environments in the transitional region between continent and ocean and are not associated with fold–thrust belts. The shape of the basin is determined by the sedimentation: high rates of sediment influx will produce a prograding deltaic wedge, but a simple basement ramp will occur in a sediment-starved basin. The stratigraphy is also variable, depending on sediment input to the evolving basin. Rarely is there the overabundance of sediments that is often seen in foreland basins. This is because the basin can continue to build seaward.

Foreland basins and the adjacent fold–thrust belt

The term foreland basin is used here in a manner that is descriptively synonymous with exogeosyncline (Kay 1951), marginal basin (Krumbein & Sloss 1963), foredeep (Aubouin 1965), foreland and foredeep trough (Price 1973), peripheral and retroarc foreland basin (Dickinson 1974), and the marginal down-flexure of Russian authors, although the genetic explanations provided by these authors are not necessarily adopted.

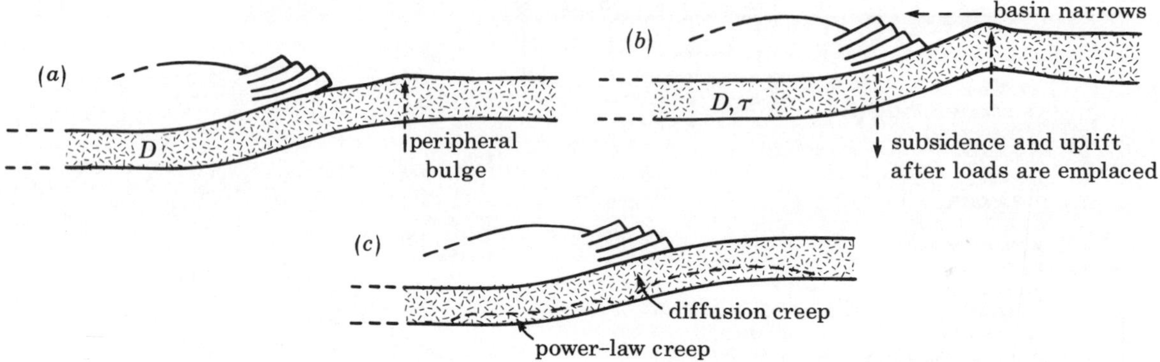

Figure 3. Diagrams illustrating flexure of the lithosphere beneath the fold–thrust belt, shown as a stack of thrust sheets. The foredeep is shown before sediment filling. In (*a*) the patterned region is that part of the lithosphere that appears to undergo elastic flexure. Its thickness is typically much less than the thickness of the thermal lithosphere. It is also assumed that the lithosphere is laterally uniform and can be characterized by its flexural rigidity, D. The peripheral bulge, which is typically a small percentage of the downwarp in amplitude, is exaggerated. It migrates to the right at a fixed distance ahead of the advancing thrust sheets. In (*b*) the patterned region is again laterally uniform but may be somewhat thicker than in (*a*). D is again the flexural rigidity and τ is the viscous relaxation time constant for the lithosphere treated as a whole. In this case the peripheral bulge tends to migrate to the left and to grow with time after a thrusting event. This leads to a coupled additional subsidence and uplift as the foredeep narrows. In (*c*) an attempt is made to outline the regions in which diffusion and power-law creep would occur were the lithosphere to have a microrheology that is the same as that determined from laboratory experiments on olivine. There will be some tendency for the basin to develop a peripheral bulge that grows and migrates to the left.

The scope of the discussion is limited to the depositional and erosional history of the basin that is directly controlled by events in the orogenic belt; that is, the flysch and molasse phases. Here, the concern is the response of the continental margin to the accretionary events associated with plate convergence and collision (Dewey & Bird 1970; Dickinson 1974) and the intraplate crustal shortening.

The working hypothesis is that foreland basins form at the site of downward-flexed lithosphere, and that this downward flexure is in response to passive loading by supra-lithospheric mass loads superimposed during formation of the fold–thrust belt. This hypothesis has more widespread application than to foreland basins only. Whenever there is lateral transfer of a rock mass over an adjacent part of the lithosphere that responds by flexure, a coupled trough is created in which sediments can accumulate. The form of this trough will depend on the timing and amount of the mass movements, the properties of the underlying lithosphere, the amount of sedimentary infill, and the role that other processes, for example subduction, under-thrusting and thermal cooling, play in warping the lithosphere at the site of the trough. The influence of each of these factors on the formation of foreland basins can be estimated from the tectonic and thermal environments in which foreland basins form.

The origin of tectonic subsidence of the foredeep by lithospheric flexure under the overthrust loads (figure 3) is analogous to the bending of a leaf spring. This, the simplest form of the model, assumes that the 'spring' is continuous and laterally uniform in its thermal and rheological properties. It is also assumed to be thermally mature, several lithospheric thermal time constants having elapsed since the last significant thermal perturbation. Such a model was shown to give a reasonably accurate account of the evolution of the Alberta Basin (Beaumont 1981) if it is assumed that the foredeep was continuously filled with sediments eroded from the adjacent Rocky Mountains and that eustatic sea level changes also occurred. It was, however, noted that the predicted average height of the Rockies was probably too large, although no quantitative estimates of palaeotopography exist. For this reason it is worthwhile considering, in a qualitative way, possible variations in the form of the tectonic subsidence, palaeotopography and basin development if the lithosphere were to have more complex properties. It is stressed that all the models proposed are fundamentally 'leaf-spring' models but that they include some second-order effects that may be detectable when other foreland basins are investigated. These complexities are briefly discussed to document the inevitable difficulties that will occur when trying to prove that any, some, or all are operative.

Lithospheric rheology: continuous uniform time-invariant elastic lithosphere

The continuous uniform time-invariant elastic lithosphere model is characterized by a lithosphere of constant flexural rigidity, D, overlying an inviscid fluid asthenosphere. The uniformity dictates that the distance between an applied load and the peripheral bulge (shown

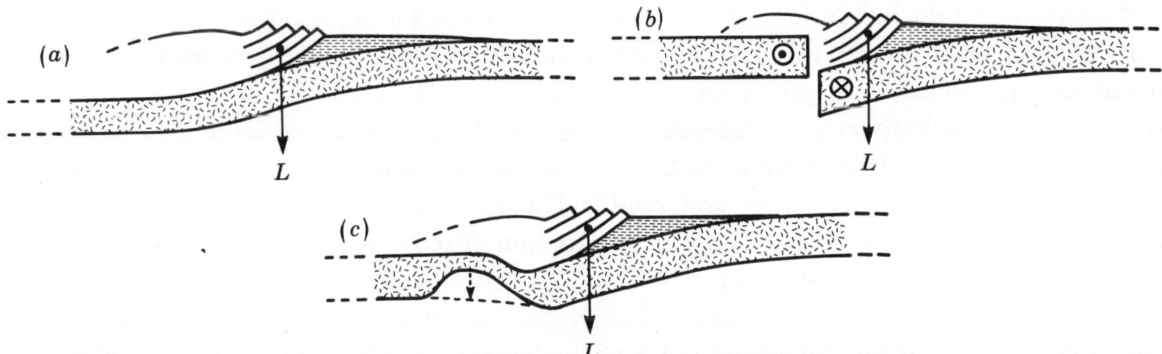

FIGURE 4. Diagrams illustrating the effects of lateral changes in properties of the lithosphere (random pattern) on the form of the foreland basin (dashed horizontal pattern), compared with a continuous uniform plate (*a*). The basin remains largely unaltered because the postulated strike-slip motions (*b*) and thermal thinning of lithosphere (*c*) occur within the orogen. The loading of the overthrusts is depicted as the cause of lithospheric flexure. The main effects of lateral changes that weaken the lithosphere will be to increase the amplitude of downwarping at the expense of reducing the width of the basin. The degree to which this occurs depends on the relative positioning of the weakened area and the fold–thrust belt.

in exaggerated form in figure 3*a*) is constant and can, in theory, be used to infer D and to provide estimates of effective lithospheric thickness. The additional growth of the basin when filled or partly filled by sediment (figure 4) must also be computed. This poses no problems if the amount of filling and sediment density distribution are known. Similarly, the effects of changes in the sea level, which lead to further onlap or offlap of the basin, can be predicted if the sea level changes are known. Pre-existing topography can also be accounted for if it is known.

We can already appreciate that the calculations are manageable for quite complex geometries but that the effects of inaccurately known quantities, such as sediment density, eustatic sea level changes and palaeotopography, will affect the model predictions. The position of the feather edge of the basin, the edge furthest from the thrust–fold belt, is very sensitive to these quantities. Consequently it should not be used to measure basin width for flexural rigidity estimates. A more accurate method is to compare overall basin shape, or the position of individual stratigraphic horizons, with model predictions. Even these require a reasonable estimate of the loading history in the fold–thrust belt. One generalization is, however, possible. If the fold–thrust belt advances basinward, then, in the absence of major sea level changes or high pre-existing topography, the feather edge of the basin will advance at the same rate. Consequently, the continuous uniform time-invariant elastic lithosphere model predicts a history of basin onlap. Furthermore, erosion of the fold–thrust belt cannot create offlap because distance from the load to the peripheral bulge is constant.

Lithospheric rheology: continuous uniform time-invariant viscoelastic lithosphere

This model casts the lithosphere as a uniform viscoelastic (Maxwell) layer overlying an inviscid fluid substratum (figure 3*b*). The elastic properties are again described by the flexural rigidity, D, but stress relaxation is permitted by viscous flow with a time constant τ. 'Time-invariant' implies that D and τ do not change with time; however, the lithospheric response to loading is time-dependent (Beaumont 1978). The response to loading is an immediate elastic flexure, the same as for the elastic models, followed by further progressive subsidence of the basin adjacent to the load and uplift and inward migration of the peripheral bulge. This, when combined with erosion of the uplifted basin strata, leads to a progressive narrowing of the basin and steepening of the dip on stratigraphic horizons. The relaxation induced by successive loads will be combined with the elastic and viscous response of subsequent and earlier loads. Therefore, the stratigraphy of the basin reflects a subtle interplay of the loading history with the viscoelastic relaxation of the lithosphere. In a general way viscous relaxation mitigates the tendency of the advancing fold–thrust belt to drive the feather edge of the basin before it. It also leads to the development of significant arches and erosional angular unconformities along the path of the inwardly migrating peripheral bulge if the amount of relaxation is large (Beaumont 1981).

When a loading phase is followed by erosion of the fold–thrust belt, the response to unloading of the 'leaf spring' is uplift of the basin. This is true for both elastic and viscoelastic models. The difference is that the immediate uplift of the basin for viscoelastic models in response to erosion has an immediate flexural wavelength that is longer than the width of the basin, simply because relaxation has had time to shrink the basin since the last loading phase. Instead of uplift and erosion that strips off stratigraphic layers much as they were deposited, as predicted by the elastic model, the viscoelastic response is therefore to truncate older stratigraphy, thereby producing a characteristic angular unconformity that exposes progressively older strata with increasing distance from the fold–thrust belt (Beaumont 1981). This is exactly as observed in the cross section of figure 1. Such an angular unconformity cannot be produced by any reasonable model that includes only loading of a continuous uniform time-invariant elastic lithosphere, although bevelling in response to an additional significant lowering of sea level can produce a similar effect. Angular unconformities do not, however, provide a conclusive case for viscoelasticity because other rheologies and lateral changes in lithospheric properties can also mimic these effects. Nevertheless, the viscoelastic model is the simplest model that produces an acceptable explanation of the Alberta Basin (Beaumont 1981).

Lithospheric rheology: inferences from laboratory experiments

Deformation diagrams for creep in olivine (see, for example, Ashby & Verrall 1977; Ranalli 1980) have been used by Beaumont (1979) to predict a first-order model of rheological zonation of the lithosphere under flexure. These models, although crude, include the effects of a uniform increase in temperature with depth in the lithosphere and a power law for the viscous creep relaxation of stress. It is unlikely that stresses induced by the distributed load of the fold–thrust belt and sediments alone will be sufficiently large to cause either brittle failure (cataclastic flow) or ductile (plastic) flow of the lithosphere if it was previously intact and without zones of local weakness. Therefore, zones of diffusion (linear) and dislocation (power law) creep similar to that shown diagrammatically in figure 3*c* can be expected. The terms diffusion and dislocation creep are used in the sense given by Beaumont (1979). Such a model is certainly more acceptable to the microrheologist than either of the previous two. The overall macro response of the lithosphere, which determines the development of the basin, is, however, likely to be very similar to the viscoelastic model. The only difference that might be detectable is that relaxation, which leads to tilting and narrowing of the basin, will rapidly decrease once the hotter regions of the lithosphere have undergone stress relaxation. Consequently, the response tends to that of a lithosphere that is apparently elastic long after loads have been added or removed. This is certainly true if thermally activated diffusion creep dominates the overall response (R. Courtney, personal communication). Were this not so, such simple rheologies as elasticity and viscoelasticity would long ago have been shown to be unacceptable. We believe that the response of the lithosphere to loading is primarily a function of its temperature distribution and that the effect of flexurally induced stress in promoting power-law creep is secondary. The geologist who is interested in basin analysis should therefore concern himself with the two simple models of lithospheric flexure and the underlying question of the thermal state of the lithosphere.

Lateral variations in lithospheric properties

There is good reason to believe that the lithosphere is not continuous and uniform beneath the whole of an orogen, as assumed in the simple models (figures 3 and 4*a*). For example the plate tectonic model explains the assembly of an orogen by subduction followed by either or both of collision and transform strike-slip motions, both of which produce lateral inhomogeneities in lithospheric structure and suggest models like those shown in figure 4*b*, *c*. Strike-slip plate motions with a component of convergence may best be represented by a broken plate model. Corresponding thermal attenuation beneath the core zone of an orogen would result in partial or total decoupling of this region. In both cases the amount of decoupling may vary with time in response to cooling of the lithosphere after the cessation of subduction and completion of suturing.

Simple elastic models in which variations in lithospheric thickness are included in the hinterland of the orogen indicate that such variations cause only second-order changes in the form of the foreland basin. The extremes (figure 4*a*, *b*) are represented by an end-loaded discontinuous but buoyantly supported beam and a point-loaded continuous buoyantly supported beam (Walcott 1970). The former predicts a somewhat narrower and deeper foreland basin than the latter for the same flexural rigidity. This reduces the predicted topography of the fold–thrust belt, as would the thermally thinned plate model (figure 4*c*), whose behaviour will lie between the two extremes. A basin of equivalent width to that predicted by a continuous uniform plate model is obtained by increasing the flexural rigidity.

It is important to ask whether the effect of lateral variations in lithospheric properties in an elastic model can mimic the effect of stress relaxation as shown by the viscoelastic model. The fundamental property of the viscoelastic model is that the plate appears weaker, thinner and to have decreasing flexural wavelength with increasing time after loading. Successive loads all show the same pattern of evolution no matter when they were added. Similar effects cannot occur in elastic models with lateral variations in properties if these lateral variations do not change with time. That is, a broken plate model will have approximately constant properties as long as the break remains. To appear weaker the plate must break, or be thermally thinned, as the fold–thrust belt evolves. To produce an angular unconformity in the basin the plate must be rewelded, or cool, before uplift and erosion of the fold–thrust belt. Although complex, such events cannot be ruled out *a priori*. Therefore it is possible, but unlikely, that time-varying lateral changes in lithospheric properties could mimic stress relaxation.

The simple concept of the fold–thrust belt advancing further onto a broken plate (figure 4*b*), so that the break recedes ever further into the hinterland until the plate appears continuous and uniform, produces an exactly opposite effect, that of an apparently increasingly strong plate when viewed from the perspective of basin stratigraphy. Post-tectonic erosion cannot produce an unconformity of the type observed because the basin strata will be stripped in much the same way as deposited.

In summary, it can be seen that lateral changes in lithospheric properties do produce second-order changes in the form of the foreland basin but that a detailed analysis, including a palinspastic reconstruction of the fold–thrust belt, will be necessary before individual lithospheric features can be demonstrated for a particular foreland basin.

Thermal maturity of the lithosphere as a key to the size of foreland basins

It is explained in the section on rifted continental margins that the flexural properties of the lithosphere are primarily determined by its internal temperature distribution. Hot, thin, thermally young lithosphere has a small flexural rigidity and wavelength, whereas old, cold, thermally mature lithosphere has a correspondingly larger flexural rigidity and wavelength. This view is confirmed by studies of the flexural response of oceanic lithosphere as it increases in age (Watts 1978). For this reason it is to be expected that, if the leaf-spring model of foreland basins is correct, the size of a foreland basin will be directly related to the thermal age of the underlying lithosphere at the time of basin formation. The thermal age is the elapsed time since the last major thermal event that thinned the lithosphere. The difference between basins that develop on thermally young and old lithosphere is illustrated in figure 5. In the limit of total thinning of the lithosphere, local, or Airy, isostatic equilibrium prevails and no foreland basin is generated.

If significant lithospheric cooling occurs during the formation of the basin (figure 5*a*) the flexural rigidity will increase, as will the width of the basin. This will accelerate the advance of the feather edge of the basin ahead of the fold–thrust belt, leading to a pattern of progressive onlap. Erosion of the fold–thrust belt and the basin can then, in theory, produce an angular unconformity because the lithosphere achieves its maximum flexural wavelength during the terminal erosion phase. This represents an additional way in which angular unconformities can arise. When invoking this model, however, evidence for a thermal event, which thins the lithosphere, is necessary.

The main point to be made in this section is that we should not expect all foreland basins to look like the Alberta Basin archetype. The reason that this basin is so pronounced is that it formed on cratonic lithosphere of the Canadian Shield that had not been subject to a major tectonic or thermal event for at least 500 Ma and probably longer. The tectonic setting of a foreland basin is therefore seen to be a critical factor in determining its final extent.

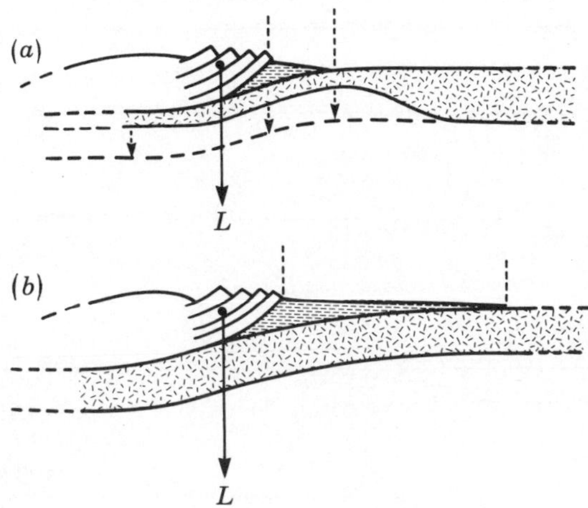

FIGURE 5. Diagrams illustrating the difference between foreland basins that form on thermally young (*a*) and old (*b*) margins. Young, hot lithosphere (less than about 30 Ma old) is flexurally weak; therefore the basin will be narrow but deep. The converse is true for thermally old (more than about 300 Ma) lithosphere. If the lithosphere cools appreciably while the basin forms, (*a*), the flexural rigidity and corresponding width of the basin will increase.

Lithospheric phase changes and topography of the fold–thrust belt

Beaumont (1981) shows the final configuration of a series of models for the development of the Alberta Basin. The evolution of the preferred model for six of the eleven time steps is shown in figure 6. The topography of the fold–thrust belt at the end of the Laramide Orogeny, *ca.* 35 Ma ago in the model, is certainly excessive in suggesting a range of mountains larger than any that currently exist, including the Himalayas. The need for such a large load is based on estimates of depth of burial of coals and shales beneath the now largely eroded Paskapoo formation, which was deposited during the Laramide Orogeny. The evidence in favour of the model load is therefore not as conclusive as it would be were no erosion to have occurred in the last *ca.* 35 Ma. Nevertheless, even minimum loads still give apparently excessive topography and there remains a need to reduce topography while preserving the load.

The topography shown diagrammatically as h_0 (figure 7*a*) would be reduced to h_1 (figure 7*b*) were the lithosphere broken or thinned or the thickness of the load reduced to reflect a load density greater than the 2400 kg m^{-3} used, but estimates suggest that such reductions would still be insufficient. The problem is not peculiar to the Canadian Cordillera. Oxburgh (1968) and Hunziker (1974), among others, have noted that for the Alpine chain, where thrust sheets tens of kilometres thick have been overthrust onto the continent, the initial topography must have been low because relatively little erosion occurred for *ca.* 30 Ma after emplacement. Richardson & England (1979) present an attractive model to explain this reduced topography by invoking amphibolite or granulite facies to eclogite facies phase transitions in the lower crust in response to the increased pressure beneath the thick overthrusts (figure 7*c*). They estimate an accompanying density change from 2850 to 3150 kg m^{-3} with a corresponding volume decrease of *ca.* 10%. Such a change within a 20 km thick region of the crust, in addition to an increase in the fold–thrust belt load density to 2600 kg m^{-3}, to include possible crystalline

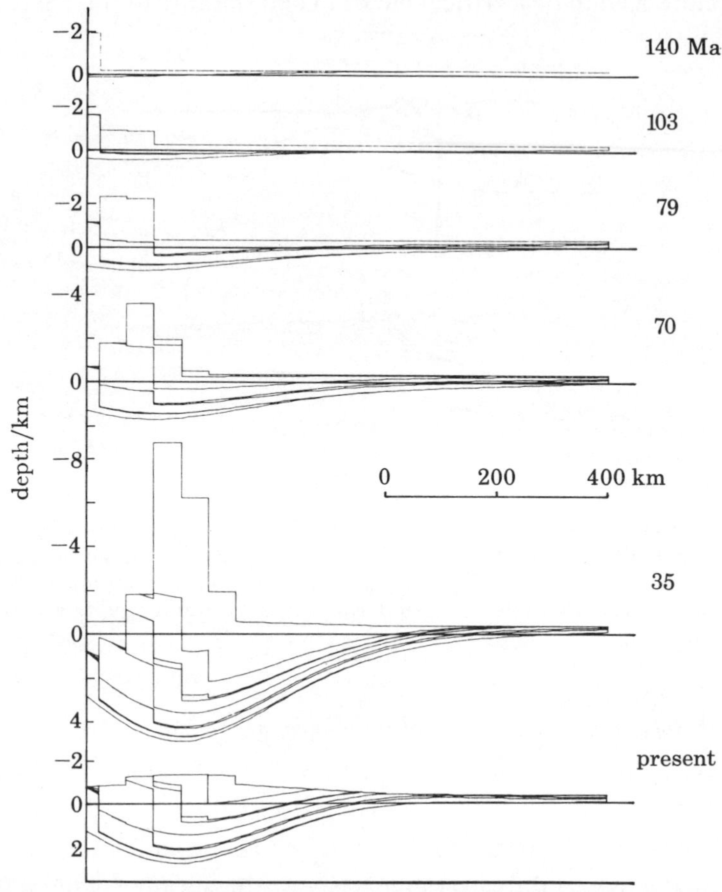

FIGURE 6. Six of eleven steps in a model of the evolution of the Alberta foreland basin (Beaumont 1981). The dates are model ages. The main points to note are: the lateral migration of the fold–thrust belt, here averaged over 50 km wide strips; that sea level changes, which widen the basin, have been included; the excessive height of the mountains at 35 Ma age; and the truncation of the basin strata during erosion between 35 Ma ago and the present. The datum is present sea level.

involvement, would reduce the average Laramide topography of the Rockies to *ca.* 2 km. The phase change is, however, temporary because relaxation of the temperature profile to the stable geotherm inverts the phase transition. This heating may take approximately 20 Ma. The time constant could vary by a factor two depending on the erosion rate, the thickness of the over-thrust sheet, and the gradient of the stable geotherm. This result suggests that the final configuration of the model (figure 6) is probably correct because thrusting ceased at the end of the Eocene Laramide Orogeny and sufficient time has elapsed for up to 10 km of erosion as proposed by the model and a return to a stable geothermal gradient.

The phase transition model (figure 7*c*) therefore provides a mechanism whereby topography can be reduced but in a temporary way. Phase transitions have often been invoked to rectify problems with models without justifying why the phase transition should occur. In this case, however, the model has a firmer basis. The forward phase change is a direct consequence of the pressure change induced by the loads, and the inversion is in response to a return to the normal geothermal gradient, both of which are predicted by equilibrium thermodynamics. No thermal pulse or inversion from a pre-existing metastable facies need be hypothesized.

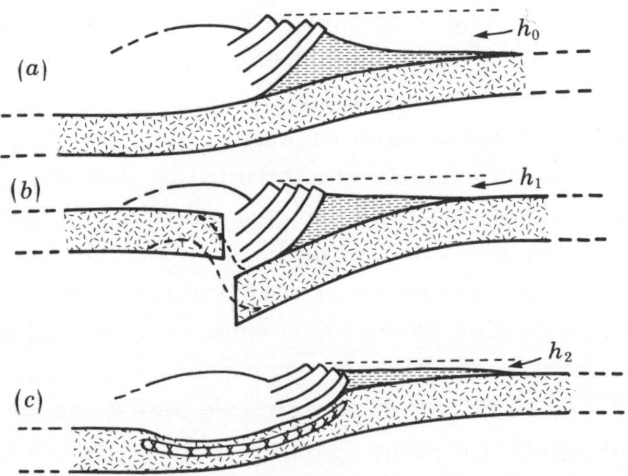

FIGURE 7. Diagrams illustrating reduction of the topography of the fold–thrust belt in response to a thinner or decoupled lithosphere, or as a result of volume changes in the crust due to phase changes (X-patterned area). (*a*) Continuous uniform lithosphere; (*b*) broken or thinned lithosphere; (*c*) phase change in lithosphere under fold–thrust belt.

It is important to note that the model (figure 6) is conservative in its load estimates. Larger loads are probably necessary to obtain the correct thickness of palinspastically restored stratigraphic units of the basin (R. A. Price, personal communication). These arguments imply the need for inclusion of phase changes in the leaf-spring model.

Lithospheric extension at rifted continental margins

Most continental margins created during rifting of a continental lithospheric plate evolve into deep sedimentary basins occupying the transitional region between oceans and continents. These rifted margins, sometimes called Atlantic-type or passive margins, are the subject of intense research interest because they may contain significant hydrocarbon resources. Of the possible mechanisms responsible for their formation and subsequent evolution, the proposal that there is significant horizontal extension during rifting which thins both the crust and subcrustal lithosphere can explain most of the first-order properties of the margins and can be tested by modelling. These properties include: the thinning of the crust landward of the ocean–continent boundary (Sheridan *et al.* 1979; Montadert *et al.* 1979; Keen & Hyndman 1979), subsidence (Sleep 1971; Watts & Steckler 1979; Watts & Ryan 1976; Keen 1979; Royden & Keen 1980), and listric faulting of the upper crust (Montadert *et al.* 1979).

Keen *et al.* (1981 *a*, *b*) and Beaumont *et al.* (1982) have recently explored the consequences of extensional models of rifting with particular reference to the rifting process, the post-rift cooling, thermal contraction and subsidence of the lithosphere, and the amplification of this subsidence by sediment and water loads. Three kinematic models of rifting, based primarily on extension and thinning of the lithosphere, were examined in detail to determine the degree to which second-order effects during the extension and rifting process dictate the evolution of the margin. To first order, details of the rifting model were found to be unimportant. Therefore, for the purposes of this paper we shall initially confine the discussion to thermomechanical models that are a direct consequence of the uniform extension model proposed by McKenzie (1978). The amount of lithospheric extension, $\beta(x)$, is assumed to be uniform with depth and to increase progressively with position, x, towards the point at which oceanic lithosphere is generated. At that point the process of tensional stress release changes from stretching of the continental lithosphere to rupture and accretion of oceanic lithosphere.

The consequences of extension are conceptually simple. The crust and subcrustal lithosphere are initially thinned by $\beta(x)$ and the space created is filled by the passive upwelling of hot asthenosphere. Extension will therefore be accompanied by isostatic elevation changes due to the replacement of crust by denser mantle lithosphere and density changes due to heating and thermal expansion. This normally results in rapid subsidence during extension. Longer-term subsidence will occur as the extended region of the margin cools by conduction and undergoes thermal contraction after extension ceases. This subsidence is analogous to that of oceanic lithosphere as it migrates from an oceanic ridge (Parsons & Sclater 1977). Extension is assumed (1) to occur on a sufficiently short geological timescale that it can be modelled as an instantaneous process without significant error (Jarvis & McKenzie 1980) and (2) to produce sufficiently small horizontal thermal gradients that horizontal heat transport can be ignored in comparison with vertical conductive cooling (Beaumont *et al.* 1982). Local isostatic equilibrium is assumed during the rifting process.

Thermal and rheological evolution of the lithosphere

An immediate consequence of the proposed rifting model is that, during the evolution of the margin, the lithosphere will undergo changes in thermal and rheological properties in both time and space as it cools. Unlike the lithosphere beneath foreland basins, no simplifying model of a uniform continuous time-invariant elastic or viscoelastic lithosphere can be assumed. This is particularly true because rheology is a strong function of temperature, as we have previously argued. A model (figure 8) for the evolution of the lithosphere beneath the Nova Scotian margin (figure 2) demonstrates how radically the lithosphere is thinned from an original thickness of 125 km during extension, and how, over the next 185 Ma after rifting, conductive cooling restores most of the original thermal regime.

For the purposes of modelling the loading response to sediment and water it is assumed that the lithosphere can be subdivided into two regions: one an elastic core in which no stress relaxation occurs; the other a hotter underlying region that releases stress rapidly in comparison with the time for evolution of the margin as a whole. This lower region is therefore treated as an inviscid fluid that is undistinguished from the asthenosphere. The boundary between the elastic and fluid regions is taken to be an isotherm, termed the relaxation isotherm, the position of which varies in space and time as the margin evolves but can be predicted by considering both the thermal and mechanical evolution of the margin.

The post-rifting history of the margin is therefore traced by a thermomechanical model in which subsidence and temperature distribution are the result of conductive cooling and in which the isostatic response to loading by sediment and water is found by assuming that those loads are supported by this elastic layer underlain by an inviscid fluid. The model is coupled in that both the space available for sediments and water and the thickness of the elastic layer, termed the rheological lithosphere, are determined by the temperature distribution. The calculations are made by using finite difference and finite element numerical models for the thermal and mechanical aspects of the problem respectively (Beaumont *et al.* 1982).

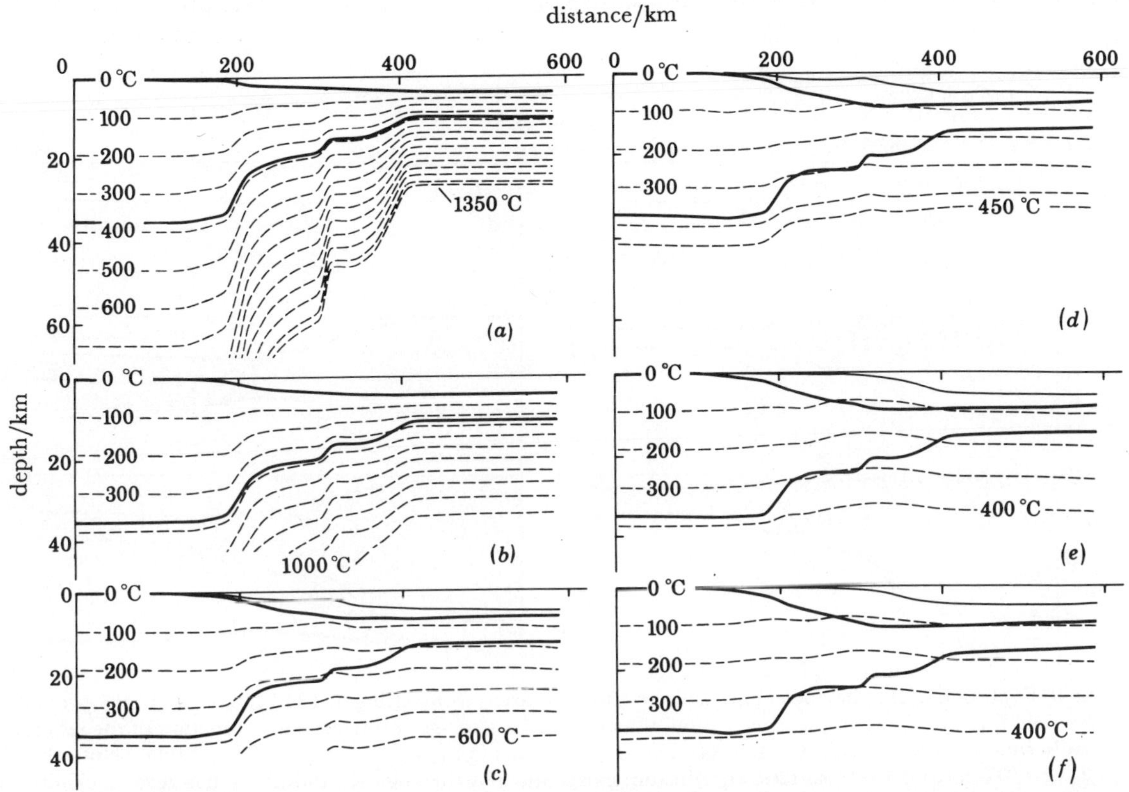

FIGURE 8. Thermal evolution of a uniform extension model of the Nova Scotian margin with a relaxation isotherm of 250 °C. (*a*) 185 Ma B.P., rifting; (*b*) 175 Ma B.P.; (*c*) 135 Ma B.P., end of Jurassic; (*d*) 100 Ma B.P., end of early Cretaceous; (*e*) 60 Ma B.P., end of late Cretaceous; (*f*) present. The lithosphere was 125 km thick before extension. Its initial thinning and that of the crust can be seen from the uplift of the 1350 °C isotherm (broken line in (*a*)) and the Moho (lower continuous bold line in (*a*)). The upper continuous bold line is the sediment–basement interface, and the continuous fine line is the sediment–water interface. Note that within *ca.* 63 Ma (the lithospheric thermal time constant for this model) after rifting, the thermal régime is largely restored to its initial state. Upwarps in the isotherms that are predicted for the present are the result of thermal blanketing by the low thermal conductivity sediments. The rheological lithosphere is that region between the sediment–basement interface and the 250 °C isotherm.

The simplicity of the rheology employed needs justification particularly by comparison with the possible rheologies suggested for the foreland basin models. Our view is that during the evolution of rifted margins, lithospheric rheology is so strongly dominated by cooling effects that it will be difficult to detect any properties that can be directly ascribed to viscous or viscoelastic relaxation. Evidence of rapid stress relaxation in the lower lithosphere will probably not be preserved in the short-term sediment record. Such relaxation is, however, allowed to occur instantaneously in the model by choosing a relaxation isotherm less than the temperature of the base of the thermal lithosphere. The same approach is implicit in the elastic lithosphere models for foreland basins because the lithospheric thickness so determined is 50–70% less than that estimated for the thermal lithosphere. Viscous flow at depth beneath continental margins will be further inhibited by cooling that 'freezes in' pre-existing stress before relaxation is complete. Emphasis is therefore placed on the spatial and temporal variations of the elastic region at the expense of more complex rheologies. This approach can also be justified on a heuristic basis by comparison with olivine microrheology (Beaumont *et al.* 1982). If simple models of this type fail to predict correctly the evolution of rifted margins, the next step will be to introduce a rheology where the temperature dependence of viscosity is explicitly included.

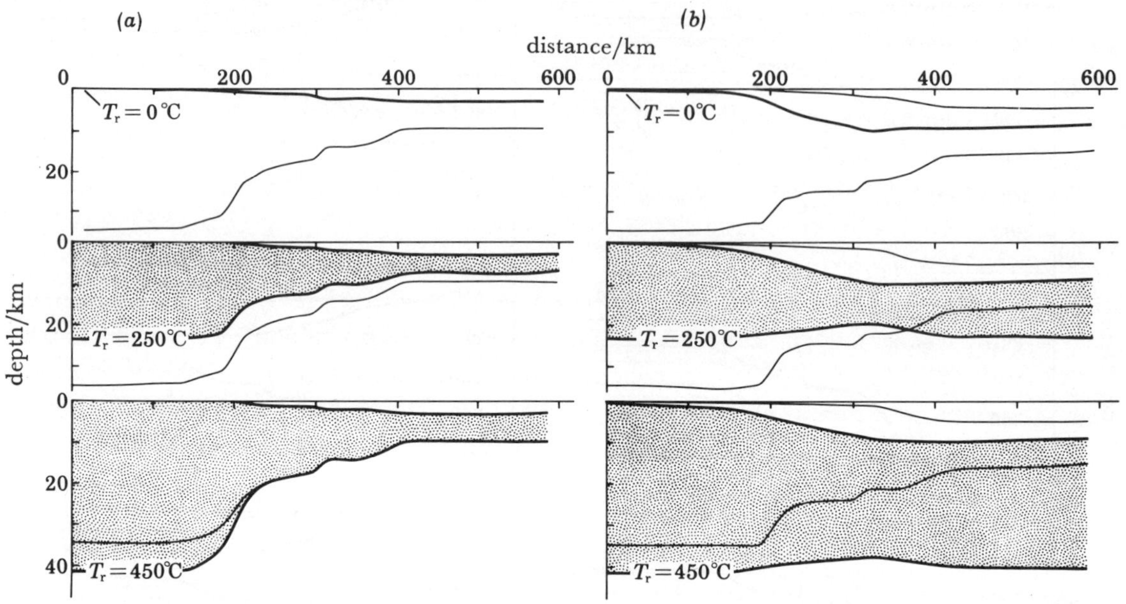

FIGURE 9. Evolution of the rheological lithosphere (stippled area) from rifting at 185 Ma B.P. (*a*) to the present (*b*) for the Nova Scotian margin cross section, illustrating the change in thickness as a function of the relaxation isotherm, T_r = 0 °C, 250 °C, 450 °C. The thin line at depth is the position of the Moho. Note that for T_r = 0 °C, local 'Airy' isostatic equilibrium exists and the rheological lithosphere has zero thickness. For finite T_r the thickness of the rheological lithosphere remains constant in the unextended region and thickens with time beneath the oceanic region. The thinning beneath the shelf edge results from upwarp of the isotherms beneath that region of the basin and depression of the surface beneath the thick sediments.

By comparison with foreland basins it is seen that the tectonic subsidence of rifted margins is not caused by supracrustal loads. Instead, rapid subsidence results from isostatic adjustment in response to what is effectively the replacement of continental crust by subcrustal lithosphere during extension. The longer-term subsidence results from thermal contraction during cooling, which re-establishes a stable geothermal gradient. The overall tectonic subsidence can be predicted if the extension $\beta(x)$ can be estimated. One method is to assume that the continental crust was uniformly thick before extension and to estimate $\beta(x)$ by measuring present crustal thickness in the belief that crustal thickness changes are solely caused by stretching. This will probably be shown to be an oversimplification but must be tested against observations.

In the same way that loading by sediments and water in foreland basins results in their lateral growth, equivalent loading of rifted margins also causes flexure of the underlying lithosphere. However, the flexural contribution to basin growth depends on the timing of loading because it is only later in the evolution of the basin that the lithosphere achieves sufficient strength to enlarge the basin significantly. This is the same thermal maturity property that determines the size of the foreland basins. In this case, however, the properties of unextended lithosphere on the continental side of a margin do not change appreciably with time (figure 9).

Flexure of this region is primarily determined by the degree of decoupling from the loaded part of extended margin. This decoupling decreases as the margin evolves but does not necessarily result in onlap of the feather edge of the basin (figure 2), particularly if the locus of sedimentation progrades seaward as the basin evolves.

Alternative lithospheric models

In an earlier model of isostasy and flexure at continental margins, Walcott (1972) demonstrated the form that flexure and gravity anomalies would take were a uniform continuous time-invariant elastic lithosphere of assumed flexural rigidity 10^{23} N m. The results suggest rather too much flexure and, as Walcott himself notes, there are good reasons for considering variations in flexural rigidity across the margin. More recently Karner & Watts (1982) have extended the analogy of the isostatic response of cooling oceanic lithosphere to rifted margins. They assume the margin to be laterally uniform and to have a lithospheric thickness that is constant for each margin studied. The results indicate that the lithosphere becomes increasingly thick for successively older margins, but does not thicken as rapidly as oceanic lithosphere. Karner & Watts (1982) interpret this result to be evidence for a continuous increase in thickness of the elastic region of the lithosphere with the time after rifting in an environment of continuous sedimentation. That such an analysis produces reasonable results with respect to isostatic response functions may be fortuitous, for, as Beaumont *et al.* (1982) show, a large part of the gravity anomaly for young margins results from transient thermal expansion that is not related to flexure of the lithosphere when loaded. This thermal component is not included in the analysis by Karner & Watts (1982).

Uplift of the adjacent continent

The deep structure of rifted margin basins remains poorly known, but the shallower structure adjacent to the continent has in many cases been drilled to basement so that the structure, lithology and age of the sedimentary sequence are well defined. Evidence from these regions and from contemporary rift zones provides a more detailed, if incomplete, test of the extension model, one that the simplest uniform extension version of the model apparently fails because no peripheral uplift to the rift zone is predicted for reasonable initial crustal thicknesses (figure 10*a*). Uplift is certainly a feature of modern rifts and young continental margins, and there is evidence (Royden & Keen 1980) that flanking highs 1–2 km high and 50–100 km wide existed at some stage during the evolution of older margins.

At least two explanations, those of lateral heat flow and depth-dependent (non-uniform) extension, have been proposed, to which we add a third, that of regional, as opposed to local, isostatic adjustment during rifting. All three variations (figure 10) cause only second-order changes to the predictions of the basic model; nevertheless, the differences among them may be sufficient that they can be tested by comparison with stratigraphy landward of the hinge line.

The depth-dependent extension model (Sclater *et al.* 1980; Royden & Keen 1980; Beaumont *et al.* 1982) predicts uplift as a result of increased extension, δ, in the lower lithosphere with respect to that, β, in the upper lithosphere. This is shown in figure 10*b*, which diagrammatically places the boundary between the two regions at the Moho, but this is not necessarily so. A more probable decoupling horizon is that corresponding to the brittle–ductile transition, which may also coincide with the level at which listric faults sole. Realistic uplifts can be achieved in this manner when $\beta \approx 1$ and δ is large. The existence of depth-dependent extension is difficult to prove beneath old margins because cooling has largely restored the lower lithosphere to its original state. The signature of the model (figure 11) is one in which uplift is concurrent with extension, depicted as an instantaneous process, and decays as the lithosphere cools and contracts. The sedimentary record would show this as an erosional unconformity on uplift, then a non-depositional interval while the high remained, followed by sediment onlap as subsidence proceeded. In the absence of erosion the original base level will be restored on cooling, for regions where only subcrustal lithosphere was thinned. Depth-dependent extension is capable of producing uplift of 1–2 km, depending on the chosen values of δ and β and on the depth of decoupling (Royden & Keen 1980).

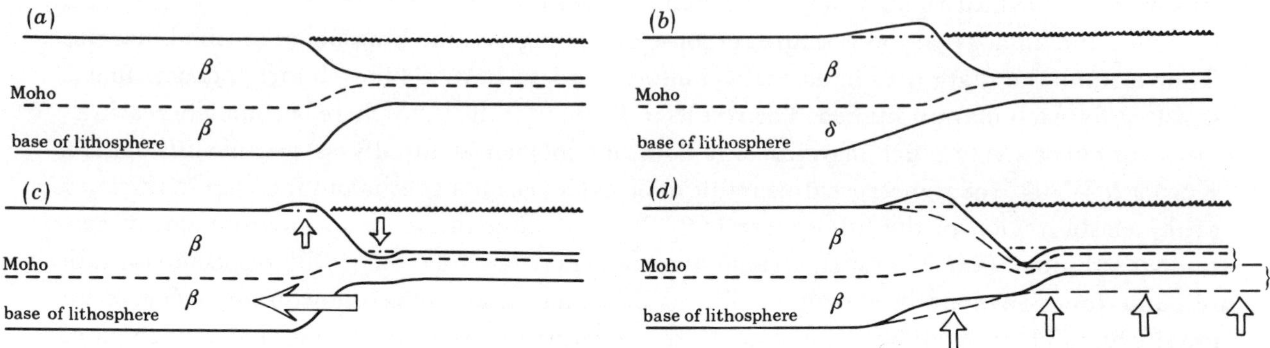

FIGURE 10. Diagrams illustrating the uniform extension model (*a*) and the three proposed models for uplift of the continent during formation of rift margin basins: (*b*) depth-dependent extension and local isostatic equilibrium; (*c*) lateral conduction of heat and local isostatic equilibrium; (*d*) uniform extension and regional isostatic equilibrium.

A second mechanism for uplift is that of lateral heat flux, a second-order effect ignored in the simple rifting models. Where the amount of extension changes rapidly across the margin, lateral temperature gradients might be sufficient to cause significant lateral heat flux. This causes enhanced cooling, contraction and subsidence of the highly extended region and heating, expansion and uplift of the unextended region (figure 10*c*). The sedimentary record would show evidence of an uplift that grew after rifting, achieving its climax after 60 Ma, and then decaying by 120–150 Ma (Beaumont *et al.* 1982) (figure 11). Deposition can, in theory, be delayed for a much longer interval than that predicted by the depth-dependent extension model. The model does, however, appear incapable of generating uplifts that exceed a few hundred metres (Beaumont *et al.* 1982; Steckler 1981).

The model that appeals to regional isostatic adjustment during rifting (figure 10*d*) extends the concept of lithospheric flexure to the rift phase. If the lithosphere retains some structural integrity during extension, lateral coupling will remain a factor in determining the final regional isostatic balance of the rifted margin. If we consider the necking of the lithosphere under tension, but in the absence of gravitational or buoyancy forces, the shape of the extended region, what we term the 'free extension shape', may be similar to that shown by the long broken lines in figure 10*d*. The depth about which necking was centred is determined by the rheology of the lithosphere and cannot be predicted without an analysis that includes the dynamics of extension and realistic rheologies, properties that are currently not well known. It is, however, highly unlikely that the necking will be symmetrical about the mid-depth of the lithosphere, or that the free extension shape will exactly coincide with that of a margin in local isostatic equilibrium. For example, figure 10*d* suggests that the free extension upper surface is at a greater depth than it would be if local isostatic conditions prevailed.

Application of buoyancy and gravitational forces to the free extension configuration will tend to raise the surface to the position of isostatic equilibrium. This is easily accommodated in regions where the lithosphere has little lateral strength. At the margin, however, where the rheological and thermal lithospheres thicken, this uplift will be flexurally resisted by the unextended region, which is already in equilibrium. Some flexural upwarp of the unextended region will occur. Correspondingly, flexural resistance to the buoyancy forces ensures that part

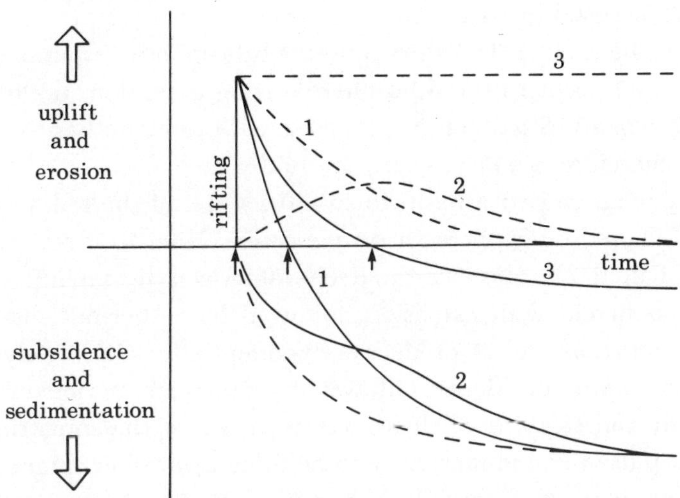

FIGURE 11. Illustrations of the form of uplift and subsidence predicted by the proposed rift models (figure 10) for a point landward of the hinge line where $\beta = 1$. Models 1 (the depth-dependent extension model) and 3 (regional isostatic adjustment during rifting) predict immediate uplift on rifting. This uplift decays for model 1 but remains for model 3 (short broken lines). The effect of lateral heat transport (model 2) is to create a smaller uplift which grows and then decays (short broken line). The solid lines illustrate the superimposed effects when flexural downwarp landward of the hinge line (long broken line) is also included. Note that models 1 and 3 predict uplift followed by subsidence, but that the subsidence of model 3 is much less. Model 2 predicts subsidence but a period of relative quiescence may exist as the thermal uplift grows. The form of the curves will vary with position across the uplifted region and with relative intensity of the competing effects.

of the thinned lithosphere remains relatively depressed. The flexural unwarp–downwarp pair balance to maintain a regional isostatic balance. The sense of the pair would be reversed were the upper surface of the free extension shape above the local isostatic depth. That this is not so may provide some insight into the rheology of lithospheric necking. By analogy with the more familiar concept of flexural downwarp, upwarps 1–2 km high and 50–100 km wide should not require excessive flexural stresses and are within the bounds of flexural rigidity estimates for the lithosphere. The signature of such flexural uplifts is that they occur concurrently with lithospheric extension and that, in the absence of erosion, they do not decay (figure 11). The sedimentary record near the peripheral bulge would be one of uplift and erosion.

One additional factor, that of flexural downwarp landward of the hinge line by sediment and water loads in the subsided areas, must also be superimposed on the simple vertical movement patterns shown as short broken lines in figure 11. In all cases this effect will accelerate onlap of the basin. The solid lines in figure 11 diagrammatically illustrate the combined effect for a point where $\beta = 1$. The long dashed line is the hypothetical flexural downwarp contribution.

The important point to note is that each of the three models predicts a sufficiently distinct pattern of non-deposition, deposition and final depth that there is hope that their relative importance can be distinguished by a careful study of the stratigraphy of the shallow regions of rift margin basins. The unknown effects of erosion and sea level change inevitably militate against this optimism.

A comparison of basin characteristics

Foreland and rifted margin basins are created in predominantly compressional and tensional environments, respectively. The key to their difference is, as previously explained, the source of tectonic subsidence, supracrustal loads versus mass replacement at depth, 'a thickening versus thinning lithosphere' during their early history, and the lithospheric response to loading. By the very nature of changes that occur in the lithosphere during extension, the lithosphere will on average appear weaker beneath rifted margins than beneath foreland basins. The properties of the two basin types are therefore to a large extent mutually exclusive.

Where it is not clear what caused a basin to subside, some of the following characteristics may prove to be useful diagnostic features for simple cases. Other features are quite similar.

1. *Subsidence history*. Rifted margins are often characterized by an initial phase of rapid subsidence, which is concurrent with extension, followed by a thermal contraction phase in which subsidence is proportional to $t^{\frac{1}{2}}$ or decays exponentially. Conversely, foreland basins exhibit episodes of very rapid subsidence followed by quiescent periods. Rapid subsidence signals the emplacement and stacking of thrust sheets at a rate that outstrips sedimentation. The basin deepens in a pulse-like manner, only to be filled as erosion catches up. Commonly, several such pulses occur, indicating more than one discrete thrusting event. Foreland basins are also characterized by episodic and terminal uplifts and erosion that appear as major unconformities.

2. *Basin shape*. Both basin types are asymmetric. They differ in that foreland basins seldom exceed a depth of *ca*. 6 km, have a characteristic convex upward flexural shape, and are bordered by a fold–thrust belt, whereas rifted margin basins can achieve depths of *ca*. 16 km, show little evidence of flexure, except landward of the hinge line, and are not bordered by a fold–thrust belt.

3. *Topography of the basin basement*. The simplest of models suggest that, when a region has been one of subdued topography before the formation of a foreland basin, the topography would undergo little deformation save that of regional flexural warping. Conversely, the process of lithospheric extension during rifting creates rugged relief by near-surface block faulting and tilting, which will be preserved as basement topography beneath the basin. Whether such listric block faulting (Proffett 1977; Montadert *et al.* 1979; Le Pichon & Sibuet 1981; Le Pichon *et al.*, this symposium) can be used to estimate extension remains controversial. Its existence, or that of a horst and graben régime, are however, first-order indicators of a tensional environment in which brittle failure of the upper crust occurred. In contrast, brittle faulting would not be expected beneath flexurally downwarped, but unextended, regions of either rift margins or foreland basins.

A pre-existing faulted upper crust can, however, be inherited. Although bevelled by erosion before basin formation, it can be reactivated during flexure, which is controlled by the underlying intact part of the lithosphere. This, together with detachments that propagate within both the older sediments and crystalline basement beneath the basin during thrusting, can lead to an irregular and faulted basement beneath foreland basins. Such a response appears more common than that exemplified by the Alberta foreland basin, where pre-Jurassic sediments remained largely unfaulted during flexure. Moreover, there is no reason why relative motion of these faulted blocks beneath either type of basin should not continue during basin evolution, thereby generating growth faults that propagate into the overlying basin sediments.

4. *Crustal structure.* If it is assumed that crustal structure is not a relic of previous events, the models predict that the crust beneath rifted margins will be attenuated whereas that beneath foreland basins remain substantially uniform. Whether this is a useful guide to the mode of basin formation depends on the degree to which changes associated with basin formation overprint and obliterate inherited structure.

5. *Gravity anomalies.* Flexural downwarp is a major feature of foreland basins, they are therefore far from local isostatic equilibrium. This is reflected in the large (*ca.* -100 mGal; -10^{-4} m s^{-2}) free air gravity anomalies over the Alberta basin. Rifted margins are seldom characterized by such long wavelength anomalies because they are closer to a state of local isostatic equilibrium.

6. *Sediment lithology.* Coarse-grained clastic sediments, which form the molasse sequences of foreland basins, are very characteristic. Interleaving finer-grained flysch sequences mark phases of rapid subsidence and basin deepening. Black shale and coal sequences are also characteristic of the foreland environment. Other organically related sediments are relatively rare, either because foreland basins fail to achieve sufficient stability for major reef growth, for example, or the influx of clastics gives an unsuitable environment. Evaporites are also relatively less common than in rifted margins, where they often occur as early post-rift deposits. This is because foreland basins spend little time in a starved shallow marine state.

Coarse-grained clastics are also a feature of the rift phase of rifted margins reflecting topographic elevation changes and block faulting. These neighbouring topographic highs decay as the basin evolves. Consequently, there is normally a progressive change to finer-grained sediments or to a deltaic sand shale environment when there is a large influx of fluvial sediment. In suitable environments the later phases of slower subsidence are often marked by reef carbonates. Deltas can also be built into foreland basins, particularly when rapid deepening of the basin is followed by rapid erosion of the fold–thrust belt. Palaeocurrent indicators may in this case provide evidence of the existence of a distal feather edge to the basin at the peripheral uplift, though the effect of a conjugate margin in young rift basins may be very similar.

7. *Thermal history of sediments.* If it is assumed that conduction dominates heat transport in both types of sedimentary basin, sediments deposited in rift basins during the early stages of their evolution will be subject to a significantly enhanced non-equilibrium geothermal gradient. In contrast, the geothermal gradient in foreland basins that form on thermally mature lithosphere will be normal throughout the evolution of the basin. The enhanced geothermal gradient in rift margin basins is approximately proportional to the amount of extension, and decays with a time constant that is proportional to the unperturbed thickness of the thermal lithosphere (Royden & Keen 1980; Royden *et al.* 1980; Beaumont *et al.* 1982). Because thermal metamorphism is directly related to temperature history, all other things being equal, equivalent levels of thermal maturation will be reached at smaller depths in rift margin basins than in foreland basins. The effect of erosion, which reduces the depth of burial and the temperature of individual strata, must also be taken into account (Beaumont 1981), as must the contributions to the geothermal gradient from radioactive heat generation in the crust and sediments (Beaumont *et al.* 1982). Some caution is needed in applying these basic ideas because

temperature gradients in both types of basin can be perturbed by local conditions, for example by the effect of salt diapirs of high thermal conductivity, or on a more regional scale by heat transport by fluid flow.

Superposition of basins

In previous sections, foreland and rift margin basins have been considered to occupy very different spatial and tectonic environments. While this is an appropriate way to discuss type examples, the cycle of opening and closing ocean basins and accompanying interactions at plate margins leads to the interchange of tensional and compressional environments, which can,

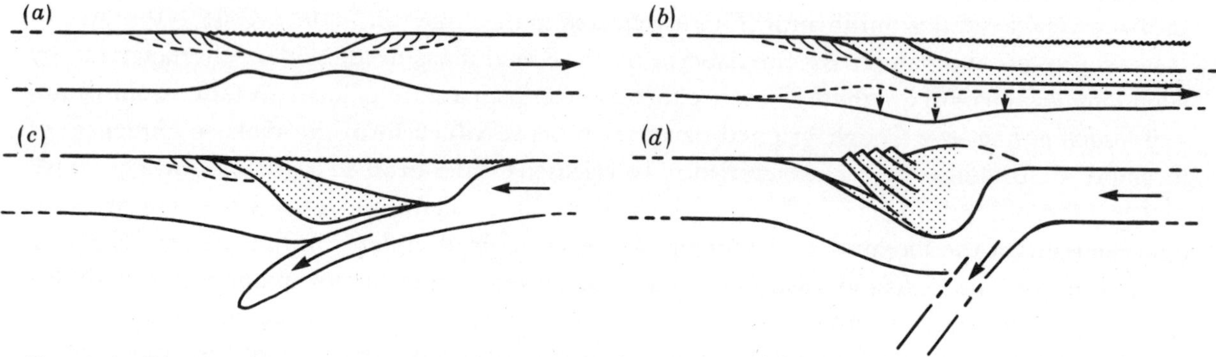

Figure 12. Diagrams illustrating the way in which a foreland basin can be superimposed on a rifted margin basin as a result of ocean closure. (*a*) Phase 1: rifting and initial subsidence. (*b*) Phase 2: drifting and thermal subsidence. (*c*) Phase 3: closure and stabilized margin. (*d*) Phase 4: collision and flexural subsidence.

in turn, lead to superposition of these basins at the same location. The simplest example, which involves only extension and compression (figure 12), illustrates this cycle for an ocean basin that existed for a timespan that was longer than several lithospheric thermal time constants. Phase 1 is that of rifting, extension and attenuation of the lithosphere. The brittle faulting of the upper crust, illustrated by listric normal faults, also serves to thin that part of the lithosphere, which acts as an intact elastic core during flexure. It is likely that these faults are not subsequently annealed and that they can be reactivated by added extension, or reversed by compression. In so far as the faulted region may represent a substantial part of the rheological lithosphere (Beaumont *et al.* 1982), we expect that the faulted lithosphere will remain flexurally weak even when thermally mature. Moreover, unfaulted zones on either side will probably become relatively stronger on cooling. Phase 2 illustrates the thickening of the thermal lithosphere during the evolution of the rifted margin basin in what is often termed its passive margin phase. Phase 3 shows the initiation of subduction at a hypothesized location sufficiently seaward that the margin remains intact during ocean closure. This is obviously a simplified view, but one that is adequate to illustrate the concept of superposition. Essentially the same development would follow were the polarity of subduction reversed. Phase 4 is that of encounter and collision with the conjugate margin, which resists subduction. The lateral forces provide the driving mechanism that disrupts the rift margin sequence and transports it landward as a succession of thrust sheets. As these are stacked onto the old rift margin, their load creates the foredeep and foreland basin. Depending on the level at which detachment occurs and on the amount of compression, the foreland sequences is superimposed on all or part of the rift margin sequence.

The foreland basin is, from this point of view, a 'successor basin', a term that we do not favour because it has no genetic connotation.

More complicated examples, in which there is more than one collision event, or there are interactions with major edifices on oceanic lithosphere, can lead to successive periods of thrusting that are recorded in the foreland sequence. If strike-slip motions are also considered, some indication of the way in which an orogen was assembled can be deduced. Purely strike-slip motions cause no overthrusting; therefore no foreland basin is created. With increasingly normal convergence the importance of overthrusts and the foreland basin increases. This view adds further credence to, for example, interpretations that suggest that the Acadian and Alleghenian Orogenies in the Appalachian fold belt were the result of largely strike-slip and normal movements in the northern and southern regions respectively.

The central Appalachians also provide an ideal example of the superposed nature of foreland sequences. At least three separate molasse wedges, separated by unconformities and corresponding to the major orogenies, have been superimposed on the Cambro–Ordovician rift sequence that corresponds to the opening of the Iapetus Ocean. Each of these sequences must be considered as a separate entity in any analysis of basin subsidence, not forgetting those parts of the stratigraphic sequence lost during uplift and erosion.

In so far as stresses acting within the lithosphere during collision are responsible for reactivation of inherited fault structures, many of the local structures of the fold–thrust belt and the foreland may reflect these stresses. We suggest that where previous basement faults exist, these will control the pattern of overthrusting, with the faults breaking upward through the overlying intact rift margin sediments. The telescoping of the margin restores much of the extension that occurred on rifting. In regions landward of pre-existing extensional faults the detachment surface will be confined to weak zones in the sedimentary sequence. This view appears to be corroborated by evidence from the Rocky Mountains in Canada (R. A. Price, personal communication) in which inferred basement involvement in thrusting does not occur eastward of the palinspastically restored position of the hinge line to the underlying rift margin sequence. Collisional stresses may also be responsible for the reactivation of horizontal motions further into the interior of the continents. It is probably no accident that the initiation of subsidence on the Hudson's Bay, Michigan and Illinois cratonic basins is coeval with the Taconic Orogeny in the Appalachians.

The implication of this discussion is that a number of factors influence the structure of foreland basins, the thermal age of the margin, the amount of overthrusting and angle of convergence, the existence and amount of reactivation of inherited basement faults, and the amount of uplift and erosion of the fold–thrust belt. The diversity of foreland basins must reflect changes in the relative importance of these factors in each case.

Conclusions

The conclusions of this descriptive comparison of foreland and rifted margin basins are simple and can be summarized as follows.

1. These basins form in respectively compressional and extensional environments within what may be regarded as lithospheric plate boundaries and as a consequence of lateral motions of the plates.

2. Tectonic subsidence of foreland basins is primarily the result of flexural downwarp of the lithosphere under supracrustal loads of the adjacent fold–thrust belt.

3. Tectonic subsidence of rifted margin basins primarily results from mass replacement at depth, during lithospheric extension on rifting, and the subsequent conductive cooling and contraction of the extended lithosphere.

4. Lateral extension of the basins beyond areas immediately influenced by the tectonic subsidence is controlled by lithospheric flexure under sediment and water loads. The flexural wavelength is primarily a function of the thermal age of the underlying lithosphere, which implies that flexure will be less important in rifted margin basins than in foreland basins.

5. The lithosphere may be modelled as a uniform elastic or viscoelastic layer supported by an inviscid fluid. Viscoelastic (Maxwell) models predict basins that shrink in size as stress is relaxed, whereas elastic models having a time-invariant flexural rigidity predict basins with constant width. More complex rheologies are considered likely to have intermediate extrinsic properties.

6. Second-order effects, such as lateral changes in lithospheric properties or some of the consequences of lithospheric cooling, will modify basin evolution.

7. Basin characteristics reflect a subtle interplay of tectonic subsidence, lithospheric flexure, sea level change, surface topography and inherited properties. All these factors must be considered in an analysis of subsidence history.

8. Changes in tectonic environment that accompany the change from opening to closing of an ocean basin, can lead to the superposition of foreland and rifted margin basins.

We thank R. A. Price for useful discussions, G. Quinlan and R. T. Haworth for comments on the manuscript, and the Natural Sciences and Engineering Research Council for financial support of this research through the Strategic Grants programme.

References (Beaumont *et al.*)

Ashby, M. F. & Verrall, R. A. 1977 *Phil. Trans. R. Soc. Lond.* A **288**, 59–95.
Aubouin, J. 1965 *Geosynclines.* New York: Elsevier.
Beaumont, C. 1978 *Geophys. Jl R. astr. Soc.* **55**, 471–497.
Beaumont, C. 1979 *Tectonophysics* **59**, 347–366.
Beaumont, C. 1981 *Geophys. Jl R. astr. Soc.* **65**, 291–329.
Beaumont, C., Keen, C. E. & Boutilier, R. 1982 *Geophys. Jl R. astr. Soc.* (In the press.)
Dewey, J. F. & Bird, J. M. 1970 *J. geophys. Res.* **75**, 2625–2647.
Dickinson, W. R. 1974 In *Tectonics and sedimentation* (ed. W. R. Dickinson) (*Spec. Publ. Soc. econ. Geol. Paleont.* no. 22), pp. 1–27.
Hunziker, J. C. 1974 *Mem. 1st Geol. Min. Univ., Padova*, no. 31.
Jarvis, G. T. & McKenzie, D. P. 1980 *Earth planet. Sci. Lett.* **48**, 42–52.
Karner, G. D. & Watts, A. B. 1982 *J. geophys. Res.* (In the press.)
Kay, M. 1951 *Mem. geol. Soc. Am.* no. 48, pp. 1–143.
Keen, C. E. 1979 *Can. J. Earth Sci.* **16**, 505–522.
Keen, C. E., Beaumont, C. & Boutilier, R. 1981*a* *Oceanologica acta* **4** (suppl.), 123–228.
Keen, C. E., Beaumont, C. & Boutilier, R. 1981*b* Presented at Hedberg Research Conference of the Am. Assoc. Petrol. Geol., Galveston.
Keen, C. E. & Hyndman, R. D. 1979 *Can. J. Earth Sci.* **16**, 712–747.
Krumbein, W. C. & Sloss, L. L. 1963 *Stratigraphy and sedimentation*, 2nd edn. San Francisco: W. H. Freeman.
Le Pichon, X. & Sibuet, J.-C. 1981 *J. geophys. Res.* **86**, 3708–3720.
McKenzie, D. P. 1978 *Earth planet. Sci. Lett.* **40**, 25–32.
Montadert, L., de Charpal, O., Roberts, D., Guennoc, P. & Sibuet, J.-C. 1979 In *Deep drilling results in the Atlantic Ocean: continental margins and paleoenvironment* (*Maurice Ewing Series*, vol. 3), pp. 154–186. Washington, D.C.: American Geophysical Union.
Oxburgh, E. R. 1968 *Proc. geol. Ass.* **76**, 1–46.
Parsons, B. & Sclater, J. G. 1977 *J. geophys. Res.* **82**, 803–827.

Price, R. A. 1973 In *Gravity and tectonics* (ed. K. A. De Jong & R. A. Scholten), pp. 491–502. New York: Wiley-Interscience.
Proffett, J. M., Jr 1977 *Bull. geol. Soc. Am.* **88**, 247–266.
Ranalli, G. 1980 *Can. J. Earth Sci.* **17**, 1499–1505.
Richardson, S. W. & England, P. C. 1979 *Earth planet. Sci. Lett.* **42**, 183–190.
Royden, L. & Keen, C. E. 1980 *Earth planet. Sci. Lett.* **51**, 343–361.
Royden, L., Sclater, J. G. & Von Herzen, R. P. 1980 *Bull. Am. Ass. Petrol. Geol.* **64**, 173–187.
Sclater, J. G., Royden, L., Horváth, F., Burchfiel, B. C., Semkin, S. & Stegena, L. 1980 *Earth planet. Sci. Lett.* **51**, 139–162.
Sheridan, R. E., Grow, J. A., Berhendt, J. D. & Bayer, K. C. 1979 *Tectonophysics* **59**, 1–26.
Sleep, N. H. 1971 *Geophys. Jl R. astr. Soc.* **24**, 325–350.
Steckler, M. S. 1981 *Trans. Am. geophys. Un.* **17**, 390.
Walcott, R. I. 1970 *Tectonophysics* **9**, 435–446.
Walcott, R. I. 1972 *Bull. geol. Soc. Am.* **83**, 1845–1848.
Watts, A. B. 1978 *J. geophys. Res.* **83**, 5989–6004.
Watts, A. B. & Ryan, W. B. F. 1976 *Tectonophysics* **36**, 25–44.
Watts, A. B. & Steckler, M. S. 1979 In *Deep drilling results in the Atlantic Ocean: continental margins and paleoenvironment* (*Maurice Ewing Symposium Series*, vol. 3), pp. 218–234. Washington, D.C.: American Geophysical Union.

Phil. Trans. R. Soc. Lond. A **305**, 319–324 (1982) [319]
Printed in Great Britain

Origin of the lithospheric tension causing basin formation

By M. H. P. Bott, F.R.S.

Department of Geological Sciences, University of Durham, South Road, Durham DH1 3LE, U.K.

Most fault-controlled basin formation within plate interiors occurs by normal faulting in response to horizontal deviatoric tension in the continental crust. It is suggested that the tension originates either from the plate boundary forces acting at trenches or as a result of isostatically compensated uplifted regions such as East Africa. The tension produced by both mechanisms is greatest in high heat-flow regions where the upper elastic part of the lithosphere is thinned and weakened. Particularly widespread tension in the continental lithosphere occurs when subduction takes place on opposite sides of a large continental mass, such as Pangaea in the early Mesozoic, where it led to widespread graben formation and, in association with hot spot activity, to continental splitting.

1. Introduction

Sedimentary basins form in many tectonic settings and complicated schemes of classification are needed to do justice to the variety of structural environments (see, for example, Bally & Snelson 1980). However, two main classes can be recognized. These are (1) the narrow basins of graben type that subside as a result of faulting in the basement, and (2) the broad basins that form by flexural downwarping without conspicuous faulting. Fault-controlled basins clearly form as a result of stressing of the continental lithosphere. Excluding some thrust-faulted basins in active mountain ranges, fault-controlled basins are generally associated with normal faulting (which may be listric), which indicates horizontal deviatoric tension. This is supported by earthquake mechanism studies in regions where graben are forming at present. In contrast, the flexural subsidence of broad basins such as the Tertiary North Sea basin is probably primarily caused by the cooling of an initially hot lithosphere with sediment loading. The subsidence is apparently not stress controlled. However, the subsidence of the North Sea is dependent on thinning of the crust by about 40 % which occurred possibly before subsidence. Similar thinning of the crust occurs beneath passive margins before their drifting stage of development. McKenzie (1978) suggested that such thinning of the crust occurs as a result of stretching of the continental lithosphere, implying an episode of strong tension before flexural subsidence. This paper examines possible sources of the horizontal deviatoric tension in the continental lithosphere that causes graben formation and more controversially may cause extreme stretching and thinning of the crust.

The deviatoric tension associated with basin formation needs to be substantial. According to Bott (1976), persistent horizontal tension of about 200 MPa (2 kbar) at least is required to form rift valleys a few kilometres deep. The tension also needs to be of a persistent and renewable type as otherwise the amount of strain that can be released would be insufficient to cause subsidence on the observed scale.

Four possible relevant sources of tension in the continental lithosphere are as follows: (1) drag on the base of the lithosphere exerted by upwelling and diverging convection currents moving faster than the lithospheric plates (see, for example, Morgan 1972); (2) plate driving forces acting on the edges of lithospheric plates, notably the 'slab pull' and 'trench suction' forces acting on subducting and overriding plates respectively at trenches (Forsyth & Uyeda 1975); (3) tensile stress associated with isostatically compensated uplifted structures underlain by crustal thickness variations or other lateral density variations in the lithosphere (Bott & Dean 1972; Artyushkov 1973; Bott & Kusznir 1979); (4) membrane stresses, which produce horizontal tension in the interior of a plate moving towards the equator or in the exterior of a plate moving towards the poles (Turcotte 1974). It is assumed here, on geological and thermodynamic grounds, that plates are driven by edge forces rather than by the drag of faster-moving underlying convection currents, so that drag on the base of the lithosphere is not regarded as a primary source of tension. It is doubtful whether the relief of membrane stress can produce enough stretching to form basins on the observed scale. Thus plate edge forces and isostatically compensated elevated features appear to be the most likely sources of the tension giving rise to basin subsidence.

2. Plate boundary forces

The most obvious source of stress in the continental lithosphere and crust is the system of plate-driving forces that are believed to act on the edges of plates at ocean ridges and trenches (Forsyth & Uyeda 1975). The edge forces acting on lithospheric plates at ridges and trenches are counteracted partly by local resistance to plate motion in the vicinity of the plate boundaries and partly by mantle drag on the base of the moving lithosphere.

At ocean ridges, the hot and consequently relatively low-density material that upwells to form new oceanic lithosphere exerts a compressive 'ridge push' force on the adjacent lithospheric plates, acting to force them apart. This produces a horizontal compression of about 30 MPa (300 bar) acting across the thickness of the lithosphere (Forsyth & Uyeda 1975).

At subduction zones, the negative buoyancy of the sinking slab exerts a pull on the plate to which it is attached, causing tension. This force is referred to as 'slab pull'. Less obviously, the subduction process may also give rise to tension in the overriding plate (Elsasser 1971; Forsyth & Uyeda 1975). This is the 'trench suction' force, which can be attributed to the lack of support for the overriding plate at the trench. The effectiveness of the slab pull and trench suction forces in giving rise to horizontal tension within the adjacent plates depends critically on the local resistances to plate motion, which may vary in the short or long term, but stress differences of at least 30 MPa can probably be produced under favourable conditions. At present, the trench suction force appears to be more in evidence along the western margin of the Pacific, where marginal basins occur, than along the eastern margin. This may be because the older dense lithosphere is being subducted in the west or it may be related to differences in absolute plate motions at the two sides of the ocean.

The edge forces acting on plates at ridges and trenches can give rise to a variety of stress régimes within the plates. A plate with ocean ridges on opposite sides would be expected to be in horizontal compression throughout from the ridge push forces acting on both edges. If an ocean ridge occurs on one side and a trench on the opposite side, then gradation from compression at the ridge end possibly to tension at the trench end would be expected, depending on the resistance near the trench (figure 1*a*, *b*). The trench suction force acting on opposite sides of a plate formed entirely of continental lithosphere would cause the whole plate to be subjected to horizontal deviatoric tension (figure 1*c*). Such a continental plate would not override the adjacent oceans as Asia and the Americas are now overriding the Pacific Ocean. Consequently, the trench suction force would not be opposed by local resistive forces to the extent that probably now applies around the Pacific. Consequently, the trench suction force would be more effective at stressing the continental plate than at the present time.

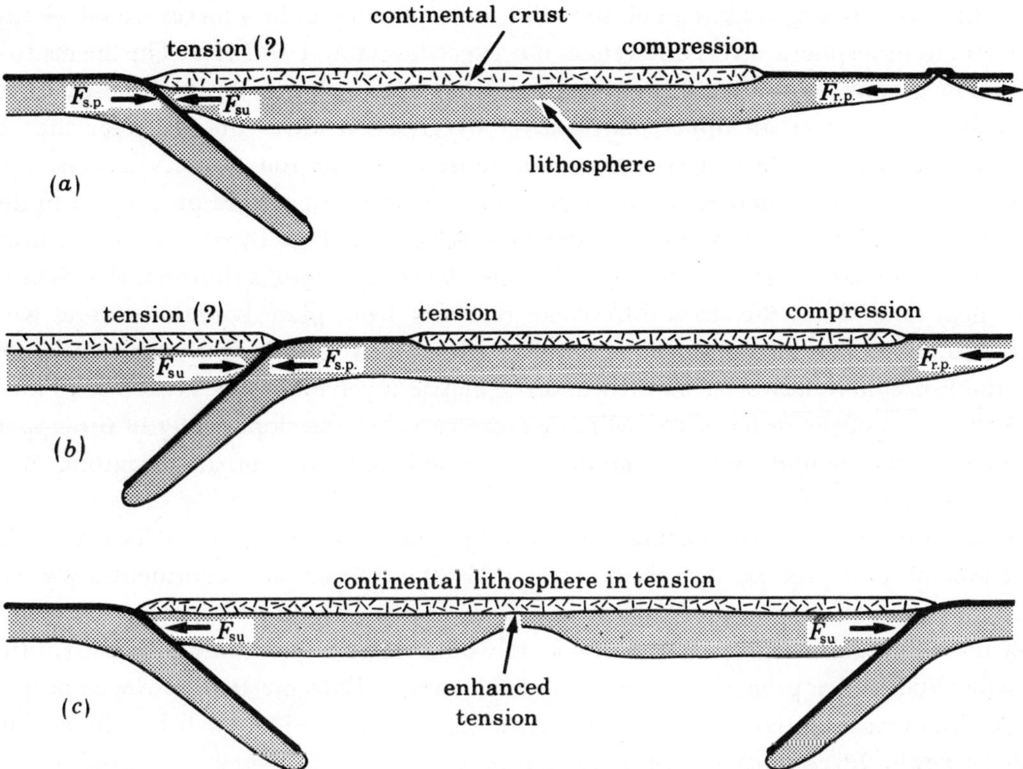

FIGURE 1. Tension in the continental lithosphere produced by plate edge forces. (*a*) Trench suction (F_{su}) at one edge and ridge push ($F_{r.p.}$) at the other edge causing a linear gradation of stress from compression near ridge possibly to tension near trench, depending on the extent of resistance near the trench. An example is the present-day South American plate. (*b*) Tension developed in a continent marginal to a narrow closing ocean as a result of the slab pull force ($F_{s.p.}$). An example is the formation of Carboniferous basins on the northern flank of the closing Hercynian Ocean. (*c*) Tension throughout a continental plate produced by trench suction at subduction zones on opposite sides. An example is Pangaea in the early Mesozoic.

During Tertiary to Recent time, all the major plates have an ocean ridge on at least one side, and thus any tension resulting from plate boundary forces is likely to be restricted to the part of the plate in the vicinity of a trench. At certain periods in the past, however, plate arrangements have been more favourable to widespread tension, such as when subduction was probably occurring on opposite sides of the Permo-Triassic supercontinent Pangaea before and during its early Mesozoic break-up. This tension may have caused the widespread formation of fault-controlled basins characteristic of the early Mesozoic and may also have given rise, with hot spot activity, to the continental break-up.

It has been suggested that tension may occur throughout a continental plate at certain rather special periods in Earth history when subduction is occurring on opposite sides, and that more generally, tension may occur in those parts of plates sufficiently close to adjacent trenches. The stresses in the vicinity of the trenches would also be expected to fluctuate on a short and long timescale as the subduction process evolves, and this may cause time variation in the subsidence, including the periodic type of subsidence that produces 'cyclic' sedimentation. Relatively simple stress systems have been considered here. More complicated situations must occur as a result of local conditions near plate boundaries, such as transform faulting and irregularities in the plate boundaries. These may produce local regions of tension of types not discussed above.

If the lithosphere is treated as an elastic solid, then plate boundary forces would be expected to give rise to lithospheric stress differences not exceeding about 40 MPa. This seems to be too low to allow graben containing 5 km of sediment to form. If, however, the lithosphere is more realistically regarded as an upper elastic layer overlying a lower ductile layer that can be treated as viscoelastic, then stress relaxation must occur in the lower viscoelastic region, causing a corresponding increase in stress in the elastic layer roughly in proportion to the ratio of lithospheric thickness to elastic layer thickness (Kusznir & Bott 1977). The stress differences produced in this way are greatest in regions where the elastic layer is thinnest, that is in regions of high heat flow. Thus the stress differences resulting from plate boundary forces would be expected to be smallest in cool shield regions and greatest in regions overlying mantle hot spots where the lithosphere has been thinned and the elastic layer may only be about 10 km thick. Stress differences of the order of 200 MPa or more can thus develop locally in the upper crust as a result of plate boundary forces, giving a possible explanation of the formation of certain types of fault-controlled basins.

Exceptionally large horizontal tensile stress might be expected to occur locally during the continental splitting process. If a crack develops in the interior of a continental plate that is subjected to general tension, then greatly increased tension would be expected to occur at the ends of the magma-filled propagating crack. It would behave like a large-scale Griffith crack filled with fluid under pressure and subjected to tension. Extreme tension developing in this way may cause substantial stretching and thinning of the continental lithosphere along the line of the newly developing passive margin.

3. Stresses associated with elevated regions

Localized tension within the continental crust occurs in regions of major surface elevation that are isostatically supported by low density rocks beneath, such as mountain ranges and plateau uplift regions (Artyushkov 1973; Bott & Kusznir 1979). The horizontal deviatoric tension is the result of the increased vertical pressure caused by the load of the surface topography and the upthrust of the compensating low density rocks beneath. This type of stress system causes the continental side of a passive margin to be in tension relative to the oceanic side (Bott & Dean 1972). Similarly, local crustal tension occurs in elevated plateau regions that are supported by low density rocks in the underlying upper mantle, such as the western United States and East Africa.

If the lithosphere is treated as wholly elastic, then the stress differences associated with an elevated plateau 2 km high that is isostatically compensated in the upper mantle are about 40 MPa. However, if the lithosphere is more realistically modelled in terms of an upper elastic layer underlain by a viscoelastic region (figure 2), then stress differences of about 200 MPa are produced in an elastic layer 10 km thick. Thus in elevated regions of high heat flow such as East Africa and the Basin and Range province of the western United States, tensile stress adequate to form substantial rift valleys is probably present. These elevated regions are the

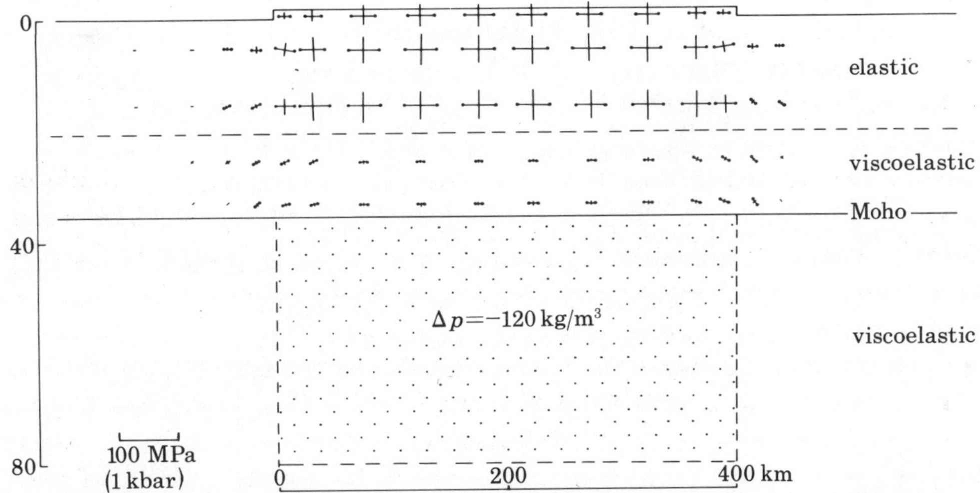

FIGURE 2. Deviatoric stresses in the continental crust produced by a compensated plateau uplift structure as a result of surface loading and upthrust of the compensating low-density mantle, assuming an upper elastic crust underlain by a viscoelastic lower crust and upper mantle. Even larger stresses are produced if the elastic layer is thinner. Dots indicate deviatoric tension. Modified from Bott & Kusznir (1979) and reproduced from Bott (1981).

location of some of the most conspicuous graben formation occurring during the Tertiary. The graben probably form as a result of the local stress system (Bott & Kusznir 1979).

4. DISCUSSION

In summary, horizontal deviatoric tension large enough to account for fault-controlled basin subsidence may occur both as a result of plate boundary forces at trenches and also in local regions of isostatically compensated elevated topography. The most widespread graben formation probably occurs at periods when a continental region is bordered on opposite sides by subduction zones. The largest tensions may occur at the time of continental splitting at the tips of the propagating crack.

Three examples of the application of these mechanisms to existing basins are as follows.

(1) *Slab pull*: the Lower Carboniferous basins of Britain formed on the northern shelf of the closing Hercynian marginal sea. It is suggested that the tension causing the formation of dominantly east–west fault-controlled basins originated as slab pull at the opposite edge of the closing ocean (Johnson 1981). Periodic variation in the stress as the subduction evolved in the closing ocean may give rise to the cyclic sedimentation so characteristic of the Carboniferous.

(2) *Trench suction*: the dominantly north–south orientated Permo-Triassic and Jurassic grabens of the North Sea and adjacent regions may have formed as a result of tension throughout Pangaea before and during break-up, originating as a result of subduction on opposite sides of the supercontinent.

(3) *Hot spot activity*: the Tertiary grabens of East Africa and the Basin and Range province are attributed to the local stress systems associated with these isostatically compensated elevated regions where the lithosphere is heated and thinned, rather than to plate boundary effects.

References (Bott)

Artyushkov, E. V. 1973 *J. geophys. Res.* **78**, 7675–7708.
Bally, A. W. & Snelson, S. 1980 *Mem. Can. Soc. Petrol. Geol.* **6**, 9–94.
Bott, M. H. P. 1976 *Tectonophysics* **36**, 77–86.
Bott, M. H. P. 1981 *Tectonophysics* **73**, 1–8.
Bott, M. H. P. & Dean, D. S. 1972 *Nature, phys. Sci.* **235**, 23–25.
Bott, M. H. P. & Kusznir, N. J. 1979 *Geophys. Jl R. astr. Soc.* **56**, 451–459.
Elsasser, W. M. 1971 *J. geophys. Res.* **76**, 1101–1112.
Forsyth, D. & Uyeda, S. 1975 *Geophys. Jl R. astr. Soc.* **43**, 163–200.
Johnson, G. A. L. 1981 *Proc. Yorks. geol. Soc.* **43**, 221–252.
Kusznir, N. J. & Bott, M. H. P. 1977 *Tectonophysics* **43**, 247–256.
McKenzie, D. 1978 *Earth Planet. Sci. Lett.* **40**, 25–32.
Morgan, W. J. 1972 *Bull. Am. Ass. Petrol. Geol.* **56**, 203–213.
Turcotte, D. L. 1974 *Geophys. Jl R. astr. Soc.* **36**, 33–42.

Phil. Trans. R. Soc. Lond. A **305**, 325–338 (1982) [325]
Printed in Great Britain

Musings over sedimentary basin evolution

By A. W. Bally†
Shell Oil Company, P.O. Box 481, *Houston, Texas* 77001, *U.S.A.*

Introduction

Field geologists and explorationists are of necessity immersed in numerous very detailed surface and subsurface observations. They are often perplexed by the choice of relatively simple geophysical models that so elegantly explain the origin and evolution of sedimentary basins. Geophysicists, on the other hand, search for a simple theme to explain the origin of sedimentary basins and, much like managers, are often impatient with lengthy detailed geological discourse that often uses fancy jargon to hide the very real difficulty that geologists have in separating important evidence from mere encyclopaedic description.

The following musings address the quality and limitations of geologic and geophysical evidence that may be used to evaluate the relative roles of stress, thermal effects and gravity loading, which have been so lucidly summarized by M. H. P. Bott in the preceding summary. The fine papers presented during this meeting, of course, have led to significant modifications of some of my earlier thoughts. Because these have been previously published elsewhere (Bally & Snelson 1980; Bally 1980), they are summarized here only for the convenience of the reader.

The definition of a sedimentary basin and the need for an ideal information package

It may be useful to define a sedimentary basin as a region that has subsided, that contains sediments in excess of 1 km, and – most important – that is *today still preserved in a more or less coherent form.* Such a definition deliberately excludes former basins whose sedimentary content is now incorporated and deformed in folded belts. The hazards of palinspastic reconstructions of folded belts are great, particularly when little, if any, information is available on the basement configuration that was originally underlying the now allochthonous sequences of folded belts. Therefore, I feel that data from folded belts are so far simply inadequate for a test of any geophysical models that describe basin origin.

The Zagros Basin of Koop (this symposium), the Franklinian Basin of Kerr (this symposium), and some of the Precambrian basins of Bickle & Eriksson (this symposium) are simplified hypothetical reconstructions of folded belts; and therefore – despite their intrinsic interest – they are of little use to test the validity of geophysical models. On the other hand, the stratigraphic evolution of folded belts is important if we are to understand their regional or plate tectonic setting and the temporal interaction of tectonics and sedimentation.

Even in basins that today are still preserved in a more or less coherent fashion we rarely have adequate and complete data. For instance, for many basins in North America we lack crustal refraction and reflexion data. Regional reflexion sections across basins are rarely published. For the basins in the U.S.S.R., we typically lack reflexion seismic data and detailed lithological and

† Present address: Department of Geology, Rice University, P.O. Box 1892, Houston, Texas 77001, U.S.A.

mechanical log suites. Only for a handful of basins in the world do we have the beginnings of a reasonably complete data bank that is also accessible to academic scientists, e.g. the North Sea, the Alberta Basin, the Paris Basin, the Aquitaine Basin, the northern Bay of Biscay, and parts of the North American Atlantic margin. Therefore, we must endeavour to obtain the most complete data base possible.

Such an 'ideal data base' necessary for the evaluation of geophysical models should include crustal refraction and reflexion profiles, conventional multichannel basinal reflexion profiles, deep stratigraphic tests with complete lithological, palaeontological and geophysical log sets, surface and subsurface maps.

Table 1. Basin classification

1. BASINS LOCATED ON THE RIGID LITHOSPHERE, NOT ASSOCIATED WITH FORMATION OF MEGASUTURES
    - 1.1. related to formation of oceanic crust
        - 1.1.1. *rifts*
        - 1.1.2. *oceanic transform-fault associated basins*
        - 1.1.3. *oceanic abyssal plains*
        - 1.1.4. *Atlantic-type passive margins (shelf, slope and rise) that straddle continental and oceanic crust*
            - 1.1.4.1. overlying earlier rift systems
            - 1.1.4.2. overlying earlier transform systems
            - 1.1.4.3. overlying earlier back-arc basins of (3.2.1.) and (3.2.2.) type
    - 1.2. located on pre-Mesozoic continental lithosphere
        - 1.2.1. *cratonic basins*
            - 1.2.1.1. located on earlier rifted grabens
            - 1.2.1.2. located on former back-arc basins of (3.2.1.) type

2. PERISUTURAL BASINS ON RIGID LITHOSPHERE ASSOCIATED WITH FORMATION OF COMPRESSIONAL MEGASUTURE
    - 2.1. deep sea trench or moat on oceanic crust adjacent to B-subduction margin
    - 2.2. *foredeep and underlying platform sediments*, or moat on continental crust adjacent to A-subduction margin
        - 2.2.1. ramp with buried grabens, but with little or no block faulting
        - 2.2.2. dominated by block faulting
    - 2.3. *Chinese-type basins* associated with distal block faulting related to compressional or megasuture and without associated A-subduction margin

3. EPISUTURAL BASINS LOCATED AND MOSTLY CONTAINED IN COMPRESSIONAL MEGASUTURE
    - 3.1. associated with B-subduction zone
        - 3.1.1. *fore-arc basins*
        - 3.1.2. *circum-Pacific back-arc basins*
            - 3.1.2.1. back-arc basins floored by oceanic crust and associated with B-subduction (marginal sea *sensu stricto*)
            - 3.1.2.2. back-arc basins floored by continental or intermediate crust, associated with B-subduction
    - 3.2. back-arc basins, associated with continental collision and on concave side of A-subduction arc
        - 3.2.1. on continental crust or *Pannonian-type* basins
        - 3.2.2. on transitional and oceanic crust or *W. Mediterranean-type* basins
    - 3.3. basins related to episutural megashear systems
        - 3.3.1. *Great basin-type* basin
        - 3.3.2. *California-type* basins

The variety among sedimentary basins

The variety among sedimentary basins, their tectonic setting and their filling history is overwhelming. It is precisely that variety that has in the past and will in future remain the big challenge for petroleum explorationists. Many authors have proposed various basin classifications and tried to find some order. One such scheme was recently proposed by Bally & Snelson

(1980). Table 1 and figure 1 illustrate the main elements of that classification. These illustrations were the outgrowth of a rather cursory study of the tectonics of the world, illustrated by maps that were displayed in the poster session of this meeting. [Those maps do not appear in this volume.]

Our primary goal was to see whether a classification that was in sympathy with and related to plate tectonics would help to establish simple rules for forecasts of hydrocarbon potential reserves. In this we failed, probably because the distribution of major hydrocarbon source bed intervals is controlled by factors (e.g. biological evolution, palaeoclimatology and regional and global palaeogeography) that often do not sufficiently coincide with the evolution of any given basin type. Nevertheless, the proposed classification provides a guide to and an overview of the

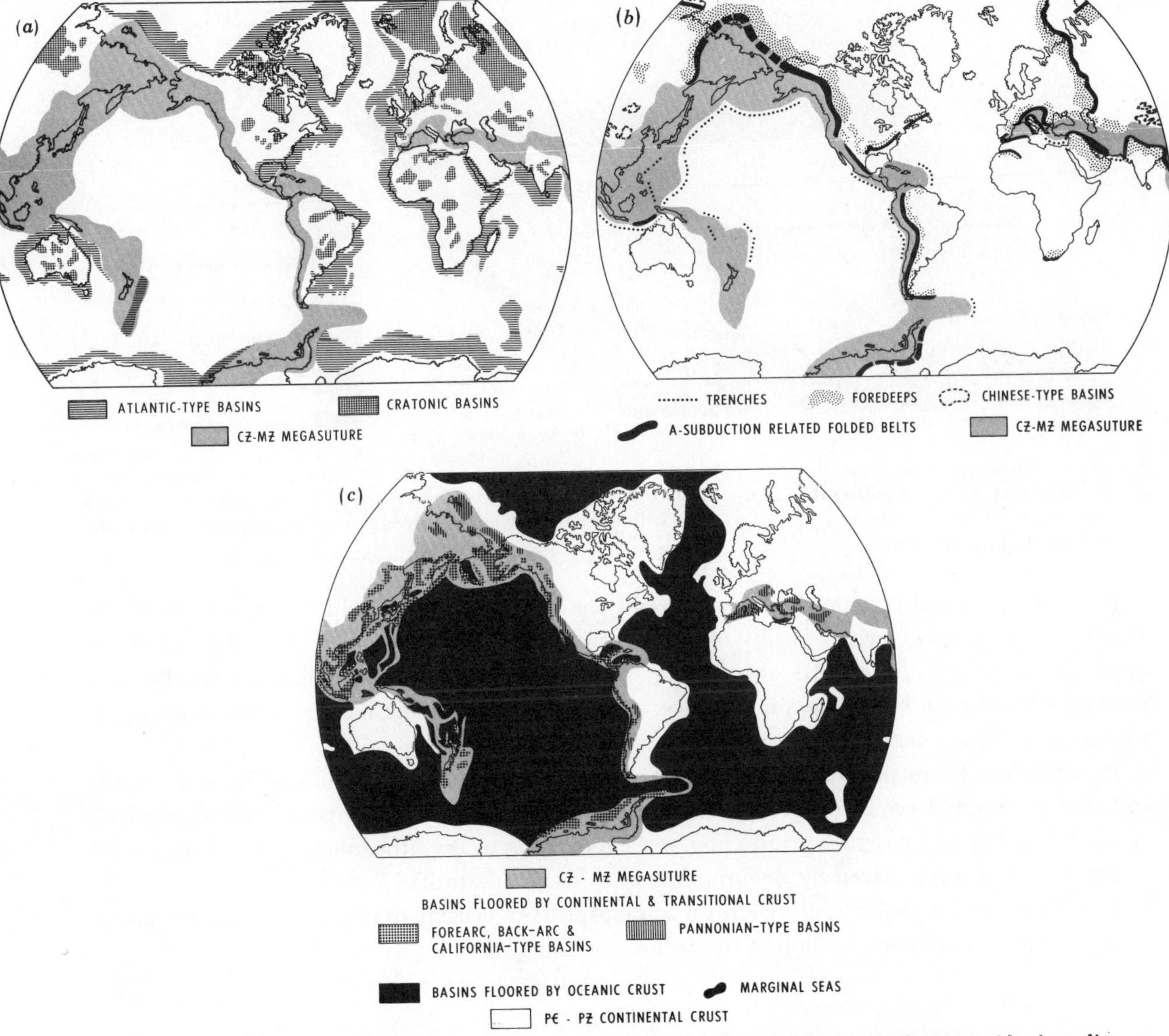

FIGURE 1. (*a*) Basins on rigid lithosphere. These are further subidivided on table 1. (*b*) Perisutural basins adjacent to subduction boundaries, but located on rigid lithosphere. These are further subdivided on table 1. (*c*) Episutural basins located on Cainozoic–Mesozoic megasuture. These are further subdivided on Table 1. A small number of preserved Palaeozoic episutural basins (e.g. Gulf of St Lawrence, Sidney Basin) are not shown. (After Bally & Snelson (1980), with permission of the Canadian Society of Petroleum Geologists.)

sedimentary basins of the world. We also hoped that our classification might provide some alternative to outdated geosynclinal concepts and jargon. There again we were not entirely successful, because we could not avoid the introduction of some new descriptive jargon.

Today it is tempting to take an idealized view of basin genesis that would initially ignore some of the more 'spooky' basin-forming mechanisms that have been proposed by our geophysical colleagues, such as those that involve phase changes at the crust mantle boundary and deeper in the Earth, or mechanisms that invoke flow within the crust, or else the flow of melts from continental to oceanic asthenosphere. Such mechanisms may well occur in Nature, but they appear to be 'spooky' because it is so difficult to map them directly or to test them with unambiguous geological and geophysical observations. At best, such concepts will appear to be permissible for a given set of data without excluding other equally permissible alternatives.

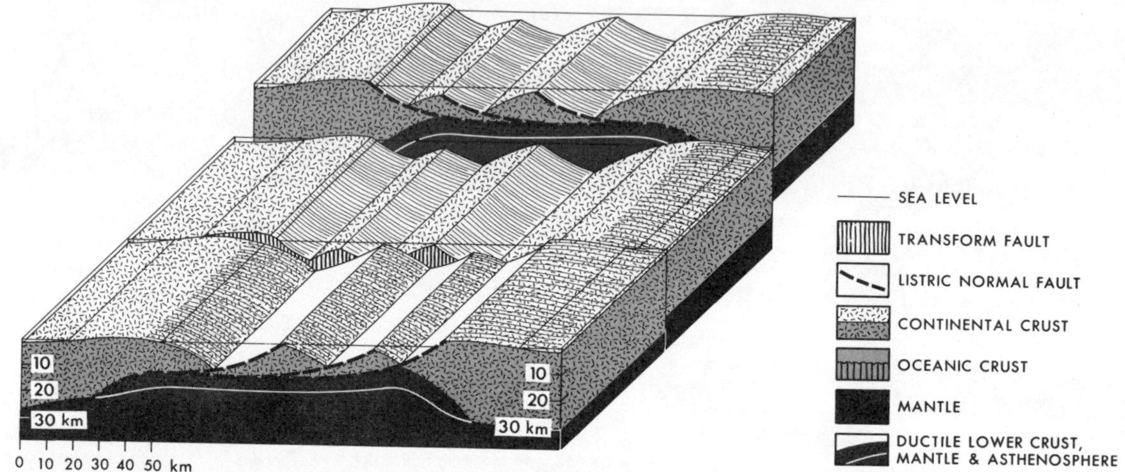

FIGURE 2. Transform zone separating two half-graben segments. This conceptual diagram shows that rift zones, in fact, may be sets of half-graben segments with changing polarity. Each segment would be separated from its neighbour by a complex transform zone. Sediments have been removed from the diagram to show the basement configuration.

A much simplified genetic view of our basin classification is that, with the exception of the perisutural basins and forearc basins, all basins appear to be initiated by a thermal event that is expressed by rifting of the brittle upper crust. Such an all-inclusive statement is naturally wide open to critical exceptions like, for instance, the number of cratonic basins that do not display an obvious rifting event that controls their early evolution.

Nevertheless, let us simply distinguish extensional basins that were initiated by a thermally induced extensional event from basins that are formed in an overall compressional régime with tectonic loading on a basement ramp and with thickening of the lithosphere. In fact, this is the contrast so well summarized by Beaumont *et al.* (this symposium).

In the following I shall dwell briefly on some aspects of crustal stretching and rifting and on matters related to documentation of the tectonic loading model.

Rifting

In the context of this meeting there is no need to recapitulate the historic evolution of various rifting concepts. The stretching model of McKenzie (1978) has triggered a wave of interesting studies (Royden *et al.* 1980; Christie & Sclater 1980; Royden & Keen 1980; Jarvis & McKenzie 1980; Le Pichon *et al.* 1981; Le Pichon & Sibuet 1981). In a nutshell, all these studies suggest that passive margins, some cratonic basins and also some back-arc-type basins may be initiated by lithospheric stretching and an associated thermal event. Subsidence following that event is due to cooling and is amplified by sediment loading. Thus, more stretching and high instantaneous strain rates lead to increased subsidence and less stretching, and low instantaneous strain rates cause less subsidence and in some cases even uplift. If the stretching occurs during a prolonged period that allows the rift system to cool, then ensuing subsidence will be diminished (Jarvis & McKenzie 1980).

The geological and geophysical evidence that is sometimes used to determine amounts of stretching is reasonably convincing but rather fragmentary. For this reason, the studies of Avedik *et al.* (this symposium) appear to be particularly useful in providing a more detailed geophysical data base for the study of the northern Bay of Biscay. The ambiguity of the data is well illustrated by different stretching estimates made by Avedik *et al.* and Le Pichon *et al.* (both this symposium) that are based on different interpretations of the same seismic section.

Very simple geometric considerations suggest that antithetic normal faulting along planar fault surfaces is difficult to achieve without associated folding or uplifting. In all cases the tilted brittle blocks have obvious bottom keels. Such a configuration is implicit in the so-called 'domino' or 'playing-card' models (see Le Pichon *et al.* this symposium). The depth of the keels is a function of the amount of tilting and the width of the rotated blocks. Unfortunately, geophysical data actually showing the existence of these keels on a crustal scale are hard to find. Thus, if the keels do not exist, one must invent some ductile flow process that eliminates them. Needless to say, the planar faults of the 'domino' model are not listric faults.

The listric fault model alleviates some of these problems. It has the obvious advantage that the beds can gradually rotate into the fault plane and that changes in bedding dips are related to changes in the curvature of the fault plane. Listric faults are often grown faults that show a divergence of beds as they dip into the fault plane (i.e. the U.S. Atlantic margin). These growth faults contrast with rotational faults with little growth and infill with more or less horizontal sediment layers (i.e. Bay of Biscay, North Sea). On a crustal scale listric normal faults also have their share of problems. These problems include excessive rotation of pointed, probably unstable, bottom keels or else compensatory faulting in the tops of the tilted fault blocks. Such faulting is commonly observed in Nature.

Reflexion seismic data may be most useful to determine the shape of listric fault planes. In some spectacular cases one can actually see the sole of the listric fault system (Montadert *et al.* 1979*a*, *b*). On other profiles, such as the ones reported from the Basin and Range province, only segments of the system are seen. A review of some of the seismic and surface evidence of a number of examples is given by Bally *et al.* 1981. Based on that summary, it would appear that for the brittle segment of the lithosphere we need better evidence for the actual shape of listric faults based on detailed seismic reflexion sections.

In recent years, after having looked at a number of seismic reflexion profiles from various areas, I have been impressed by the dominant asymmetry of rift systems. Half-grabens and systems of multiple half-grabens are the rule, and reflexion seismic evidence for symmetrical grabens is virtually absent. The main theme always seems to be rotation into a main master-fault system with some subsidiary compensatory faulting. The polarity of the master fault may change within a graben, and two segments of opposing master-fault systems appear to be separated by a transform fault (see figure 2). The transform segments are often complex and difficult to map. In hindsight, it is puzzling why the concept of more or less symmetrical graben systems has dominated the geological literature for decades. Is it possible that the geophysical tools and geological methods that were used to depict the graben configuration were simply too crude to resolve the shallow structure and the question of symmetry?

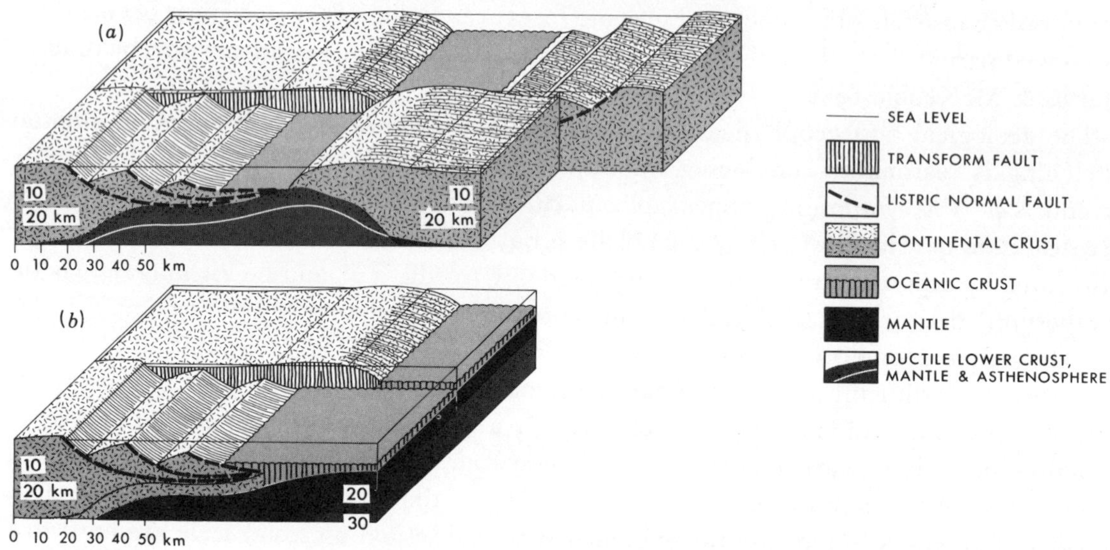

FIGURE 3. The development of a passive margin, with fault blocks rotating toward the ocean. (*a*) Rifting stage; (*b*) spreading (drifting) stage.

The asymmetrical nature of graben systems may be of some consequence in the assessment of changes occurring along Atlantic-type margins and in the comparison of conjugate margins across a spreading ocean. Figures 3 and 4 show the evolution of half-graben systems into passive margins. They suggest that the polarity of the bounding listric fault systems may change, i.e. the listric faults may dip towards the ocean (e.g. Bay of Biscay) or else away from the ocean (e.g. Baltimore Canyon). The same diagrams also suggest that relatively abrupt changes in basement depth (e.g. from Georges Bank to Baltimore Canyon) probably occur across transform fault systems that offset major half-graben systems. Another consequence of the asymmetrical nature of half-graben systems is that the genesis and evolution of conjugate margins across a spreading ocean may differ substantially. All these concepts need more verification.

Bott (this symposium) discusses the global origin of large tensile stresses. On a more restricted regional scale, it may also be important to look into the demonstrability of pre-existing zones of weakness. For instance, listric normal faulting in the Western Cordillera appears to be semi-concordantly superposed on an earlier gently westward dipping listric décollement – thrust fault system. The suggestion there is that listric normal faults prefer to sole out in the basal sole fault system of earlier thrust faults (Price 1981). By analogy, a quick glance at the disposition of the Triassic graben systems of the Appalachians shows a crude parallelism with the strike of the foreland folds. This again strongly suggests that late Triassic listric normal faults merge into the sole of a late Palaeozoic thrust fault system, which, as the COCORP data across the Appalachians suggest (Cook *et al.* 1979; Brewer *et al.* 1981), may underlie much if not all of the Piedmont. The observation that fault blocks of the U.S. Atlantic shelf are rotated towards the continent would in turn suggest that the bounding listric faults are soling into a decoupling system that has an easterly vergence and relates to the Palaeozoic folded belts of Africa.

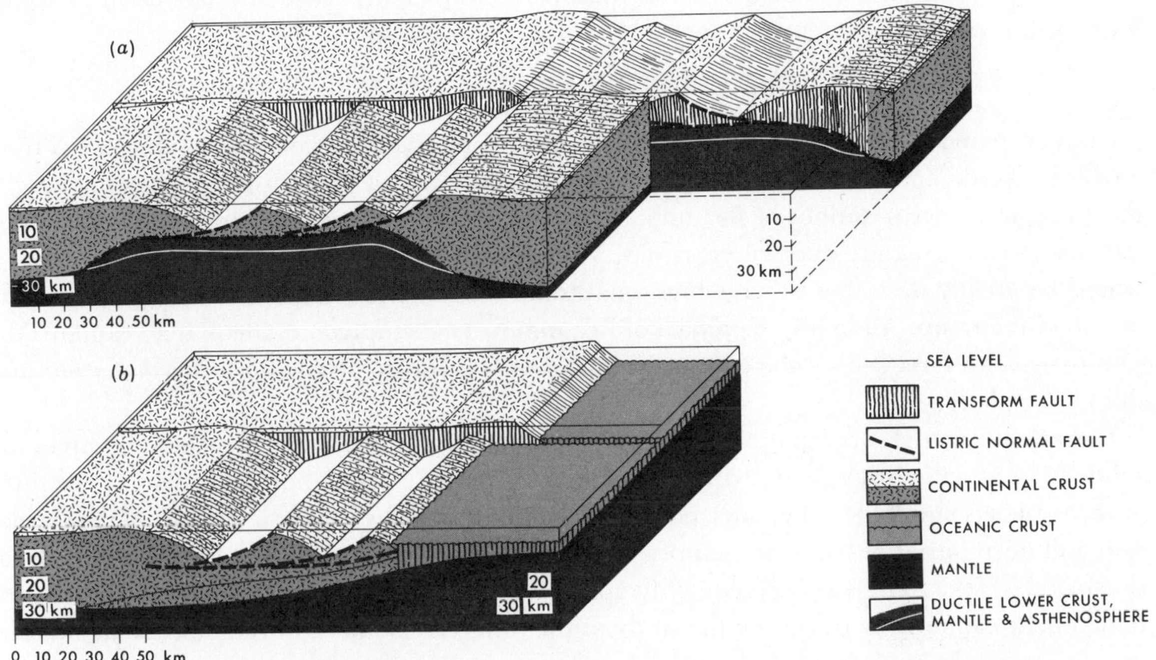

FIGURE 4. The development of a passive margin with listric fault blocks rotating toward continent. (*a*) Rifting stage; (*b*) drifting stage.

In conclusion, the stretching model appears to be a reasonably viable model for the formation of many basins. The evidence documenting planar or else listric fault blocks is far from conclusive, although in my judgement reflexion profiles and mechanical considerations (Hafner 1951) so far favour a low-angle listric fault model. Either way, estimates of stretching based on reflexion seismic data will be limited by the general absence of detailed regional reflexion lines, with key beds in the stretched crust that are calibrated by drilling. In all cases it is important to differentiate growth faults that are associated with one or more sets of updip converging beds from fault systems that show no growth and slow infilling by horizontal beds.

When analysing the subsidence of basins that suggest a thermal extensional origin, it may be wise to isolate the effects of sediment loading, as is done by Watts *et al.* and Beaumont *et al.* (both this symposium) and the effects of demonstrable minimum stretching (Avedik *et al.*) from the effects of stretching permissible but not shown by actual data (Le Pichon *et al.*) as well as the effects of other 'spooky' but possible mechanisms such as thermally induced phase changes (Turcotte) and ductile flow in the lower crust (Bott).

Foredeeps

Earlier it was pointed out that the perisutural basins of our classification are formed in a compressional, i.e. collisional, régime. The Carboniferous Permian basins of the U.S. continental interior (Dewey & Pitman, this symposium), the restored Zagros Basin (Koop), the restored southern part of the Franklinian Basin (Kerr), some of the Carboniferous sedimentary basins of northern Europe (Clarke), and the Western Canada Basin (Porter *et al.* and Beaumont *et al.*) all formed in a collisional régime.

Only for western Canada, Wyoming and some parts of the Appalachians are published reflexion seismic sections available that permit observation of the common ramp that underlies both the foredeep and the adjacent folded belt. The most extensive reflexion seismic data set is from western Canada and, together with surface and subsurface data (Price 1981), forms the base of the structural interpretations of Beaumont *et al.* Unfortunately, most of the published seismic sections in that area are over 20 years old. With advances in modern reflexion seismology, it would be greatly desirable to have high-quality data available for the study of the structural details of the ramp. All in all, the model of Beaumont *et al.* simply and adequately explains the subsidence of a foredeep by the mechanical loading (by thrust sheets) on a thermally old and thick lithosphere.

Beaumont's model is compatible with a general mountain-building concept that involves the subduction of sialic lithosphere, with extensive decoupling of the sedimentary skin in the form of thrust sheets and folds. The inner portion of such folded belts is characterized by the formation and deep burial of basement nappes that leads to regional metamorphism and the formation of gneiss folds. These are subsequently uplifted and eroded. The uplift history of the interior folded belts can now be traced by the study of the different methods of radiometric age dating and fission tracks analysis.

Obviously then, future studies on foredeep subsidence should be supported by high-quality seismic reflexion data and a comparison of the uplift history in the mountains with the sedimentation and subsidence in the adjacent foredeep. In effect, the foredeep stratigraphy directly reflects the subsidence and uplift history of the basement ramp that underlies both the foredeep and the adjacent folded belt.

A number of foredeeps, i.e. the Palaeozoic Arkoma Basin and the St Lawrence lowlands, and parts of the Tertiary Alpine Molasse basin, show considerable normal faulting in the underlying ramp. Much of the normal faulting occurs at about the same time as the thrust faulting. The significance of these normal faults is still obscure, but it is not likely that they are associated with a thermal event or else with significant crustal stretching.

An important variation on the theme of mechanical loading is offered by the development of foreland uplifts and associated basins that involve the basement, i.e. the Colorado–Wyoming Rocky Mountains and some of the Palaeozoic foreland uplifts of Oklahoma and West Texas (Dewey & Pitman, this symposium). Relatively little is known about these uplifts, but the Cocorp line across the Wind River Mountains and reflexion profiles obtained by the oil industry indicate that these foreland blocks are bounded by reverse faults and oblique-slip fault systems (Brewer *et al.* 1979). This in turn suggests a common decoupling level near the crust–mantle boundary. Where basement uplifts are in effect thrust over an adjacent block, basins develop that appear to be explainable by simple loading. More quantitative studies, preferably based on relatively complete data packages, are needed to confirm this suspicion. A better documentation is particularly urgent for the Palaeozoic basins studied by Dewey & Pitman. My recollection of these basins is that seismic reflexion data show little, if any, support for the late Palaeozoic extensional–thermal events postulated by these authors.

What little we know about the basins of interior China (e.g. Tarim and Dzungarian Basins) Bally *et al.* 1980) suggest that they are formed in a compressional, collisional context and that their subsidence is due to loading by reverse faulting and overthrusting of thick sialic sheets on the platforms that underlie these basins (i.e. the Chinese basins of our classification).

The global synchronism of subsidence and uplift episodes

A number of authors, mostly following in the footsteps of Sloss (1972), have made a convincing case for the worldwide correlatibility of major sedimentary sequences that are separated by unconformities. The concept has gained further support by the worldwide study of stratigraphy as displayed on reflexion seismic profiles undertaken by Vail and his colleagues at Exxon (Vail *et al.* 1977). There is little doubt in my mind that in a first approximation, the major sequences are indeed correlatable. I am disturbed, however, that we lack in-depth stratigraphic and palaeontological documentation to support the postulated correlations.

The issue of documentation is of some importance in the context of basin origin and evolution. Vail and his associates interpret their cyclical sequences in terms of eustatic sea level changes. They are convinced that from the reflexion seismic record of subsiding basins, they can extract a signal that – if correlated – is uniquely characteristic of eustatic sea level changes. From looking at subsidence plots such as those published by Keen (1979) and by Watts & Steckler (1979), it would appear that the influence of sea level changes on subsidence is minimal and therefore that most of what we see on reflexion seismic lines is in fact a record of subsidence. Watts (this symposium) makes a good case for the tectonic origin of 'coastal' onlap on which many of Vail's curves are based.

In other words, it appears to a number of authors and myself that the major stratigraphic sequences that we see and correlate in many parts of the world are correlatable tectonically induced subsidence cycles, separated by ubiquitous unconformities. The latter may well in some cases reflect a lowering of the sea level.

A simple glance at a Lower Cretaceous isopach of North America indicates substantial subsidence occurring at the Atlantic margin, in the Gulf of Mexico, in the foredeep of the Western Cordillera, and in the Great Valley of California. We thus see more or less synchronous subsidence that on Atlantic-type margins may be caused by more or less advanced cooling and sediment-loading that is dependent on the opening time of the adjacent ocean, while in the Rocky Mountain foredeep subsidence is dominated by mechanical loading by thrust sheets of a lithospheric ramp; and, finally, in the Great Valley we see subsidence that may either be associated with a fore-arc basin or else by a back-arc basin. The point is that plate tectonics is responsible for different modes of simultaneous subsidence that are dependent on the plate tectonic setting. For the Western Cordillera, this context has been well described by Coney (1979).

If, however, instead of focusing on the sequences themselves, we look at the unconformities separating them, it turns out that they are typically continent-wide, as shown by characteristic subcrop maps that are based on subsurface control. The apparent structural origin of major arches associated with these subcrops remains obscure. It is not likely that they are composites of peripheral bulges associated with sedimentary loads of a number of adjacent basins, because

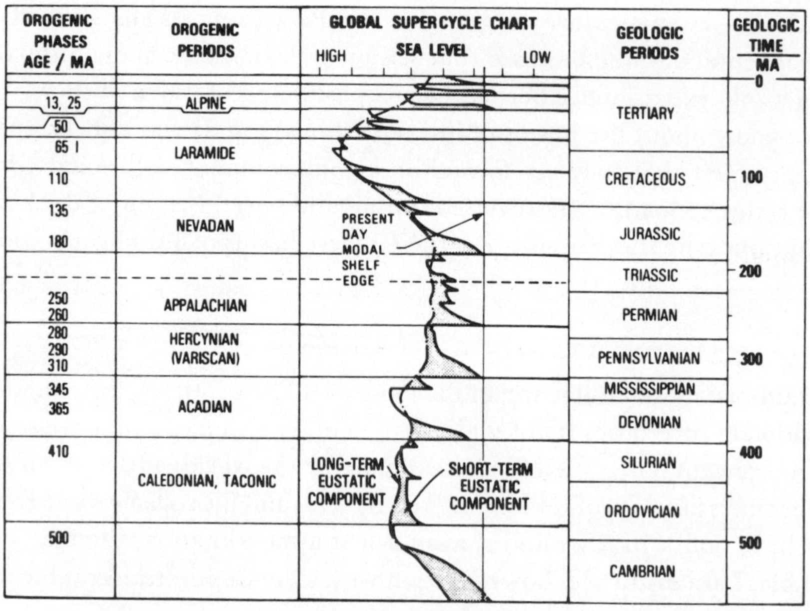

OROGENIC PHASES AGE / MA	OROGENIC PERIODS	GLOBAL SUPERCYCLE CHART SEA LEVEL (HIGH – LOW)	GEOLOGIC PERIODS	GEOLOGIC TIME MA
				0
13, 25	ALPINE		TERTIARY	
50				
65	LARAMIDE			
110			CRETACEOUS	100
135	NEVADAN	PRESENT DAY MODAL SHELF EDGE	JURASSIC	
180				
			TRIASSIC	200
250, 260	APPALACHIAN		PERMIAN	
280, 290, 310	HERCYNIAN (VARISCAN)		PENNSYLVANIAN	300
345, 365	ACADIAN		MISSISSIPPIAN	
			DEVONIAN	
410	CALEDONIAN, TACONIC	LONG-TERM EUSTATIC COMPONENT; SHORT-TERM EUSTATIC COMPONENT	SILURIAN	400
			ORDOVICIAN	
500			CAMBRIAN	500

FIGURE 5. Global supercycle chart and correlation with orogenic activity. (From Vail (1977).)

they extend well beyond, and often are discordant with, the outline of single basins. Regional subcrops appear to be preferentially formed at the inception of a major collisional phase, and they often immediately precede the formation of foredeeps (i.e. the pre-middle Devonian subcrop of North America, the pre-Cretaceous subcrop of the Western Interior basins of North America (Bally 1980), and the pre-Eocene subcrop of the Molasse Basin of the Alps). One gains the impression that somehow, with the inception of a major collision, extensive cratonic portions of a continent are being warped or tilted.

Whatever the reason for the extensive supraregional subcrops, we are still confronted with their approximate worldwide correlatability. A major problem becomes evident when we compare a chart by Vail (1977) with one by Schwan (1980) (see figures 5 and 6). With regard to unconformities, the obvious question is: what is an orogenic phase? Schwan (1980), following the Stille tradition, equates the unconformity with an orogenic phase, while Vail (see also Johnson 1971) has his orogenic phases bracketed by major unconformities or, in other words, correlated with his sequences. As expressed elsewhere (Bally 1980), and in support of the Johnson (1971) concept, my own suspicion is that the sequences reflect continuing plate tectonics, while the major unconformities mark the slow-down and ultimately the cessation of one major plate tectonic régime and a plate reorganization into a new plate tectonic régime. However, to demonstrate this point convincingly, one would have to undertake a well documented study of the structural palaeogeographic characteristics that precede the formation of supraregional unconformities. A precise statement of the palaeontological brackets that straddle the unconformity would also be necessary. For the Mesozoic and Cainozoic, similar age brackets would be needed for discordances that separate parallel packages of magnetic stripes, as well as for the inception and cessation of igneous events (see Schwan 1980).

Clearly all of this would entail a major international cooperative effort, and I am planning to explore the feasibility of such an effort.

A few concluding remarks

The papers in this meeting focused only on a limited number of basin types: passive margins, cratonic basins and foredeeps. A quick glance at table 1 will show that a large number of basin types were not discussed, most notably forearc and backarc basins. The latter were discussed in some detail during a 1980 Royal Society Discussion Meeting on extensional tectonics associated with convergent plate boundaries (Vine & Smith (eds) 1981). Furthermore, the basins that are associated mega-shear systems, i.e. the oblique-slip related basins, were discussed in a recently published collection of papers (Ballance & Reading 1980). In oblique-slip systems both compressional and extensional basins may form, and it may be reasonable to presume that either lithospheric stretching or mechanical loading may be responsible for the formation of individual basins. The problem is that we so often lack critical reflexion and refraction data to tell us what the precise basement configuration in oblique-slip related basins is.

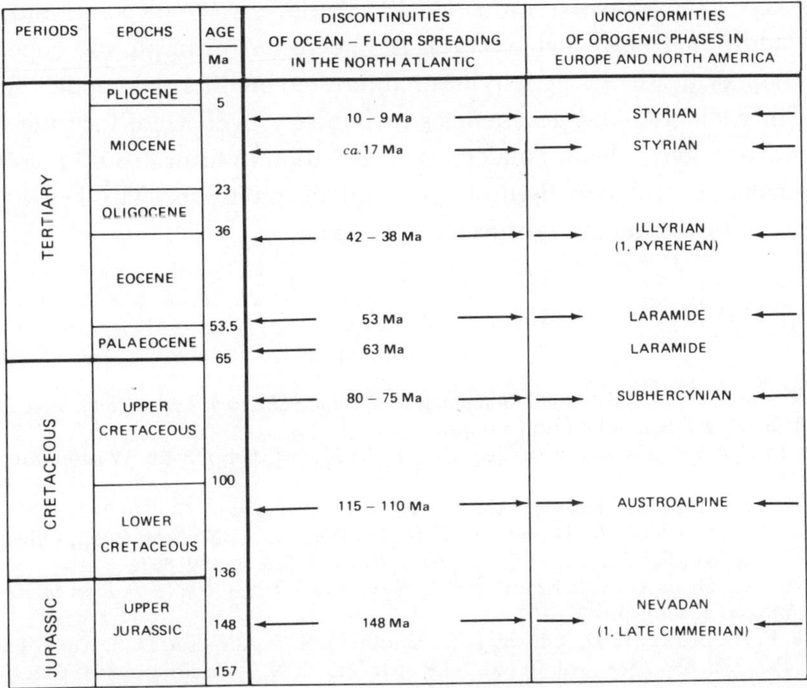

Figure 6. Outline of discontinuities in the course of plate movements in the north Atlantic Ocean and of orogenic events in bordering continents showing time parallels. (After Schwan (1980), with permission of the American Association of Petroleum Geologists.)

A number of the proposed basin-forming mechanisms make assumptions with regard to processes that might occur in the middle and lower crusts. There detailed refraction, wide-angle reflexion, and more conventional reflexion profiles might provide additional information. Another research avenue, which to my knowledge has not yet been adequately explored, involves surface geological studies in basement thrust sheets that bring to the surface the deeper crustal portions of obvious passive margins, e.g. the Eastern Alpine thrust sheet, the Insubric and Ivrea Zones of the Alps, and the main basement sheet of the High Himalayas. Typically the work in these areas is aimed at studying the age and origin of the basement, the effects of orogenic metamorphism, etc. If, however, these basement sheets are the exhumed deeper portions of former passive margins, they should also reveal some of the effects related to the origin of these margins, e.g. stretching in the lower crust and the soles of listric normal fault systems. Of course, the Himalayas would offer a particularly attractive target for such a study, because there the deeper portions of an ancient passive margin to the south at the Yalu–Tsanpo suture can be compared with the deeper portions of the ancient active margin of the Trans-Himalaya to the north (Bally *et al.* 1981).

It looks as if the ability of our geophysical colleagues to invent models is moving much faster that the capacity of geologists and geophysicists to acquire the critical high-quality information to unambiguously support these models. At issue, of course, is the possibility for research institutions to acquire expensive crustal reflexion and refraction profiles and to have access to detailed industry reflexion profiles across sedimentary basins. With research funds becoming scarcer on both sides of the Atlantic, a close dialogue and cooperation with industry becomes increasingly desirable. In this context it will become more important to view sedimentary basins in their totality, including their hydrological and hydrodynamic characteristics, their finite groundwater reserves, the potential and risks of subsurface waste disposal, and, of course, the exploration for additional fossil fuel reserves. A fine recent example of cooperation was the creative interaction of academic geophysicists interested in thermal models for sedimentary basin genesis with geologists and geochemists in industry, who were focusing on maturation studies of hydrocarbon source beds. Analogous cooperation in future could involve geochemists interested in diagenetic problems, hydrologists and geophysicists. The key to successful cooperation is going to be a common subsurface data bank.

References (Bally)

Ballance, P. F. & Reading, H. G. (eds) 1980 *Sedimentation in oblique-slip mobile zones* (*Spec. Pub. int. Ass. Sedimentol.* no. 4). Oxford: Blackwell Scientific Publications.

Bally, A. W. 1980 In *Dynamics of plate interiors* (*Geodynamics Series*, vol. 1, pp. 5–20. Washington, D.C.: American Geophysical Union.

Bally, A. W. & Snelson, S. 1980 Realms of subsidence. *Mem. Can. Soc. Petrol. Geol.* **6**, 9–94.

Bally, A. W., Allen, C. R., Geyer, R. B., Hamilton, W. B., Hopson, C. A., Molnar, P. H., Oliver, J. E., Opdyke, N. D., Plafker, G. & Wu, F. T. 1980 *U.S. geol. Surv. Open-File Rep.* no. 80–501.

Bally, A. W., Bernoulli, D., Davis, G. A. & Montadert, L. 1981 *Oceanologica Acta* (*S. P. Proc. Int. Geol. Congr., Geology of Continental Margins Symp.*), pp. 87–101.

Brewer, J. A., Cook, F. A., Brown, L. D., Oliver, J. E., Kaufman, S. & Albaugh, D. S. 1981 In *Thrust and Nappe Tectonics* (*Spec. Pub. geol. Soc. Lond.*, vol. 9) (ed. K. R. McClay & N. J. Price), pp. 501–511. Oxford: Blackwell Scientific Publications.

Brewer, J. A., Smithson, S. B., Oliver, J. E., Kaufman, S. & Brown, L. D. 1979 *Tectonophysics* **62**, 165–189.

Christie, P. A. F. & Sclater, J. G. 1980 *Nature, Lond.* **283**, 729–732.

Coney, P. J. 1979 *Mem. geol. Soc. Am.* **152**, 33–50.

Cook, F. A., Albaugh, D. S., Brown, L. D., Kaufman, S., Oliver, J. E. & Hatcher, R. D., Jr 1979 *Geology* **7**, 563–567.

Hafner, W. 1951 *Bull. geol. Soc. Am.* **62**, 373–398.

Jarvis, J. G. & McKenzie, D. P. 1980 *Earth planet. Sci. Lett.* **48**, 42–52.

Johnson, J. G. 1971 *Bull. geol. Soc. Am.* **82**, 3263–3298.

Keen, C. E. 1979 *Can. J. Earth Sci.* **16**, 505–522.

Le Pichon, X., Angelier, J. & Sibuet, J.-C. 1981 *Tectonophysics*. (In the press.)

Le Pichon, X. & Sibuet, J.-C. 1981 *J. geophys. Res.* **86**, 3708–3720.

McKenzie, D. 1978 *Earth planet. Sci. Lett.* **40**, 25–32.

Montadert, L., Roberts, D. G., de Charpal, O. & Guennoc, P. 1979*a* *Initial reports of the Deep-Sea Drilling Project*, vol. 48 (ed. L. Montadert & D. G. Roberts), pp. 1025–1060. Washington, D.C.: U.S. Government Printing Office.

Montadert, L., de Charpal, O., Roberts, D. G., Guennoc, P. & Sibuet, J.-C. 1979*b* In *Deep drilling results in the Atlantic Ocean* (*Maurice Ewing Ser.*, vol. 3) (ed. M. Talwani, W. W. Hay & W. B. F. Ryan), pp. 164–18. Washington, D.C.: American Geophysical Union.
Price, R. A. 1981 In *Thrust and nappe tectonics* (*Spec. Pub. geol. Lond.*, vol. 9) (ed. K. R. McClay & N. J. Price), pp. 427–448. Oxford: Blackwell Scientific Publications.
Royden, L. & Keen, C. E. 1980 *Earth planet. Sci. Lett.* **51**, 343–361.
Royden, L., Sclater, J. G. & von Herzen, R. P. 1980 *Bull. Am. Ass. Petrol. Geol.* **64**, 173–187.
Schwan, W. 1980 *Bull. Am. Ass. Petrol. Geol.* **64**, 359–373.
Sloss, L. L. 1972 In *Twenty Fourth Session, Int. Geol. Cong., Section* 4, pp. 24–32.
Vail, P. R. 1977 Sea level changes and global unconformities from seismic sequence interpretation. A report of the JOIDES Subcommittee on the Future of Scientific Ocean Drilling, Woods Hole, Mass. (unpublished).
Vail, P. R., Mitchum, R. M., Jr, Todd, R. G., Widmier, J. M., Thompson, S., III, Sangree, J. B., Bubb, J. N. & Halelid, W. G. 1977 In *Seismic stratigraphy – applications to hydrocarbon exploration* (*Mem. Am. Ass. Petrol. Geol.* no. 26) (ed. C. E. Payton), pp. 49–212.
Vine, F. J. & Smith, A. G. (eds) 1981 *Phil. Trans. R. Soc. Lond.* A **300**, 217–442.
Watts, A. B. & Steckler, M. S. 1979 In *Deep drilling results in the Atlantic Ocean* (*Maurice Ewing Symposium Ser.*, vol. 3) (ed. M. Talwani, W. W. Hay & W. B. F. Ryan), pp. 218–234. Washington, D.C.: American Geophysical Union.

Discussion

R. A. PRICE. In western North America there are a number of places where you can see two different crustal levels in contact, separated by a normal fault with a shallow dip. The metamorphic rocks beneath the fault show evidence of rapid quenching. A wide variety of rocks are found above the fault, from well consolidated rocks to fanglomerates containing detritus from the metamorphic rocks. These structures are probably the flat parts of listric normal faults. Frequently the metamorphic rocks form a dome that rises well above the surrounding basins, which contain the rocks overlying the faults. It is not obvious how to use this geometry to estimate the amount of extension.

A. W. BALLY. Listric normal faults may sole at different levels. For instance, in the Western Cordillera some faults appear to be rooted in the Moho and therefore will offset the basement overlying the Moho. Other faults, however, are rooted at the sediment–basement interface, at the base of the overlying décollement thrust sheets. Price has described some that are involved in mantled gneiss domes. These too appear to be located at an old sediment–basement interface. I am puzzled by these features and wonder particularly why the domes themselves should be uplifted. There is little doubt in my mind that some kind of broad regional uplift affected the Western Cordillera of the U.S. as suggested by Oligocene levels that climb from the Western Plains into the Colorado Mountains, and then descend across the Sierra Nevada into the Great Valley. Thus the whole intervening region has been regionally uplifted and been intersected by a complex system of listric normal faults that are likely to be linked by transform segments. What is needed now is a detailed study of the geomorphology, the geology and the structures as seen on reflexion seismic profiles, because the surface geology of the cross section of Nevada that I showed on my slide is only known in outline.

D. G. ROBERTS. In the Bay of Biscay granites have been dredged from the tops of the listrically faulted blocks. Therefore these faults cannot sole out against the basement sediment interface, but must be rooted in a plane within the crystalline basement. Why should this behaviour occur?

A. W. BALLY. I have no difficulty seeing listric normal faults within the basement. The COCORP line shows that the Appalachian thrusts are all rooted in a large subhorizontal thrust which essentially moves basement over basement. Listric normal faults that merge with older décollement systems of the Appalachian type must be mostly intrabasement faults. I can easily imagine a Precambrian shield having a similar internal structure, with basement structures stacked on top of each other. On a reflexion seismic line all fault planes would be within the basement. Precambrian basement may contain many planes of weakness corresponding to older décollement systems which could provide soles for listric normal fault systems.

M. F. Osmaston. Could the normal faults in the Baltimore Canyon area postdate the continental separation, and have been produced by the thermal upwarping of the margin? They would then have to be downthrown towards the continent to the west.

A. W. Bally. I do not believe that the normal faults in the Baltimore Canyon area postdate the continental separation. I believe that the faulting I see on the Atlantic offshore sections forms part of a 'Triassic' graben system that is well known on land. On land the rifting starts in late Triassic and ceases at the end of the lower Jurassic. In the Grand Banks (offshore Newfoundland) rifting starts at the same time but the cessation is considerably later. Thus different parts of rifted margins have different histories. The influence of an extended period of extension on thermal subsidence has been investigated by Jarvis and McKenzie. If the extension is rapid, the faults are only active for a short period of the time, and sediment layers deposited after the grabens have formed will be horizontal.

Reactivation of such faults will displace younger beds, and new grabens will form over the old ones. Therefore the breakup unconformity may be in some cases at the base of horizontal sediments that fill a graben and reactivation of the extension may result in the formation of other unconformities.